John Thornton

Human Physiology

John Thornton

Human Physiology

ISBN/EAN: 9783337365370

Printed in Europe, USA, Canada, Australia, Japan

Cover: Foto ©berggeist007 / pixelio.de

More available books at **www.hansebooks.com**

HUMAN PHYSIOLOGY

BY

JOHN THORNTON, M.A.

AUTHOR OF 'ELEMENTARY PHYSIOGRAPHY' 'ADVANCED PHYSIOGRAPHY ETC.
HEAD MASTER OF THE CENTRAL HIGHER GRADE SCHOOL, BOLTON

WITH 272 ILLUSTRATIONS, SOME COLOURED

THIRD EDITION, REVISED

LONGMANS, GREEN, AND CO.
39 PATERNOSTER ROW, LONDON
NEW YORK AND BOMBAY
1899

PREFACE

THIS work has been mainly prepared for lay students preparing for the Second or Advanced Stage Examination of the Science and Art Department. Such students are supposed to have acquired some preliminary knowledge of the subject, and I have endeavoured in the following pages to furnish them with an opportunity of deepening and extending that knowledge. I have not tried to produce a superficial cram book, but have endeavoured to furnish precise and accurate information on such parts of histology and anatomy as are required, as well as to give a reasoned account of the physiological processes of the human body.

Attendance at a course of lectures and demonstrations given some years ago by Dr. A. Gamgee, F.R.S., first gave me a real interest in, and insight into, the subject, while continued study and experience in teaching have only served to deepen that interest and insight. In writing the book the chief authorities have been consulted. In Anatomy I have made use of Gray, Quain (edited by Schäfer and Thane), and Testut. In Physiology I have consulted Beaunis, Foster, Halliburton, Hermann, Langlois and Varigny, Landois and Stirling, and Waller. To the student who wishes to continue his study I would strongly recommend Dr. Waller's admirable work as his next book, to be followed by Professor Foster's Text-book or that of Landois and Stirling. Care should be taken to obtain the latest edition of these standard works.

Most of the illustrations have been derived from works published by Messrs. Longmans & Co., and for permission to use these I am very grateful. A large number has been employed, partly to supplement the text, partly to illustrate objects not easily obtained, and partly to take the place of detailed description. The student is strongly advised to give considerable attention to the figures and their description, as his progress will thereby be greatly facilitated.

A set of progressive questions has been added to the book, not merely to let the student see the nature of the examination he may have to undergo, but with a view of calling attention in another way to certain important points connected with his study.

I am indebted to my brother, G. Thornton, M.D., for reading over the proof sheets, and for some valuable suggestions.

JOHN THORNTON.

Bolton : *December* 1893.

In this Second Edition a few corrections and alterations have been made in most chapters, and a Glossary of the chief technical words has been inserted.

J. T.

Bolton : *July* 1896.

CONTENTS

HUMAN PHYSIOLOGY

CHAPTER I

HISTOLOGY, OR THE MICROSCOPICAL STRUCTURAL ELEMENTS OF THE BODY

1. The Microscopic Elements of the Body.—A simple examination and dissection of the human body, or that of any other higher animal, soon shows that it is made up of easily separated portions having distinct forms, such as the heart, lungs, stomach, brain, &c. Such parts are called *organs*, and each organ is found to have its special work or function in the animal economy. A department of study called *descriptive anatomy* furnishes an account of the several organs, members, and regions of the body, pointing out their form, structure, and mutual connections. Much of this knowledge may be obtained from the dead body, and the student is supposed to have already acquired by this means and from the earlier chapters of an elementary work on animal physiology a rudimentary acquaintance with the build of the human body and with the position, form, and function of its chief organs. Dissection further teaches that the body of any of the higher animals is made up of different kinds of materials, such as muscle, fat, cartilage, nervous matter, &c. These various constituent materials are spoken of as textures or *tissues*, and while some of the organs are formed mainly of one elementary texture or tissue, other organs or parts of the body are compounded of several tissues. The portion of anatomy that treats of the minute structure of the tissues is called *Histology* (Gk. *histos*, a web ; *logos*, a discourse). Histology requires the use of the microscope, and by its aid we quickly learn that the separable structural

B

elements of the various tissues are for the most part either minute corpuscles named 'cells' or elongated threads named 'fibres.' But since many of the fibres, if not all, are either modified cells or are derived directly or indirectly from cells, and since all the tissues contain cellular elements and originate as collections of cells, the cell may be regarded as the histological unit of the body. To the minute anatomy of the tissues we shall devote our first chapter, as a sound knowledge of this is essential to the proper understanding of much that follows. Human physiology, in fact, may, from one point, be regarded as the knowledge of the structure, properties, and functions of cells and their derivatives.

As we shall presently see, the cells of the body differ widely in form, appearance, and structure in the different tissues entering into the composition of the organs, while corresponding to these differences of structure and constitution will be found differences of function or properties. In other words, morphological differentiation (Gk. *morphe*, form) is accompanied by physiological differentiation. Thus the cells of muscle, or motor cells, possess the power of contraction, and do the mechanical work of the body ; the cells of the alimentary canal either secrete fluids from the blood and pour it out on the food to digest it and render it soluble, or they are specially adapted for absorbing into the blood-stream the food so prepared. The liver cells, among other functions, further elaborate the food-stuffs absorbed into the blood so that they may serve the more easily as nutriment for the cells of the various tissues. For every cell needs to assimilate food in order to live, and every cell must be supplied with oxygen for the oxidation of the matter by which its energy is produced. The cells must also be able to get rid of the waste products of their activity. The lungs are the organs of the body where a thin layer of cells in the cavities allows the passage of oxygen into the blood, to be carried to all the tissues, and where also a discharge of carbon dioxide, CO_2, one of the products of cell activity, takes place ; while the cells of the kidneys perform the special duty of freeing the blood from the nitrogenous waste passed into it. There is, therefore, a physiological division of labour in the human body, and the parts work together harmoniously because the organs and tissues are co-ordinated and regulated by a complex system of cells and fibres known as the Nervous System.

Although the modern conception of a cell differs from that which prevailed some few years ago, the doctrine that the bodies of all animals and plants consist either of a single cell or of a number of cells and their products, and that all cells proceed from pre-existing cells (*omnis cellula e cellula*), is now regarded as the basis of biological science. It is often spoken of as the *cell theory*, and may be regarded as 'the greatest discovery in the natural sciences in modern times.' To the study of the animal cell we now therefore proceed.

2. **The Animal Cell.**—An animal cell is a minute corpuscle of living substance, or protoplasm, those of the human body seldom exceeding $\frac{1}{300}$th of an inch in diameter, while many are but one-tenth this size. It consists of two distinct parts, the main substance of the cell, called *protoplasm*, and

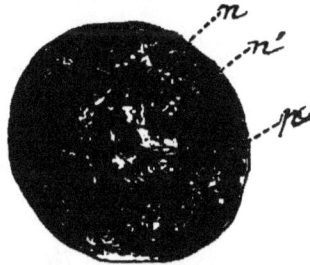

Fig. 1.—Diagram of a Cell, the Protoplasm of which is composed of Spongioplasm and Hyaloplasm. Highly magnified.

p, protoplasm; *n*, nucleus; *n'*, nucleolus.

a tiny body embedded in the protoplasm, called the *nucleus*. Seen under a low power of the microscope, the protoplasm appears either to be a homogeneous substance or to contain granular matter and vacuoles (globular cavities containing a watery fluid); but viewed with very high powers, part of the protoplasm is seen to be fibrillated and to form a fine network. This network is known as the *reticulum* or *spongioplasm*, and the rest of the protoplasm occupying the meshes of the network is a clear substance termed *hyaloplasm*. The arrangement of the network and the size of the meshes vary in different cells, the nodes or junctions of the fibrils probably giving rise to some of the granular appearances before mentioned. It must, however, be remembered that the protoplasm often includes actual granules of nutritive material or of matter stored up for nutrition, and that the vacuoles may contain glycogen or other substances in solution.

The external layer of protoplasm in some cells becomes altered so as to form a cell-membrane or cell-wall, but this is more frequently found in vegetable than in animal cells.

Within the protoplasm, and generally near the middle of the cell, is the minute body of somewhat firmer consistence called the *nucleus*. It is probably a portion of the cell-protoplasm somewhat altered in chemical nature, and its chief office is to preside over the nutritive activity and reproduction or division of the cell. It is believed by many biologists that the nucleus is also the conveyer of hereditary properties. The

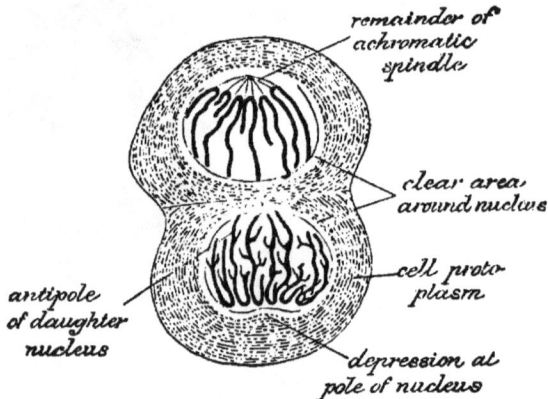

FIG. 2.—A dividing Cell. Formation of Chromatin Network in Daughter-nuclei. (Rabl.) The upper daughter-nucleus is at an earlier stage than the lower one. The cell-protoplasm is now completely divided.

nucleus is spherical or elongated in shape, often bounded by a membrane, and consists of a clear substance known as the nuclear matrix or *nucleoplasm*, pervaded by a network of fibres differing in nature from those of the cell-substance. The intranuclear network may show enlargements called *nucleoli* (fig. 1). Nuclei are often recognised by treating a cell with acetic acid, which destroys the cell-protoplasm and thus renders the nucleus more visible. The intranuclear network stains with certain dyes, and is hence called *chromoplasm* (Gk. *chroma*, colour), while the matrix or nucleoplasm is achromatic. Histologists distinguish between a resting or non-dividing nucleus and a dividing nucleus. When cells are about to divide, the nucleus undergoes certain definite changes, the protoplasm outside the nucleus taking part in the process by means of the so-called *attraction spheres*. These consist of a wheel-like arrangement of fine fibrils starting from a central particle called the *attraction particle*, and the twin spheres are connected by the delicate system of fibrils denominated an *achromatic spindle*. The fibrillar network of the nucleus becomes arranged into definite patterns (skeins, stars, rosettes, &c.), finally separating into two groups of chromatic fibres and giving origin to daughter-nuclei, around which the protoplasm of the cell ultimately arranges itself and divides to form two separate cells. This indirect process of cell-division is spoken of as *karyokinesis* (Gk. *karuon*, kernel, nucleus; and *kinesis*, movement). A cell may therefore be described as a minute mass of protoplasm con-taining a specially for-med constituent known as the nucleus, the pro-

Fig. 3. - An Amœboid Pale Corpuscle of the Newt, showing a Double Nucleus with Reticulum of Chromoplasm, and the Protoplasm showing Spongioplasm and Hyaloplasm.

toplasm consisting of two substances, a network called *spongio-plasm* and a clear fluid called *hyaloplasm*, and the nucleus being formed of a fluid called *nucleoplasm*, and a system of fibres called *chromoplasm*. Protoplasm is a living substance, and forms the 'physical basis of life.'

3. **Composition and Properties of Protoplasm.**—Proto-plasm is a semi-fluid transparent substance that swells up in but does not mix with water. It usually contains granular matter and vacuoles ; but these are not essential to its nature. It consists of 80 to 85 per cent. of water with 15 to 20 per cent. of solids, chiefly proteids, the chief proteid being plasten. Proteids are substances containing carbon, hydrogen, nitrogen, oxygen, and sulphur ; and they are distinguished by certain

reactions (see Appendix). Carbohydrates like glycogen, as well as minute quantities of inorganic substances like calcium, are also present. The vital properties of protoplasm may be studied either in the amœba, a microscopic unicellular animal found in stagnant water and damp earth, or in the white blood-corpuscle, which is an example of a cell retaining its simple primitive form in an adult higher animal. The chief of these physiological characters are :—

(1.) **Power of Movement.**—On examining an amœba with a high power of the microscope, it is seen to consist of an irregular mass of protoplasm with a nucleus (or nuclei), the protoplasm usually containing granular matter and vacuoles. Its shape soon changes ; processes called pseudopodia are extended and retracted ; at times the whole mass seems to follow a process and the animal changes its position. Colourless blood-corpuscles may be noticed to exhibit similar movements, which are hence termed *amœboid.*

(2.) **Power of Response to Excitation, or Irritability.**—Besides the apparent spontaneous movements just described, protoplasm can be excited by external agencies called *stimuli*. Such stimuli are heat,

FIG. 4.—Human Colourless Blood-corpuscle, showing its successive changes of outline within ten minutes when kept moist on a warm stage. (Schofield.)

electricity, foreign particles, &c., and the protoplasm exhibits its excitability by movement or contraction of its mass when a suitable stimulus is applied.

(3.) **Power of Assimilation and Growth.**—The amœba and other similar protoplasmic cells possess the power of taking in material and using it as food. Such matter becomes modified by chemical changes in the protoplasm of the cell, so that part of it becomes converted into protoplasm itself, while the part that is not assimilated is rejected. The power of digestion and nutrition possessed by protoplasm is accompanied with a power of respiration. Oxygen is absorbed and drawn into chemical combination, but it is afterwards exhaled, for the most part united with carbon as carbon dioxide.

(4.) **Power of Secretion.**—Protoplasmic cells can elaborate new substances from the food supplied, some of which are stored for future use and some of which are excreted. In higher animals special cells in organs called secreting glands are set apart to prepare material known as enzymes or ferments. Living protoplasmic cells are thus the seat of continuous chemical changes, the cell either building up its own substance from the food supplied, or forming materials like glycogen, fat, ferments, &c., within its substance ; while, on the other hand, these constructive processes are always accompanied by destructive changes, of which oxidation forms the chief. The whole series of chemical changes within a cell, or within a body composed of cells, beginning with assimilation and ending with excretion, is termed *metabolism.* The process of building up living material, or constructive metabolism, is called *anabolism* ; the breaking down of material into simple products in the protoplasm, or destructive metabolism, is called *katabolism*.

(5.) **Power of Reproduction.**—When an amœba, by the growth of its protoplasm, has reached a certain size, its power of reproduction is manifested by fission, a simple division of the cells preceded by a division of the nucleus taking place. Two independent amœbæ are thus produced, each of which may afterwards undergo similar direct division. A more complex process, called karyokinesis or indirect division, is more common both in vegetable and animal cells. This has already been referred to.

The body of man, as well as that of every higher animal, is developed from a single cell, termed the *ovum*, which after fertilisation first divides into two, each of which again divides. By further division, eight, sixteen, thirty-two, and so on, cells are formed. After cell-multiplication has proceeded for some time, the cells become arranged in three layers ; the outermost layer is termed the *epiblast*, the innermost the *hypoblast*, and the middle layer the *mesoblast*. From these three layers all the tissues and organs of the adult are formed. From the epiblast are formed the epidermis and nervous system ; the hypoblast will form the digestive and absorbent membranes of the digestive and respiratory systems ; the mesoblast will give origin to the blood-corpuscles, the connective tissues, the bones and muscles of the body. In this process of development

FIG. 5.- Ovum. Magnified 350 times.

a, vitelline membrane, *zona pellucida* or *zona radiata*. *b*, external border of the yolk and internal border of the vitelline membrane. *c*, germinal vesicle and germinal spot.

from the three primary embryonic layers there is not merely cell-multiplication, but the cells become modified and metamorphosed from their primitive condition. Some become thin and flat, and cohering at their edges form the blood-vessels ; others become elongated and threadlike to form the fibres of muscular tissue ; others become separated by intercellular substance derived from the cells themselves, and in this intercellular substance fibres such as are found in connective tissue may arise, or calcareous matter be deposited, as in bone. To the study of the varieties of tissue thus produced in the adult we now proceed.

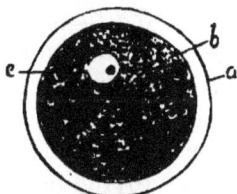

4. **Classification of Tissues.**—Having described the structure and properties of the protoplasmic units called animal cells, we now proceed to study the different kinds of material, or tissues, which are formed of or arise from these cells. The number of distinct tissues or textures forming the various organs of the human body has been reduced to five : (1) *epithelial* or surface-limiting tissues ; (2) *connective* or supporting tissues ; (3) *muscular* or contractile tissues ; (4) *nervous* or sensory tissues ; (5) *blood* and *lymph* particles, or nutritive tissues. Each of these classes, as will be seen, may be sub-

divided ; but the various kinds are not always sharply distinguished, forms of transition between them being often recognisable.

5. **Epithelial Tissues.**—Epithelium is a tissue covering the external and internal free surfaces of the body, and composed of cells placed side by side, with a small amount of cementing intercellular substance.   It thus forms—(1) the outer surface of the skin, where it is known as *epidermis* ; (2) the covering of mucous membranes, *i.e.* those membranes that line the passages and cavities of the body that communicate with the exterior, including the ducts and tubes of glands opening into these cavities ; (3) the terminal parts of the organs of special sense ; (4) the inner surface of serous membranes, *i.e.* the membranes forming what appear closed sacs in the thorax and abdomen ; (5) the inner surface of the heart, blood-vessels, and lymphatics ;[1] (6) the inner lining of the ventricles of the brain and the central canal of the spinal cord.

Epithelial cells consist of protoplasm and a nucleus, and when multiplying undergo division by karyokinesis in the way already described.   Having different functions, as protective, secreting, &c., and being exposed to diverse conditions, they present varieties of structure and form.   As regards *arrangement*, epithelial cells may be *stratified*, forming several superposed layers ; or *simple*, forming a tissue of one layer.   Where there are but two or three layers fitting into one another the term *transitional* may be used.   As regards *shape*, epithelial cells are classified as *squamous* or flat, *columnar* or cylindrical, and *cubical*.   No blood-vessels pass into epithelial tissue, the cells deriving their nourishment by imbibition of the plasma exuded into the subjacent tissue, but in many parts nerve fibrils, as fine filaments, exist among the epithelial cells.

6. **Stratified Epithelium.**—This epithelium forms the external surface of the body, or epidermis, and is also found lining the cavity of the mouth, pharynx, œsophagus, and the anterior surface of the cornea.

The *epidermis* consists of (a) the superficial horny layer of flattened cells ; (b) the *stratum lucidum*, formed of dense horny scales, showing traces

---

[1] The term *endothelium* is applied by some authors to the single layer of flattened cells which line the internal free surface of the serous sacs, blood-vessels, and lymphatics (4 and 5 above).

of a nucleus ; (c) the *rete mucosum* or Malpighian layer, consisting of several layers of nucleated cells, the deeper ones of which become columnar as they rest on the underlying corium or dermis. It is in the deepest part of a stratified epithelium that new cells are formed by cell-division. Cells are pushed outwards from below, and as they near the surface they not only become compressed and changed in shape, but they undergo a change in chemical composition, their protoplasm becoming converted into a horny substance, *keratin*, the nucleus also being often involved. Before this conversion occurs there is usually a deposit of granular material within

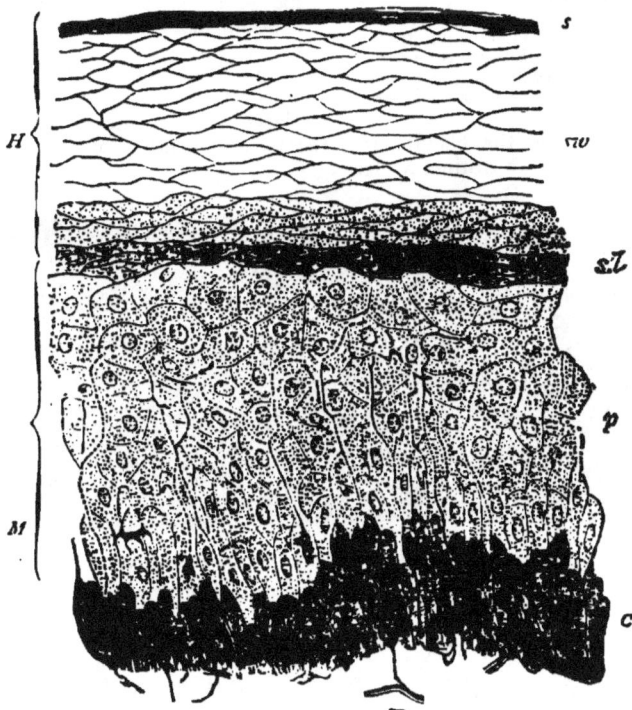

FIG. 6. Section of Epidermis from the Human Hand. Highly magnified.
(Ranvier.)

H, horny layer, consisting of *s*, superficial horny scales ; *sw*, swollen-out horny cells ; *s.l*, stratum lucidum ; *M*, rete mucosum or Malpighian layer, consisting of, *p*, prickle-cells, several rows deep, and *c*, elongated cells forming a single stratum near the corium ; *n*, part of a plexus of nerve-fibres in the superficial layer of the cutis vera. From this plexus fine varicose nerve-fibrils may be traced passing up between the cells of the Malpighian layer.

the cells, and the upper layer of cells of the rete mucosum has been termed the *stratum granulosum*. In the deeper layers of a stratified epithelium the surfaces of the cells are not close together, but a radiating system of fibrils connects the protoplasm of adjacent cells, and between these lie very fine channels. The exact purpose of these intercellular bridges and canals is not fully understood.

When such cells, known as *prickle-cells*, are isolated, the broken fibrils give them a spiked or dentated appearance. In coloured races the colour is due to pigment-granules, contained chiefly in the cells of the deeper layer of the rete mucosum. Pigment epithelial cells also occur on the internal surface of the choroid coat and iris of the eyeball. The older superficial cells of the epidermis are continually being removed by wasting and friction, a constant renewal taking place from below.

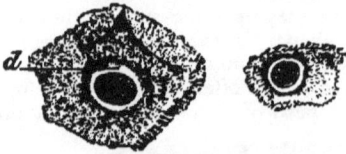

FIG. 7.—Two ' Prickle-cells from the deeper part of the Epidermis. (Ranvier.)
d, space around the nucleus, probably caused by shrinking of the latter.

7. **Transitional Epithelium.**— This variety is found in several places. A squamous transitional epithelium lines the bladder and ureters, and consists of three or four layers of cells with distinct nuclei.

FIG. 8.—Section of the Transitional Epithelium lining the Bladder. (E. A. S.)
a, superficial, b, intermediate, and c, deep layer of cells.

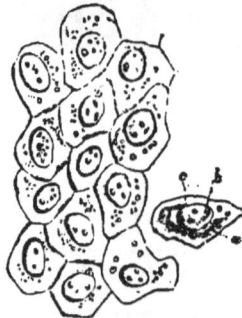

FIG. 9.—Pavement Epithelium, scraped from a Serous Membrane.
a, cell-body ; b, nucleus ; c, nucleoli. (Henle.)

The inner or most superficial layer consists of somewhat flattened cells, each of which overlies two or three pear-shaped cells, which form the second layer. A third layer fills up the intervals between the cells of the second layer. A columnar transitional epithelium of three or four layers of cells is found on the lining membrane of the larynx, trachea, and large bronchi, but these will be described with the corresponding simple epithelium below.

8. **Simple Squamous or Pavement Epithelium.**—A single layer of

simple scaly epithelium, consisting of flattened polygonal cells fitting together at their edges, lines (*a*) the air-cells of the lungs, certain looped tubules of the kidney, and the inner surface of the iris and choroid ; (*b*) the free surfaces of serous membranes—as the pleura, pericardium, peritoneum, arachnoid—and the interior of the heart, blood-vessels, and lymphatics. It is to the epithelial tissue of the latter division (*b*) that the term *endothelium* is sometimes applied. The outline of these simple squamous epithelial cells is readily brought out by treatment with silver nitrate, as this salt darkens the small amount of intercellular substance.

9. **Simple Columnar Epithelium.**—This kind of epithelium consists of cells of cylindrical or prismatic form set upright on a surface, and, except in the cases mentioned above, the cells form but one layer side by side. Owing to mutual compression

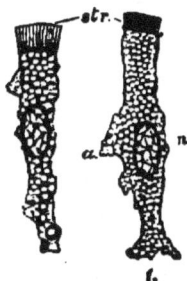

FIG. 10.—Columnar Epithelium Cells of the Rabbit's Intestine. (E. A. S.)

The cells have been isolated after maceration in very weak chromic acid. The protoplasm is reticular and vacuolated ; the striated border (*str.*) is well seen, and the bright disc separating this from the cell-protoplasm ; *n*, nucleus with intranuclear network ; *a*, a thinned out wing-like projection of the cell, which probably fitted between two adjacent cells.

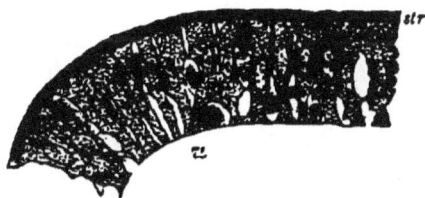

FIG. 11.—A row of Columnar Cells from an Intestinal Villus of the Rabbit. (E. A. S.)
str, striated border : *w*, wander-cells between the epithelium cells.

and the frequent presence of lymphoid or wander-cells, their shape is often very irregular. Columnar epithelium is found lining the alimentary canal from the lower end of the œsophagus to the anus, as well as on the free surface of the trachea and largest air-tubes. In some cells of columnar epithelium, as those of the small intestine, the free edge is finely striated. Columnar cells often undergo a modification of shape owing to the secretion in their interior of *mucin*, the chief organic constituent of mucus. The mucus distends the upper part of the cell until a rupture occurs and the mucus is discharged on the surface. Such mucus-secreting cells are called *goblet* or *chalice* cells, and the columnar epithelial cells of the intestinal and respiratory tracts may at any time undergo this transformation into goblet-cells temporarily, while the epithelial cells of mucus-secreting glands have this function permanently.

10. **Ciliated Epithelium.**—This variety of epithelium consists of cells, usually columnar, having at their free ends

FIG. 12.—Simple Columnar Epithelium from the Mucous Membrane of the Intestine, with goblet-cells pouring out their contents. (Klein and Noble Smith.)

fine hair-like processes called *cilia*. Each cell bears a brush of cilia, about $\frac{1}{8500}$ of an inch long in the windpipe, and during life or for a short time after removal from the body the cilia exhibit a rapid whip-like movement, the cells of the surface of a membrane moving simultaneously or in quick succession in one direction. Bending swiftly in one direction and returning to the upright position more slowly, these cilia set in motion in a definite direction the fluid which bathes the ciliated surface. It is thus that mucus is moved along the bronchial tubes and trachea to the pharynx. Ciliated epithelium is found in the respiratory region of the nose, on the upper half of the pharynx, along the Eustachian tubes, in the lower part of the larynx, except over the vocal cords, in the trachea and bronchial tubes, lining the

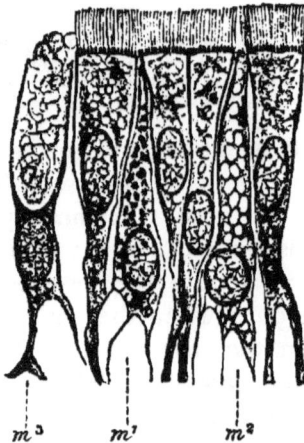

FIG. 13.—Ciliated Epithelium Cells from the Trachea of the Rabbit. Highly magnified. (E. A. S.)

m^1, m^2, m^3, mucus-secreting cells, lying between the ciliated cells, and seen in various stages of mucin-formation.

ventricles of the brain and the central canal of the spinal cord, and on the mucous lining of the uterus and Fallopian tubes. As to the nature and cause of ciliary movement little is definitely known, except that the source of motion is contained in the ciliated cell itself and is independent of any connection with nerve fibres. Ciliary movement may be readily studied by taking a piece of the gill of a salt-water mussel, mounting it in sea water, and examining it with a power of about 200 diameters.[1]

11. **Sensory Epithelium.**—Various forms of modified epithelial cells are connected with the terminations of certain sensory nerves to form the receptive end-organ for different kinds of vibrations. Examples of such sensory epithelium are the rods and cones of the retina, the auditory hair-cells, &c. These will be best studied in connection with the different senses.

12. **Functions of Epithelium.**—Putting aside the varieties of sensory epithelium, it will be seen that the functions of epithelium are either *protective* or *secretive*. Thus the layers of epithelium forming the epidermis, the epithelium lining the air-passages and that lining the eyelids, serve mainly as a protective covering. But the epithelial cells of the salivary glands, those of the gastric glands which secrete gastric juice, those of the liver, of the sweat-glands, &c., are composed of protoplasm which is the seat of active chemical operations. New substances are formed from the blood and poured out as secretions to fulfil important functions, or discharged from the body as excretions. ' A secreting epithelium may be regarded as a partition between the blood, or, more properly speaking, between the lymph on the one side and the lumen of the secreting gland on the other. From the lymph the materials are taken by the secreting cells and then worked up into the components of the secretion, and finally discharged on the other side into the lumen, and thence by the ducts of the secreting gland to their destination. The amount of secretion is in some cases, as in that of the kidney, very largely influenced by the amount of blood reaching the organ, and by the blood pressure ; this again is dependent on the size of the blood-vessels, which is regulated by the vaso-motor nerves that supply their muscular tissue.'—Halliburton. Ciliated epithelium, besides being protective, also aids by its movements in propelling fluids and minute particles from the body.

[1] See note on Microscope in Appendix.

13. **Basement Membrane.**—Beneath the epithelium of certain parts appears a fine homogeneous membrane spoken of as *basement membrane* or *membrana propria.* That at the base of most mucous membranes and beneath the epithelium of secreting glands, however, is found to consist of a very thin layer of flattened cells belonging to the variety of tissue called connective tissue, now to be treated of.

14. **Connective Tissues.**—The term *connective tissues* includes a number of tissues which at first sight appear to differ greatly, but which are, nevertheless, properly grouped together owing to their common function, origin, histological and chemical similarities. Their common function is that of connecting and supporting the other tissues, and they have a common origin, as they are all derived from the mesoblast in the developing body. The histological similarities or points of structure in common are three microscopic elements which exist in varying proportion in each, *viz.* a ground-substance or matrix, cells, and fibres. All varieties of connective tissue that contain white fibres yield gelatine on boiling. Moreover, these tissues may under certain circumstances replace each other or merge into one another. There are three chief varieties of connective tissue, the first variety being separated into six kinds. The following table gives a list of these tissues, with a few preliminary remarks about each :—

I. Connective tissues proper.

(1.) **Areolar Tissue.**—This is distributed as an irregular meshwork or loose connective tissue through all parts of the body. It is found beneath the skin and mucous membranes, it forms a sheath for the muscles and the organs generally, binds the parts of organs together and different organs one to another. It contains cells and fibres (both white and yellow) embedded in a ground-substance.

(2.) **White Fibrous Tissue.**—This forms the chief part of tendons and ligaments, is found in the true skin and the denser fasciæ binding down the muscles. It consists of bundles of white parallel fibres, the other elements being relatively unimportant. The fibres run for the most part in one or two directions, instead of interlacing in every direction as in areolar tissue.

(3.) **Yellow Elastic Tissue.**—This is the variety of connective tissue in which elastic fibres preponderate. It is found in the ligamentum nuchæ of quadrupeds, in the ligamenta subflava between the arches of adjacent vertebræ, in the walls of the trachea and its branches, and with other textures in the walls of the blood-vessels.

(4.) **Retiform or Adenoid Tissue.**—This is found chiefly in lymphatic glands and allied structures.

(5.) **Adipose Tissue or Fat.**—This is developed from areolar tissue, the protoplasm of the cells being for the most part replaced by fat, the fibrous element almost disappearing.

(6.) Mucous or jelly-like connective tissue, found in the vitreous humour of the eye and in the body during its early development.

The above varieties of connective tissue are often referred to as *the fibrous connective tissues par excellence.*

II.—Cartilage or gristle, of which there are three varieties :—

(1.) Hyaline cartilage.

(2.) White fibro-cartilage.

(3.) Yellow fibro-cartilage.

III.—Bone and dentine.

In all the forms of connective tissue it will be noticed that the intercellular substance is greatly developed and the cells are much less numerous than in epithelium, epithelium having but little intercellular material.

15. **Areolar Tissue.**—The most widely distributed kind of connective tissue has a more or less open texture, and appears to the naked eye to consist of fine transparent threads and films intercrossing in every direction, and leaving, especially when stretched, open spaces or areolæ between them. Hence the name given to it, areolar tissue. Its universal distribution in and around the other tissues and organs of the body is thus described in Quain's 'Anatomy' :—

'**Areolar Tissue.**—If we make a cut through the skin and proceed to raise it from the subjacent parts, we observe that it is loosely connected to them by a soft filamentous substance of considerable tenacity and elasticity, and having, when free from fat, a white fleecy aspect ; this is the substance known as *areolar tissue.* In like manner the areolar tissue is found underneath the serous and mucous membranes which are spread over various internal surfaces, and serves to attach these membranes to the parts which they line or invest ; and as under the skin it is named "subcutaneous," so in the last-mentioned situations it is called "subserous" and "submucous" areolar tissue. But on proceeding further we find this substance lying between the muscles, the blood-vessels, and other deep-seated parts, occupying, in short, the intervals between the different organs of the body where they are not otherwise insulated, and thence named "intermediate ; " very generally, also, it becomes more consistent and membranous immediately around these organs, and under the name of the "investing" areolar tissue, affords each of them a special sheath. It thus forms inclosing sheaths for the muscles, the nerves, the blood-vessels, and other parts. Whilst the areolar tissue might thus be said in some sense both to connect and to insulate entire organs, it also performs the same office in regard to the finer parts of which these organs are made up ; for this end it enters between the fibres of the muscles, uniting them into bundles ; it connects the several membranous layers of the hollow viscera, and binds together the lobes and lobules of compound glands ; it also accompanies the vessels and nerves within these organs, following their branches nearly to their finest divisions, and affording them support and protection. This portion of the areolar tissue has been named the " penetrating," " constituent," or " parenchymal."

‘ It thus appears that the areolar is one of the most general and most extensively distributed of the tissues. It is, moreover, continuous throughout the body, and from one region it may be traced without interruption into any other, however distant; a fact not without interest in practical medicine, seeing that in this way dropsical waters, air, blood, and urine, effused into the areolar tissues, and even the matter of suppuration, when not confined in an abscess, may spread far from the spot where they were first introduced or deposited.

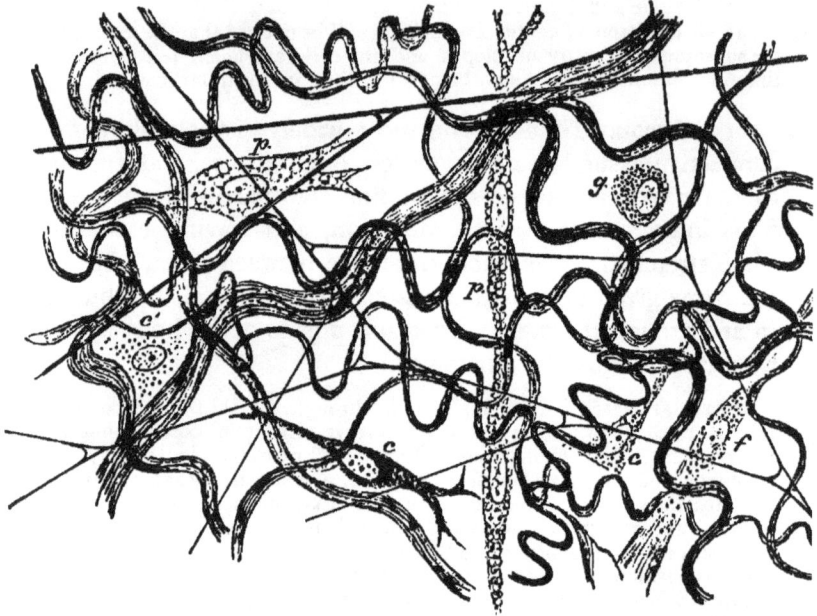

Fig. 14.—Subcutaneous Areolar Tissue from a young Rabbit.   Highly magnified.
(E. A. S.)
The figure shows the appearance of the tissue examined perfectly fresh.
The white fibres are in wavy bundles, the elastic fibres form an open network. *p, p*, vacuolated cells (plasma-cells); *g*, granular cell; *c, c*, branching lamellar cells; *c'*, a flattened cell of which only the nucleus and some scattered granules are visible; *f*, fibrillated cell.

‘ On stretching out a portion of areolar tissue by drawing gently asunder the parts between which it lies, it presents an appearance to the naked eye of a multitude of fine, soft, and somewhat elastic threads, quite transparent and colourless, like spun glass; these are intermixed with fine transparent films, or delicate membranous laminæ, and both threads and laminæ cross one another irregularly and in all imaginable directions, leaving open interstices or areolæ between them.   These meshes are, of course, more apparent when the tissue is thus stretched out; it is plain also that they are not closed cells, as the term " cellular tissue," which was formerly used to denote the areolar tissue, might seem to imply, but merely interspaces, which open freely into one another : many of them are occupied by the fat, which, however, does not lie loose in the areolar spaces, but is enclosed in its own vesicles. A small quantity of colourless transparent fluid of the nature of lymph is also present in the areolar tissue, but, in health, not more than is sufficient to moisten it.

' On comparing the areolar tissue of different parts, it is observed in some to be more loose and open in texture, in others more dense and close, according as free movement or firm connection between parts is to be provided for.'

Examined under the microscope, the transparent threads are seen to consist of wavy bundles of very fine parallel fibres (*white fibres*) running in various directions, with a few single branching fibres of another kind, known as *elastic fibres*. Besides these fibres several varieties of connective-tissue cells may be seen: (1) *flattened connective tissue corpuscles*, with an oval nucleus, and often showing branching processes, *c, c* ; (2) *plasma-cells*, with the protoplasm markedly vacuolated, *p, p* ; (3) *granular cells*, or cells containing distinct granules, *g*. In addition to the fixed connective-tissue cells, leucocytes identical with white blood-corpuscles and lymph-corpuscles are often plainly visible (wander-cells). Cementing the white fibres together and forming the matrix or basis of the tissue between the bundles, is a clear homogeneous material containing mucin, called the *ground-substance*. It is difficult to see in the fresh tissue, but, like the intercellular cement of epithelium, it is stained brown by silver nitrate. It is in depressions of the ground-substance on the surface of the white fibres or in the spaces—cell-spaces—of the ground-substance uniting the bundles that the fixed connective-tissue cells are found. Both the cells and the spaces in which they lie may intercommunicate by their branches, and thus bring into connection the superficial and deeper parts of the tissue. Blood-vessels, lymphatics, and nerves are conveyed in areolar connective tissue to the parts in which they are to be distributed. A few capillaries are distributed to the tissue itself, and numerous lymphatic networks are found in it. It is doubtful whether any nerves terminate in it, for it may be cut in a living animal without giving any pain, unless the instrument meets nerves passing through it to other parts.

C

16. **White Fibrous Tissue.**—Besides forming the wavy
bundles in areolar tissue, *white fibres* arranged in parallel
bundles form compact bands or cords, the ligaments by which
the bones are connected together at the joints, and tendons by
which muscular fibres are affixed to the bones. They also
form fibrous membranes, such as the periosteum covering the
bones, the perichondrium covering cartilages, the dura mater
lining the skull, and the fasciæ or aponeuroses enveloping and
binding together the muscles. In the true skin and mucous

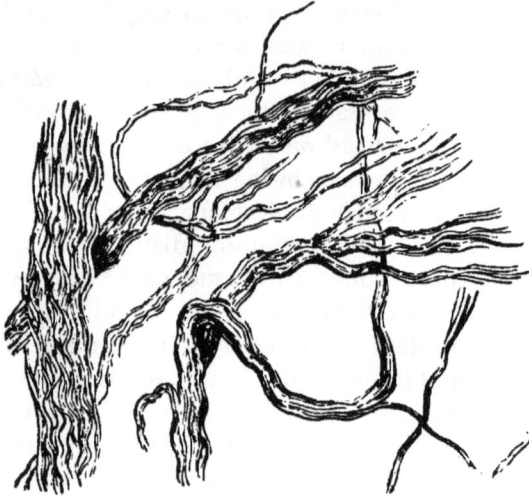

FIG. 15.—Bundle of White Fibres of Areolar Tissue, partly unravelled.

membrane the bundles of interlacing white fibres form a close
felt work.

Compact white fibrous tissue has a shining pearly aspect.
It is exceedingly strong and pliant, but quite inelastic. Examined with a high power, a small bundle of white fibres, when
teased with a needle, is seen to consist of very fine transparent
fibres or filaments from $\frac{1}{30000}$ to $\frac{1}{23000}$ of an inch in thickness.
These fine filaments do not occur singly, but are cemented into
small bundles by a small quantity of the mucin ground-substance. Each filament of a bundle seems parallel with its
neighbouring filaments, neither branching nor uniting with
others. Though transparent when seen with transmitted light,

the filaments in mass appear white. Acetic acid causes them to swell and become almost invisible. Boiled with water they are converted into gelatine.

In fibrous tissue (tendon and ligament) the cells, termed

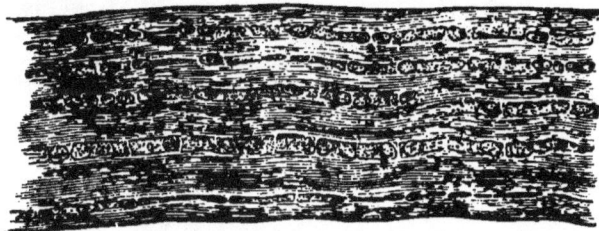

FIG. 16.—Tendon of Mouse's Tail, stained with logwood, showing chains of cells between the tendon-bundles. 175 diameters. (E. A. S.)

'tendon-cells,' are all flattened and arranged in rows, parallel to the bundles of fibres. The ends of these nucleated tendon-cells lie close together, the nuclei of adjacent cells often being in close proximity, as may be noticed in the figure. Fibrous tissue receives blood-vessels that run in tendons and ligaments between the longitudinal fibrous bundles, sending communicating branches across to form an open network. Lymphatics are abundant both in the enveloping sheaths of tendons and in the penetrating areolar tissue. Many tendons, ligaments, and fibrous sheaths possess nerve fibres, those of tendon often terminating in a special manner (par. 49).

FIG. 17.— Elastic Fibres of Areolar Tissue.

17. **Yellow Elastic Tissue.**—Elastic tissue is that variety of connective tissue in which elastic fibres abound. These elastic fibres, already referred to as occurring in areolar tissue, are also found in small and variable proportions among the white fibrous bands and sheaths already mentioned, and between the bundles of white fibres of the dermis and mucous membranes. They form a fenestrated membrane in the middle coat of large blood-vessels, and are found in large numbers in the walls of the pulmonary alveoli. They also occur closely massed together in the elastic

C 2

ligaments which extend between the arches of the vertebræ and in the ligamentum nuchæ of the neck. When seen in quantity the elastic fibres have a yellowish appearance. They are distinguished from the white fibres not only by their elasticity and colour, but also by their sharper out-line, by joining or anastomosing in their course so as to form a network, and by a tendency to curl up at the ends when broken across. Elastic fibres are best seen after treating a tissue with acetic acid. This causes the white fibres to swell and become indistinct, but the elastic fibres remain unaffected. Elastic fibres do not yield gelatine on boiling, but they yield or are composed of a different substance, elastin. Their size varies from $\frac{1}{8000}$ to $\frac{1}{24000}$ of an inch, large fibres occurring in the ligamenta sub-flava of the vertebræ and very fine ones being found in the vocal cords.

18. Retiform or Adenoid Tissue.—This is a variety of areolar or connective tissue found in the spleen, in the tonsils, in lymphatic glands, and in many mucous membranes. It is com-

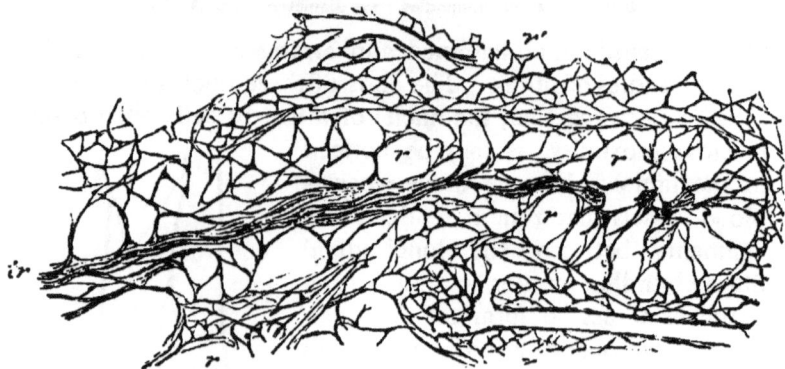

FIG. 18.– Reticulum from the Medullary part of a Lymphatic Gland. (E. A. S.)
tr, end of a trabecula of fibrous tissue ; *r*, *r*, open reticulum of the lymph-path, con-tinuous with the fibrils of the trabecula ; *r'*, *r'*, denser reticulum of the medullary lymphoid cords. The cells of the tissue are not represented, the figure being taken from a preparation in which only the connective tissue fibrils and the reticulum are stained.

posed of a very fine network or reticulum of fibrils continuous with white fibres of ordinary connective tissue, but having few or no elastic fibres. Around the fibrils of the network are wrapped the fixed cells of the tissue, so that it seems to be formed, when these cells are not cleared away, of stellate cells and their anastomosing processes. The meshes of the network are occupied with lymph and by numerous corpuscles which closely resemble lymph-corpuscles. These are known as *lymphoid cells*, and the tissue containing them is known as lymphoid or adenoid (gland-like) tissue. In the spleen the interstices of the retiform tissue are mostly occupied by blood instead of lymph. Retiform tissue thus forms the stroma or framework of lymphoid tissue. (See fig. 18.)

19. **Adipose Tissue, or Fat.**—Fatty tissue is distributed almost generally through the body, though collected more abundantly in certain positions. It forms in the healthy subject a layer beneath the skin, is gathered in quantity around the kidneys, fills up the furrows on the surface of the heart, exists in abundance in the marrow of the bones, but it is absent from the brain and lungs. Examined by the microscope, fat is seen to consist of cells or vesicles collected into lobules,

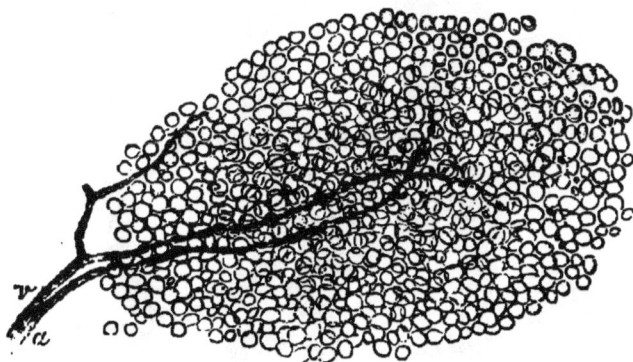

Fig. 19.—A small Fat-lobule from the Subcutaneous Tissue of the Guinea-pig. Magnified about 20 diameters (E. A. S.)

a, small artery distributed to the lobule ; *v*, small vein. The capillaries within the lobule are not visible.

these lobules being collected into clusters, which appear like grains to the naked eye, the cells and lobules being held together by a small amount of areolar tissue in which the blood-vessels ramify. Each lobule has its afferent artery, capillaries, and efferent vein, but no nerves. The cells in well-nourished bodies are round or oval, $\frac{1}{300}$th to $\frac{1}{800}$th inch in diameter. They are ordinary connective-tissue cells in which oily matter has been secreted from the blood, leaving the original protoplasm of the cell as a mere envelope. The nucleus of the protoplasm may often be detected flattened out in this envelope. Fat consists of stearic, oleic and palmitic acids united with glycerine. The fat is fluid during life, but becomes solid after death, often forming needle shaped crystals. During starvation the fat may be absorbed and used up in the body, and then the cells become ordinary connective-tissue cells again. Adipose tissue serves as a protective packing material, prevents the heat of the body from passing away too rapidly, as it is a bad conductor, and furnishes a store of substance rich in carbon and hydrogen for use in the body.

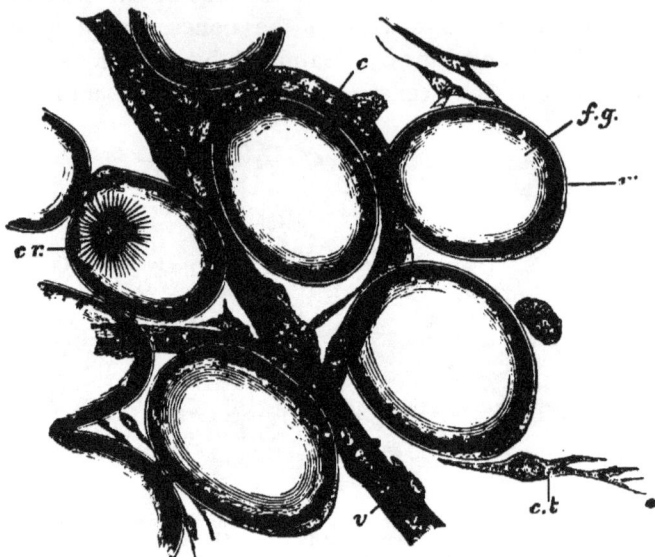

FIG. 20. —A few Cells from the margin of the Fat-lobule represented in the preceding figure. Highly magnified. (E. A. S.)

f.g, fat-globule distending a fat-cell ; *n*, nucleus ; *m*, membranous envelope of the fat-cell ; *cr*, bunch of crystals within a fat-cell ; *c*, capillary vessel ; *v*, venule ; *c.t*, connective-tissue cell ; the fibres of the connective tissue are not represented.

20. Cartilage.—Cartilage or gristle is a tough bluish-white substance, opaque in mass, but translucent in thin slices. Though of firm consistence it is very elastic, yielding readily to pressure or torsion, but recovering its shape when the constraining force is removed. On prolonged boiling it yields an albuminoid substance named *chondrin*, which is closely allied to gelatine. No nerves have been found in cartilage and it is devoid of sensibility. It is also non-vascular, and derives its nourishment by imbibition of lymph which exudes from the neighbouring capillary vessels, *i.e.* from the vessels of the perichondrium or from those of the bone and synovial membrane in articular cartilage. All cartilage, except articular cartilage, is surrounded by a vascular membrane consisting chiefly of white fibrous tissue, and called the *perichondrium*.

When a thin slice is examined under the microscope cartilage is seen to consist of nucleated cells and a firm ground-substance or matrix. The matrix is either without distinct structure, *i.e*, homogeneous like ground glass, or fibrous. Two varieties of fibrous are found, one in which the fibres are white like those of white fibrous tissue, and the other in which the fibres are yellow like those in yellow elastic tissue. Cartilage may therefore be divided thus : —

(1.) Hyaline . . . . . $\begin{cases} \text{Costal} \\ \text{Articular} \\ \text{Temporary} \end{cases}$

(2.) Fibro-cartilage . . . $\begin{cases} \text{White} \\ \text{Yellow.} \end{cases}$

(1.) *Hyaline cartilage* (Gk. *hualinos*, of glass) occurs in the adult, forming the costal cartilages, the nasal cartilages, investing the ends of bones

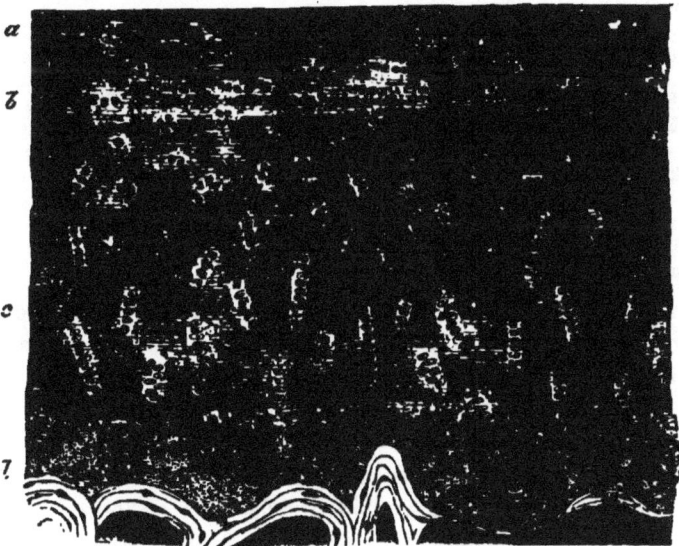

FIG. 21.—Vertical Section of Articular Cartilage covering the lower end of the Tibia, Human. Magnified about 30 diameters.

a, cells and cell-groups flattened conformably with the surface ; *b*, cell-groups irregularly arranged ; *c*, cell-groups disposed perpendicularly to the surface ; *d*, layer of calcified cartilage ; *e*, bone.

at the joints, entering into the structure of the larynx, trachea, and bronchi. A temporary form exists in the fœtus and young animals which is destined to be replaced with bone by the deposition of lime salts. Examined under the microscope, hyaline cartilage is seen to consist of rounded, oval, or irregular cells, lying in what appears a homogeneous matrix resembling ground glass. In articular cartilage the cells are scattered in groups of two, four, or eight through the matrix, the arrangement of the cell groups being vertical near the surface of the bone, but horizontal near the outer surface of the cartilage where it joins the synovial membrane (see fig. 21). The protoplasm of the cells under high powers shows fibrils and granules, and its nucleus has the usual intranuclear network. The part of the matrix next the cartilage-cell forms a capsule, and as the cells multiply by indirect division a new capsule soon forms around each, the primary capsule disappearing. It is thought by some observers that fine channels in the matrix pass from one cartilage cell to another. In the costal carti-lages the cells are often larger and contain a little fat, while the matrix is more distinctly fibrillated.

(2.) *White fibro-cartilage* is found where great strength and a certain amount of rigidity are required. It occurs (*a*) connecting cartilage form-ing part of the intervertebral discs ; (*b*) as interarticular discs between the ordinary articular cartilages of certain joints, *e.g.* the knee-joint ; (*c*) as marginal cartilages round the rim of the shoulder and hip-joints ; (*d*) as lining the groove in which certain tendons of muscle glide. Under the microscope white fibro-cartilage shows a matrix composed of many fibres, in which cartilage-cells appear.

(3.) *Yellow fibro-cartilage*, or elastic cartilage, is found in the epiglottis, the cornicula of the larynx, in the outer ear, and in the Eustachian tube. It is more flexible and tough than the hyaline cartilage, and the matrix in which the cells lie is composed mainly of fine elastic interlacing fibres.

Cartilage has several important uses. It assists in binding bones together and yet allowing a certain degree of movement ; it acts as a buffer to deaden shocks ; it reduces friction at joints ; it serves to keep open and maintain the shape of tubes, as in the trachea ; it forms attachments for muscles and ligaments ; and it serves to deepen joint cavities. The costal or rib cartilages form an important part of the framework of the thorax and impart elasticity to its walls.

21. Bone—Physical Properties and General Structure.—

With the exception of enamel, bone is the hardest structure of the body, and yet it possesses considerable toughness and elasticity. In its fresh state it is pinkish-white externally, though redder within. On cutting through a bone the eye easily sees two kinds of bony tissue, a hard and *compact* tissue like ivory forming the outside shell, and a looser tissue internally, with fibres and thin plates, having the appearance of lattice-work and hence called *cancellous* tissue, a gradual transi-tion from one kind to the other being apparent. In a long bone the large round end consists of the cancellous tissue with a thin coating of compact tissue ; in the hollow shaft of the long bone the sides are almost entirely of compact tissue. In tabular or flat bones we find two layers or plates of compact tissue at the surface and a spongy texture known as diploë between. A fresh or living bone is covered by an outer tough fibrous membrane called the *periosteum*. From a close network of blood-vessels in the periosteum branches find their way through small openings on the surface of the bone, mainly to nourish the compact tissue, running in the bone through tiny channels called Haversian canals. By stripping the periosteum from the surface of living bone, small bleeding-points indicating the entrance of the periosteal vessels may be seen, and a longitudinal section of a long bone not only shows that the substance of the bone must be well supplied with blood, as it is seen to exude in all parts, but the interior of such a bone is seen to be

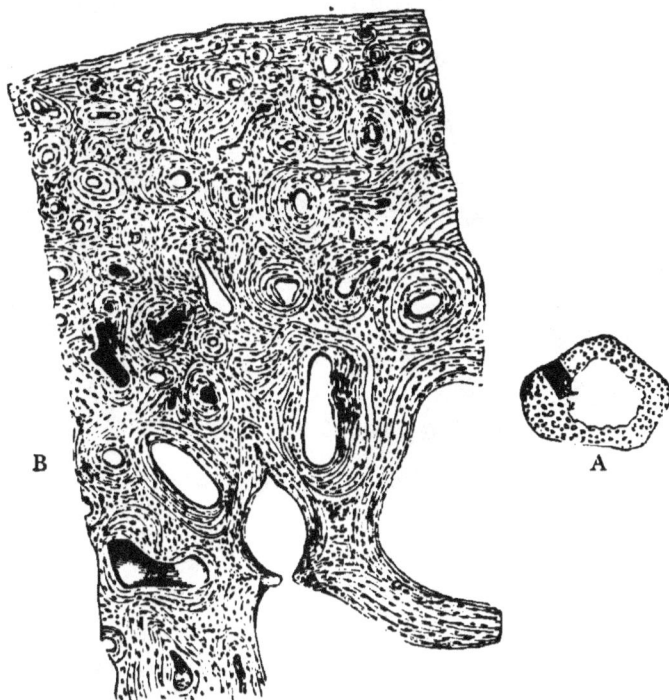

FIG. 22.—A. Transverse Section of a Bone (Ulna) deprived of its earth by acid.
(Sharpey.)
The openings of the Haversian canals are seen. Natural size. A small portion is shaded to indicate the part magnified in Fig. B.
B. Part of the Section A, magnified 20 diameters.
The lines indicating the concentric lamellæ are seen, and among them the lacunæ appear as little dark specks.

occupied by a cylindrical cavity filled with marrow, the marrow being surrounded by a vascular areolar envelope lining the medullary cavity, and termed the *endosteum*. Besides the blood-vessels that pass from the periosteum, long bones have a nutrient artery entering at some part of the shaft which, passing into the medullary canal, breaks up in the marrow. Other small vessels reach the articular extremities to supply the cancellous tissue, for during life the spaces of the cancellous tissue contain marrow and blood-vessels.

22. **The Periosteum.**—The periosteum adheres very firmly to the bone and invests every part except at the joints where it is covered with cartilage. Its chief uses are to support the

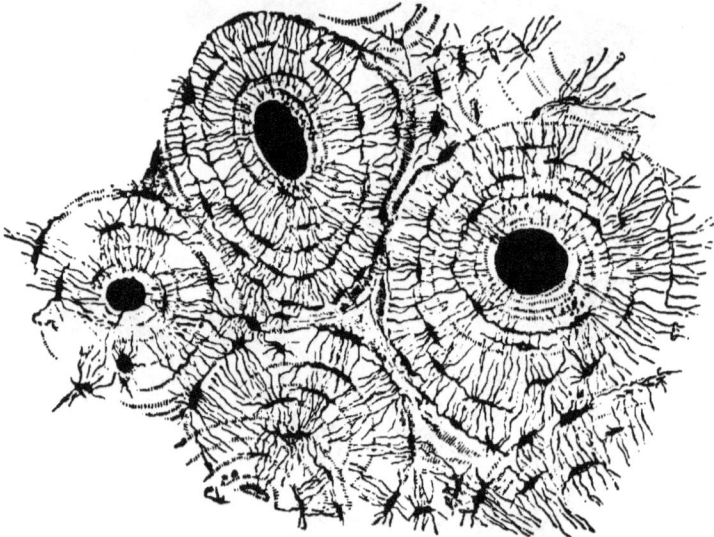

FIG. 23.—Transverse Section of Compact Tissue (of Humerus). Magnified about 150 diameters. (Sharpey.)

Three of the Haversian canals are seen, with their concentric rings; also the lacunæ, with the canaliculi extending from them across the direction of the lamellæ. The Haversian apertures had become filled with air and débris (from the grinding), and therefore appear black in the figure, which represents the object as viewed with transmitted light. (In a longitudinal section the Haversian canals would be seen uniting with one another.)

vessels going to the bone and to afford attachment for the tendons and ligaments where they are fixed to the bone. It consists of two layers : (*a*) an outer layer, chiefly of white fibres, in which the blood-vessels ramify before entering the compact bony tissue ; (*b*) an inner layer consisting largely of elastic fibres and granular cells called *osteoblasts* or bone-forming cells. Fine nerves and lymphatics are found in the periosteum and accompany the arteries into the bone. Since bony tissue is nourished in great part by vessels that leave the periosteum, it is apt to die if this membrane be removed.

23.—Minute Structure of Bone. —Besides the spaces in the cancellous or spongy tissue of bone formed by the reticulated spicules of bone, compact bony tissue is permeated by anastomosing channels called Haversian canals. Fig. 22, A, represents a cross section of a long bone near the middle, and on examining this with a hand-glass the openings of these longitudinal passages may be seen. Their average diameter is about $\frac{1}{500}$th of an inch, though wider near the medullary cavity and less near the outside of the bone.

Fig. 22, B, shows a small part of A magnified. Here the Haversian canals are seen to be surrounded by an appearance of concentric rings. This appearance is caused by the transverse section of thin tubes or cylinders of bony tissue (spoken of as concentric plates or *lamellæ*) fitting inside one another and surrounding the Haversian canal. Besides the concentric lamellæ there are other thin layers of bony tissue arranged parallel to the surface as at *a*, while others run between the Haversian sets. Compact bone is in fact made up of these bony lamellæ closely applied and lying in various directions. The Haversian canals are minute channels through the compact tissue, the bony lamellæ around them having a concentric arrangement. In a longitudinal section the Haversian canals are seen to be but short, as they soon unite with others, and thus form a network of tubes in the compact bone.

All over the section little dark specks are seen among the lamellæ. These are in reality minute cavities named *lacunæ*. With a higher power these lacunæ are more dis-

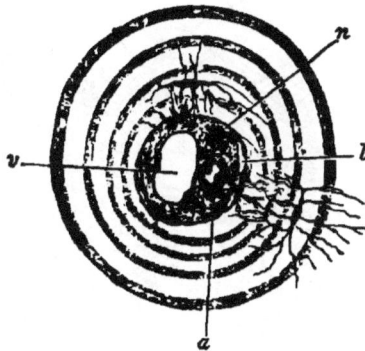

FIG. 24. Section of a Haversian Canal, showing its contents. Highly magnified. (E. A. S.)

a, small arterial capillary vessel ; *v*, large venous capillary ; *n*, pale nerve-fibres cut across ; *l*, cleft-like lymphatic vessel : one of the cells forming its wall communicates by fine branches with the branches of a bone-corpuscle. The substance in which the vessels run is connective tissue with ramified cells ; its finely granular appearance is probably due to the cross section of fine fibrils.

tinctly seen, while extending from them are minute tubes named *canaliculi*. The canaliculi pass across the lamellæ and connect neighbouring lacunæ with one another and with the Haversian canal.

Most of the Haversian canals contain two small blood-vessels, nerve-fibres, and a lymphatic vessel, and they communicate externally with the periosteum and internally with the marrow. Each Haversian canal, with its concentric lamellæ, lacunæ, and canaliculi, form what is known as a Haversian system.

The lacunæ are situated between the lamellæ, and each lacuna is occupied by a bone-cell, a flattened nucleated cell that sends processes into the canaliculi. The plasma of the blood that passes through the capillary walls passes into the dense bony substance by means of the lacunæ and canaliculi, the bone-cells playing, no doubt, an important part in the nutritive process.

The lamellæ of which we have spoken are formed of fine fibres decussating so as to form a close network. By peeling off thin films from the surface of a bone that has been softened in acid and examining it under the microscope, this fibro-reticular structure can be made out. In many places the various lamellæ may be seen to be bolted or held together by larger tapering fibres known as *perforating fibres*. Bone, in fact, is a variety of connective tissue, and consists of (1) cells or branched bone-corpuscles, (2) interlacing and decussating fibres, (3) a ground-substance in which calcium salts are deposited. The lamellæ are made up of the fine fibres just mentioned lying in calcified ground-substance, while the cells occupy the lacunæ between the lamellæ, and send processes into the canaliculi.

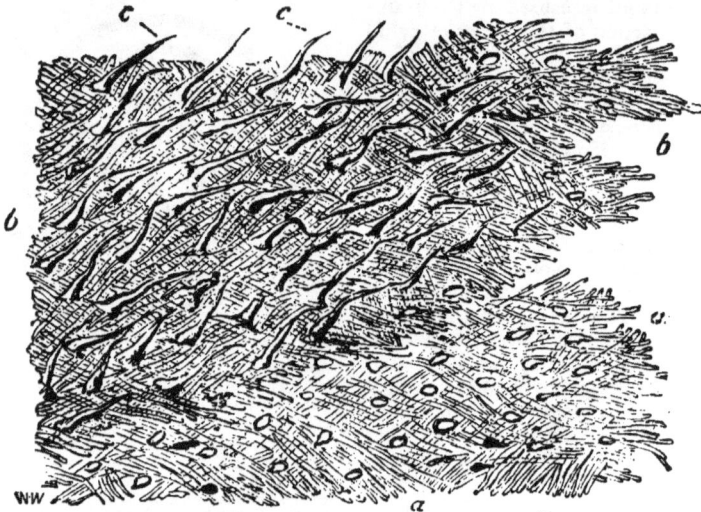

FIG. 25. - Lamellæ torn off from a Decalcified Human Parietal Bone at some depth from the surface. (Sharpey.)

a, lamellæ, showing decussating fibres ; *b, b*, thicker part, where several lamellæ are superposed ; *c, c*, perforating fibres. Apertures through which perforating fibres had passed are seen especially in the lower part, *a, a*, of the figure. Magnitude as seen under a power of 200 but not drawn to a scale. (From a drawing by Allen Thomson.)

24. Chemical Composition of Bone.—Bone consists of mineral and organic matter in intimate combination. The former gives hardness and rigidity, the latter tenacity. The mineral matter can be removed by maceration in dilute acids, and the bone thus decalcified is softer, tougher, and more flexible. The organic or animal part may be burnt out, when the bone will retain its original form but be white and brittle. The animal part may be resolved into gelatine on boiling. The

mineral or inorganic matter forms about two-thirds of dry bone, and the animal matter one-third. In undried bone, without separation of marrow or blood, there is nearly 50 per cent. of water. The chief organic constituent of bone is collagen, which is changed into gelatine on boiling with water ; the chief inorganic constituent is calcium phosphate, $Ca_3(PO_4)_2$. The following table gives the percentage composition of dry bone :—

Organic matter		33·30
Mineral matter Calcium phosphate . . .		51·04
,, carbonate . . .		11·30
,, fluoride		2·00
Magnesium phosphate . . .		1·16
Sodium salts		1·20

25. The Marrow. There are two kinds of marrow found in bone, yellow and red. Yellow marrow fills the medullary cavity and some of the larger cancelli of long bones. It consists chiefly of fat-cells with blood-vessels and a small amount of connective tissue. Red marrow is found in the cancellated tissue of long bones, in the diploë of the bones of the cranium, in the bodies of the vertebræ, the sternum and the ribs. Red marrow consists of delicate connective tissue, numerous small blood-vessels, and a large number of round nucleated cells called *marrow-cells*, some large cells known as *giant cells* or *myeloplaxes*, and a number of small nucleated cells of a reddish tint. These coloured cells are termed *erythroblasts*, and resemble nucleated coloured corpuscles found in the embryo. They multiply by karyokinetic division, and are believed to be the cells from which ordinary red blood-corpuscles are developed in the adult, their transformation into biconcave discs being accompanied by a disappearance of the nucleus. They are thought to get into the circulation by passing into the capillaries of the marrow, the walls of which are exceedingly thin, or even imperfect in some places. (See par. 56.)

26. **Secreting Glands**—Secreting glands are organs in which nucleated cells, spread over a surface in the form of an epithelium, separate or secrete from the blood, or rather from the lymph exuded from the neighbouring capillaries, certain raw materials to form the substances they discharge. The secreting cells are supported by a small amount of connective tissue forming a basement-membrane (par. 13), which may either be continuous or form a mere network, and this basement-membrane (or the cells themselves) is surrounded by lymph that has passed out of a plexus of capillary blood-vessels in close proximity. The essential parts, therefore, are active cells and capillary blood-vessels in close relation to supply the fluid

from which certain constituents are drawn to form the secretions. A cell charged with its selected or converted contents delivers up its secretion either by exudation or by the bursting and destruction of the cell itself.

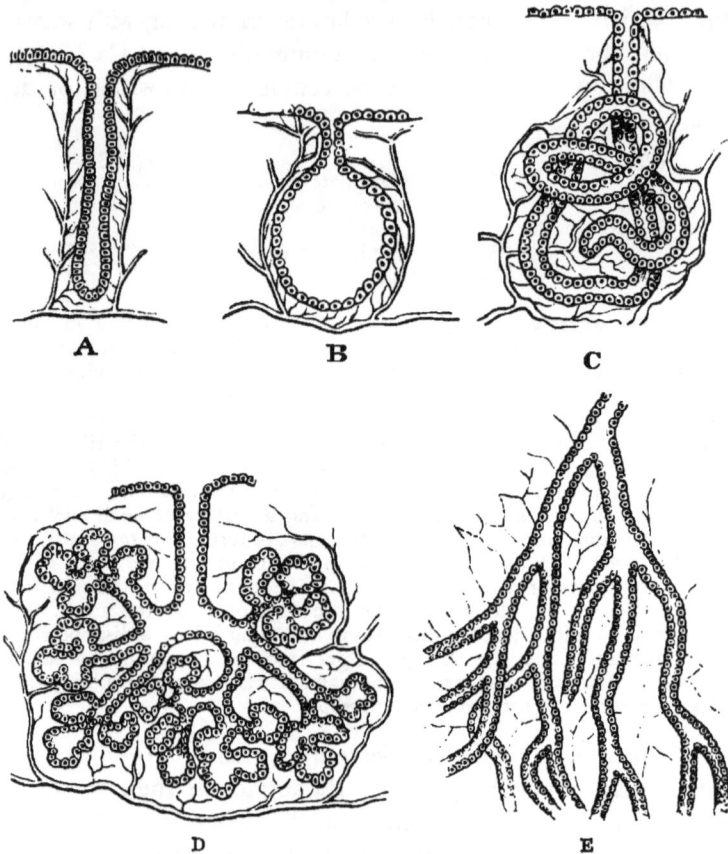

FIG. 26. –Diagrammatic Plan of Varieties of Secreting Glands.
A. Simple gland.   B. Sacculated simple gland.   C. Simple convoluted tubular gland.
D. Racemose gland.   E. Compound tubular gland.

The simplest form of a secreting surface is a plain and continuous one, as in various parts of certain mucous membranes, but in what are usually called secreting glands an increase of surface is obtained by a recession or inversion of the membrane in the form of a cavity. The tube or cavity, of whatever shape, terminates in a blind extremity or extremities, and is lined by secretory epithelial cells having on their outer surface a plexus of capillary blood-vessels.

In the simplest form a single involution takes place, constituting a *simple gland* ; this may be either in the form of an open tube (fig. 26, A) or the walls of the tube may be dilated so as to form a saccule (fig. 26, B). These are named the *simple tubular* or *saccular* glands. Or, instead of a short tube, the involution may be lengthened to a considerable extent, and then coiled up to occupy less space. This constitutes the *simple convoluted tubular* gland, an example of which may be seen in the sweat glands of the skin (fig. 26, C).

If, instead of a single involution, secondary involutions take place from the primary one, as in fig. 26, D, the gland is then termed a compound one. These secondary involutions may assume either a saccular or tubular form, and so constitute the two subdivisions—the *compound saccular* or *racemose* gland, and the *compound tubular*. The racemose gland in its simplest form consists of a primary involution which forms a sort of duct, upon the extremity of which are found a number of secondary involutions, called saccules or alveoli, as in Brunner's glands. But, again, in other instances, the duct, instead of being simple, may divide into branches, and these again into other branches, and so on; each ultimate ramification terminating in a dilated cluster of saccules, and thus we may have the secreting surface almost indefinitely extended, as in the salivary glands (fig. 26, D). In the *compound tubular* glands the division of the primary duct takes place in the same way as in the racemose glands, but the branches retain their tubular form and do not terminate in saccular recesses, but become greatly lengthened out as in the kidney (fig. 26, E). In the simple gland the blind termination of the duct is often called the 'fundus,' but in the compound glands we have duct, intermediate portions, and terminal recesses termed 'alveoli' or 'acini.' Racemose glands have often a lobular structure, the lobules being held together by the branching ducts and interlobular connective tissue. Their alveoli or acini are sometimes almost filled by the secreting cells, only a

very small cavity being left in the centre to communicate with
the excretory duct that ultimately opens into a cavity lined by
mucous membrane or on the surface of the skin.   Besides
blood-vessels glands have lymphatic vessels that begin in the
lymphatic space around the alveolar cells.   Nerves are found
in the connective tissue, and it has been asserted that in some
cases they terminate in the secreting cells.   The varying
appearance of gland cells when loaded with their special
products and when empty, as well as the influence of the
nervous system on the secretory process, will be described later
(figs. 111, 114, &c.).   The pressure frcm behind, or *vis a tergo*,
of the accumulating secretion is believed to be the main cause
of its discharge, though the contraction of the muscular tissue
in the wall of the duct where it exists doubtless gives some aid.

27. **Mucous Membrane.**—Mucous membranes are the soft
and highly vascular membranes that line the internal passages
communicating with the exterior.   They are moistened with a
transparent slimy alkaline fluid termed *mucus*, which is discharged
from secreting cells on the surface of the membrane or in
special glands communicating with the surface.   Mucus
consists of 95 per cent. of water, and contains a small quantity
of a peculiar compound (probably globulin with animal gum)
termed mucin.   Mucin has a ropy consistence, and is precipitated with alcohol.   The other solids are mineral salts and
organic bodies.

The various mucous membranes will be described in detail
when treating of the organs of which they form a part, but the
following general remarks from Gray's 'Anatomy' may be now
studied with advantage :—

'Mucous membranes line all those passages by which the
internal parts communicate with the exterior, and are continuous with the skin at the various orifices of the surface of the
body.   They are soft and velvety, and very vascular, and their
surface is coated over by their secretion, mucus, which is of a
tenacious consistence, and serves to protect them from the
foreign substances introduced into the body with which they
are brought in contact.

'They are described as lining the two tracts—the gastropulmonary and the genito-urinary ; and all, or almost all, mucous membranes may be classed as belonging to and continuous with the one or the other of these tracts. .

' The external surfaces of these membranes are attached to the parts which they line by means of connective tissue, which is sometimes very abundant, forming a loose and lax bed, so as to allow considerable movement of the opposed surfaces on each other. It is then termed the submucous tissue. At other times it is exceedingly scanty, and the membrane is closely connected to the tissue beneath ; sometimes, for example, to muscle, as in the tongue ; sometimes to cartilage, as in the larynx ; and sometimes to bone, as in the nasal fossæ and sinuses of the skull.

' In structure a mucous membrane is composed of CORIUM or DERMIS and EPITHELIUM. The epithelium is of various forms, including the squamous, columnar, and ciliated, and is often arranged in several layers. This epithelial layer is supported by the corium, which is analogous to the derma of the skin, and consists of connective tissue, either simply areolar, or containing a greater or less quantity of lymphoid tissue. This tissue is usually covered on its external surface by a transparent structureless basement-membrane, and internally merges into the submucous areolar tissue. It is only in some situations that the basement-membrane can be demonstrated. The corium is an exceedingly vascular membrane, containing a dense network of capillaries, which lie immediately beneath the epithelium, and are derived from small arteries in the submucous tissue.

' The fibro-vascular layer of the corium contains, besides the areolar tissue and vessels, unstriped muscle-cells, which form in many situations a definite layer, called the muscularis mucosæ. These are situated in the deepest part of the membrane, and are plentifully supplied with nerves. Other nerves pass to the epithelium and terminate between the cells. Lymphatic vessels are found in great abundance, commencing either by cæcal extremities or in networks, and communicating with plexuses in the submucous tissue.

' Embedded in the mucous membrane are found numerous

D

glands, and projecting from it are processes (villi and papillæ) analogous to the papillæ of the skin. These glands and processes, however, exist only at certain parts, and it will be more convenient to defer their description to the sequel, where the parts are described as they occur.'

_____ .. _____

CHAPTER II

MUSCULAR AND NERVOUS TISSUE

28. WE soon learn that the movements of the body are produced by means of muscles in the limbs and trunk, and that some of the internal organs are also made up in great part of muscular tissue. A special property of this tissue is its power of *contraction* under an excitation or stimulus ; and since we find that all the muscles are supplied with nerves, we come to the conclusion that contraction normally takes place as the result of a nervous impulse arriving at a muscle. The close connection of these two tissues leads us to treat them together in a separate chapter. Besides, both the muscular and nervous systems intervene so often in the great functions of circulation, respiration, digestion, and secretion that some preliminary acquaintance with these systems is necessary before treating of these functions.

The two chief varieties of muscular tissue are—(1) *unstriped* or *smooth muscular tissue*, called also involuntary because it is found in organs whose contraction is not under the control of the will; (2) *striped* or *striated muscular tissue*, called also voluntary because it is found in the skeletal muscles, which are under the control of the will. Striped muscle is also found in certain parts of the internal ear and in the upper half of the pharynx. Intermediate in structure and properties is the striated muscle of the heart. Ordinary striped muscle contracts most quickly; next comes the muscle of the heart ; and plain or unstriped muscle contracts most slowly.

29. **Unstriped or Plain Muscular Tissue.**—Unstriped

muscle is found in the muscular coat of the alimentary canal below the middle of the œsophagus, in the trachea and bronchi, in the middle coat of the arteries, the coats of many veins and the larger lymphatics, in the bladder and ureters, and in the ducts of glands. It also occurs in the iris and ciliary muscle, and in the true skin, especially between the bases of the papillæ. When it contracts in the skin under the influence of cold or fear, or any other stimulus, it· causes the papillæ to become unusually prominent, thus producing the peculiar roughness termed 'goose skin.' (See par. 161.)

Plain or non-striated muscular tissue is composed of spindle-shaped nucleated cells, somewhat flattened, and having a length seldom more than $\frac{1}{500}$ of an inch and a breadth about one-eighth of the length. The oval nucleus exhibits the intranuclear network and one or two nucleoli. The cell-substance is longitudinally but not transversely striated, and each cell seems to have a delicate sheath. Between the fibres there is a small quantity of cementing substance. The fibres are collected together into bundles ensheathed in the connective tissue. The small blood-vessels run in the connective tissue, and from these capillaries pass between the individual fibres.

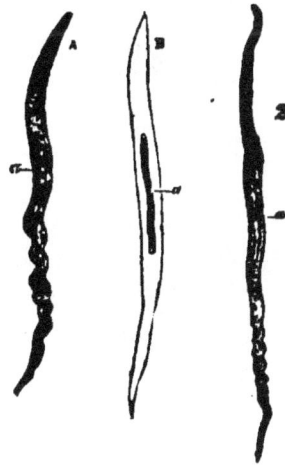

FIG. 27.—Muscular Fibre-cells from Human Arteries. Magnified 350 diameters. (Kölliker.)
a, a, nucleus; B, a cell treated with acetic acid.

Non-medullated nerves are supplied to plain muscular tissue from the sympathetic or ganglionic system, and this tissue responds but slowly to a stimulus, the contraction lasting several seconds and spreading as a wave from fibre to fibre.

30. **Cardiac Muscle.**—The heart, though an involuntary muscle, contains fibres that present transverse marks or striæ. But the fibres are smaller than those of voluntary muscle, and

the striæ are not so well marked. The fibres consist of quadrangular cells joined end to end, and many of these have branches uniting with neighbouring fibres. Each cell has a clear oval nucleus near the centre, with one or more nucleoli (fig. 28). The cells have no investing sheath. Like the skeletal muscles, the heart has an abundant supply of blood-vessels. It has also a rich supply of lymphatic vessels, occupying the interstices of the muscular network. Its nerve-supply will afterwards be considered.

31. **Voluntary or Striated Muscle.**—Each voluntary muscle

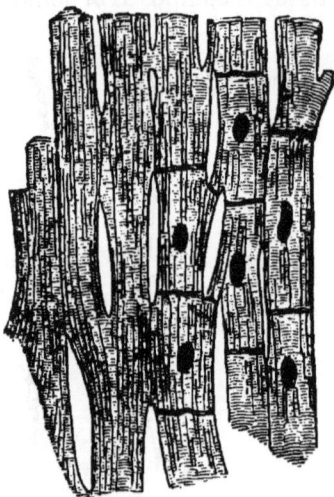

FIG. 28.—Muscular Fibres from the Heart, magnified, showing their cross striæ, divisions, and junctions. (Schweigger-Seidel.)
The nuclei and cell-junctions are only represented on the right-hand side of the figure.

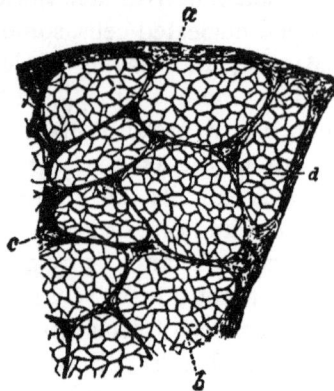

FIG. 29.—Transverse Section from the Sterno-mastoid in Man. Magnified 50 times.

a, external perimysium ; *b*, fasciculus ; *c*, internal perimysium ; *d*, fibre.

in reality forms an organ, composed chiefly of a mass of contractile fibrous tissue called muscular, and of other tissues and parts that may be regarded as accessory. Such a muscle as the biceps or gastrocnemius, for instance, consists of a mass of red flesh surrounded by a sheath of areolar connective tissue called the external *perimysium*, and from the inner surface of this divisions pass off into the interior and inclose bundles of fibres named fasciculi. The fasciculi are of various sizes, each fasciculus having its sheath of perimysium, the tissue extending even between the individual fibres as *endomysium*. Besides serving to bind together the fibres and fasciculi, the areolar tissue just described serves to conduct and support the blood-vessels and nerves in their ramifications in the muscle. This areolar tissue between the fibres and bundle is also continuous with that of the tendon in which the muscle terminates, a further connection between muscle and tendon being effected by the fibres of the tendon becoming united with the sarcolemma

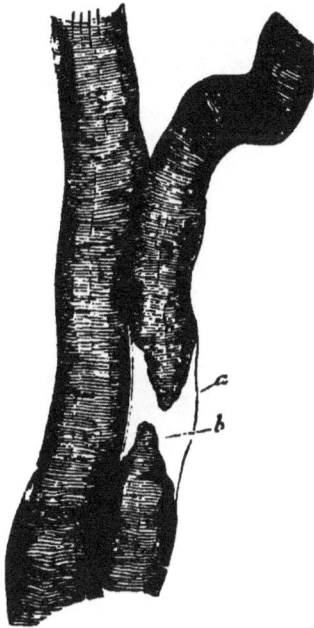

FIG. 30.—Two Human Muscular Fibres. Magnified 350 times. In the one, the bundle of fibrillæ (*b*) is torn, and the sarcolemma (*a*) is seen as an empty tube.

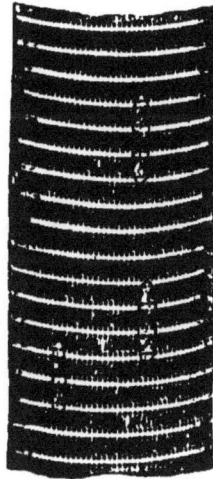

FIG. 31.—Muscular Fibre of a Mammal, examined fresh in serum. Highly magnified. (E. A. S.)
This figure was drawn with the surface layer of muscular substance accurately focussed, the lateral portions having been added by gradually sinking the focus. The nuclei are seen on the flat at the surface of the fibre, and in profile at the edges.

or sheath of the muscular fibre. In the body of the muscle the fibres, which are a little over an inch in length, often fail to reach an end ; they are then connected with the fibres of the connective tissue by their sarcolemma. When a muscle contracts, therefore, the fibres which end directly in tendon pull on the tendon, while those that do not so end pull on the tendon indirectly by means of the connective tissue in the body of the muscle, for this tissue is continuous with that of the tendon.

32. Muscular Fibres.—We now turn to the individual muscular fibre ; for the structure, composition, and properties of the fibres constitute knowledge of muscular tissue. By careful teasing a shred of muscle with a needle single fibres may be isolated and examined under the microscope. A muscular fibre is usually cylindrical in shape, on the average $\frac{1}{500}$th in diameter and a little more than an inch long. Each fibre has a transparent elastic sheath called the *sarcolemma* (distinct from the areolar tissue previously mentioned), inclosing a contractile substance. The contractile substance is marked by alternate dim and light stripes running across the fibre, which accounts for the name *striped* or *striated muscle*. A dark line, known as ' Dobie's line ' or ' Krause's membrane,' can be seen with very high powers passing through the clear band, while through the centre of the dim band a clearer line, called ' Hensen's line,' is sometimes visible in an extended condition. On carefully focussing the surface of a fibre, rows of apparent granules are seen at the boundaries of the light streaks, with fine longitudinal lines uniting the granules. These lines are

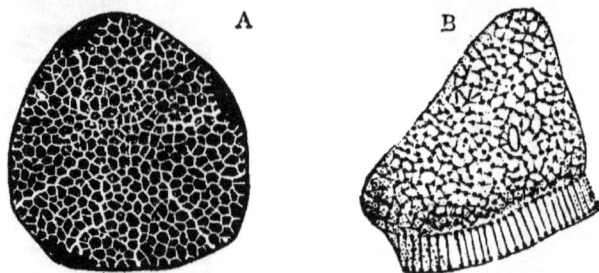

FIG. 32. Transverse Sections of Muscle Fibres. (E. A. S.)

A. Transverse Section of a Mammalian Muscular Fibre showing Cohnheim's areas. Alcohol preparation. Three nuclei are visible under the sarcolemma.
B. An isolated Disk of Leg-muscle of a Beetle treated with dilute acid.

The disk is seen partly on the flat, partly in profile, and exhibits the net-like appearance of the sarcoplasm in the transverse section of the fibre ; the meshes represent the areas of Cohnheim.

most conspicuous in the muscles of insects. They indicate the longitudinal elements or muscle-columns (sarcostyles) which compose a fibre, and are probably due to an interstitial substance between the columns termed *sarcoplasm*. If a *transverse* section of a fibre be examined with a high power, it is seen to be subdivided into small angular parts, *the areas of Cohnheim*. These represent sections of the muscle-columns of which the fibre is composed. After death, or on being hardened with certain reagents, as chromic acid or alcohol, a fibre may easily be split up into smaller threads or *fibrillæ*,[1] a term that is here used to signify the same as the term muscle-columns or sarcostyles above. Fibres may also, with suitable

---

[1] The term *fibrillæ* or *fibrils* is used by some authors to indicate still finer elements, which are supposed to constitute the muscle-columns themselves.

reagents, be split up into transverse discs. Each fibre thus appears to be composed of a mass of fibrils (sarcostyles) embedded in interfibrillar substance or sarcoplasm, and composed of alternate segments of dim and clear substance. The columns of muscular substance are the actual contractile elements of the tissue.

Besides the tubular sheath of sarcolemma and the striated substance, a muscular fibre shows reticulated nuclei surrounded by a variable amount of granular protoplasm. A nucleus with its adjacent protoplasm is called a muscle-corpuscle.

33. **Blood-vessels of Muscular Tissue.**—The muscles are abundantly supplied with blood-vessels. The arteries, accom-

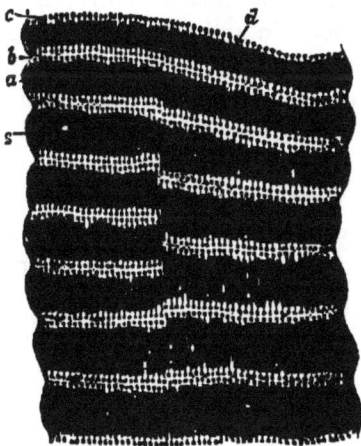

FIG. 33.—Living Leg-muscle of Water Beetle. Highly magnified. (E.A.S.)

s, sarcolemma ; *a*, dim stripe ; *b*, bright stripe ; *c*, row of dots in bright stripe, which are enlargements or thickenings on the longitudinal septa of sarcoplasm. These septa are represented by the longitudinal lines, *d*. The continuity of these lines through the bright stripe is difficult to see in the fresh fibre, but after treatment with acid it becomes quite distinct.

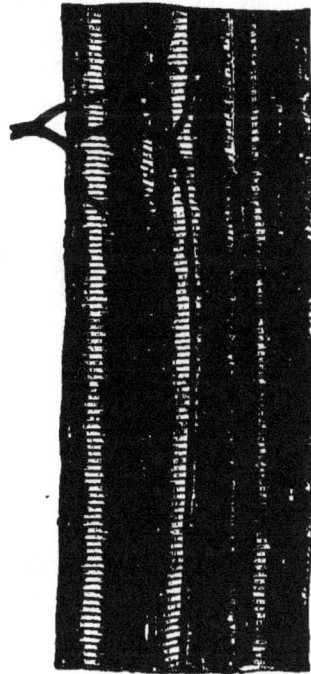

FIG. 34.—Capillary Vessels of Muscle. Moderately magnified. (E. A. S.)

panied by the associated veins, enter at various points, branch in the areolar tissue between the fasciculi, and at length terminate in capillaries that form an oblong network around the fibres on the outside of the sarcolemma. These capillaries are very small, and from them is exuded the fluid by which the muscular tissue is nourished. There is also in the connective tissue lymph spaces that form the commencement of the lymphatic vessels of the tissue, so that every muscular fibre is surrounded by capillary blood-vessels and lymph spaces, and the nutriment passes from the blood by means of the lymph through the sarcolemma to the substance of the fibre, waste products passing from the fibre into the blood in the opposite direction.

During muscular action the blood-supply of a muscle is increased, for the arteries are so arranged that they do not undergo compression during contraction, while the venous blood and lymph are squeezed out as the muscle contracts.

34. **Nerves of Muscular Tissue.**—Voluntary muscular tissue receives a supply of nerve-fibres, chiefly medullated, from the cerebro-spinal system. The nerves branching in the connective tissue first form plexuses, and then gradually divide until a single nerve-fibre enters each muscular fibre, the primitive sheath of the nerve-fibre fusing with the sarcolemma, while the axis cylinder of the nerve passes through it and ends in a terminal ramification called an *end-plate* on the substance of the fibre (fig. 35). Each fibre appears to receive one end-plate about its middle, and as the fibres of a muscle are but about an inch in length, the end-plates are distributed over the length of the muscle ;

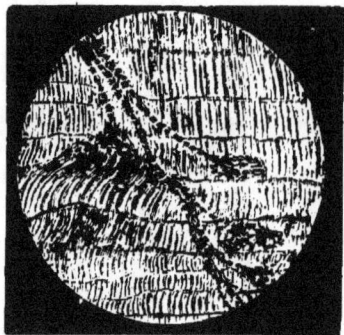

FIG. 35.—Motor Nerve-endings in Snake's Muscle. (Waller.)

so that a nervous impulse along the different fibres going to a muscle reaches the different parts about the same time, and sets up a simultaneous contraction in the organ.

Plain muscle is supplied with non-medullated nerve-fibres on which knob-like endings have been described in some cases.

35. **Chemical Composition of Muscle.**—From a perfectly fresh muscle the contractile semi-fluid substance of the fibres can be squeezed out from the sheaths of sarcolemma, and it is then called *muscle-plasma*. Muscle-plasma

is a slightly alkaline viscid fluid which, if exposed to the ordinary temperature of the air, coagulates, separating after some time into a transparent *muscle-clot* and a watery fluid, *muscle-serum*. Muscle-clot consists of a substance called myosin, which is a proteid substance belonging to the class called globulins, as it is coagulable by heat and soluble in saline solutions. The formation of myosin from muscle-plasma is probably due to a ferment, just as in the case of the formation of fibrin from blood-plasma. In the formation of myosin by the clotting of muscle-plasma an acid called sarco-lactic acid is developed, no acid being formed during the formation of fibrin in blood clotting. The serum contains three different proteid bodies, some extractives, and two pigments; hæmoglobin and myohæmatin.

After a muscle has been removed from the body some time, or even in the body after general death, the muscular tissue becomes dead. When a muscle dies it becomes rigid and more opaque, loses its irritability, and undergoes chemical changes. *Rigor mortis* (the stiffness of death) is the condition of the muscles that follows the death of man or other animal, and it is due to the coagulation of the muscle-plasma within the sarco-lemma sheaths of the muscular fibres. The onset of this rigidity in man varies from some minutes to a few hours, though it is usually complete in four or five hours after death. It is accompanied by evolution of CO_2 and heat. After a time—one to five days—it passes away and the muscles become soft and flaccid. The earlier it occurs the sooner it passes off. The cause of the disappearance is said to be due to the putrefaction that begins.

A living muscle derives nutriment for building up the complex coagu-lable plasma from the lymph that passes out of the blood—the lymph holding in solution the various kinds of food-material taken up in the ali-mentary canal and oxygen absorbed in the lungs—and continually gives off CO_2 and nitrogenous waste. When in action it consumes an increased quantity of oxygen, produces more carbonic acid, as well as some other products, and acquires an acid reaction. On passing into a condition of rigor mortis the complex living molecule breaks up, a solid clot of myosin forms, a sudden increase of CO_2 occurs, and sarcolactic acid appears.

36. **Muscular Contraction.**—By removing the gastrocne-mius muscle from a recently killed frog, whose tissues retain their vitality longer than those of a warm-blooded animal, the living muscular substance may be submitted to experiment. The muscle is left attached to the femur above and to the *tendo Achillis* below, with a length of the sciatic nerve that goes to the muscle exposed. Such an isolated muscle with its attached nerve is called a *muscle-nerve preparation*. On stretching such a muscle it is found to be extensile and elastic. Its elasticity is slight but perfect, that is to say a small force will extend it, but it returns exactly to its original length when the force is removed. The elongation of a muscle is not proportional to the weight used, but diminishes in proportion as the weight

increases.   But the most important property of muscle in
the living state is its contractility or power of shortening when
irritated by a stimulus, its volume or bulk remaining the same.
Normally the muscle contracts in response to a stimulus from
the central nervous system, but other kinds of excitation or
stimuli may be applied.   These may be mechanical (pricking
or pinching), thermal, chemical, or electrical, and the stimulus
may be applied to the nerve going to the muscle or to the
muscular tissue direct.   It should be noted that a healthy muscle
when at rest is contracted or retracted to a small degree, this
small amount of contraction being known as *muscular tone*,
and is due to the influence derived from the central nervous

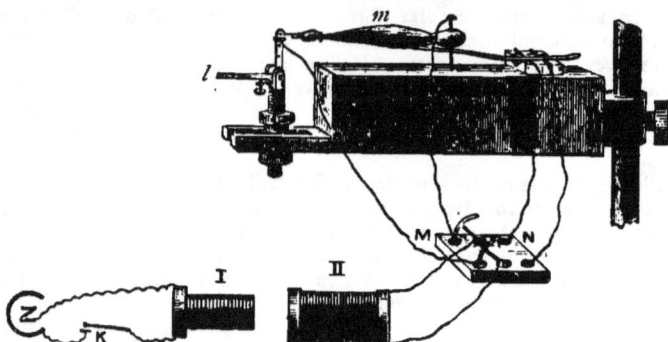

FIG. 36.—(Waller.)

l, lever cut short passing to drum ; M, wires from secondary coil of battery to muscle, *m* ;
N, wires across which the nerve of the muscle is laid. (For a drum see fig. 73.)

system, for on cutting its nerve the muscle becomes longer
and more relaxed.   That the muscular fibres have in themselves
the power of contraction is proved by the use of the drug
curare, which paralyses the motor nerve endings in muscle,
whilst the muscular tissue still responds to direct stimulation.
Ammonia also destroys the nerve and yet stimulates muscular
tissue.

To study the phenomena of contraction a graphic representation (called a *muscle-curve*) of the contraction and relaxation of the muscle is obtained by a *myograph*.   This instrument
consists of a drum covered with a coat of lampblack revolving
at a definite rate, against which the point of a light lever attached
by a thread to a muscle can be brought.   When the drum is

moving and the muscle is quiet a horizontal white line is drawn on the paper ; but when the muscle contracts the lever rises, falling again during relaxation (see fig. 36 and 37).

Fig. 37 represents a muscle-curve obtained by sending a single induction shock from an electrical apparatus through the sciatic nerve to the gastrocnemius muscle of a frog, the time occupied in tracing the curve being indicated by the small curves below, each of which represents $\frac{1}{100}$th of a second.

A study of the curve of a simple muscular contraction or twitch shows three phases :—(1.) A phase termed the *latent period*, during which no apparent change takes place. This interval is said to be taken up by the propagation of the impulse along the nerve and by the preparatory changes in the muscle. It becomes longer as the muscle becomes fatigued.

(2.) A phase of shortening or contraction, during which the lever rises. The height of contraction diminishes as the muscle

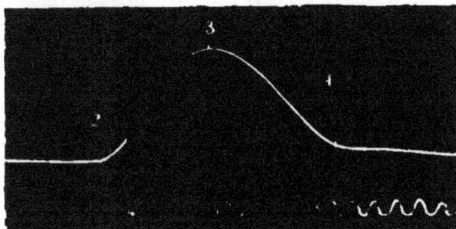

FIG. 37.—A Single Muscular Contraction. (Frog's Gastrocnemius.)

From 1 to 2 is the latent period ; from 2 to 3 the period of shortening ; from 3 to 4 the period of relaxation.

ishes as the muscle tires, though the period remains the same.

(3.) A phase of relaxation of the muscle or return to its original length, during which the lever falls. This period also becomes longer as the muscle tires.

From the time-marking below it is seen that the latent period in a fresh muscle occupies about $\frac{1}{100}$th of a second, the phase of contraction $\frac{4}{100}$, and the phase of relaxation $\frac{5}{100}$ second. A single muscular twitch is thus completed in about $\frac{1}{10}$th of a second, but it must be noted that the time varies, being much longer in a fatigued muscle.

By sending in a second shock before relaxation sets in we shall get a second curve added to the first, and on increasing the rapidity of the shocks each succeeding contraction will start from some part of the preceding one and raise the lever to a greater height. When the shocks increase more rapidly still, the individual shocks, visible at first, fuse together into one continuous curve, the lever having attained a maximum height, where it remains until the muscle is exhausted if the shocks continue.

A muscle in a state of continuous contraction is said to be in a condition of spasm or *tetanus*. Tetanus may occur as the result of disease (lockjaw) or be produced by poisons such as strychnine. A fatigued and exhausted muscle cannot act owing to the accumulated products of action, and if left to

FIG. 38.—Composition of Tetanus.
Stimuli caused by a spring interrupting primary circuit by vibrating in and out of a mercury cup ; the vibration frequency is increased by shortening the spring.

rest, may recover by the circulating blood bringing fresh nourishment and carrying off the accumulated waste products. Every contraction of a voluntary muscle in the living body is considered to be tetanic in character, a sudden jerk being in reality a tetanic contraction of short duration. A good example of a clear case of tetanus is seen when a person takes hold of the handles of a strong galvanic battery in action. Cramp or tetanus is produced in the muscles of his fingers by the rapidly repeated contractions in the muscles bending his fingers, and he cannot let go.

Besides doing mechanical work during contraction, the muscle produces heat and undergoes an electrical change. In other words it sets free energy, the energy being obtained from that stored up in the muscular tissue and derived from the food.

37. Summary of Changes occurring during Muscular Contraction.—When a muscle contracts the following changes occur :—

(*a*) Structural :—
    (1.) The whole muscle shortens, thickens, and hardens.
    (2.) The muscle-columns or sarcostyles of the fibres contract, the dark discs encroaching on the clear intervals.
    (3.) The blood supply is increased, while venous blood and lymph are squeezed out.

(*b*) Physical :—
    (1.) There is a slight rise of temperature, the venous blood from an active muscle being warmer than that from a muscle at rest.
    (2.) The electrical conditions of the muscle are changed, a negative variation of the natural muscle-current taking place.
    (3.) The extensibility of the muscle is increased—that is to say, a given weight stretches a contracted muscle more than the same muscle at rest.

(*c*) Chemical :—
    (1.) The chemical changes occurring in resting muscle become more active during contraction.
    (2.) Carbon dioxide (CO_2) is suddenly evolved in greater quantity, and the amount of oxygen absorbed is increased, but not in proportion.
    (3.) Sarcolactic acid ($C_3H_6O_3$) is produced, and the muscle becomes acid to litmus paper.
    (4.) Sugar is produced, presumably from the muscle glycogen.

38. General Arrangement of the Nervous System.—The nervous system, whose general function is to guide, regulate, harmonise, and co-ordinate the other functions of the body, consists of (1) masses of nervous matter situated within the bony cranium and spinal canal, forming what is known as the *cerebro-spinal system* ; (2) cords of nervous matter termed *nerves*, extending from the cerebro-spinal system to the muscles, sense organs, and other organs of the body ; (3) smaller masses of nervous matter called *ganglia*, lying along nerves situated in the neck, thorax, and abdomen, and constituting the so-called *sympathetic system.*

The cerebro-spinal system consists of the brain, divided into *cerebrum* or large brain and *cerebellum* or small brain. The brain, with its twelve pairs of cranial nerves, will be treated of in a later chapter. Connected with the brain is the spinal cord, which is united with a portion of the brain in the cranial cavity known as the *medulla oblongata.* Passing from the spinal cord are 31 pairs of spinal nerves, 8 cervical, 12 dorsal, 5 lumbar, 5 sacral, and 1 coccygeal. The sympathetic system consists

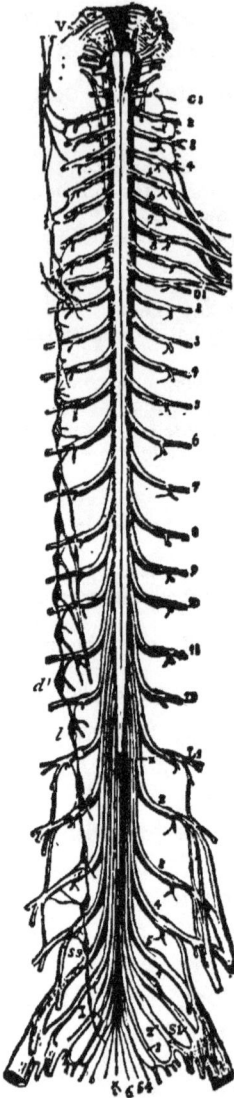

in the main of a double chain of small swellings, or ganglia, on each side of the front of the spinal column, connected with each other and with the spinal nerves. Cords, sometimes called sympathetic nerves, pass from these ganglia to the viscera and blood-vessels, the movements of which are involuntary. These nerves on their way to the viscera sometimes unite with each other to form interlacing networks or *plexuses*, on many of which collateral ganglia are found. The sympathetic system is not really an independent system, as was once supposed, for we now know that its ganglia have a distinct relation to certain fibres that leave the spinal cord, and that its nerves are under the control of some part of the central system. Nervous substance is found, under the microscope, to consist of two different structural elements, *fibres* and *cells*. The fibres are found in the nervous cords and also in certain parts of the brain and spinal cord ; the cells are mainly confined to the cerebro-spinal centre and the ganglia, although some are present at the terminations of the nerves of special sense. A nervous cord or nerve is found to be composed of bundles of nerve-fibres running side by side, bound together and inclosed by a sheath of connective tissue. According as the nerve-fibre has a sheath of medullary substance or is without the sheath, it is spoken of as a *medullated* fibre or *non-medullated* fibre. According to their function, nerves and nerve-fibres are classified as (1) *afferent* or *centripetal*, (2) *efferent* or *centrifugal*. Nerve-fibres which convey impulses *from* a centre to some part of the body are *efferent* nerve-fibres ; nerve-fibres which convey impulses *to* a centre from some part of the body are *afferent* fibres. The term *motor* is sometimes used instead of efferent, and the term *sensory* instead of afferent ; but these terms are not strictly interchangeable, for impulses passing along fibres to the central system may give rise to effects that do not result in sensation, *e.g.* reflex action ; and impulses passing along fibres from the central nervous system often produce effects other than movement, *e.g.* secretion or inhibition. A nerve-trunk, as the vagus, which contains both kinds of fibres, is spoken of as a *mixed nerve*. We also speak of *somatic* and *splanchnic* nerve-fibres. Somatic (Gk.

FIG. 39.—The Spinal Cord and its Nerves (the Spinal Nerves), together with the Sympathetic Chain on one side

v, pons Varolii, below which is the medulla oblongata ; c 1 to 8, the cervical nerves ; D 1 to 12, the dorsal nerves ; L 1 to 5, the lumbar nerves ; s 1 to 5, the sacral nerves ; 6, the coccygeal nerve ; ×, the terminal fibre of the cord ; a to x, the sympathetic chain, showing the connection with the spinal nerves.

soma, body) nerves supply such body structures as the muscles, bones, skin, &c. ; splanchnic (Gk. *splanchna*, viscera) nerves are distributed to the viscera —heart, lungs, alimentary canal, &c. The viscera of the body are innervated by nerve fibres which leave the spinal cord with a medullary sheath, run forward into the sympathetic or more distal ganglia, and there lose their sheath as they issue from the ganglion as non-medullated fibres.

Before describing the microscopical structure of nerve-fibres and nerve-cells we will return to the spinal cord, to obtain some general ideas of the functions of nervous tissue and the close connections that exist between the central nervous system and the various other tissues of the body. Examination of a portion of the spinal cord shows that it is cylindrical in shape and separated into right and left halves by median fissures. Nerves are given off from each half, each nerve arising by two roots, a posterior and anterior root. The posterior root has a slight swelling or ganglion just before it unites with the anterior root, while still in the spinal canal. The section of the cord shows the interior to contain grey matter arranged somewhat like the letter H, and this grey matter is composed mainly of cells (fig. 40). The external white matter is composed of fibres. By help of the

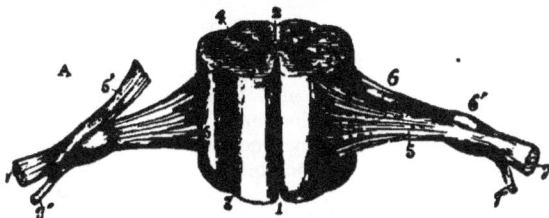

FIG. 40.—Section of Spinal Cord, showing mode of origin of Spinal Nerves.
1, Anterior median fissure ; 2, posterior median fissure ; 3, antero-lateral impression ; 4, postero-lateral groove ; 5, anterior root ; 6, posterior root ; 6', ganglion on posterior root ; 7, united mixed nerve.

annexed diagram (fig. 41) we can learn something about the distribution of the nerves springing from a typical region of the cord. The diagram represents one lateral half only, the corresponding half being similar. From the horns of the grey matter arise the two roots of the spinal nerve N. Experiments have shown that the nerve-fibres of the posterior root P are *afferent* fibres and that the fibres of the anterior root A are *efferent*, the former conveying impulses to the cord and the latter carrying impulses from the cord. The nerve-trunk N must therefore be a mixed nerve. The mixed trunk undergoes many divisions, but the greater portion of these, represented by N', pass to the skeletal muscles M, or to the sensory cells of the skin in a definite part of the body, S. An important branch of the mixed nerve-trunk N is the small branch V, known as the *ramus communicans*. The fibres from this pass into one of the ganglia of the sympathetic chain Σ, where some of the fibres appear to become connected with the nerve-cells of the ganglion. From the sympathetic ganglion fibres pass out to supply the viscera (or internal organs), these fibres often passing on their way through collateral ganglia. The greater part of these sympathetic or splanchnic fibres probably carry impulses from the central system to the plain muscular fibres of the viscera (*m*), while others are afferent, and carry impulses to the central system from the sensory cells (*s*) of the internal organs. Some terminate in other ways, *x*.

One portion of the ramus communicans in the thoracic region is the grey ramus communicans, *r V*. It is shown in the diagram running back from the ganglion Σ to the spinal cord, and giving off a branch *v m*, which runs in connection with the spinal nerve to the muscular tissue of the blood-vessels, *m'*. The nerves that thus regulate the calibre of the blood-vessels are spoken of as *vaso-motor* nerves. The upper end of the spinal cord, known as the medulla oblongata or spinal bulb, lies in the cranium, and may be regarded as the lowest division of the brain (fig. 42). It is pyramidal in shape, with its broad end upwards, and surmounted by a part of the brain called the pons Varolii. On the anterior surface of the lower part certain fibres coming up from the spinal cord may be seen to decussate with (cross) each other before going to the highest parts of the brain (decussation of pyramids). From its front and sides may be seen issuing the sixth to the twelfth cranial nerves. These nerves have their ultimate or deep origin in particular parts of the grey matter of the bulb. In the posterior part of the medulla is found a diamond-shaped space termed the *fourth ventricle*, which is continuous below with the central canal of the spinal cord, and the roof of which is formed by a thin membrane. The pointed lower end of the fourth ventricle, from its shape resembling that of a pen-nib, is called the *calamus scriptorius*. In the grey matter forming the floor of the fourth ventricle are found groups of cells forming the nerve-nuclei of the cranial nerves just mentioned. One of these numbered X, and termed the *vagus* or pneumogastric, is very important, being intimately associated with the functions of circulation, respiration, and digestion. The importance of the medulla as a nerve-centre for controlling and regulating various offices is the reason that we have introduced this short account of it at this stage. By a *nerve-centre* we must understand a ganglion cell or group of cells capable of receiving, modifying, and discharging nerve impulses, and thus acting for the performance of some function.

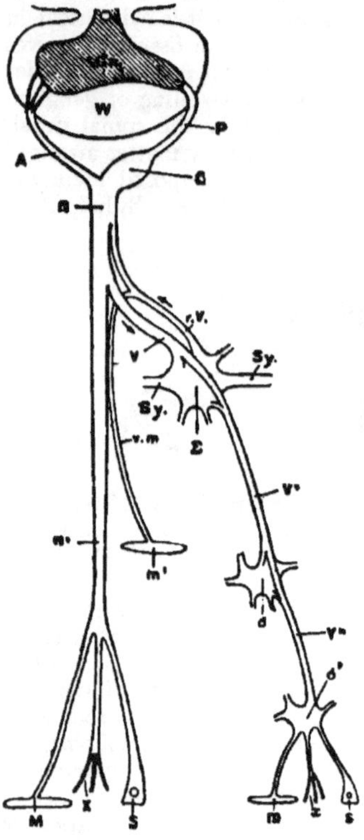

Fig. 41.—(After Foster.)

39. **Nerves.**—A nerve is a whitish cord formed of a number of nerve-fibres bound together by connective tissue. In a small nerve we may find a single bundle, or *funiculus*, of fibres in a tubular covering, but in larger nerves, as the sciatic, several bundles, or *funiculi*, are united together by a common covering

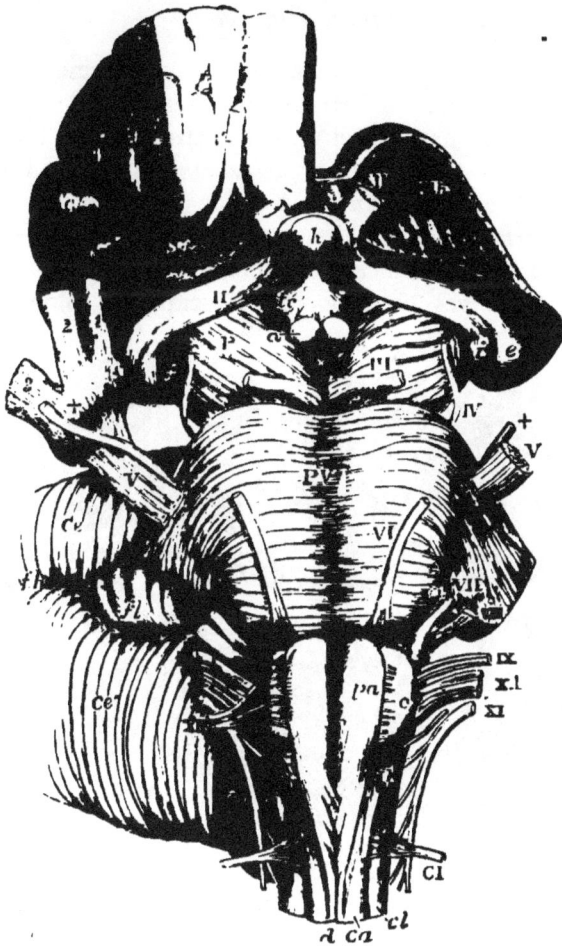

Fig. 42.—View from before of the Medulla Oblongata, Pons Varolii, Crura Cerebri, and other central portions of the Encephalon. Natural size. (Allen Thomson.)

On the right side the convolutions of the central lobe, or island of Reil, have been left, together with a small part of the anterior cerebral convolutions; on the left side these have been removed by an incision carried between the thalamus opticus and the cerebral hemisphere.

I′, the olfactory tract, cut short and lying in its groove; II, the left optic nerve in front of the commissure; II′, the right optic tract; *Th*, the cut surface of the left thalamus opticus; C, the central lobe, or island of Reil; *Sy*, fissure of Sylvius; » », anterior perforated space; *e*, the external, and *i*, the internal corpus geniculatum; *h*, the hypophysis cerebri or pituitary body; *tc*, tuber cinereum with the infundibulum; *a*, one of the corpora albicantia; P, the cerebral peduncle or crus; III, close to the left oculo-motor nerve; *x*, the posterior perforated space.

The following letters and numbers refer to parts in connection with the medulla oblongata and pons. PV, pons Varolii; V, the greater root of the fifth nerve; +, the lesser or motor root; VI, the sixth nerve; VII, the facial; VIII, the auditory nerve; IX, the glosso-pharyngeal; X, the pneumogastric nerve; XI, the spinal accessory nerve; XII, the hypoglossal nerve; C I, the suboccipital or first cervical nerve; *pa*, pyramid; *o*, olive; *d*, anterior median fissure of the spinal cord, above which the decussation of the pyramids is represented; *ca*, anterior column of cord; *r*, lateral tract of bulb continuous with *cl*, the lateral column of the spinal cord.

E

of connective tissue. This common sheath, formed of white and elastic fibres, passes between the funiculi of nerve-fibres and supports the fine blood-vessels distributed to the nerve. It has received the name of *epineurium*. Each funiculus has also a sheath of connective tissue of a lamellar nature, called the *perineurium*. The nerve-fibres are also separated from one another by delicate connective tissue called *endoneurium*, which serves to support the capillary blood-vessels

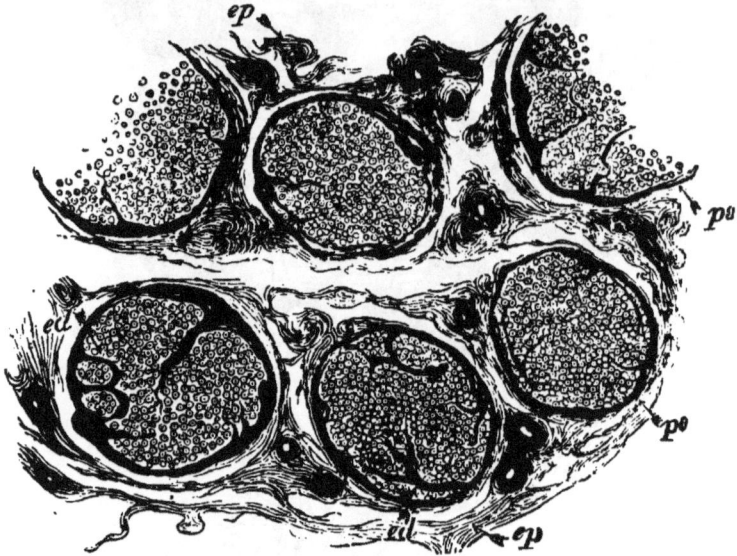

FIG. 43.—Section of a Part of the Median Nerve (Human). Drawn as seen under a low magnifying power. (From Landois, after Eichhorst.)

ep, epineurium, or general sheath of the nerve, consisting of connective-tissue bundles of variable size separated by cleft-like areolæ, with here and there blood-vessels; *pe*, lamellated connective-tissue sheaths (perineurium) of the funiculi; *ed*, interior of funiculus, showing the cut ends of the medullated nerve-fibres, which are embedded in the connective tissue within the funiculus (endoneurium).

nourishing the nerve-fibres. Lymphatic vessels are also found in this tissue. The branches of a nerve and single fibres passing to their distribution are invested with a delicate continuation of the perineural sheath known as the *sheath of Henle*. (The nerve-fibres in the brain and spinal cord are held together by a special kind of tissue termed *neuroglia* (par. 43).) Nerve-fibres from one funiculus often pass into another, but in these communications the medullated fibres remain individually distinct and never anastomose together. Most nerves contain two kinds of fibre, though the fibres of the cerebro spinal system are chiefly medullated fibres, while in those of the sympathetic system non-medullated fibres preponderate. Nerve-trunks themselves receive nerve-fibres (*nervi nervorum*), which ramify chiefly in the epineurium.

40. **Medullated Nerve-fibres.**—Medullated or white nerve-fibres form the white part of the brain and spinal cord and the greater part of the cerebro-spinal nerves.

The fibres vary in size even in the same nerve, but more so in different parts of the nervous system. Some are $\frac{1}{1500}$ inch (17 μ) in diameter, but others only $\frac{1}{12000}$ inch ($2\,\mu$), the smallest being those that pass from the spinal nerves to the sympathetic system. Examined fresh, these fibres appear as glassy threads with a double contour (fig. 44), but when treated with certain reagents each medullated fibre is found to consist of:

(1.) An axis cylinder in the centre, continuous from its origin to its distribution, and showing a fibrillar structure under high powers.

(2.) A medullary sheath, or white substance of Schwann, not continuous, but showing gaps known as the 'nodes of Ranvier.' It is of a fatty nature, and stains dark with osmic acid.

FIG. 44.—White or Medullated Nerve-fibres, showing the sinuous outline and double contours.

(3.) A primitive sheath outside—the neurilemma—continuous but dipping down at the node of Ranvier. This is absent from the nerve fibres in the brain and spinal cord. On the inner surface of the primitive sheath, usually midway between each node, is a clear oval nucleus surrounded by a little protoplasm (fig. 45).

The axis cylinder is the conducting and important part of the nerve-fibre, being often seen continuous with one of the processes of a nerve-cell. At the central and peripheral ends both sheaths disappear, but the axis cylinder continues.

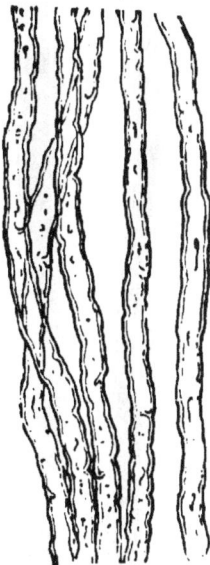

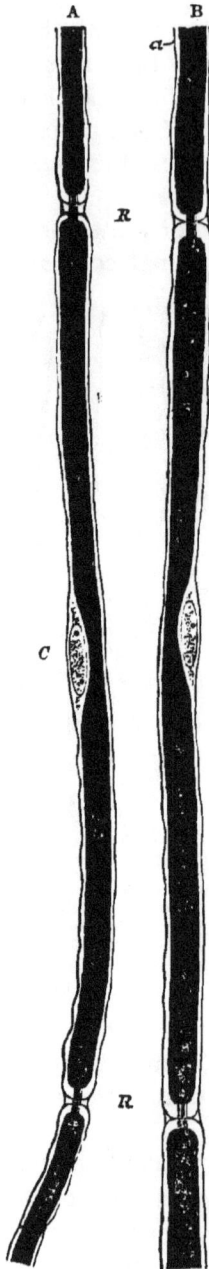

41. Non-medullated Nerve-fibres.—

These grey fibres occur chiefly in branches from the sympathetic ganglia, though found to some extent in the nerves of the cerebro-spinal system. They are transparent and soft, having nuclei at intervals but no medullary sheath (fig. 46). The nuclei appear to lie beneath a fine primitive sheath, though it is difficult to see this sheath separate. Many of these fibres branch and unite with neighbouring fibres in their course to form a network. Medullated fibres never show this phenomenon. The branches of the olfactory nerve consist of non-nucleated fibres with a distinct nucleated sheath.

42. Nerve-cells. —

Nerve-cells are found in the grey matter of the brain, spinal cord, sympathetic ganglia, the ganglia of the posterior roots of spinal nerves, and in small ganglia on the course of some other nerves. While the chief function of a nerve-fibre is to conduct nervous impulses, nerve-cells may not only conduct and transfer impulses, but they may modify or even originate nervous impulses of various kinds.

The nerve-cells are microscopic bodies of various shapes and sizes in different parts of the nervous system. They give off *processes* or *poles*, and may be unipolar (*i.e.* with one pole), bipolar, or multipolar. A typical multipolar nerve-cell is such as is found in the anterior horn of the grey matter of the spinal cord. Such a cell is seen to be a nucleated

FIG. 45.—Portions of two Nerve-fibres stained with osmic acid (from a young Rabbit). 425 diameters.
R, R, nodes of Ranvier, with axis cylinder passing through. *a*, primitive sheath of the nerve ; *c*, opposite the middle of the segment, indicates the nucleus and protoplasm lying between the primitive sheath and the medullary sheath. In A the nodes are wider, and the intersegmental substance more apparent than in B. (Drawn by J. E. Neale.)

mass of protoplasm having numerous branching processes, mostly ending as fine twigs. One process becomes continuous with a nerve-fibre, and is known as the axis-cylinder process. In

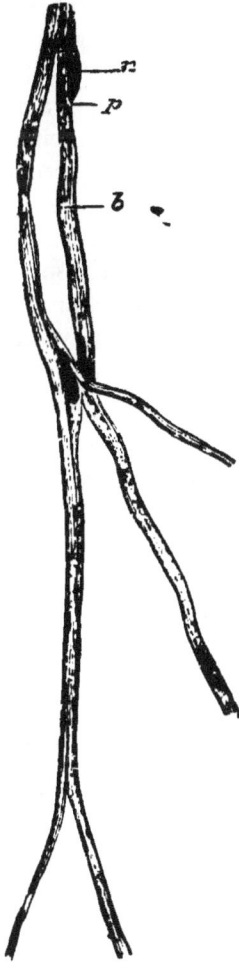

FIG. 46.—Portion of the Network of Non-medullated Fibres of Remak, from the Pneumogastric of the Dog. (Ranvier.)

n, nucleus; *p*, protoplasm surrounding it; *b*, striation caused by fibrils.

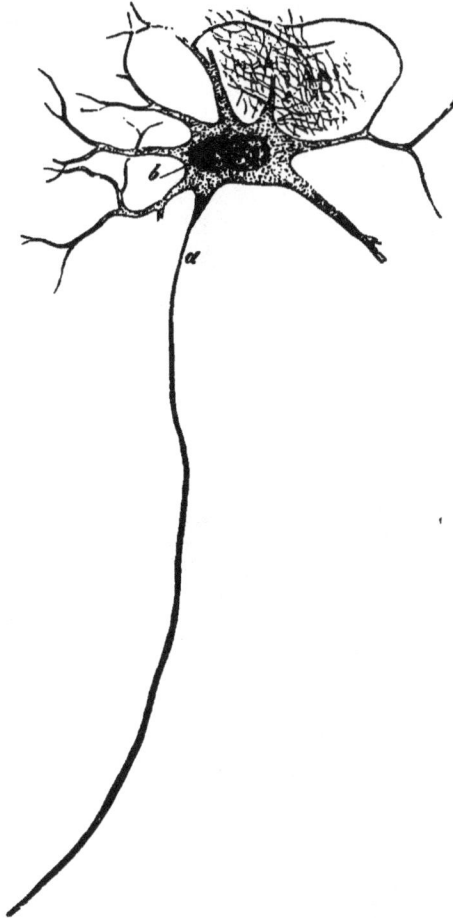

FIG. 47.- Multipolar Nerve-cell from Anterior Horn of Spinal Cord (Human). (Gerlach.)

a, axis-cylinder or nerve-fibre process; *b*, pigment.

the case of the sympathetic ganglia all the processes of the cell appear to become nerve-fibres, one process being often continuous with the fine medullated fibre entering the ganglion, and the other being continuous with the non-medullated fibres that pass to their peripheral distribution. In the spherical cells of the spinal ganglion there is but one process, which, after winding about outside the cell, joins a nerve-fibre passing through the ganglion on its way to the spinal cord, cells and process forming a T-shaped junction with the fibre. In the ganglia the nerve-cells have a sheath continuous with the sheath of the nerve-fibres they join.

43. Supporting Tissue of the Nerve Centres.

A supporting substance is found among the grey and white matter of the brain and spinal cord termed *neuroglia*. It is composed of cells with numerous fibre-like processes (glia-cells), which fill up the interstices of the proper nervous elements, fibres and cells. Further support is given by ramifications from the ciliated epithelium-cells lining the central canal of the spinal cord and the ventricles of the brain, except the fifth. Support is also given by the connective-tissue sheath of the brain and spinal cord (pia mater), which sends in delicate partitions and carries the blood-vessels for the nervous tissue.

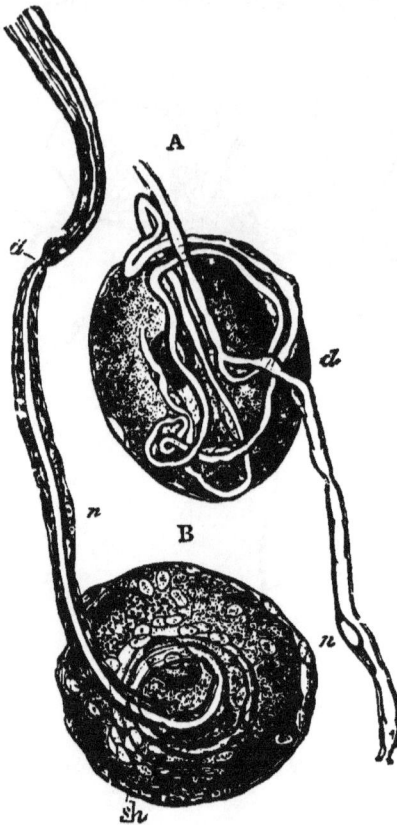

Fig. 48. Two Nerve-cells from a Spinal Ganglion (Human). (Retzius.)

sh, nucleated sheath ; *n, n*, nuclei of the primitive sheath of the nerve. From each cell a fibre can be seen to arise, and, after a convoluted course on the surface of the nerve-cell, to bifurcate (opposite *d*); from which point the divisions pass either in the opposite direction to one another, as in A, or at first in the same direction, as in B. The nuclei of the sheath of the nerve-cell are all represented in B, but only those seen in profile have been represented in A.

44. Reflex Action.

The central nervous system of man, the cerebro-spinal system, contains in addition to nerve-fibres a number of nerve centres localised in certain parts, yet having communications with each other and influencing each other upon suitable occasions. These nerve centres of the brain and spinal cord may be regarded as consisting of closely connected groups of ganglion-cells acting together and endowed with the power of regulating some function of the body. Thus we speak of a respiratory centre, a vaso-motor centre, a centre for defæcation, &c.

Speaking physiologically, the whole of the processes of the nervous centres are divided according to the mode of reaction into *reflex* acts and *voluntary* acts.

A reflex action is the immediate efferent response to afferent impulses, independently of the will. The spinal cord possesses this power of responding to afferent impulses in a remarkable degree, and may in fact be regarded as formed in part of a number of reflex centres. It also acts as a conductor of impulses, *via* the spinal nerves, to and from the brain, as well as to and from different parts of the cord. As we shall afterwards learn, the brain itself may also become a centre of reflex action. The mechanism of a reflex action involves the following elements :—(1) A sentient surface or peripheral sense organ ; (2) an afferent nerve-fibre ; (3) a cell or group of cells on the afferent tract ; (4) a connection of these cells (commissural fibres) with (5) efferent nerve-fibres ; (6) a muscle, gland, secreting cell, or other cells capable of some response to the stimulus from the nerve centre involved in 3, 4, and 5. These are the component parts of a *reflex arc*.

An appropriate stimulus applied to the sentient surface passes along the afferent fibres and leads to a discharge of efferent impulses which set the muscle or other responding cell in action. Thus a frog whose brain has been removed will often remain at rest, but if its skin be irritated by a prick or other stimulus its limbs will be set in motion by the contraction of certain muscles.

The reflex process thus set up is illustrated diagrammatically in the annexed figure from Dr. Waller. From the sentient surface (1) an afferent impulse passes along (2) to the posterior root of the spinal cord, the nerve-fibres of the posterior root ending in minute filaments among the small cells of this part of the cord (3). In some unexplained way this impulse passes across the grey part of the cord to the large cells of the anterior root (5), the cells of this part being connected by their axis-cylinder process with the efferent fibres (6). These convey the stimulus to the fibres of the muscle (7), which accordingly contract. Where the brain is concerned in the action the circuit is longer through s and M.

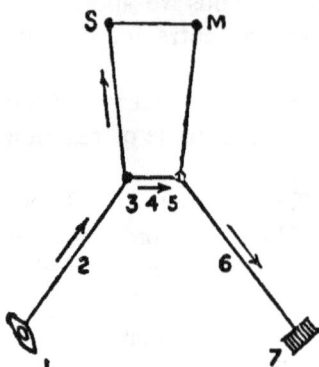

FIG. 49.—Reflex Action.

It should be noted that the centre in the cord does not merely reflect the afferent impulses into efferent impulses ; the reflex act often seems to be adapted for a purpose, and to be of a useful nature. The brainless frog's movements on being irritated by an acid are directed to removing the irritating substance. Winking on irritation of the conjunctiva or at the sudden approach of a missile indicates a purpose. Sneezing, coughing, and vomiting are typical reflex actions of a similar nature.

There is a variety of actions occurring rhythmically in the body to which the term *automatic* is sometimes applied, because they appear to arise in the nerve centres themselves without any external impulse (automatic = self-acting, literally). Such an action is breathing. Breathing is brought about by the action of certain muscles which expand the chest and so cause air to rush into the lungs, followed by the action of another set of muscles which contract the chest and drive out air, and this appears due to the rhythmical discharge from nerve centres in the medulla of efferent impulses at the rate of sixteen to eighteen times per minute. But such discharges are not really self-caused ; the centres are no doubt stimulated by the condition of the blood. Keeping the term automatic to refer to those motor reactions that occur in series and are due to rhythmic discharges from nerve centres, we may give as instances of automatism, besides breathing, the activities of the centres that keep up a constant or tonic contraction in the arterial walls and in sphincter muscles, as well as those which provide for regular contraction and relaxation of organs. All these centres are doubtless set in action by impressions from the organs they thus regulate. though the afferent impulses may not always be easy to describe.

In a typical reflex action afferent impulses reach a nerve centre and are transmitted by the irritable protoplasm of the centre, not simply turned aside, or reflected, into efferent impulses. It is true that there is often a close relation between the strength of the stimulus applied to the afferent nerve and the magnitude of the efferent impulse, as shown in muscular movement, *i.e.* a slight impulse will usually give rise to slight muscular movement and a strong impulse to more forcible movement. But a very slight stimulus may also be so intensified in a nerve centre, where the arrangement of the reflex mechanism is complex, that the result is the powerful contraction of many muscles. Witness the convulsive fit of coughing when a small particle passes into the larynx. The condition of the centre also modifies greatly the resulting reflex action. Thus strychnia heightens the excitability of the cord to a great degree, and in a brainless frog poisoned with this drug a slight touch on the skin causes violent convulsive contractions in most of the muscles of its body, a discharge of energy from the centre spreading in nearly all directions. In *simple* cases of reflex action between the part stimulated and the muscle or muscles set in motion, the action is such that the movement appears to be the result of an efferent impulse in a motor nerve that is the companion to the sensory nerve ; in *complex* cases of reflex action co-ordinated movements are set up that appear adapted to accomplish a purpose, as when a brainless frog attempts to wipe off the irritating acid on its flank. It should also be noted that a stimulus applied to the surface of the skin or other terminal sensory organ leads to stronger and more complex movement than when the stimulus is applied to the sensory nerve directly. Reflex movements being immediate, unchosen motor responses to stimulation of afferent nerves may be performed unconsciously or consciously. When the iris contracts under the influence of a strong light, the excited retina leads to the transference of an afferent impulse along the optic nerve to a centre in the brain, and an efferent impulse is then sent along the third cranial nerve to the circular fibres of the iris, and their contraction follows without our knowledge.   Sneezing, on the other hand, is a reflex action set up by irritation of the nasal branch of the fifth nerve conveying impulses to the respiratory centre in the medulla that lead to the action of expiratory muscles, and of this we are usually conscious at the time or may become conscious subsequently.   Finally, we may note that reflex actions may often be *inhibited* or arrested by an effort of the brain (will), though in some cases the stimulus may be so powerful as to defy inhibition.

45. Voluntary Actions in contrast with Reflex Actions.— A large class of conscious efferent impulses appear to arise in the brain without any immediate afferent impulses.   They proceed from the will or choice of the individual.   These volitional or voluntary impulses differ from the immediate motor responses to stimuli that constitute reflex acts by seeming to be free, irregular, spontaneous, or even chosen in spite of exciting impressions.   There is no doubt some preceding stimulus, but between the sensation and the motion the brain compares and judges.   Drawing a plan, writing a theme, or solving a problem involve conscious voluntary acts.   But it is easy to see that there is difficulty in fixing the boundary between voluntary and reflex acts, and that acts that are reflex ordinarily may be modified by the influence of the will.

46. Velocity of Nerve Current.—The rate of transmission of nerve impulses along a nerve-fibre has been found to be about 100 feet per second (light travels 186,000 miles per second, and electricity 200,000 miles per second).   This rate has been directly measured by stimulating one of the nerves for the small muscles of the hand near its origin in the neck and at another point in the forearm (the nerve coming near the surface at these places), and observing by the help of electric signals the difference of time between the contraction of the muscles when the nerve is stimulated at the distant position and the near position.   Knowing the rate of conduction in nerve-fibres, it is possible to determine the time occupied by a nerve centre in its work.   This is done by stimulating a sensory or afferent fibre of the spinal cord, and observing the length of time between throwing in a shock and the commencement of the muscular contraction on the motor or efferent side.   Taking from this whole time the time required to pass along the known length of nerve-fibres and the time lost in the muscular nerve-ending, the time occupied in the nerve centre in the reflex act is found to vary from ·006 second to ·06 second according to the condition of the nerve centre and the path taken. This period is known as *reflex-time*.

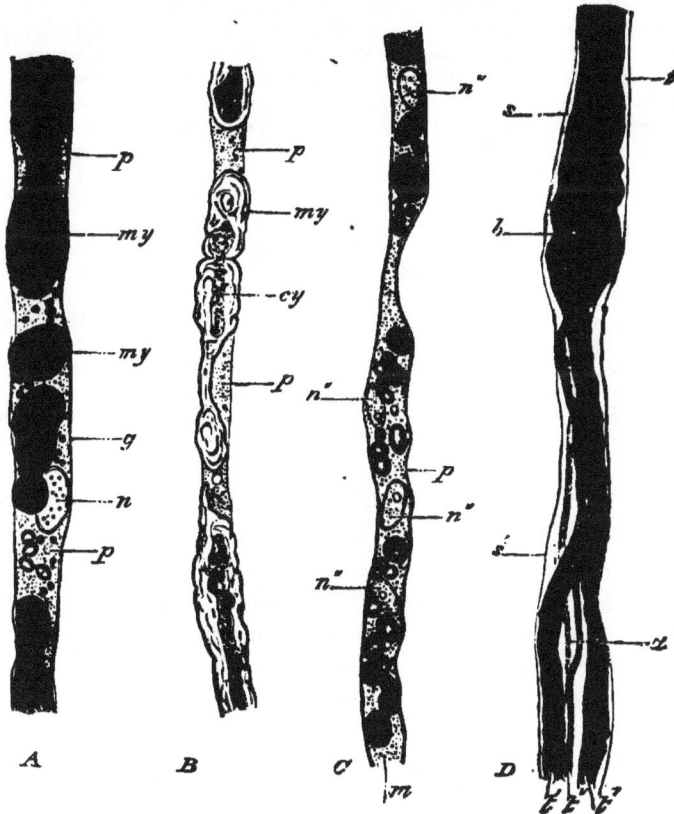

FIG. 50.—Degeneration and Regeneration of Nerve-fibres in the Rabbit. (Ranvier.)
A, part of a nerve-fibre in which degeneration has commenced in consequence of the section, fifty hours previously, of the trunk of the nerve higher up ; *my*, medullary sheath becoming broken up into drops of myelin ; *p*, granular protoplasmic substance which is replacing the myelin ; *n*, nucleus ; *g*, primitive sheath. B, another fibre in which degeneration is proceeding, the nerve having been cut four days previously ; *p*, as before ; *cy*, axis-cylinder partly broken up, and the pieces inclosed in portions of myelin. C, more advanced stage of degeneration, the medullary sheath having almost disappeared, and being replaced by protoplasm in which, besides drops of myelin, are numerous nuclei, which have resulted from the division of the single nucleus of the internode. D, commencing regeneration of a nerve-fibre. Several small fibres, *t't''*, have sprouted from the somewhat bulbous cut end, *b*, of the original fibre, *t* ; *a*, an axis cylinder which has not yet acquired its medullary sheath ; *s,s'*, primitive sheath of the original fibre. A, C, and D are from osmic preparations ; B from an alcohol and carmine preparation.

47. **Degeneration of Nerves.**—When a nerve is cut the part supplied by the nerve becomes paralysed, loss of motion or sensation. or both, according as the nerve is sensory or motor, or mixed, being the result. Excitability also gradually becomes less, and disappears entirely in a few

days. Soon after section a gradual degeneration begins in the peripheral fibres thus cut off from the cells to which they belong, and which govern their nutrition. The white medullary sheath breaks up into small granular oily-looking masses and at length disappears. The axis cylinder also breaks up, and this also is absorbed, so that all that is left of the severed nerve is the primitive sheath with protoplasmic contents. This degeneration takes place along the whole of the severed nerve simultaneously, and the process lasts about ten or twelve days. (Degeneration of a nerve after section is often called, from its discoverer, *Wallerian degeneration.*)

The cut end connected with the centre does not degenerate, but in the course of two or three months regeneration of the nerve has begun by the growth from the central stump of new axis cylinders, a repetition of the process of the development of the nerve. These pass along the old track and form the axis cylinders of new nerve-fibres, which in time become surrounded with a medullary sheath, and reaching the periphery may restore the lost excitability of the part. From what has just been said it is evident that section of a spinal nerve results in degeneration of all the nerve-fibres leading to the periphery because all these have their trophic centres near the cord—the afferent fibres being connected with cells in the ganglion of the posterior root, the efferent fibres with cells in the anterior horn of grey matter in the cord (par. 51).

48. Functions of Nerves.

—We have already said that nerves are divided into two great classes according as they conduct impulses from a nerve centre (efferent) or to a nerve centre (afferent). From what has been said it will be understood that the function of a nerve is ascertained by determining its mode of origin, both superficial and deep, its mode of distribution to muscle or gland or sense organ, and its physiological behaviour. If it spring from a tract of matter where other afferent or efferent nerves arise, its general function may be surmised; if it be distributed to a muscle or gland it is evidently motor or efferent; if it terminate in a sense organ in such a way as to be adapted to receive particular impressions, it must be sensory or afferent. But experiment and observation of the nerve under varying conditions will be the best guide. It may be stimulated in its course and the effects observed, or it may be divided and the peripheral and central ends stimulated separately. Stimulation of the end going to the periphery would show whether this led to contraction of muscle, secretion of gland, arrest or increase of action. Stimulation of the end connected with the nerve centre might show whether a sensation of any kind was set up. Finally, examination of the cut end under the microscope would show in what direction degeneration of the nerve-fibres had occurred, and from this we should learn the direction of its cell origin, the portion remaining in connection with the cell undergoing little or no change. In one or more of these ways the functions of the nerve-fibres have been ascertained, and the nerves divided into groups according to their functions. These functional groups of nerve-fibres may be thus arranged :—

I. *Efferent* or centrifugally conducting nerve-fibres carry impulses from a centre to the periphery.

(*a*) Motor fibres to striped skeletal muscles.

(*b*) Motor fibres to cardiac muscle (accelerator) (par. 83).

(*c*) Motor fibres to unstriped muscle, *e.g.* the intestine.

(*d*) Secretor-motor fibres to gland-cells, *e.g.* the salivary glands (par. 114).

(*e*) *Trophic* fibres to tissues generally to regulate their metabolism and nourishment (?).

(*f*) Vaso-motor fibres constricting arteries (par. 84).

(*g*) Vaso-inhibitory fibres dilating arteries (par. 84).

(*h*) Cardo-inhibitory fibres checking or arresting action of cardiac muscle (par. 83).

[Inhibitory nerve-fibres are those which check, arrest, or suppress a motor or secretory act already in progress.]

II. *Afferent* or centripetally conducting nerve-fibres carry impulses from the periphery or outlying parts to a nerve centre.

(*a*) Sensory fibres in the narrow sense, conveying general impressions of pain, heat, pressure, and common sensation.

(*b*) Sensory fibres of the special senses, viz., optic, auditory, olfactory, gustatory, and tactile.

(*c*) Musculo-sensory fibres.

(*d*) Excito-motor fibres ministering to reflex action.

III. *Intercentral* nerve-fibres connecting ganglionic centres with each other.

It will be seen that all efferent nerve-fibres are not motor in the sense of leading to muscular movement, as some lead to the secretory activity of cells merely, while others actually inhibit or arrest motion. Nor are all afferent fibres sensory in the strict meaning of this word, as some do not excite sensation or consciousness, for in the case of the excito-motor nerves that lead to reflex action impulses may be conducted to a nerve centre and transferred thence to an efferent fibre whose peripheral end is set in action without the mind being aware of the process.

49. Nerve-endings or Peripheral Distribution of Nerves.—

We have already spoken of the way in which nerves terminate in voluntary muscles and described the motorial end-plates of these organs (par. 34). In the involuntary muscles, such as are found in the muscular layers of the hollow viscera, the nerves, which are composed for the most part of non-medullated fibres, form complicated networks or plexuses near their termination, ganglion-cells being seen at the junctions as in the figure. From these plexuses anastomosing branches pass off and penetrate between the involuntary muscular tissue, finally ending in fibrils that come into close relation with individual muscle-cells.

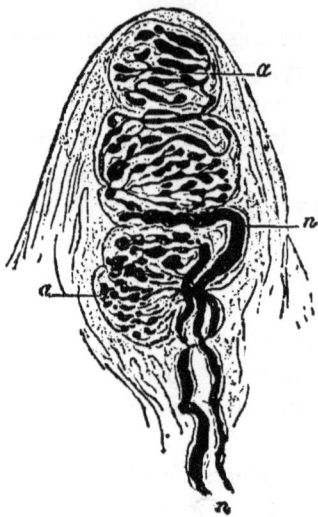

FIG. 51.—Tactile Corpuscle within a Papilla of the Skin of the Hand, stained with chloride of gold. (Ranvier.)

n, two nerve-fibres passing to the corpuscle ; *a, a*, varicose ramifications of the axis cylinders within the corpuscle.

A special plexiform ending, probably sensory in function, is found in many tendons near their junction with muscles. The tendon-bundles are somewhat enlarged at this point, and the nerve-fibres passing to this part divide and spread between the smaller bundles so as to form a branched expansion almost similar to the motorial end-plates of the voluntary muscles. The whole structure in a tendon thus provided with an expanded nervous plexus is known as the *organ of Golgi*.

Most of the afferent or sensory nerve-fibres that go to the special organs of sense (eye, ear, &c.) end in cells, as will be described when treating of these organs. In other cases ordinary sensory nerves, including those devoted to tactile sensations, end in special organs, of which there are three chief varieties, *tactile corpuscles or touch bodies, end bulbs*, and *Pacinian corpuscles*. *Free* nerve fibrils among the deeper layers of the epidermis have been already referred to (par. 6).

Tactile corpuscles are small oval bodies about $\frac{1}{300}$ in. long found in the papillæ or among the epithelium of the skin of the hand, foot, lip, and nipple. The main substance of the corpuscle is composed of connective tissue, and within the capsule of this one or more nerve-fibres wind about, and then

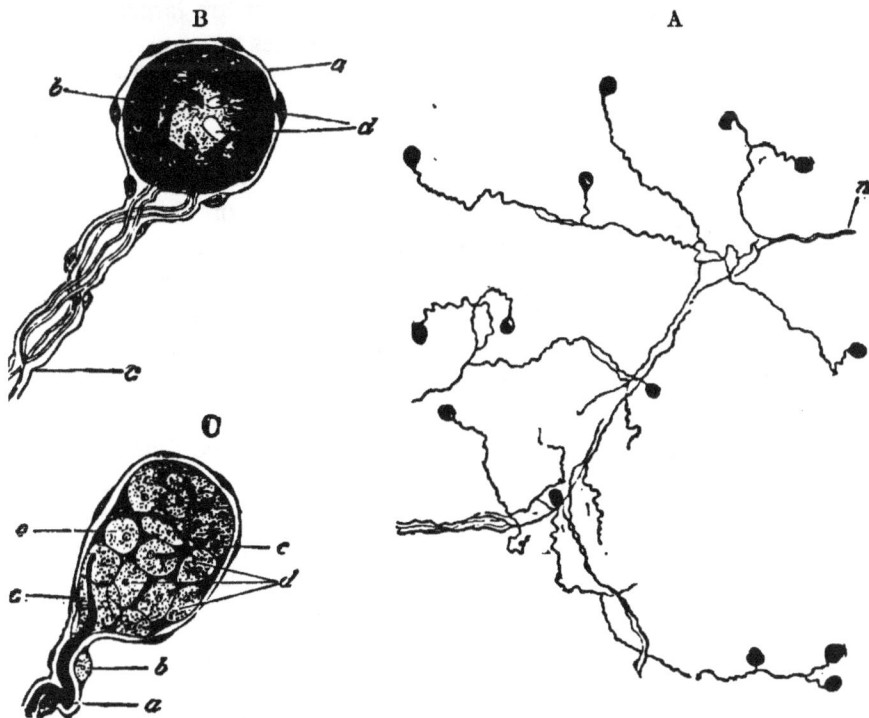

FIG. 52.—End bulbs from the Human Conjunctiva. (Longworth.)

A, ramification of nerve-fibres in the mucous membrane, and their termination in end bulbs, as seen with a lens; B, an end bulb more highly magnified; *a*, nucleated capsule; *b*, core, the outlines of its component cells are not seen; *c*, entering fibre branching and its two divisions passing to terminate in the core at *d*; C, an end bulb treated with osmic acid, showing the cells of the core better than B; *a*, the entering nerve-fibre; *b*, capsule with nuclei; *c, c*, portions of the nerve-fibre within the end bulb, the ending of the fibre is not seen; *d, e*, cells of the core.

ramify within it, the axis cylinders of the nerve-fibres terminating in enlargements. As these bodies are only met with in highly sensitive parts, they are thought to be connected with the sense of touch.

End bulbs are spheroidal corpuscles about $\frac{1}{600}$ in. in diameter composed of a connective tissue capsule, and a core of cells in

which the axis cylinder of a nerve-fibre terminates after twisting
or expanding.   They are found in the papillæ of the skin of
the lips, in the conjunctiva of the eye, and in the mucous
membrane of the cheeks.   Similar bodies of a larger size are
also found in the synovial membrane of certain joints.

The Pacinian bodies or corpuscles of Vater are larger and
more complex bodies than tactile corpuscles or end bulbs.
They are oval bodies $\frac{1}{10}$ to $\frac{1}{15}$ in. long, and $\frac{1}{20}$ to $\frac{1}{30}$ in. broad,
attached to some of the cerebro-spinal and
sympathetic nerves, especially on the cutaneous nerves of the hand and foot.   They
also occur in the abdomen on the nerves
of the solar plexus and the nerves of the
mesentery.   Each Pacinian corpuscle is
attached by a stalk to the nerve, and is
found to consist of several concentric layers
of connective tissue which enclose the
prolonged end of a nerve-fibre (fig. 54).
On reaching the core the nerve-fibre loses
its medullary sheath, the axis cylinder
terminating in an arborisation or bulbous
enlargement.   The physiological office of
Pacinian corpuscles is not clear.

Other modes of nerve termination will
be described in the account of the special
senses.

FIG. 53.—A Nerve of the
Middle Finger, with
Pacinian bodies at-
tached.   Natural size.
(After   Henle   and
Kölliker.)

50. **Electrical Phenomena of Nerves** —
In living uninjured tissue at rest the electrical state of all parts is similar; but in a
divided nerve the injured part or cut end becomes electro-
negative to the uninjured portion, and a current passes from
the longitudinal surface to the cut end.   By causing a nervous
impulse to pass down such a nerve by any method of stimulation
—mechanical, thermal, or electrical—the electrical difference
between the injured and uninjured surfaces undergoes a diminution, or a *negative variation*.   This negative variation or
diminution of previously existing electrical difference is
indicated by a galvanometer, and is made use of in detecting

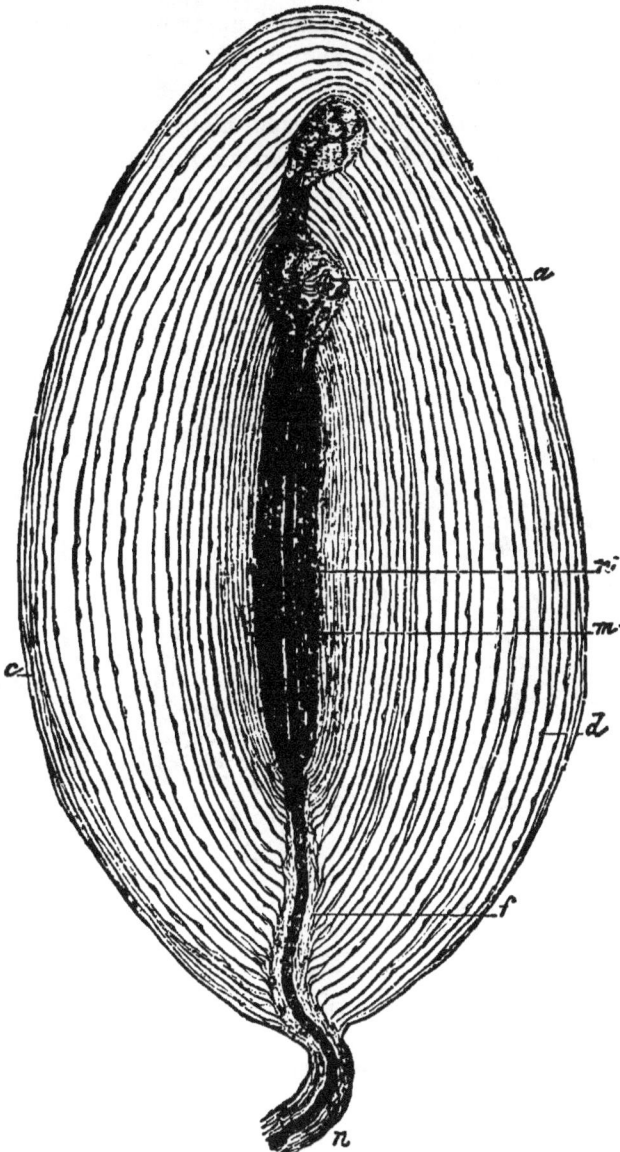

FIG. 54.—Magnified view of a Pacinian Body from the Cat's Mesentery. (Ranvier.)
n, stalk with nerve-fibre enclosed in sheath of Henle passing to the corpuscle; *n'*, its
continuation through the core, *m*, as a pale fibre; *a*, termination of the nerve in the
distal end of the core. In the corpuscle here figured the termination is arborescent;
a', lines separating the tunics of the corpuscle, often taken for the tunics themselves;
f, channel through the tunics, traversed by the nerve-fibre; *c*, external tunics of the
corpuscle.

F

whether nerve impulses are passing along nerve-fibres or not.
It also shows that nerves conduct in *both* directions, though in
one only is any effect manifest.

Sending a *constant* electric current along nerve-fibres throws
them into a state called *electrotonus*, modifies the conductivity
of the fibres, and increases the natural nerve current in the

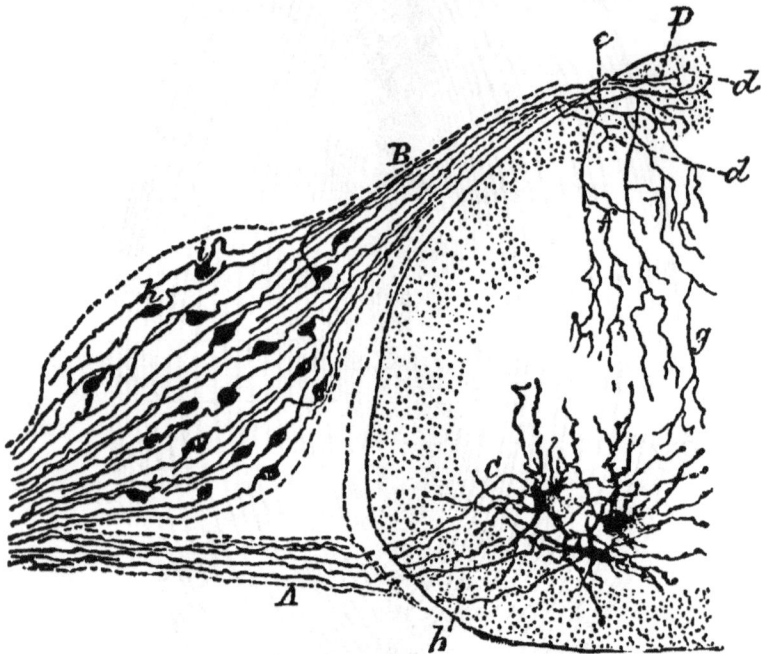

Fig. 55. —Transverse Section of the Spinal Cord of a Chick on the 9th day of
incubation, prepared by Golgi's method. (Ramón y Cajal.)

A, axis cylinders of anterior root-fibres issuing from large cells of the anterior horn, *C*.
B, posterior root-fibres passing from the bipolar cells of the spinal ganglion into the
posterior column of the spinal cord, *D*, where they bifurcate, *d*, and become longitudinal ; *e, f, g*, collaterals from these fibres, passing into the grey matter.

same direction. No chemical change has been detected in
nerve-fibres as the result of nervous impulses passing, nor has
any *thermal* change been observed in them, though the activity
of nerve *centres* has been found to be accompanied by a
development of heat.

51. **Origins or Roots of Nerves.**—A spinal nerve has its
central extremity connected with the cord at a point termed

its origin or root, or rather at two points of origin, an anterior
and a posterior root. But as the fibres can be traced for some
distance in the cord the place of attachment to the surface is
only the superficial root, and thè deep origin is concealed. On
tracing this out it is found that nerve-fibres usually originate
as prolongations of one of the processes of a nerve-cell. Researches in the development of nerves, in fact, show that nerve-
fibres have grown out from nerve-cells. The efferent fibres of
the anterior roots arise from the àxis-cylinder processes of the
cells in the anterior horn of grey matter in the cord, and the
afferent fibres of the posterior root arise from cells in the
posterior ganglion, and grow into the cord, where they bifurcate
and give off fine branches (see fig. 55), as well as towards the
periphery.

CHAPTER III

BLOOD

52. BLOOD, as it exists in the living body, is, in one sense, a
tissue consisting of cellular elements called corpuscles, and an
intercellular matrix called plasma. It flows as a fluid through
a system of closed tubes —arteries, capillaries, and veins—and,
traversing all the parts of the body, serves as the medium of
nourishment to all the other tissues. Into it, in a way shortly
to be described, are passed at certain parts of the circulatory
system the various kinds of food that have been duly digested,
as well as a supply of oxygen derived from the air. As it
passes through the minute capillary blood-vessels it gives up
to the surrounding tissues some of these new materials for their
nutrition, and receives in return from these tissues certain
waste products which have to be removed, the lymph, an
almost clear fluid, lying outside the capillaries and bathing the
elements of each tissue, serving as the medium of this exchange
between the blood and other tissues. All tissues take up
oxygen from the blood and give up carbonic acid to the blood,
though often at various rates ; but the tissues of many organs
either consume and assimilate particular materials from the

blood or throw off into the blood-stream special waste products. It is thus evident that the composition and characters of blood must vary in different parts of the body, and at different times, as will be more fully noted when dealing with the several organs. But the combined action of all the tissues keeps the whole mass of blood in the body at an average uniform composition, so that there are certain general properties of the blood as a whole which we may now proceed to describe.

53. Physical Properties of Blood. --The *colour* of blood varies from a bright scarlet-red in the arteries to a dark purplish-red in the veins. Oxygen or the air makes it red; want of oxygen renders it dark. Even in thin layers blood is *opaque*, as is easily seen by trying to read print through a thin layer placed on glass. This opacity is due to the fact that blood owes its colour to certain fine red particles floating in a clear pale fluid called the *plasma*. Blood has a slightly saline *taste*, due to the salts dissolved in the fluid part; it emits a peculiar *odour*, which varies with each animal species. It has an alkaline *reaction*, due to the presence of disodic phosphate and bicarbonate of soda. Its alkalinity may be shown by placing a drop for a few seconds on a specially prepared glazed red litmus paper, and then wiping it off, when a blue spot will be apparent. Blood is slightly heavier than water, its mean *specific gravity* being about 1050, when water is taken as 1000. The *temperature* of the blood varies within narrow limits, 97·8° F. to 100° F., according to the part of the body—near the surface or deep-seated—from which it is drawn. A special property of blood, partly physical and partly chemical, is its power under certain conditions of passing from a fluid condition into a jelly-like mass, which then separates into a fluid called *serum*, and a solid clot floating in the serum. The *coagulation of the blood*, as this phenomenon is called, is caused by the formation of threads of solid fibrin which, contracting, entangle the corpuscles and form a clot. Put in a tabular form we have :—

$$\text{Living Blood} = \left\{ \begin{array}{l} \text{Plasma} \\ \text{Corpuscles} \end{array} \right\} = \left\{ \begin{array}{l} \text{Serum (appearing as a clear fluid).} \\ \text{Fibrin threads} + \\ \text{Corpuscles} \end{array} \right\} = \text{Clot.}$$

or thus :

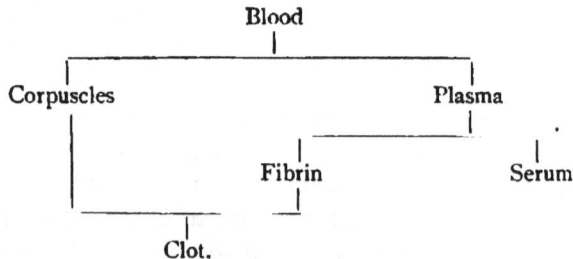

The phenomena of clotting are further treated of in par. 60.

54. Microscopic Appearance of Blood.— When examined by the microscope, either within the minute vessels or spread out on a piece of glass, blood is seen to consist of an enormous number of small bodies or *corpuscles* floating in a clear fluid, the *plasma* or *liquor sanguinis*. These corpuscles are seen to be of two kinds, white or colourless corpuscles and red corpuscles. By diluting a small but carefully measured quantity of blood with a comparatively large quantity of some indifferent fluid, as a 5 per cent. solution of sodium sulphate, the actual number of corpuscles may be counted. It is found that, on the average, the number of corpuscles in a cubic millimetre exceeds 5,000,000. The number of white corpuscles is far less than the number of red ones, in the proportion on the average of 1 white to 350 red. The relative numbers of the two kinds, however, vary with age, situation, state of health, and other conditions. The proportion of white to red is greater in children than in adults, during digestion than while fasting. In the splenic vein there is 1 white to 70 red, but in the splenic artery only 1 white to 2,260 red ; in the hepatic vein there is 1 white to 170 red, but in the portal vein only 1 white to 740 red. Each red corpuscle consists of a *stroma* or framework of proteid material, having within it a red colouring matter termed hæmoglobin. This gives the colour to the corpuscle and to the blood itself. Shaking up venous blood with air or oxygen turns it bright red ; passing in carbon dioxide gas darkens it in colour again. (See Glossary, Hæmoglobin.)

55. The Colourless Corpuscles. —The white or colourless corpuscles of the blood (often called *leucocytes*) are usually somewhat larger than the red ($\frac{1}{2500}$th of an inch in diameter compared with $\frac{1}{3200}$th of an inch) and always much fewer in number. They are essentially similar to the animal cells called leucocytes (see Glossary) found in other parts of the body, and to the simple organism called the amœba.

Colourless corpuscles or leucocytes are nucleated masses of protoplasm without cell-wall and of nearly globular shape. Several varieties exist in human blood. The most numerous variety has a lobed nucleus, and the surrounding protoplasm contains numerous fine granules (see fig. 57, 1). Another less numerous variety contains several large granules, and the granules of these coarsely granular corpuscles stain deeply with acid aniline dyes like eosin. A third variety of leucocyte in the blood is the *hyaline corpuscle*, the protoplasm of which is quite clear of granules. A fourth variety is a small corpuscle called a *lymphocyte*. This has a large nucleus and a thin layer of clear protoplasm. Colourless corpuscles have the remarkable property of spontaneously altering their form, that is, they are capable of amœboid movement and change of shape. These amœboid movements consist of alternate protrusions and contractions of processes of protoplasm (fig. 4). The addition of water or dilute acid renders the presence of a nucleus or nuclei in these cells more apparent. White blood-corpuscles have the power of taking into their interior small solid particles or granules, and also foreign bodies like small bacteria, while in some cases distinct vacuoles may be seen in their protoplasm. Another result of their power of amœboid movement is the power they possess under certain circumstances of passing through the walls of the minute

Fig. 56.—Human Blood as seen on the warm stage. Magnified about 1000 diameters. (E. A. S.)

r, r, single red corpuscles seen lying flat ; *r′,r′*, red corpuscles on their edge and viewed in profile ; *r″*, red corpuscles arranged in rouleaux ; *c, c*, crenate red corpuscles; *p*, a finely granular pale corpuscle : *g*, a coarsely granular pale corpuscle—both have two or three distinct vacuoles, and were undergoing changes of shape at the moment of observation ; in *g*, a nucleus also is visible. '

Fig. 57.—Colourless Corpuscles treated with Water and with Acetic Acid. (E. A. S.)

1, first effect of the action of water upon a white blood-corpuscle ; 2, 3, white corpuscles treated with dilute acetic acid ; *n*, nucleus.

blood-vessels into the surrounding tissue. This process, called ' emigration of white corpuscles ' or *diapedesis*, often appears active in the neighbourhood of an inflamed area, where they may accumulate to form new tissue or degenerate and form an abscess, which contains a large quantity of what are known as pus-cells. In other cases we may find the emigrated leucocytes in connective tissue as *wander-cells*, or they may possibly take a part in the natural growth and repair of tissue. As to those that remain in the blood-stream, it is certain, since there is a constant influx, that they disappear in some way, possibly by the substances composing them being dissolved in the plasma during circulation. That they give rise to red corpuscles is very doubtful.

As to the *origin* of the colourless corpuscles there is little doubt. It will shortly be seen that white corpuscles are continually being poured into the blood-stream from the lymphatic system, and that the corpuscles of lymph are far more numerous in the lymphatic vessels after they have passed through the lymphatic glands. These glands, as already shown, consist partly of a special kind of connective tissue, arranged as a delicate network, and known as *lymphoid* or *adenoid tissue*. In this tissue, which is found in the spleen and certain mucous membranes as well as in lym-

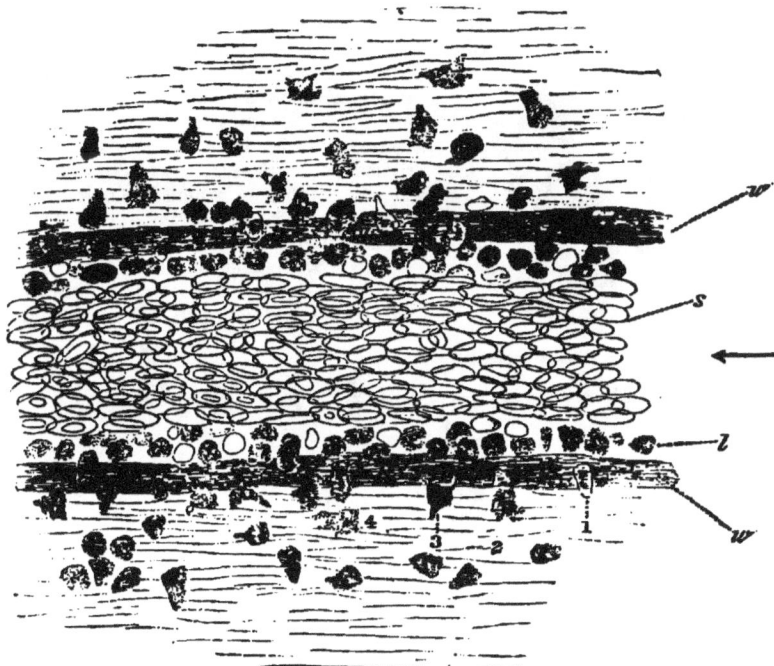

FIG. 58.— Migration of the White Corpuscles from a Small Vein.
w, wall of vessel ; *l*, leucocyte-layer on inner side of the vessel wall ; 1, 2, 3, and 4, leucocytes in different stages of migration ; *s*, red blood-corpuscles (axial stream).

phatic glands, multitudes of colourless nucleated cells are found undergoing karyokinetic cell-division in the same way as other animal cells. Here, then, is the chief source of supply. The white corpuscles multiply in the adenoid tissues of the lymphatic glands, to pass into the lymph-stream, and from it into the blood. Here they appear to be of use in carrying absorbed nutriment to the tissues, their numbers being greatly increased after a meal. Emigrated leucocytes also serve in protecting the organism against bacteria, &c. that are the source of diseases, as they have the power in many cases of enclosing and devouring such injurious objects. Leucocytes acting in this way are termed phagocytes (Gk. *phagein,* to devour). (See par. on Inflammation.)

56. Red Corpuscles.—Human red corpuscles are circular biconcave discs, having a diameter of $\frac{1}{3200}$th of an inch (7μ), and a thickness one-fourth of the diameter, red in mass but yellow when seen singly. Examined under a high power of the microscope, they show when lying flat a dark centre with a light rim, or *vice versa*, according as the edge or centre is brought into focus. There is no nucleus. Seen edgewise the shallow depression on both surfaces plainly appears. When blood is shed the red corpuscles tend to adhere together at their surface, forming rouleaux like rolls of coin. Red corpuscles are extensible and elastic, readily gliding past each other in the blood-vessels. Though they never protrude processes of protoplasm (pseudopodia), like white corpuscles, certain reagents alter their shape. Thus dilute salt solution causes them to show thorny prominences, so that they are then described as *crenated*. When water is added to blood under the microscope, the corpuscle absorbs a quantity, bulges out, and at last becomes decolorised. Each red corpuscle, in fact, consists of a delicate colourless framework or *stroma*, in the meshes of which the colouring matter, called *hæmoglobin*, is contained. This experiment with water is considered by Schäfer to prove that 'the corpuscle is composed of a membrane or external envelope with coloured fluid

FIG. 59.—1. Human Blood Corpuscles, side and front view.
2. Frog's Blood Corpuscles, side and front view. (Waller.)

The corpuscles are drawn to the same scale, viz. ×1000. 1μ or $\frac{1}{1000}$mm. of actual length is represented by 1 mm.; the human corpuscle as sketched above is 7 mm. in diameter, *i.e.* its actual diameter is 7μ; the frog's corpuscle is 20μ by 11μ. The corpuscles of mammalia resemble those of man, but differ in size. Average magnitudes are: elephant, 9μ; dog, rat, 7μ; cat, horse, ox, 6μ; goat, sheep, 5μ. But the corpuscles of man, averaging 7μ, may range from as low as 5μ to as high as 9μ.

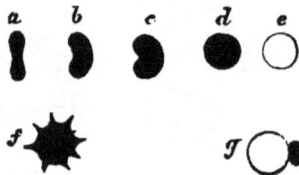

FIG. 60.

a e, successive effects of water upon a red blood-corpuscle : *a*, corpuscle seen edgeways, slightly swollen ; *b, c*, one of the sides bulged out (cup form) ; *d*, spherical form ; *e*, decolorised stroma ; *f*, a thorn-apple-shaped corpuscle (due to exposure) ; *g*, action of tannin upon a red corpuscle.

contents, for the above reaction is precisely the same as would occur by osmosis with a bladder of the shape of a corpuscle filled with a strong solution of albuminous substance and placed in water.' Tannic acid causes . the hæmoglobin to be extruded from red corpuscles in drop-like masses.

Hæmoglobin forms 90 per cent. of dried red corpuscles. It is a complex albuminous substance, containing C.H.O.N.S.Fe, the percentage of iron being 0·4. Its chief use is to act as an oxygen carrier from the lungs to the tissues, and to this end it forms a loose compound with oxygen (in addition to the oxygen it always contains), as the red corpuscles pass through the lung capillaries. It thus exists in two forms, oxy-hæmoglobin and reduced hæmoglobin. Both forms crystallise, and the two can be distinguished by their colour and by their different absorption spectra when examined in solution with the spectroscope. Reduced hæmoglobin is often referred to simply as hæmoglobin. Hæmoglobin can be readily split up into a coloured pigment called *hæmatin* and a proteid body called *globin*. Now one of the constituents of the bile, which is a secretion from the blood formed by the liver, is a pigment called *bilirubin*. The composition of bilirubin indicates that it is a derivative of hæmatin, and as no other source of hæmatin is known than the hæmoglobin of the red corpuscles, it appears that a continuous destruction of red corpuscles must go on in the liver. Red corpuscles have also been seen in various states of disintegration in the spleen pulp. If destruction is thus going on continually—the life of a red corpuscle has been estimated at from three to four weeks—it becomes important to ascertain the *origin* of these bodies. Careful inquiry has shown that after birth the chief source of supply of red corpuscles is the *red marrow of bone* that fills the

FIG. 61.—Coloured Nucleated Cells from the Red Marrow of the Guinea-pig. (E. A. S.—

internal cavities of many bones, particularly the ribs. Here is found the great seat of formation of red corpuscles, for in it are found certain peculiar nucleated red corpuscles called *erythroblasts*, which gradually lose their nuclei and pass into circulation as ordinary red corpuscles. These early forms of the red corpuscles multiply by karyokinetic division, and this division becomes very rapid after much loss of blood. A similar formation of red corpuscles is said to take place in the spleen after great loss of blood, but of this there is some doubt. To sum up, we may say that red corpuscles are minute biconcave non-nucleated discs, consisting of a framework or stroma, which contains hæmoglobin ; that the chief function of these bodies is the carrying of oxygen by their hæmoglobin from the lungs

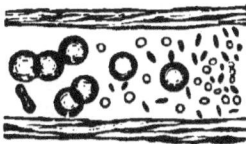

FIG. 62.—Blood Corpuscles and Platelets within a Small Vein of the Rat's Mesentery. (Osler.)

FIG. 63.—Network of Fibrin, shown after washing away the corpuscles from a preparation of blood that has been allowed to clot. (E. A. S.)

Many of the filaments radiate from small clumps of blood-tablets.

to the tissues ; that it is their fate to be broken up in the liver and spleen after a short period ; that the hæmoglobin thus set free from these bodies is of secondary use in forming the bile pigments ; and that new red corpuscles arise in the adult from division of a special class of cells in the red marrow of bones.

57. **Blood-platelets.**—A third microscopic element of blood, of very minute, variously shaped particles, is known as *blood-platelets*. They may be noticed free within the living blood-vessel, though they often become massed together after the blood is drawn, the fibrin filaments which then form radiating from the groups of platelets thus gathered together. The

real nature and function of these particles are by no means determined. Filaments of fibrin often seem to radiate from small clumps of blood-platelets in a portion of clot examined under the microscope.

58. Gases of the Blood.—When blood is collected over mercury to keep it from contact with the air, about 60 vols. of gases per 100 vols. of blood can be extracted from it by a mercurial pump. These gases are collected in a graduated eudiometer tube, and are found to consist of oxygen, carbon dioxide, and nitrogen.

The amount of CO_2 in the total volume thus extracted from the blood is estimated by introducing into the eudiometer tube a small ball of fused caustic potash fixed on a platinum wire. After some hours this absorbs the CO_2, and the diminution in volume shows the amount of CO_2. The amount of oxygen is then estimated by placing in the tube, thus deprived of its CO_2, a piece of phosphorus, which absorbs the oxygen. After the ball is removed the diminution in volume now indicates the amount of oxygen. The remaining gas is nitrogen. As a result of careful experiments, the average total gases in human blood at the normal temperature and pressure are a little over 60 vols. per cent., distributed thus :—

	O	CO$_2$	N
Arterial blood	20	40	1·4
Venous blood	8 to 12	46	1·4

All measured at 760 mm. pressure and 0° C.

It will be noticed that while the amount of oxygen in arterial blood is nearly constant, the amount in venous blood varies considerably. It varies with the vein from which it is drawn, and also with the same vein according to the activity of the organ from which the vein proceeds. Thus in the case of a gland at rest through which the blood flows slowly, the issuing blood has less O than from an active gland ; but the blood issuing from an active muscle is, in spite of the rapid flow, more venous than from the same muscle at rest. But the student must not forget that arterial blood does not give up its whole oxygen and always contains carbon dioxide, nor that venous blood does not give up all its CO_2 and always contains some oxygen.

The oxygen which comes off from blood in the vacuum

of the mercurial pump exists in the blood in two distinct states :—(1) a very small amount is simply absorbed by the blood, about as much as is absorbed by water under ordinary atmospheric pressure ; (2) the larger portion is in a state of *loose chemical combination* with the hæmoglobin of the red corpuscles, thus forming oxy-hæmoglobin. That the greater portion of the oxygen is in a state of chemical combination is evident from the fact that it is not subject to the law of absorption of gases by a liquid, for it does not leave the blood in the mercurial pump gradually as the pressure is reduced, but comes off suddenly at a certain very low pressure ; that it is but *loosely* combined is evident from the fact that it is thus able to be set free at all by reduction of pressure, and because it is liberated by boiling the blood and by the simple process of passing H, N, or CO through the blood.

The carbon dioxide is mainly contained in the plasma of the blood, a very small portion being simply absorbed; the largest portion is in feeble chemical combination with the sodium salts of this fluid, sodium carbonate and sodium phosphate. It is also very probable that the red corpuscles contain a portion of the CO_2 in loose chemical union. The nitrogen which exists in the blood appears to be simply absorbed, as the quantity is about the same as water would dissolve.

59. **Composition of Blood.**—The composition of living blood by weight may be thus given :—

Living blood $\begin{cases} \text{Plasma or liquor sanguinis } \frac{3}{4} \text{ or 60 to 65 p.c.} \\ \text{Corpuscles (red and white) } \frac{2}{5} \text{ or 35 to 40 p.c.} \end{cases}$

Plasma is composed, roughly, of 90 per cent. water and 10 per cent. solids. (Plasma also contains a small portion of the gases of the blood dissolved in it.) Of the solids of plasma about 8 per cent. are proteid [1] in nature, the chief proteids being paraglobulin (serum-globulin), serum-albumin, and fibrinogen. Fibrinogen is the substance that produces fibrin. It only forms 0·2 to 0·4 per cent. of the proteids of the plasma. The other solids of plasma are inorganic or mineral salts and a class of bodies called extractives, because they are extracted by such

[1] See appendix, ' Proteid.'

substances as alcohol and ether. The chief extractives are fats, lecithin, cholesterin, urea, and a sugar termed dextrose or glucose.

Serum is plasma minus the elements of fibrin, and has, therefore, the same composition as plasma, except that it contains no fibrinogen. Corpuscles, when moist, contain a far greater proportion of solids than plasma, the approximate composition being 62 water and 38 solids per cent. Moist *red* corpuscles contain from 60 to 70 per cent. water and from 30 to 40 per cent. solids. Of the dry solids of red corpuscles 90 per cent. is hæmoglobin, 8 per cent. proteid, and 2 per cent. other substances. The chemical nature of white corpuscles is not fully understood. The chief portion of their protoplasm is proteid in nature, and the nucleus contains a substance called nuclein. Of the mineral salts contained in the blood a point worthy of special notice is that nearly all the sodium salts are contained in the plasma (or serum), and the potassium salts in the corpuscles. Less marked is the preponderance of the chlorides in the plasma and of the phosphates in the corpuscles.

Taking the blood as a whole, 100 grammes consist of about 80 grammes of water and 20 grammes of solids, some in solution and some in suspension; while the same amount contains 60 cubic centimetres of gases, partly in solution and partly in loose chemical combination. Of the 20 grammes of solids, 10 consist of hæmoglobin, 8 consist of the proteids— paraglobulin, serum-albumin, and fibrinogen, and 2 grammes are inorganic salts and the extractives. The chief inorganic salts are sodium chloride (in the plasma), potassium phosphate (in the corpuscles), sodium bicarbonate, and sulphates. Of the extractives, fat forms a variable proportion, sugar about one-tenth (·1) per cent. and urea only about ·03 per cent.

60. Coagulation of the Blood.—Shortly after blood has been shed it becomes viscid and then rapidly sets into a jelly. A few minutes later drops of a faint straw-coloured liquid appear and form a thin layer on the surface of the jelly, and this is soon followed by layers of the same fluid at the sides, and finally at the bottom of the vessel. The jelly continues shrinking from the sides of the vessel, more fluid becomes apparent, till at last a clot floats in the clear straw-coloured liquid. The appearance of this clear fluid, which is called *serum*, is due to the shrinking and condensation of certain fibrillar constituents of the clot, called *fibrin*, and the consequent squeezing out of the fluid from the interior of the contracting clot. If we examine a small part of the clot with a microscope, we soon learn that it consists of a fine network of fibrin fibrils, in the meshes of which are entangled the red and white corpuscles of the blood. Fibrin is thus formed from the blood-plasma of living blood, and the clot or crassamentum consists of filaments of fibrin with the blood-corpuscles entangled in its meshes. The interior of the clot is darker and firmer than the outside.

Serum is plasma fibrin or the elements of fibrin.

Human blood becomes viscid in two or three minutes, forms a jelly five or six minutes later, produces serum in a few more minutes, the clot continuing to shrink for several hours. By surrounding freshly drawn blood with a freezing mixture coagulation is delayed, the heavier red corpuscles sink towards the bottom, and the lighter white corpuscles that do not disintegrate remain near the top. On removing the blood from the influence of the low temperature, coagulation occurs, and the clot that forms has its upper portion formed of whitish fibrin and colourless corpuscles only. This pale upper layer is termed the *buffy coat*. The blood of the horse and inflammatory blood generally coagulates slowly and forms a buffy coat. Blood does not clot during life in the uninjured blood-vessels, as the living tissue of the inner coat either prevents the formation of the fibrin 'ferment' that changes the fibrinogen of the plasma into fibrin or removes

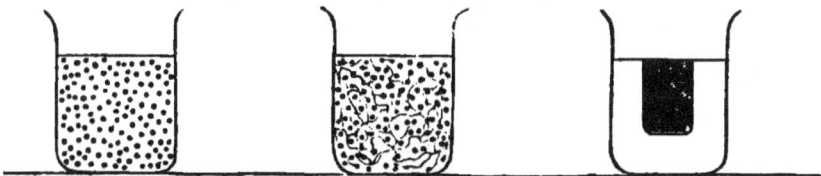

Fɪɢ. 64.—Diagram to illustrate the Process of Coagulation.

I. Fresh.	II. Coagulating.	III. Coagulated.
(Corpuscles and plasma.)	(Birth of fibrin.)	(Clot and serum.)

Plasma minus fibrinogen equals serum   Corpuscles plus fibrin equal clot.

the ferment body as soon as it is formed. Injury to the inner coat of the blood-vessels may produce coagulation within the vessels.

Both red and white corpuscles are specifically heavier than the plasma (the red are heavier than the white), and hence if shed blood be kept from coagulation, as may be done by surrounding horse's blood with a mixture of ice and salt, the corpuscles have time to sink almost completely, and plasma may thus be obtained separately. Such plasma, under favourable conditions, coagulates in the same manner as the entire blood, first forming a jelly and then separating into a colourless shrunken clot and serum. The corpuscles, therefore, are not an essential part of the clot, though it is possible, as the colourless corpuscles especially begin to disintegrate in shed blood, that the corpuscles play some part in the act of coagulation. Indeed, the theory of coagulation now generally held is thus stated by Halliburton :—'When the blood is within the vessels, one of the constituents of the plasma, a proteid of the globulin class called fibrinogen, exists in a soluble form. When the blood is shed, fibrinogen is converted into the comparatively insoluble substance, fibrin. This change is brought about by the activity of a special unorganised ferment, called the fibrin-ferment. This ferment does not exist in healthy blood contained in healthy blood-vessels, but is one of the products of the disintegration of the white corpuscles, and probably also of the blood-tablets, that occurs when the blood leaves the vessels or comes into contact with foreign matter.' To this it may be added that the presence of a certain quantity of a soluble calcium salt seems to be a necessary condition for the due action of fibrin-ferment on fibrinogen, as in the complete absence of such salt no fibrin is formed.

The facts that support the theory above given may be thus stated. It is certain that coagulation is due to the appearance of threads of fibrin which contract and entangle the corpuscles so as to form a clot; for, if the blood, as soon as it is shed, be whipped with a bundle of twigs, stringy masses of fibrin, buff-coloured, on washing in water, adhere to the twigs, and the defibrinated blood will not clot. Fibrin is a proteid body. The amount of fibrin in blood is in reality very small, only about ·1 per cent. of the blood. That the fibrin is derived from the plasma is evident from the fact that plasma free from corpuscles can easily be made to clot and produce fibrin. Such plasma can be obtained either by receiving the blood of the horse (horse's blood coagulates slowly) into a tall narrow glass surrounded by a freezing mixture, and allowing the corpuscles some hours to subside, or by receiving the blood from an artery into one-third of its bulk of a saturated solution of a neutral salt like magnesium sulphate. The vessel is set aside in a cool place for some hours until the corpuscles subside, and the upper layer of clear plasma mixed with the salt is then drawn off. On removing the salt by dialysis coagulation of the plasma occurs; or coagulation may be caused by diluting the salted plasma with much distilled water. To show that fibrin is formed under certain circumstances from the fibrinogen of the plasma and not from the other proteids, the following experiments have been made. By adding to plasma an equal vol. of a saturated solution of sodium chloride a white flaky sticky precipitate of fibrinogen is thrown down. This is then redissolved in a dilute saline solution and reprecipitated, until a solution of fibrinogen is obtained free from other proteids. This solution will not coagulate when left to stand, but if a small quantity of 'fibrin ferment,' obtained as described below, be added, it readily clots and yields normal fibrin. A solution of paraglobulin or serum-albumin, when thus acted on, yields no fibrin. To obtain the substance which induces the change of fibrinogen into fibrin, and which is called 'fibrin ferment' from the nature of its action, serum is mixed with twenty times its volume of alcohol, and allowed to stand two or three weeks. The insoluble precipitate thus produced is filtered off, dried carefully at a low temperature, and then extracted with distilled water. This aqueous extract contains the ferment in solution, for on adding it to a solution of fibrinogen, or to such a fluid as hydrocele fluid, which contains fibrinogen but no fibrin, coagulation at once begins. That the ferment is derived from colourless corpuscles (and blood-platelets?) is surmised from the observed breaking-up of the leucocytes when blood is shed, and from the fact that they have been seen to collect on foreign bodies introduced into blood-vessels before the clot forms on these bodies. The need for a minute quantity of a neutral salt before fibrin forms has also been proved. Hence the belief in the theory we have adduced.

61. Circumstances influencing Coagulation.—1. Coagulation is *hastened* by :—(a) A temperature a little above that of the body, as 100° to 120° F. (b) Contact with foreign matter. The greater the surface of contact the more rapid are the changes which induce fibrin-formation. Thus the introduction of a thread or wire leads to clotting around the object so put in. Injury or disease of the vascular wall acts in the same way. Brisk agitation, by multiplying the points of contact, hastens coagulation. (*c*) Dilution with a *small* quantity of water. 2. Coagulation is *hindered* or *prevented* by:—(*a*) A low temperature ; a temperature of melting ice ($32°$ F. or $0°$ C.) delays coagulation almost indefinitely. (*b*) Contact with the living uninjured lining membrane of the blood-vessels prevents coagulation. As the tissues continue to live some time after the death of an animal, the blood may remain fluid in the vessels after an animal has been dead some hours. If the jugular vein of a horse be tied in two places and cut out from the body, the blood will remain fluid for a day or more. (*c*) The addition of a sufficient quantity of a neutral salt, *e.g.* magnesium sulphate, prevents coagulation. On diluting the mixture with a considerable quantity of water, however, coagulation sets in. (*d*) The addition of *much* water to the blood delays coagulation considerably. (*e*) Heating the blood to a temperature of $133°$ F. ($56°$ C.) coagulates fibrinogen, the antecedent of fibrin, and so prevents a clot being formed. Blood coagulates more slowly in a smooth than in a rough vessel, in a vessel smeared with oil than in one not so treated, in a shallow vessel than in a deep one.

62. **Quantity and Distribution of the Blood in the Body.**— From observations made on executed criminals it has been estimated that the total quantity of blood in the human body is about $\frac{1}{13}$th of the body weight. This does not appear to vary much, the tissues serving to regulate both the average composition and the average quantity. Even in a starving body the quantity of blood remains nearly constant for a considerable time, the fat, muscles, and other tissues suffering the waste that must go on (par. 110). The distribution of the blood may be thus approximately given :—

One-fourth in the heart, great blood-vessels, and lungs.
    ,,   ,,   liver.
    ,,   ,,   skeletal muscles.
    ,,   ,,   other organs.

A man weighing 11 stone, or 154 lbs., would therefore have about 12 lbs. of blood—*i.e.* about 9 pints—in his body, a quarter of which would be in the thoracic viscera, a quarter in the liver, a quarter in the muscles, and a quarter in the other organs.   The heart itself will only hold about ten ounces.

63. Lymph.—Lymph is a transparent, slightly yellow albuminous fluid, in which float lymph-corpuscles identical with the colourless corpuscles of the blood.   It is contained in spaces between the cells of the tissues and in a system of vessels that have their *origin* in nearly all the tissues of the body (par. 138).   Passing from these lymph-spaces of the tissues into lymphatic capillaries, the lymph is pressed onwards into larger vessels, most of which converge and empty their contents into the thoracic duct which terminates in the left subclavian vein.   On its way it is subjected to the influence of the lymphatic glands, where an increase in the number of colourless corpuscles occurs, so that the number of these bodies varies in different parts of the lymphatic system.   Lymph is largely the excess of blood-plasma which has exuded through the walls of the blood capillaries and has not been needed for the nutrition of the tissues.   It also contains a number of colourless corpuscles which have emigrated into the tissues from the blood capillaries.   Some recent observations appear to show that it is also in part produced by active secretive processes of the cells of the capillaries and tissues.   Its composition appears to be nearly the same as that of blood minus the red corpuscles, *i.e.* it is nearly like plasma with leucocytes.   It has, however, a greater proportion of water, a greater proportion of carbonic acid and urea, derived from the waste of the tissue, and a smaller proportion of fibrin and other proteids, than plasma.   Like blood, or plasma, it coagulates when shed owing to the formation of fibrin, but the clot is less pronounced owing to the smaller proportion of fibrin which it yields (0·04 to 0·08 per cent. of fibrin, compared with 0·2 to 0·4 per cent. in blood-plasma).   The fluid that exudes from the clot may be called lymph serum, and like blood-serum it contains the proteids paraglobulin and serum-albumin, though in much less proportion.

	1,000 parts of human blood-plasma contain	1,000 parts of human lymph contain
Water	902·90	986·34
Solids	97·10	13·66
(a) Proteids :—		
Fibrin	4·05	1·07
Other proteids . . .	78·84	2·30
(b) Extractives	5·66	1·31
(c) Inorganic salts . . .	8·55	8·78

More CO_2 can be extracted from lymph than from plasma, but it contains hardly any oxygen.   Lymph for examination is obtained from the thoracic duct of an animal that has not been fed for a considerable time.

64. Chyle is the fluid contained in the lymphatic vessels of the intestine (lacteals) during digestion.   It differs from lymph in being an opaque milky-looking fluid, owing to the large number of minute fatty granules it contains (fig. 65, *a*), the fat in this form being spoken of as the ' molecular basis' of the chyle. A few fat-globules like those of milk are also found. While in the intestine the fat is emulsified or finely divided by the action of the bile, but the ' molecular basis' condition of the fat is not found until it has passed through the lacteals. Chyle also contains a greater propor-tion of solids than lymph. In a fasting animal the fluid contained in the lacteals does not differ from ordinary clear lymph. Lymph examined under the microscope shows colourless corpuscles floating in a clear fluid, and a very small number of fat-globules ; chyle shows colourless corpuscles and numerous minute fat-granules. When chyle coagulates it yields a soft white clot.

The *quantity* of lymph in the body has been estimated at about one quarter of the body weight. The total amount of lymph—*i.e.* lymph mixed with chyle—that passes into the blood of a well-fed animal through the thoracic duct in twenty-four hours is believed to be almost equal to that of the whole blood in the body. One half of this daily amount comes from the lacteals, and therefore in consider-able part from the food ; the other half comes from the other parts of the body. The above figures can-not be regarded as exact, for the estimates can hardly be said to be based on sufficient data. They serve, however, to impress the fact that the amounts of lymph in the body and passing daily into the blood are very large. All the tissues of the body derive their nutrient material and excrete their waste products into the lymph. The oxygen and food absorbed into the blood pass through the capillary walls into the lymph bathing the cells and other tissue elements ; the carbonic acid, urea, and other waste pro-ducts of tissue activity pass, for the most part by diffusion, in the opposite direction, from lymph into the blood, in order to be removed, a portion, however, of these waste products reaching the blood indirectly through the thoracic duct.

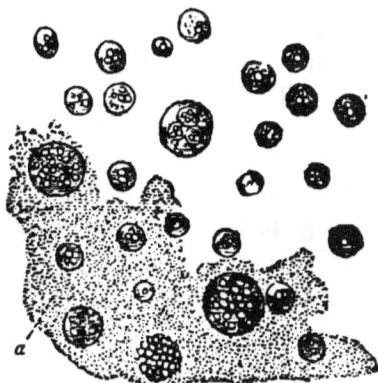

FIG. 65.— Chyle from the Lacteals.

The circulation of the lymph is described in chapter ix.

CHAPTER IV.

THE HEART AND THE CIRCULATION.

65. General Statement.— Circulation is the incessant move-ment of the blood through a central pumping organ, called the heart, and a system of closed tubes, called blood-vessels, the whole constituting the circulatory apparatus. The study of the circulation comprises an account of the arrangement and structure of this apparatus, the mode of action of its various parts, the speed and pressure of the blood at different parts of the circuit, and the control exercised by the nervous system over the heart and the blood-vessels.

The circulatory apparatus consists of : — (*a*) A central hollow muscular organ, termed the heart, which is separated into two distinct sides by a longitudinal fixed division, each side being separated into two cavities by a transverse movable partition or valve. The upper cavity on each side is called an auricle, and the lower cavity a ventricle. (*b*) A system of tubes or vessels connected with the heart, of three kinds : (1) arteries—vessels with relatively stout and elastic walls, which convey blood away from the heart ; (2) veins—vessels with thinner and more flaccid walls, that carry blood to the heart ; (3) capillaries—fine vessels connecting the arteries to the veins, and having walls so thin that a portion of the blood-plasma and some colourless corpuscles escape into the surrounding tissue as the blood passes through them.

To get a general idea of the blood-current we may begin with the blood in the right auricle. The contraction of this cavity drives the blood through the auriculo-ventricular opening into the right ventricle, the valves between auricles and ventricles being so disposed that the blood can only pass in one direction. From the right ventricle the blood is forced along the pulmonary artery into the lungs, in the capillaries of which it gives up CO_2 and water and takes in oxygen. From the lungs the blood is collected by the four pulmonary veins and taken to the left auricle, from which it passes into the left ventricle. This part of the circulation is often spoken of as the **pulmonary circulation.** Driven by the contraction of the left ventricle into the great artery called the aorta, the blood bright red in colour and laden with nutritive materials, traverses the arterial ramifications, and passes into the numerous capillaries distributed in the various tissues of the body. Here it gives up oxygen and certain nutritive materials to the tissues, receives certain products of waste, and, changed to a darker colour, passes from the capillaries into the veins. All the veins from every part of the body, except those from the lungs and the heart itself, ultimately join together into two large veins, the lower or inferior vena cava and the upper or superior vena cava, and these pour their contents into the right auricle of the heart. This is termed the greater or **systemic circulation**. A particular part of this circulation is often spoken of as the **portal circulation**. It consists of the passage of the blood through the portal vein and liver. The portal vein is formed by the union of the veins from the stomach, spleen, pancreas, and intestines, and contains the blood laden with certain of the products of digestion. This portal vein passes, along with a branch of the aorta, to the liver, where it is broken up into a second set of capillaries, from which the hepatic veins arise and join the inferior vena cava. Strictly speak-

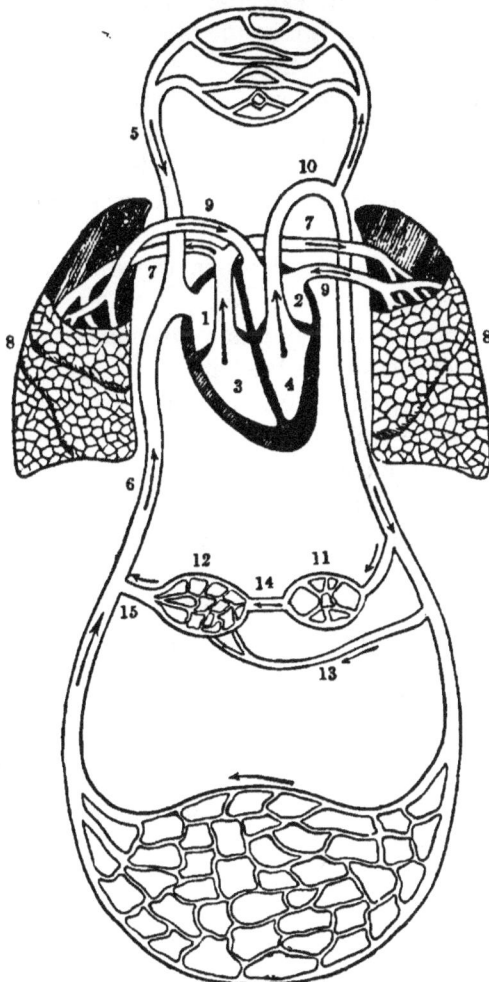

Fig. 66.—Diagram illustrating the Circulation.

1, right auricle ; 2, left auricle ; 3, right ventricle ; 4, left ventricle ; 5, vena cava superior ; 6, vena cava inferior ; 7, pulmonary arteries ; 8, lungs ; 9, pulmonary veins ; 10, aorta ; 11, alimentary canal ; 12, liver ; 13, hepatic artery ; 14, portal vein ; 15, hepatic vein.

ing, this is not, however, a special circulatory system, but only a special part of the systemic system. We thus see that from one point of view we may regard the circulatory system as formed of two distinct parts, the heart being looked on as a double organ. The right heart leads to the pulmonary system, its ventricle being the agent for forcing the blood through the lungs; the left heart leads to the systemic circulation, its ventricle forcing the blood through the systemic circulation. But in truth there is only one circulation, a drop of blood, to pass through a complete circuit, making two journeys from and to the heart. The *cause* of the circulation is the difference of pressure which exists in the different parts of the circuit, for the blood, as in the case of other liquids, flows in the direction of least pressure (see Pressure of Blood, par. 77). The pressure is greatest in the pulmonary and systemic systems at the beginning of the arterial systems, and it is least at the venous terminations in the auricles. The propulsive action of the heart keeps up this difference of pressure and thus produces the blood-current.

The mean time which the blood requires to make a complete circuit is a little over half a minute.

Before treating of the various organs and facts of circulation in detail, the accompanying diagram (fig. 66) will help to illustrate what has been said. (The student must note that a diagram is not an actual representation of an object, but merely a figure to illustrate facts in a general or schematic way.)

66. **The Heart.**—The heart is a hollow muscular organ of a conical form, placed obliquely in the thorax behind the sternum and costal cartilages and between the lungs, having its broad end or base upwards. Its pointed end or apex is directed downwards, forwards, and to the left, so as to terminate in the interval between the fifth and sixth ribs, about $3\frac{1}{2}$ inches from the middle line of the sternum and $1\frac{1}{2}$ inch below the left nipple. It thus projects further into the left than into the right half of the chest. The great blood-vessels proceeding from its base help to suspend and keep it in position. Both the heart and the roots of the great blood-vessels are enclosed in a membranous sac or bag, called the *pericardium*. The pericardium is a fibro-serous membrane consisting of two layers, an external fibro-serous layer and an internal serous layer. The fibrous layer is a strong dense membrane, forming a conical bag,

attached below to the diaphragm, and continued above in the
form of tubular prolongations along the great blood-vessels, and

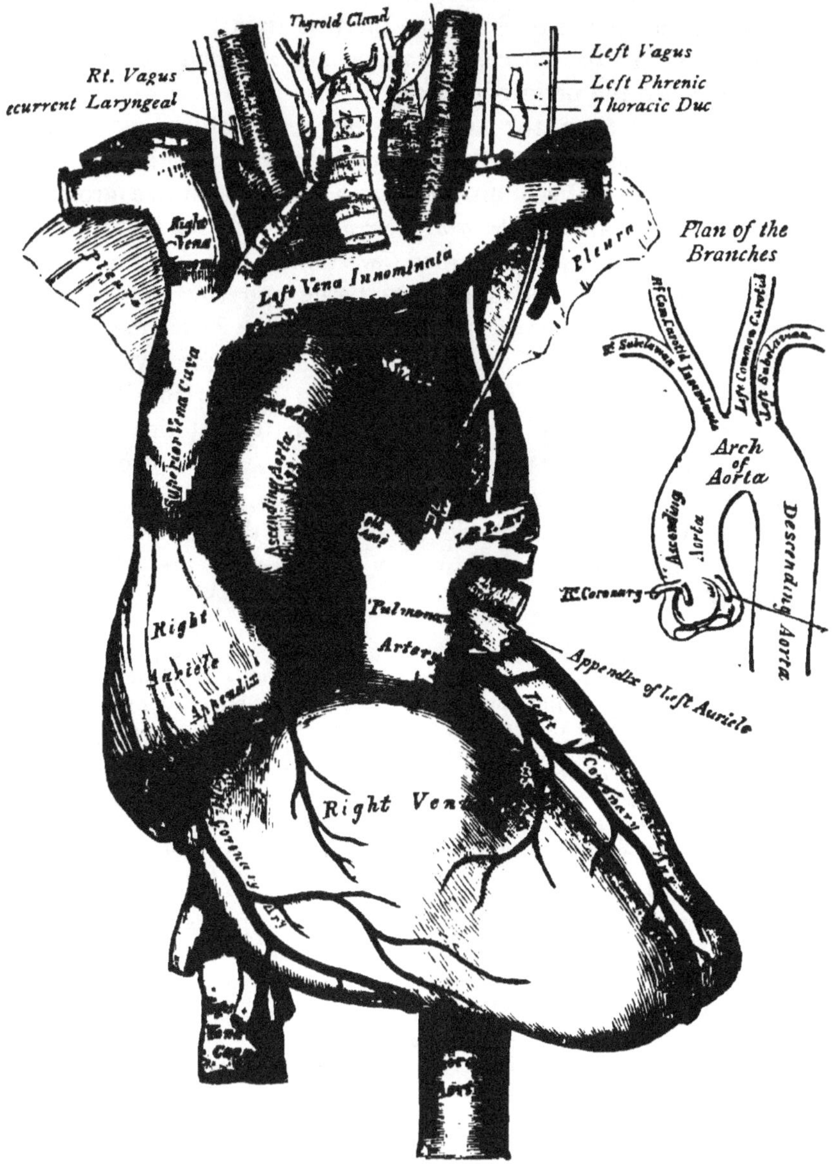

FIG. 67.—The Arch of the Aorta and its Branches.
(Blood-vessels coloured according to the kind of blood in them.)

gradually passing into their external coats. The serous layer of the pericardium not only lines the fibrous layer, but is also reflected above on the surface of the heart. Between the two serous layers of the pericardium there is thus formed a completely shut narrow space, containing a very small quantity of *lymph*, called the pericardial fluid.[1]

Examined from the outside, the heart, which measures about ˙5 inches in length and 3½ inches in breadth at the broadest part in the adult, exhibits a deep transverse groove called the *auriculo-ventricular furrow*, and in its lower part two longitudinal grooves, called *interventricular furrows*. In these furrows, embedded in fatty tissue and covered by the pericardium, lie the chief blood-vessels of the heart itself, the *coronary* arteries and veins, together with the lymphatic vessels and nerves. These grooves on the surface indicate the four cavities or chambers found in the interior of the heart—viz. two upper cavities called *auricles*, and two lower cavities called *ventricles*. Looked at *in situ*, the front convex surface of the heart is made up chiefly of the right ventricle, and the posterior surface is mainly occupied by the left ven tricle (fig. 67).

67. Chambers of the Heart.— The interior of the heart is divided by a fixed longitudinal muscular septum into two chief cavities, a right and left cavity. Each of these right and left halves is again subdivided by a trans verse constriction into an upper chamber called an *auricle* and a lower one called a *ventricle*. There is free communication between auricle and ventricle of each side of the heart, the auriculo-ventricular opening, however, being so guarded by membranous flaps as to allow the blood to pass from auricle to ventricle, but not in the opposite direction. There is no direct communication between the left and right sides or halves of the heart. The two auricles are the receptive cavities and the two ventricles the propulsive cavities of the heart.

The Right Auricle occupies the right and anterior portion of the heart. Its walls are thin, and appear of a quadrilateral form, prolonged at the upper

---

[1] This fluid, like that contained in the closed sac lining the abdomen (the peritoneum), and like that contained in the sacs investing the lungs (the pleuræ), used to be called *serum*, and hence the membranes forming these sacs are often termed *serous membranes*. The ' serous fluid ' in these closed cavities is, however, really lymph, and by means of minute openings (stomata) between the cells of the simple epithelial layer forming the inner surface of the membrane, a communication is formed with the subjacent lymphatic vessels. (See fig. 99.)

left corner into a tongue-shaped portion called the *right auricular appendix*, which overlaps the root of the aorta. On opening the cavity its interior is seen to be lined by a delicate smooth membrane, the *endocardium*. This

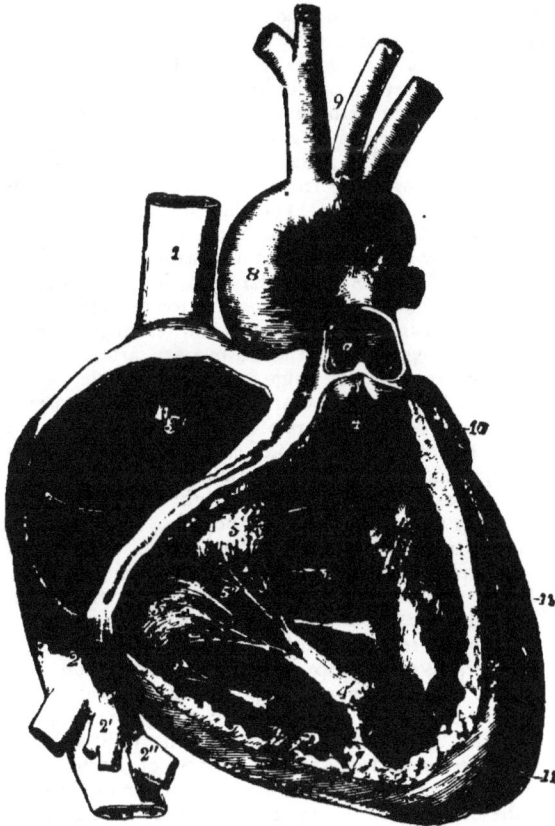

FIG. 68.—Interior of the Right Auricle and Ventricle, exposed by removal of the greater part of their right and anterior walls. (Allen Thomson.) ½.

1, superior vena cava ; 2, inferior vena cava ; 2′, hepatic veins ; 3, septum of the auricles ; 3′, fossa ovalis, the Eustachian valve is just below ; 3″, aperture of the coronary sinus with its valve ; +, +, right auriculo-ventricular groove, a narrow portion of the adjacent walls of the auricle and ventricle having been preserved ; 4, 4, on the septum, the cavity of the right ventricle ; 4′, large anterior papillary muscle ; 5, infundibular, 5′, right, and 5″, posterior or septal segment of the tricuspid valve ; 6, pulmonary artery, a part of the anterior wall of that vessel having been removed, and a narrow portion of it preserved at its commencement where the pulmonary valve is attached ; 7, the aortic arch close to the cord of the ductus arteriosus ; 8, ascending aorta, covered at its commencement by the auricular appendix and pulmonary artery ; 9, placed between the innominate and left common carotid arteries ; 10, appendix of the left auricle ; 11, 11, left ventricle.

membrane lines the other cavities also, and is continuous with the middle and inner coats of the blood-vessels. Within the cavity are to be seen : —
(1) The opening of the superior vena cava, the direction of the orifice being

downwards and forwards. This is the only large aperture of the heart without a valve. The superior vena cava returns the impure venous blood from the upper part of the body. (2) The opening of the inferior vena cava below and to the right, the direction of the orifice being upwards and inwards. This vessel returns the blood from the lower part of the body, and its opening is slightly protected by a crescent-shaped membrane, known as the Eustachian valve, the convex border of which is attached to the vein. (3) The dilated portion of the right coronary vein, known as the *coronary sinus.* This vein returns the blood from the substance of the heart, and its orifice is protected by a semicircular fold of membrane known as the *coronary valve.* (4) The *fossa ovalis*, an oval depression in the posterior wall of this auricle which corresponds to an opening between the two auricles which exists in the fœtus. (5) The auriculo-ventricular opening, an oval or funnel-shaped aperture of communication between the auricle and ventricle, surrounded by a fibrous ring, and guarded by the tricuspid valve.

The Right Ventricle is of a triangular form, its walls forming the chief part of the anterior surface of the heart. Its lower angle does not quite reach the apex of the heart, while its upper and left angle is prolonged into a conical form to the beginning of the pulmonary artery. When opened the septum of the heart is seen to bulge into this ventricle (fig. 68), and its inner surface is marked by muscular bundles, *columnæ carneæ,* some of which, named *musculi papillares,* are attached at their bases to the ventricular wall, and at the other end are prolonged into small tendinous cords (*chordæ tendineæ*), which are joined to the edges of the segments of the auriculo-ventricular valve. This valve on the right side consists of three triangular flaps, and is therefore termed *tricuspid.* The flaps, the anterior one of which is the largest, are composed mainly of fibrous tissue covered by endocardium, and are continuous at their bases, joining the fibrous ring spoken of above. During the *systole* or contraction of the ventricle the segments of the valve are driven up towards the auricle so as to meet, and thus prevent the blood returning into that cavity. The chordæ tendineæ attached to the margins of the flaps, through the contraction of the papillary muscles, keep the valve from yielding too much towards the auricle. Three semilunar valves guard the orifice of the pulmonary artery, which takes its rise at the upper and left angle of the right ventricle. These valves consist of three semicircular folds formed of fibrous tissue, covered by a prolongation of the endocardium on one side, and of the inner coat of the artery on the other side. They are attached by their convex margin to the wall of the artery at its junction with the ventricle, while their straight border is free and directed upwards. In the centre of each free margin is a little nodule, known as the *corpus* Arantii, the three meeting in the centre when the valve is closed. Above and behind each semilunar valve the wall of the pulmonary artery bulges to form the *sinus* of Valsalva. During the contraction of the ventricle the valves are pressed back towards the walls of the artery, but when the ventricle relaxes, during diastole, the pressure of the column of blood in the artery forces them inwards and downwards so as to bring their margins together and close the opening.

The Left Auricle occupies the left and posterior part of the base of the heart. It is slightly smaller than the right, but its walls are a little thicker. Its interior is covered with smooth endocardium. Behind are seen four orifices in its walls, two on each side. These are the openings of the pulmonary veins, which bring the blood that has been oxygenated in the lungs into this cavity, those from the left lung entering close together, sometimes

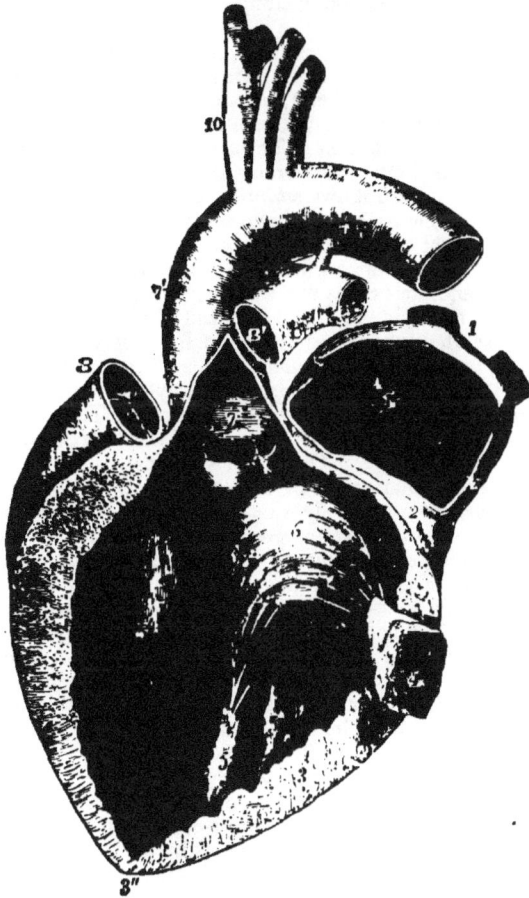

FIG. 69.—The Left Auricle and Ventricle opened and a part of the wall removed so as to show their interior. (Allen Thomson.) ½.

The commencement of the pulmonary artery has been cut away, so as to show the aorta ; the opening into the left ventricle has been carried a short distance into the aorta between two of the semilunar flaps ; and part of the auricle with its appendix has been removed. 1, right pulmonary veins cut short ; 1', placed within the cavity of the auricle on the left side of the septum, on the part formed by the valve of the foramen ovale, of which the crescentic border is seen ; 2', a narrow portion of the wall of the auricle and ventricle preserved around the auriculo-ventricular orifice ; 3, 3', cut surface of the wall of the ventricle, seen to become very much thinner towards 3'', at the apex ; 4, a small part of the wall of the left ventricle which has been preserved with the left papillary muscle attached to it ; 5, 5, right papillary muscles ; 5', the left side of the septum ventriculorum ; 6, the anterior or aortic segment, and 6', the posterior or parietal segment of the mitral valve ; 7, placed in the interior of the aorta near its commencement and above its valve ; 7', the exterior of the great aortic sinus ; 8, the upper part of the conus arteriosus with the root of the pulmonary artery and its valve ; 8', the separated portion of the pulmonary trunk remaining attached to the aorta by 9, the cord of the ductus arteriosus ; 10, the arteries arising from the aortic arch.

indeed terminating by a common opening. The pulmonary veins have no valves.

The *left ventricle* is the thickest part of the heart, and its walls form the apex projecting beyond the right ventricle. It forms a small part of the left side of the front of the heart, but a considerable part of its posterior surface. On opening it its walls are seen to be about three times thicker than those of the right ventricle. They are thickest at the broadest part of the ventricle. Its lining membrane is continuous with that of the auricle, and its inner surface shows *columnæ carneæ, musculi papillares*, and *chordæ tendineæ*. The fleshy columns are smaller but more numerous than those of the right ventricle, and those forming the papillary muscles are collected into two groups, which are larger than those of the right ventricle. The tendinous cords are thicker, stronger, but less numerous than those on the right side of the heart, and they act in the same way, being attached at their upper end to the margins of the two flaps of the *mitral* or *bicuspid* valve, which guards the auriculo-ventricular opening. This valve, as its name implies, consists of only two segments, but they are thicker and stronger than those of the tricuspid. They are continuous at their bases, and are attached to the walls of the heart in the same way as those on the opposite side. The larger segment is to the front. Close to the auriculo-ventricular orifice of the left side is the opening leading into the aorta, the artery which distributes the pure blood to all parts of the body. The aortic orifice is protected by three semilunar valves, which are formed and which act in a manner similar to those of the pulmonary artery, though they are thicker, larger, and stronger. In two of the pouches known as the sinuses of Valsalva of the aorta, arise the coronary arteries which supply blood to the substance of the heart. The right coronary artery arises from the right side of the aorta on a level with the free margin of one of the semilunar valves, runs obliquely in the auriculo-ventricular groove, and distributes branches to the aorta, pulmonary artery, right auricle, and right ventricle. The left coronary artery arises from the left sinus of Valsalva, and gives off branches to the left auricle, both ventricles, and to the interventricular septum (see fig. 70). The two coronary arteries have fine anastomoses with each other on the surface of the heart. The coronary veins for the most part join to form a common trunk, which empties itself into the lower part of the right auricle, its terminal part expanding to form a sinus, the orifice of which is guarded by the coronary or Thebesian valve. The coronary circulation shows the shortest way by which blood can pass from the left to the right side of the heart.

68. Working of the Heart.

68. **Working of the Heart.**—The human heart of an adult in middle life contracts at the rate of about seventy times per minute, the parts acting in a regular or rhythmical sequence. The action is marked by an alternate contraction and relaxation of its muscular walls, the total movement being called a 'cardiac revolution' or 'cardiac cycle.' Each cardiac cycle consists of three acts : (1) a short contraction of the auricles ; (2) a longer contraction of the ventricles as soon as the auricular contraction is over ; (3) a pause in action nearly

equal in time to that occupied by the two contractions, until the auricles again contract. The contraction of any part is called its *systole* (Gk. *systello*, to contract); its relaxation its *diastole* (Gk. *diastello*, to dilate). Both auricles contract together and both ventricles contract together. The exact time occupied by each phase of a cardiac cycle has been found approximately by register-

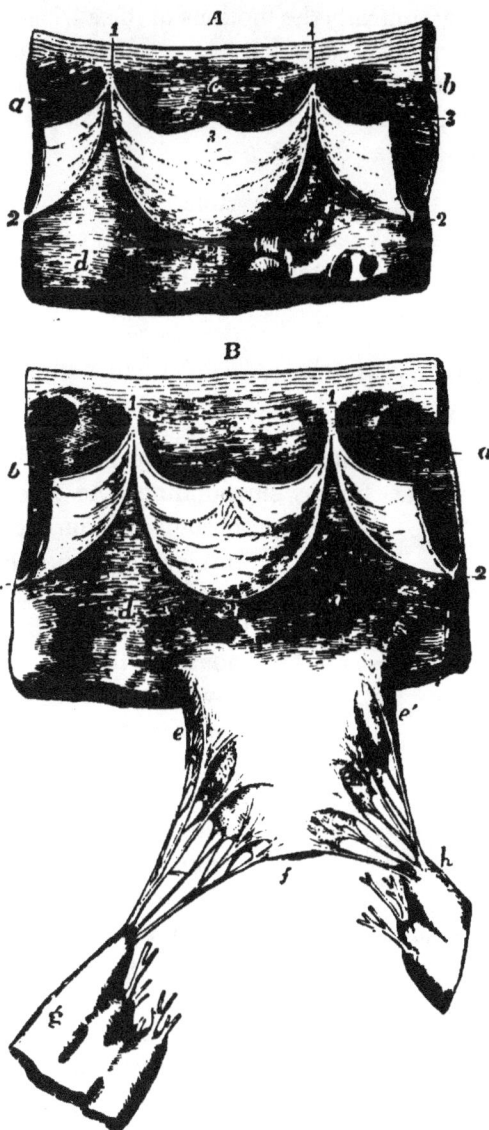

FIG. 70.— Views of Parts of the Semilunar and Mitral Valves, as seen from within the Ventricle. (Allen Thomson.)

A, portion of the pulmonary artery and wall of the right ventricle with one entire segment and two half-segments of the valve. *a, b, c,* sinuses of Valsalva opposite the segments; *d, d',* inner surface of the ventricle; 1, 2, curved attached border of the segments; 3, corpus Arantii, at the middle of the free border.

B, portion of the aorta and wall of the left ventricle with one entire segment and two half-segments of the aortic valve, and the right or anterior segment of the mitral valve. *a, b, c,* sinuses of Valsalva opposite the segments; in *a* and *b* the apertures of the coronary arteries are seen; *d, d',* the inner surface of the wall of the ventricle; 1, 2, and 3, as before; *e, e',* the base of the anterior segment of the mitral valve; *f,* its apex; between *e,* and *e',* and *f,* the attachment of the branched chordæ tendineæ to the margin and outer surface of the valve-segment; *g,* right, *h,* left papillary muscle: the cut chordæ tendineæ are those which belong to the posterior segment and the small or intermediate segments

ing graphically the motions of the auricles and ventricles directly communicated to light levers brought into contact with their surfaces (see par. 72). Taking the heart-beat at seventy times per minute, each cycle would occupy about $\frac{8}{10}$ of a second, made up as follows :—

Auricular systole $= \frac{1}{10}$ of a second.
Ventricular systole $= \frac{3}{10}$ „   „
General pause in action $= \frac{4}{10}$ „   „

or thus :—

Auricular systole $\frac{1}{10}$ + Auricular diastole $\frac{7}{10}$ = $\frac{8}{10}$ of a sec.
Ventricular systole $\frac{3}{10}$ + Ventricular diastole $\frac{5}{10}$ = $\frac{8}{10}$ „

The auricular systole begins with contraction of the muscular fibres surrounding the great veins, and passes through the auricles as a peristaltic wave, thus sending the blood into the dilating ventricles. Regurgitation into the great veins is prevented by contraction of the mouths of these vessels, and by the coronary valve on the right side. The so-called Eustachian valve is imperfect, and of little use in the adult. The systole of the ventricles immediately follows that of the auricles ; they change from a rounded to a more conical form, their walls become tense and hard, the apex of the heart is slightly tilted upwards,· and the heart itself twists somewhat on its long axis from left to right. The ventricular cavity is thus greatly diminished, the segments of the valves meet in a nearly horizontal position, the chordæ tendineæ, aided by the contracting papillary muscles, holding fast their margins and surfaces 'like a taut sail.' The pressure within the ventricles being now greater than that in the arteries, the semilunar valves at the entrance of the great arteries are forced open, and pressed towards but not close to the arterial walls, and the blood enters these vessels. During this ven-

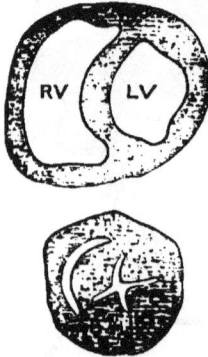

FIG. 71.— Transverse Section through the Middle of the Ventricles of a Dog's Heart in Diastole and in Systole. (After Ludwig.)

tricular systole, an ear placed against the walls of the chest hears a dull sound, the 'first sound' of the heart, and the eye may notice and the hand feel an impulse against the chest-wall in the fifth left intercostal space. As soon as the ventricular contraction ceases, ventricular diastole begins, and proceeds during the greater part of the auricular diastole, which began at the close of the auricular contraction. At the beginning of the ventricular diastole the semilunar valves close, owing to the negative pressure in the ventricle and the

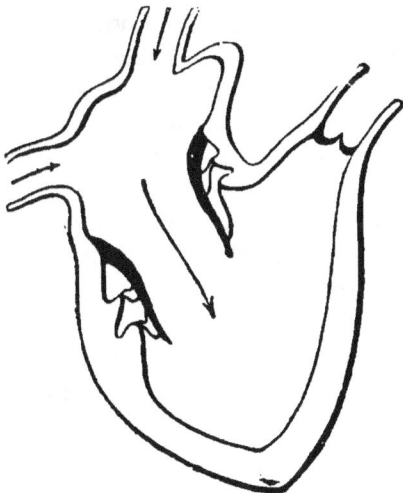

FIG. 72 (A). —A diagram showing the Position of the Valves of the Heart while the walls are relaxed. Diastole.

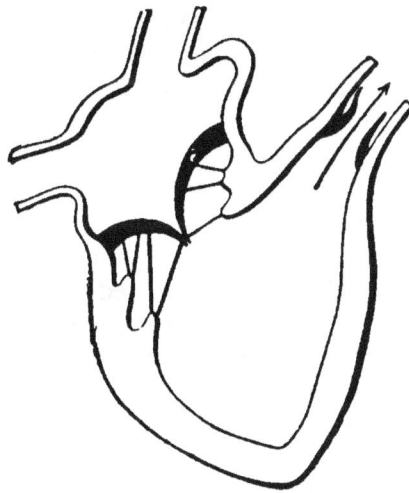

FIG. 72 (B). - Diagram showing the Position of the Valves of the Heart during the contraction of the Ventricles. Systole.

elastic reaction of the arterial walls ; a second sharp sound is heard ; the heart returns to its oblique position, the auriculoventricular valves open, and blood begins to pass from the auricles into the ventricles. After a general pause or passive interval, equal in time to the two systoles, during which the auricles are filling with blood, the cycle is completed, and auricular contraction again begins to expel the blood into the ventricles that had not already passed therein. 'It is very important to notice the cardiac cycle is not a see-saw between auricles and ventricles,' but the blood flows from the great veins into the auricles during ventricular systole, and goes through the auricles into the ventricles during the pause.

69. Impulse and Sounds of the Heart.—As already remarked, the heart during action communicates a stroke or beat to the chest-wall, and this cardiac impulse, as it is called, can be both seen and felt in the left side. This impulse has been found to occur at the same time as the ventricular systole. It is usually best felt at a point in the fifth left intercostal space, midway between the left edge of the sternum and a vertical line drawn through the left nipple. It is caused by the sudden pressure of the lower anterior tense part—not the apex or extreme lip—of the ventricle against the chest-wall during ventricular contraction. It is common to call this impulse the 'apex beat,' though the name is not strictly correct. Besides the sensation which can be perceived by the eye and felt by the hand, when the ear is placed against a person's chest over the region of the heart, two distinct sounds can be heard, a muffled dull sound, known as the 'first sound,' followed almost immediately by a shorter and sharper sound, known as the 'second sound' of the heart. Their character is imitated by pronouncing the syllables *lubb-dup*. The first sound is found to begin with the ventricular systole, and to continue during the greater part of the contraction. But though it occurs at the same time as the impulse of the heart, it is not caused by the heart's impulse, as it can be heard quite clearly when the chest-wall is removed. It is most probably a composite sound due to two chief causes : (1) the vibration of the tricuspid and mitral valves, when they are suddenly put into tension at the beginning of the ventricular systole ; (2) the sound produced by the contraction of the muscular fibres forming the walls of the ventricles. In proof of this it may be said (1) that injury or disease of the auriculo-ventricular valves alters the character of the first sound (this is often observed in the case of the mitral valve) ; (2) the first sound can still be heard in a modified form in the excised and empty heart of an animal, where the valves cannot become tense. The second sound is undoubtedly caused by the vibration set up in the semilunar valves of the aorta and pulmonary artery, when they become tense at their sudden closure on the completion of the ventricular systole ; for curved needles have been introduced into the aorta and pulmonary artery so as to hook back one or more of the semilunar valves, and the second sound has then ceased. Moreover, modification of the second sound has been frequently found, after death, to be associated with injury or disease of these valves.

70. Summary of Events occurring during a Cardiac Revolution.— The average frequency of the heart-beats is seventy-two per minute, which allows about $\frac{8}{10}$ second for each cardiac cycle. The events occurring during such a cycle, and the time-relations of these events, may be shown graphically as below :—

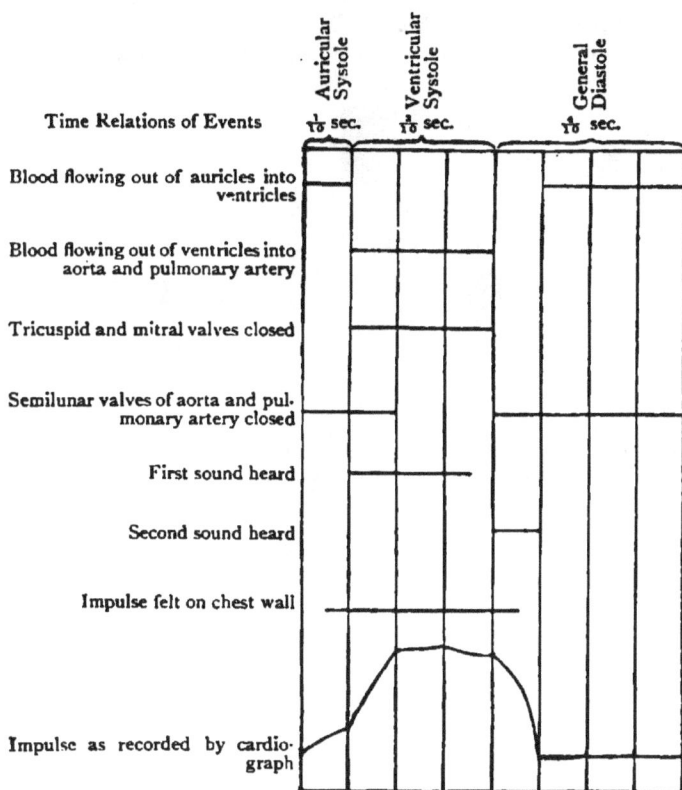

Time Relations of Events	Auricular Systole $\frac{1}{10}$ sec.	Ventricular Systole $\frac{4}{10}$ sec.	General Diastole $\frac{3}{10}$ sec.
Blood flowing out of auricles into ventricles			
Blood flowing out of ventricles into aorta and pulmonary artery			
Tricuspid and mitral valves closed			
Semilunar valves of aorta and pulmonary artery closed			
First sound heard			
Second sound heard			
Impulse felt on chest wall			
Impulse as recorded by cardiograph			

Dividing the time of a cardiac cycle into three periods to correspond with auricular systole, ventricular systole, and the period of rest or passive interval, we may tabulate the events thus : —

First Period $\frac{1}{10}$ *sec.*	*Second Period* $\frac{3}{10}$ *sec.*	*Third Period* $\frac{4}{10}$ *sec.*
1. Auricular systole. 2. Completion of the filling of the ventricles.	1. Ventricular systole. 2. Closure of the mitral and tricuspid valves. 3. Opening of the semilunar valves of aorta and pulmonary artery. 4. Propulsion of blood into the aorta and pulmonary artery. 5. Impulse of the heart against the chest-wall. 6. Gradual filling of the auricles with blood. 7. First or long dull sound heard during greater part of this period. 8. Short silence following first sound.	1. General diastole of heart, blood flowing into both auricles and ventricles. 2. Semilunar valves of aorta and pulmonary valves closed. 3. Second or short sharp sound heard at the beginning of this period only.

71. Frequency of Heart-beat.—The number of cardiac pulsations in the adult varies from 65 to 75 per minute, the average being 72. At birth

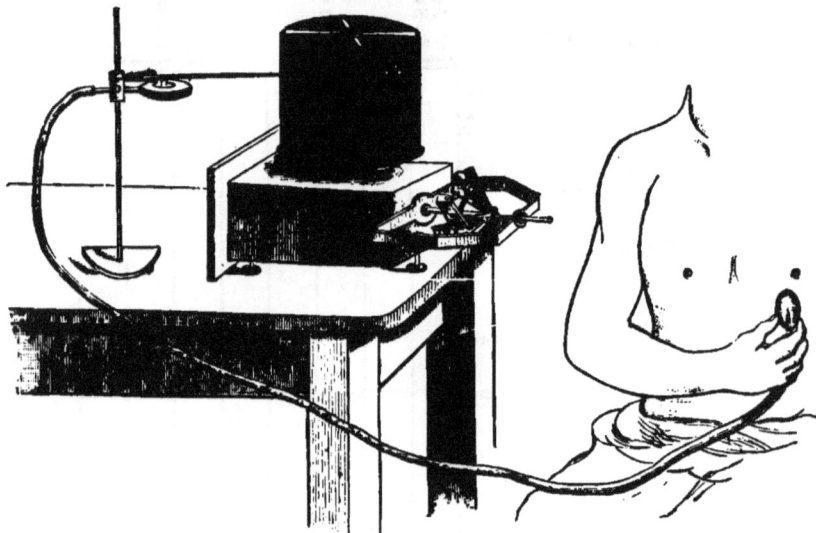

Fɪɢ. 73.—The Cardiograph as applied to Man. (Waller.)

the number is 140, at the end of the first year 120, at the end of the second year 110, at the end of the fifth year 100, at twenty-one 75. It is slower in man than in woman, and becomes slightly increased in old age. Active exercise and active digestion increase the number of beats, but it is important to notice that variations in the frequency of the beats are mainly due to shortening or prolongation of the general diastolic pause, while the duration of the act of contraction remains nearly constant. Increased frequency thus increases the working time of the heart and diminishes its resting time. Increased resistance to the outflow of blood at first increases the number of heart-beats, but if continued or increased beyond a certain point diminishes the number ; with diminished resistance in the arteries the heart is accelerated. Hence in normal conditions it has been said that ' the rate of the beat is in inverse ratio to the arterial pressure.'

72. The Cardiograph.— This is an instrument by means of which the impulse of the heart is magnified and recorded on a surface made to travel at a uniform rate by clockwork arrangement (fig. 73). Its chief parts are : (1) a tambour or drum, to be applied to the spot where the impulse is felt most strongly ; (2) an air-tight tube connecting this tambour with a second tambour, so that the impulse on the elastic membrane of the first tambour is communicated through the enclosed air to the membrane of the second tambour ; (3) a long light lever resting on the second tambour so that the lever is free to move and record its movements on a suitable register-ing surface ; (4) a registering surface, which is usually a cylinder covered

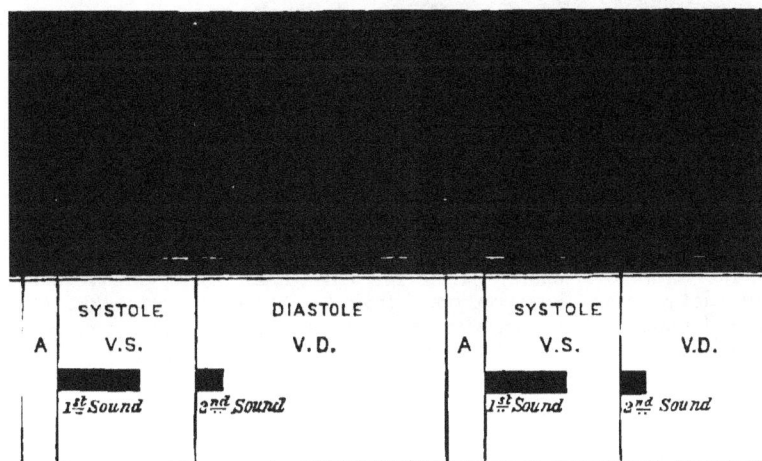

FIG. 74.—A Tracing from the 'Apex-beat' of Man. (Waller.)
Relation of sounds to contraction indicated below the tracing.

with smoked paper ; (5) clockwork causing the cylinder to revolve with a definite velocity, and having a time-marker to indicate seconds, tenths of a second, &c.

A cardiograph placed on the spot where the impulse of the heart against the chest-wall is most strongly felt furnishes a tracing of the impulse very similar to one furnished by placing the instrument on the ventricular wall of the contracting heart of an animal, in which the chest-wall has been removed and the beating of the heart kept up by artificial respiration. The tracing, called a *cardiogram,* is caused by the pressure of the anterior surface of the right ventricle near the apex against the chest wall, and this pressure varies with the changes in the front-to-back dia-meter of the ventricle, and with the changing position of the heart's apex during ventricular contraction and relaxation. Observation and experi-ment show that the cardiac impulse in the fifth left intercostal space occurs at the same time as the systole of the ventricle, and the curves or tracing

H

furnished by the cardiograph enable us to fix the relative duration of each of the phases of a cardiac cycle. Moreover, by listening to the heart and recording its sounds by means of an electric signal, while a cardiogram is being registered, observers have been able to fix the precise place of the sounds in the events forming a cardiac cycle. The annexed figure 74 of a cardiogram or tracing furnished by the 'apex-beat' of a heart beating sixty times a minute has, therefore, been interpreted according to the directions given below it. The small rise preceding the larger one occurs at the auricular systole, and is caused by this contraction producing a slight pressure in the ventricles.

By placing small air-bags connected with tubes and levers within the chambers of the heart, the changing pressure *within* these cavities has been recorded, and the curves thus obtained assist in interpreting those obtained by the cardiograph. The curve of pressure within the ventricle shows that at the end of the systole there is a negative pressure or suction-power in this cavity. This suction-power of the ventricle is mainly due to the active dilatation of the ventricles caused by the elastic relaxation of their forcibly contracted walls, and it is doubtless of some service in carrying on the venous circulation (par. 81).

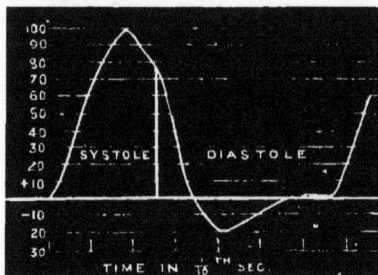

FIG. 75   Intraventricular Pressure of Dog's Heart. (Waller.)

Positive during systole, negative during the first part of the diastole. The numbers and horizontal lines indicate pressure in mm. Hg.

CHAPTER V

BLOOD-VESSELS AND CIRCULATION

73. **Three kinds of Blood-vessels.**—Excluding the central organ, the heart, the blood vascular system includes three kinds of vessels—arteries, capillaries, and veins. Before describing the circulation of the blood through these vessels, we proceed to say a few words about their distribution, and to give an account of their structure.

74. **Structure of Arteries.**—The arterial system commences with the aorta, which springs from the left ventricle of the heart. After ascending a little over two inches it curves backwards to the left, descends along the vertebral column,

and, after passing through the diaphragm into the abdomen, divides opposite the fourth lumbar vertebra into the right and left common iliac arteries. From the arch of the aorta are given off the three large trunks for the supply of the head, neck, and upper limbs. The names of the chief branches may be learnt from figs. 67 and 77. The arteries usually run in protected situations, and as they proceed in their courses they divide and subdivide into vessels of smaller and smaller calibre until they pass into the capillaries. ' A branch of an artery is less than the trunk from which it springs, but the combined area or collective capacity of all the branches is greater than the calibre of the parent vessel immediately above the point of division. From this it is plain that, since the area of the arterial

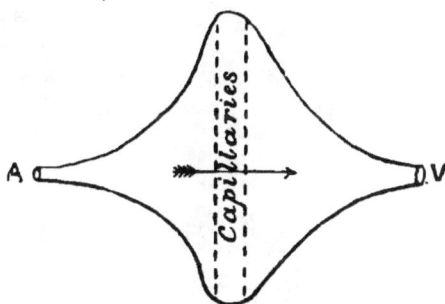

Fig. 76. – Diagram of the Sectional Area of the Blood-vessels in the different parts of the Vascular System. A, aortic orifice ; V, venous orifice.

system increases as the vessels divide, the capacity of the smallest vessels and capillaries will be greatest ; and as the same rule applies to the veins, it follows that the arterial and venous systems may be represented, as regards capacity, by two cones whose apices (truncated, it is true) are at the head and whose bases are united in the capillary system. The effect of this must be to make the blood move more slowly as it advances along the arteries to the capillaries, like the current of a river when it flows into a wider and deeper channel, and to accelerate its speed as it returns from the capillaries to the venous trunks.' Arteries occasionally unite with one another ; they are then said to anastomose or form anastomoses.

The Structure.—Besides a sheath of connective tissue, arteries (except the minute ones) have walls consisting of three coats : (*a*) an internal coat, or *tunica intima* ; (*b*) a middle coat, or *tunica media* ; and (*c*) an external coat, or *tunica adventitia*.

The *internal coat* of a large artery may, in part, be raised

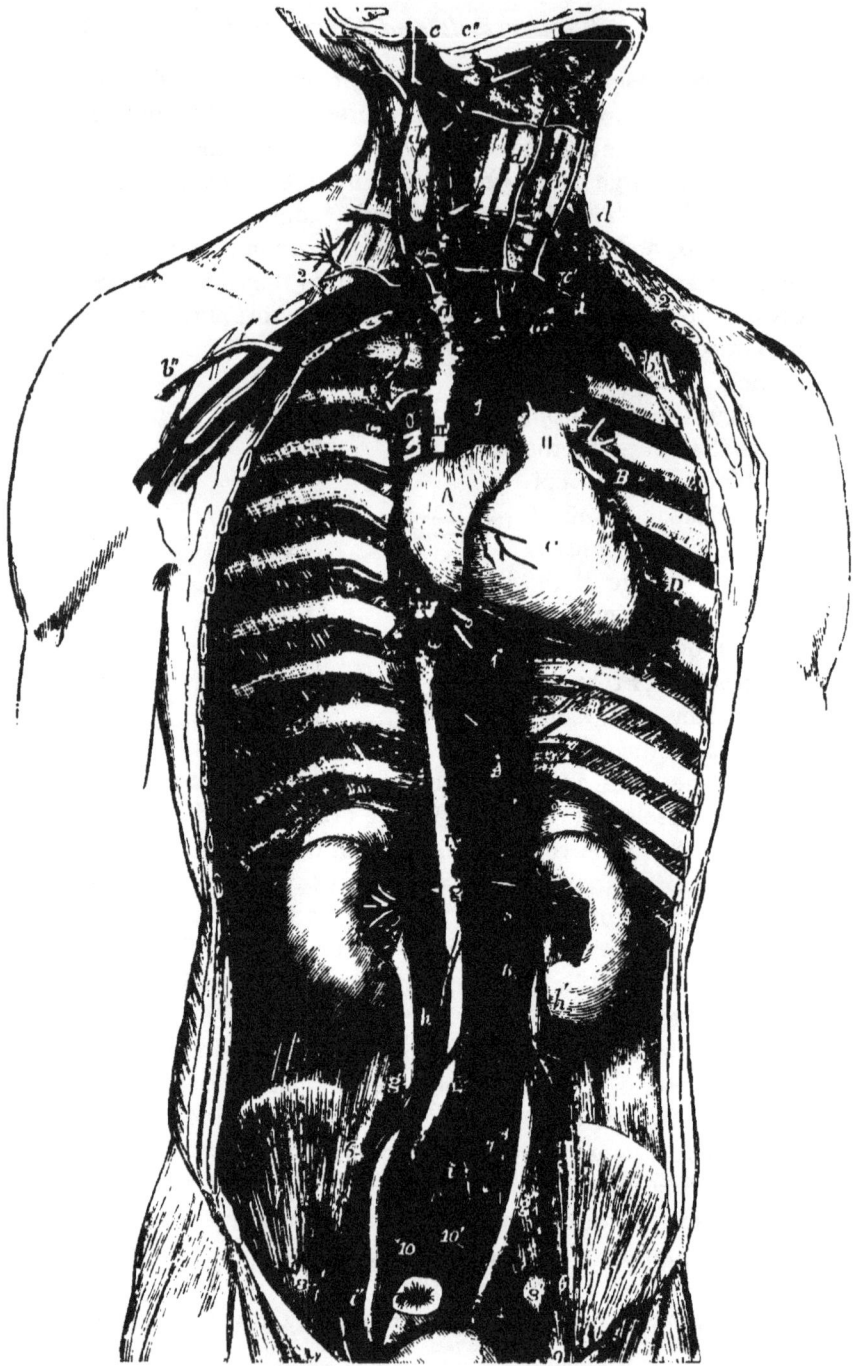

with a needle as a fine transparent membrane, but when examined in section under the microscope it is seen to consist of three parts : (1) a layer of flattened epithelial cells (the endothelium), (2) a subepithelial layer of delicate connective tissue, (3) a layer of elastic fibres forming an interlacing network.

The *middle coat* consists of unstriped muscular fibres ar-

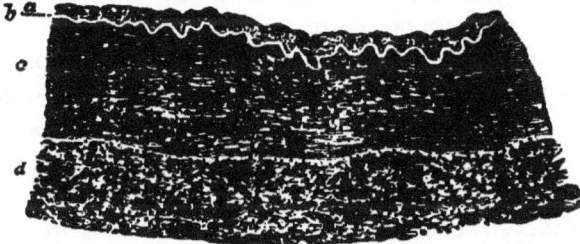

FIG. 78.—Transverse Section of part of the Wall of the Posterior Tibial Artery (Man). 75 diameters. (E. A. S.)

a, epithelial (endothelial) and subepithelial layers of inner coat ; *b*, elastic layer (fenestrated membrane) of inner coat, appearing as a bright line in section ; *c*, muscular layer (middle coat); *d*, outer coat, consisting of connective-tissue bundles. In the interstices of the bundles are some connective-tissue nuclei, and, especially near the muscular coat, a number of elastic fibres cut across.

ranged circularly round the vessel, with a certain portion of elastic fibres, the elastic element being very largely developed in the aorta and some other large arteries. The thickness of the arterial wall is mainly due to the middle coat. In the smaller arteries (arterioles) the muscular element bears a greater proportion to the elastic element than in the larger arteries.

_____

FIG. 77.—General View of the Heart and Blood-vessels of the Trunk, from before, in a Male Adult. (Allen Thomson.) ⅓.

A, right auricle ; B, left auricular appendix ; C, right ventricle ; D, part of left ventricle ; I, I, aorta ; II, trunk of pulmonary artery dividing into its right and left branches, and connected to the aorta by the cord of the ductus arteriosus ; III, superior vena cava ; IV, IV, inferior vena cava.

1, innominate artery and right carotid ; 1′, left carotid ; 2, 2, right and left subclavian arteries ; 3, intercostal vessels ; 4, inferior phrenic arteries ; below 4, the cœliac axis and superior mesenteric artery ; 5, renal arteries ; 6, 6′, spermatic arteries ; below 6, the inferior mesenteric ; 7, 7′, right and left common iliac arteries ; 8, 8′, external iliac arteries ; 9, left epigastric and circumflex iliac arteries ; 10, 10′, internal iliac arteries ; and between these two figures, the middle sacral artery ; 11, 11, femoral arteries prolonged from external iliacs.

a, right innominate vein ; *a′*, the left. *b*, *b′*, right and left subclavian veins ; *b″*, the cephalic vein of the right arm ; *c*, *c′*, internal jugular veins ; *c″*, right facial vein ; *d*, *d*, external jugular veins, formed by the posterior auricular and part of the temporomaxillary ; *d′*, anterior jugular veins with the transverse branch joining them ; *e*, azygos vein arching over the root of the right lung ; *f*, hepatic veins ; *g*, renal veins ; to the sides are seen the kidneys and the suprarenal bodies ; *g′*, right, *g″*, left ureter ; *h*, *h′*, spermatic veins ; *i*, *i*, common iliac veins ; *i′*, *i′*, external iliac veins ; *k*, *k*, femoral veins passing up into external iliac veins.

The *external coat* consists chiefly of fine white connective tissue, with a variable amount of elastic tissue arranged longitudinally. It is this coat that is loosely connected with the arterial sheath of similar connective tissue.

It will thus be seen that the large arteries possess considerable strength and elasticity. After death arteries are usually found dilated and empty (the ancients believed they conveyed air), so that when cut across they show an open orifice owing to their thick elastic walls, while the veins usually collapse.

Vessels and Nerves of Arteries.—The coats of all the larger arteries contain small vessels, both arterial and venous, called *vasa vasorum* (the vessels of the vessels). The small nutrient arteries arise from some branch or from a neighbouring artery, ramify in the sheath, and are distributed mainly in the *outer coat*. Minute veins return the blood to the veins accompanying the arteries. Lymphatic vessels are also present in the outer coat. Nerves are also distributed to the arteries, some filaments of which penetrate to the muscular tissue of the middle coat, and thus regulate, by causing contraction or relaxation, the amount of blood sent to any part. (See par. 84.)

75. Capillaries.—The smallest arteries gradually terminate in a network of vessels, which from their minute size are called capillaries (Lat. *capillus*, a hair). Interposed between the smallest branches of the arteries and the commencing veins, they constitute a network of vessels, which in any particular part remain of the same diameter throughout, though the size of the meshwork, or closeness of the capillaries, varies in different regions. The network is very close, and the supply of blood therefore great, in the lungs, the choroid coat of the eye, in muscular tissue, in the skin and mucous membranes, and in glands and secreting structure ; but there are wide meshes and comparatively few vessels in the ligaments and tendons. Cartilage contains neither blood-vessels nor nerves, its nutrition being effected by imbibition from adjoining tissues. Other non-vascular parts are the nails, the hair, and the cornea.

The size of the capillaries varies between $\frac{1}{3500}$ and $\frac{1}{2000}$ of an inch. Some smaller than these are found in the brain, and some larger in the marrow of the bones. They may almost be regarded as 'minute tubular passages hollowed out in the connective tissue which binds together the elements of a tissue.' The structure of the capillaries is very simple, for their walls consist of a single layer of elongated epithelial cells joined edge to edge by cement substances. The outline of the cells can be rendered apparent by staining with nitrate of silver, while the nuclei can be brought into view by staining with logwood. It is in the capillaries that the interchange of material takes place between the blood and the elements of the tissue outside the

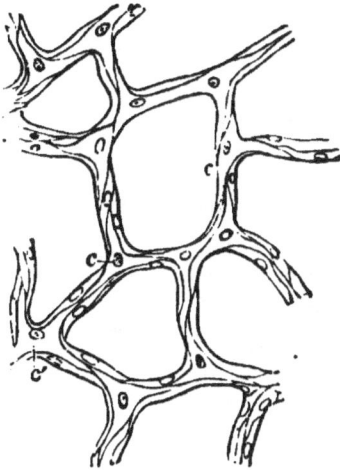

FIG. 79.—Capillary Vessels from the Bladder of the Cat. Magnified. (After Chrzonszczewsky.)

The outlines of the cells are stained by nitrate of silver.

FIG. 80.– Capillary Blood-vessels in the Web of a Frog's Foot, as seen with the microscope. (After Allen Thomson.)

The arrows indicate the course of the blood.

capillaries, the lymph playing the part of intermediary. Some interchange also takes place in the minute arteries and veins. The walls of the vessels on either side of the capillaries, however, soon become too thick for any such interchange.

76. Distribution and Structure of Veins.—The venous system begins in the small vessels springing from the capillaries, and joining to form larger and larger trunks, at last convey the blood to the heart, the two venæ cavæ conveying the blood from the systemic system to the right auricle, and the four pulmonary veins bringing the blood from the lungs to the left auricle (fig. 69). The veins are arranged in a superficial and a deep set, the latter generally accompanying the arteries, and being named *venæ comites*. Connections or anastomoses between veins are more common than in the arteries. It is important to note that the total capacity of the veins di-minishes as they approach the heart, but being more numerous and larger than the arteries, their total capacity (except in the pulmonary circulation,

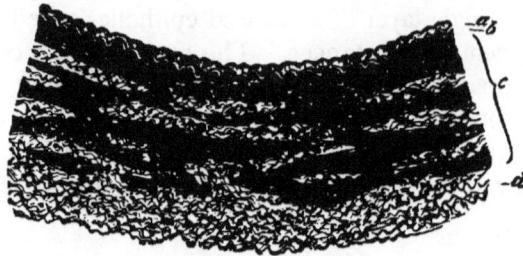

FIG. 81. —Transverse Section of part of the Wall of one of the Posterior Tibial Veins (Man). (E. A. S.)

a, epithelial and subepithelial layers of inner coat ; *b*, elastic layers of inner coat ; *c*, middle coat, consisting of irregular layers of muscular tissue alternating with connective tissue, and passing somewhat gradually into the outer connective tissue and elastic coat, *d*.

where the veins do not exceed in capacity the pulmonary arteries) is two or three times greater than that of the arteries.

The veins have thinner walls than arteries and usually collapse when cut across, though their walls have a general resemblance to those of arteries. Thus in most veins of moderate size three coats may be distinguished : an *internal* coat of three layers, as in the arteries ; a *middle* coat, thinner and less muscular and with less elastic fibres than those of the arteries ; an *external* coat of connective tissue, often thicker than the middle coat, and in some cases, as in the trunks of the hepatic vein and venæ portæ, containing muscular tissue. Most of the veins are provided with valves which are adapted to prevent a reflux of blood. These are formed of semilunar folds of the internal coat, strengthened by included connective tissue, like the semilunar valves of the aorta and pulmonary artery, but their free margins are turned in the opposite direction, *i.e.* towards the heart. Usually two folds or flaps are placed opposite each other, and the wall of the vein immediately on the cardiac side of the valve

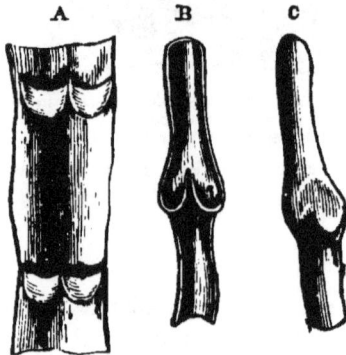

FIG. 82. · Diagram showing Valves of Veins. (Sharpey.)

A, part of a vein laid open and spread out, with two pairs of valves ; B, longitudinal section of a vein, showing the apposition of the edges of the valves in their closed state ; C. portion of a distended vein, exhibiting a swelling in the situation of a pair of valves.

is dilated into a pouch or sinus, so that when the blood tends to flow backwards a knot or swelling appears wherever a valve is placed. This can readily be seen in the superficial veins of the forearm if we press along its surface in a direction opposite to the venous flow. Single folds are found in some of the smaller veins, and in some cases over the opening of small entering branches. An imperfect fold forms the Eustachian valve, and a more complete crescentic fold covers the opening of the chief coronary vein. Valves are not found in all veins. They are absent in those measuring less than $\frac{1}{12}$ inch in diameter, and generally in those not subject to muscular pressure. Thus there are no valves in the superior and inferior vena cava, in the trunk and branches of the portal vein, in the hepatic and renal veins, in the pulmonary veins, in the veins within the cranium and

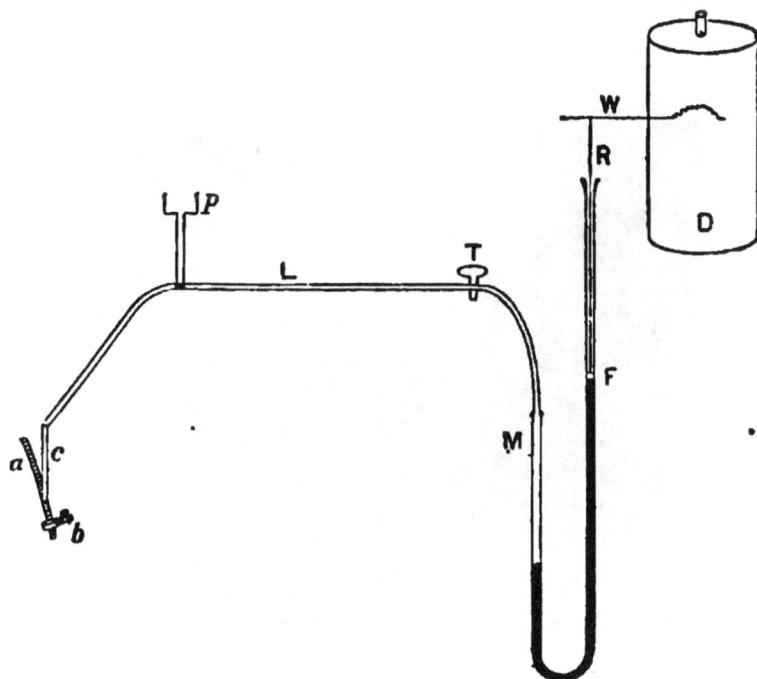

FIG. 83.—Scheme of Mercurial Kymograph for taking Blood-pressure.

a, artery; *b*, clip; *c*, cannula; P, pressure syringe for forcing in sodium bicarbonate solution; L, lead tube; T, stop-cock; M, manometer tube; F, float; R, rod, W, writing-style; D, drum on which pressure curve is recorded.

spinal canal, and in those of the bones. In the azygos and intercostal veins they are few in number; but in the veins of the limbs, especially the legs, valves are numerous.

77. **Blood-pressure.**—If an artery be punctured or cut across, the blood escapes with such force that it is sent to a considerable distance, and this occurs not evenly, but in jerks corresponding to the heart-beats, the flow between the jerks being much less. The larger the artery and the nearer the heart, the more marked is the force of the jets. Further, when the artery is severed the flow is almost entirely from the proximal or heart end, unless there is collateral communication in the neighbourhood. In

the case of a divided vein the blood flows chiefly from the distal cut end — that end in connection with the capillaries- escaping with but little force, and not in jets but in a continuous stream. When an artery is tied or ligatured the vessel swells up on the side of the heart, and the pulse-beat may be felt right up to the ligature, while on the side of the capillaries the vessel becomes empty and shrunk and no pulse is felt. In the case of ligature of a vein, the vein swells up on the distal side of the ligature, while on the proximal or cardiac side there is collapse unless there is collateral anastomosis with the vein. No pulse is felt on either side of the ligature in the case of the vein. These facts not only indicate the course of the blood-stream and show that, as regards the capillaries, it is moving in different directions in the two kinds of vessel, but they also

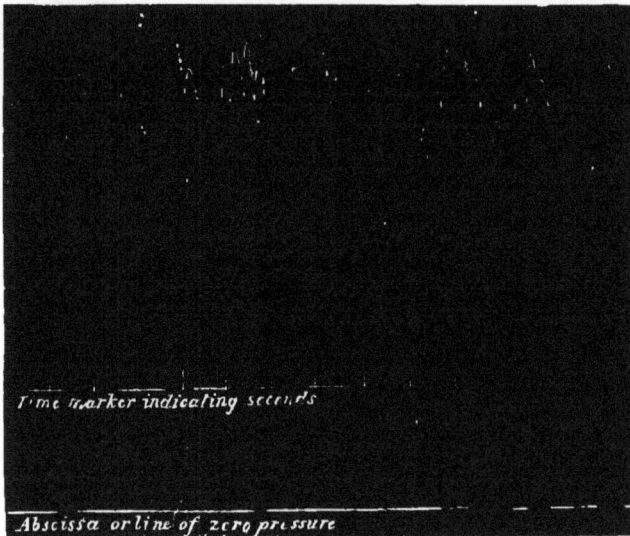

FIG. 84.—Portion of a Blood-pressure tracing from the Carotid of a Rabbit. (Waller.)

The small undulations are due to the heart-beat ; the larger undulations upon which they are superposed are due to the movements of respiration. The height above the zero line is one-half of the blood-pressure expressed in terms of mercurial pressure.

indicate a different kind of stream in the vessel and a considerable pressure on the walls of an artery. The existence of this pressure in the artery may be illustrated in other ways. If the interior of the carotid artery of a living animal be placed in connection with a vertical glass tube, the blood rises by leaps to a height of three feet (in the case of a rabbit), remaining about this height, called the mean height, with small oscillations corresponding to the heart-beats. If a similar tube be placed in the jugular vein, the column of blood will be found to be hardly an inch high.

A more accurate method of measuring the blood-pressure is found in the use of an instrument called the *manometer*, combined with a recording apparatus moving at a definite rate, the whole apparatus being termed a *kymograph* or ' wave-writer (fig. 83).' The manometer is a U-tube with limbs about eighteen inches in length half filled with mercury. On the surface of the mercury in the open limb there rests a float, attached to a fine rod bearing at its upper end a writing-style to record its movements on the revolving cylinder. The other limb of the manometer is connected to a rigid but flexible tube (as lead) to a small glass tube called a *cannula*, which is to be inserted into the blood-vessel. The cannula, tube, and connected limb of the manometer are filled with a strong solution of sodium bicarbonate (which prevents coagulation), this liquid being forced into these tubes so as to raise the mercury in the open limb to nearly the anticipated height. On removing the clip which has prevented the blood from entering the cannula inserted into the artery, the blood is put into connection with the manometer. Pressing through the cannula and tube, the blood forces down the mercury in one limb of the manometer and raises it in the other. The difference of level at which the mercury stands in the two limbs gives the pressure. This level is recorded on the revolving cylinder by the pin attached to the float. The height of the column, however, varies, and these variations are also recorded. When the animal is quiet the variations rise and fall with each heart-beat and each act of respiration.

Fig. 84 gives a tracing of the pressure obtained by a mercurial kymo-

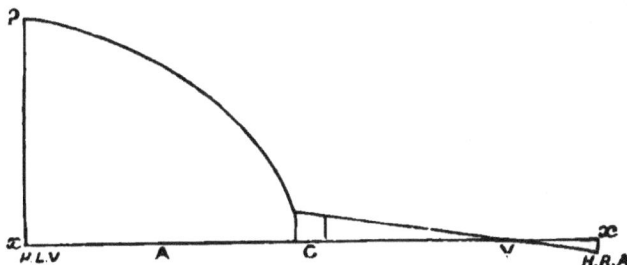

FIG. 85.—Diagram indicating the Relative Height of Blood-pressure in the different Regions, *x x* being the line of Zero Pressure (Atmospheric Pressure).

H. L. V., left ventricle ; A, arteries ; C, capillaries ; V, veins ; H. R. A., right auricle where pressure is negative or below atmospheric pressure.

graph. The mean pressure in the aorta is about 200 mm., or nearly 8 inches of *mercury*, in the carotid 6 to 7 inches, and in the bronchial artery 4 to 5 inches (multiply by 13 for height of blood column).

The pressure of blood in the capillaries cannot be found by the manometer, as these vessels are too minute for any cannula, but it has been estimated by the amount of compression required to drive the blood out of the capillaries and prevent its return. The following results have been obtained by the use of the kymograph and the method just indicated :—
(1) The blood-pressure is highest in the left ventricle during systole and at the beginning of the aorta, falling as we pass from the aorta to the arterial branches, but very gradually. (2) A rapid and sudden fall takes place in the minute arteries and capillaries. (3) The blood-pressure in the veins is less than that in the capillaries, though it is greater in the smaller veins nearer the capillaries than in the larger veins approaching the heart. (4) The blood-pressure in the large veins near the heart is negative—that is, below that of the atmosphere—except during forced expirations. The

continued decline of blood-pressure from the root of the aorta through the arteries, capillaries, and veins to the right auricle is shown graphically in the annexed diagram (fig. 85).

Moreover, it should be noted that the blood-pressure curve of an artery shows small oscillations corresponding to the heart-beats; due to a wave of distension called the pulse, that these pulse-waves become less marked as we get towards the capillaries, and that in the capillaries and veins the pulse entirely disappears.

To explain the high pressure of blood and the pulse in the arteries, the comparatively small pressure and steady flow in the capillaries, and the uniform pulseless current in the veins, we must bear in mind the following facts :—

(1.) That at each systole of the ventricle the muscular action of the heart drives about five ounces of blood into the already full arteries.

(2.) That the passage of the blood into the veins is hindered by the friction of the mass of blood against the walls of the minute arteries and capillaries ; for, although the sectional area of the minute arteries far exceeds that of the aorta, and the united sectional area of the capillaries is several hundred times that of this vessel, yet the assistance to the flow of the viscid blood through these narrow tubes due to this increase of total capacity is more than counterbalanced by the increase of friction resulting from the greater extent of the surface of the walls. The resistance is greatest in the small arteries, where the blood is moving faster than in the capillaries, and where the contractility of their muscular walls enters into action, though the friction in the capillaries is an item in the resistance. The resistance thus offered in the small arteries and capillaries to the flow of the blood is often spoken of as ' the peripheral resistance.'

(3.) That there is a considerable length of elastic tubing (the arteries) between the muscular pump and the peripheral resistance, beyond which there is a ready passage for the blood through the venous system.

It has already been remarked that the blood-current is the result of a difference of pressure in the vascular system, and that experiment shows a gradually decreasing pressure from the root of the aorta to the mouths of the great veins. The heart produces this difference of pressure and thus causes the blood-stream. It also keeps up this difference of pressure, for at each systole of the ventricles a certain quantity of blood is forced into the great arteries, an equal amount flowing from the arteries through the capillaries into the veins. But the flow in this last case is 'continuous, while the supply from the ventricles is intermittent. If the arteries had rigid walls the flow would be intermittent, as in such a case each stroke of a pump forces out at the distal end just as much as is forced in, the flow ceasing when the propelling force ceases. Though the arteries have elastic walls, the outflow at their distal end would still be intermittent were there no resistance to the flow of the fluid. This can easily be shown with a length of rubber tubing connected with a pump. But if the tubes be elastic, and if there be considerable resistance at their further end, the sudden supply of fluid from the pump distends their walls and calls into action their elasticity. This elasticity exerts a pressure on the blood, forcing it onwards, and when the strokes of the pump are repeated before the tube has time to empty itself further distension takes place and stronger elastic reaction is set up. By repeating the strokes with sufficient rapidity the distension of the tube becomes so great that the elastic force called into action is just sufficient to force through the resistance an amount of liquid between two strokes equal to that sent in at each stroke, and the outflow becomes continuous. This is just what happens in the systemic and pulmonary circulations. The peripheral resistance causes the force of the ventricular systole to be spent in keeping the arterial system in a state of great distension, and this is manifested as the blood-pressure about the measurement of which we have spoken ; the elastic reaction of, or pressure exerted by, the arterial walls is continually being expended in overcoming the resistance and driving the blood onwards through the minute arteries and capillaries into the venous system (hence the diminution of pressure as we pass from the heart) ; the outflow becomes continuous, and at such a rate that just as much blood passes from the arteries to the veins during a cardiac revolution as enters the aorta (and pulmonary artery) at each short ventricular systole. The interrupted inflow from the heart becomes, therefore, a continuous current in the capillaries and veins because of the rapidity of the heart-beats, the peripheral resistance in the arterioles and capillaries, and the great elasticity of the arterial walls. It must be noted, however, that the arterial elasticity is not a fresh force, but only plays the part of a reservoir for the heart's force, converting its intermittent force into a continuous force and producing the constant flow.[1]

The arterial blood-pressure varies with variation of either of its two chief factors, the contractions of the heart and peripheral resistance. Pressure becomes greater with greater force of the heart or with greater peripheral resistance ; pressure becomes less with diminished heart force or diminished peripheral resistance. Thus violent exertion of any kind increases the number of heart-beats, and therefore its total force in a given time, and a temporary increase of pressure results. A sudden shock or experimental stimulation of the pneumogastric nerve weakens or checks the action of the heart, and a temporary diminution of pressure results. Increase of peripheral resistance may be caused by stimulation of the muscular coat of the smaller arteries ; shock may lead to a relaxation of their wall. But in the healthy body these variations limit and neutralise each other, so that a fairly constant pressure is maintained.

78. **The Arterial Circulation.**—We have already described the structure of the arteries, and have shown that they are elastic and contractile tubes, the elasticity being the marked feature of the larger arteries, and the contractility the chief feature of the smaller arteries (par. 74). We have also seen that, in virtue of the elasticity of the walls of the larger arteries, the intermittent gushes of blood produced by the action of the heart are changed into a continuous stream, that portion of the energy of the heart's action which is used up or rendered potential in dilating the arteries being expended or rendered kinetic in the elastic recoil. We have further seen that the blood in the arteries exerts considerable pressure on their walls, and that this pressure diminishes as we approach the capillaries, the decrease being due to the fact that the pressure is being used up in driving the fluid along against resistance. As there

[1] See note, page 424.

is great resistance as the blood flows into the peripheral region (the small arteries and capillaries), much of the pressure is expended in driving the blood through this region, and therefore a great fall of pressure results here, though enough remains to send the blood along the veins to the heart.

The muscular tissue of the smaller arteries consists of fibres circularly disposed, and is of great service—(1) in regulating the supply of blood to an organ, a greater supply being required during activity than in repose ; (2) in enabling the arterial system to accommodate itself to the amount of blood in the body ; (3) in arresting bleeding (hæmorrhage) when an artery is cut, as its contraction in combination with the formation of fibrin tends to close the orifice.

The muscular tissue of the arteries is regulated by nerve-fibres called *vaso-motor nerves* (par. 84).

79. **The Pulse.**—A special feature of the blood-current in the arteries is the wave of blood produced by the intermittent injection of several ounces of blood into the already distended aorta. This wave of blood, which must be distinguished from the onward flow of the blood itself, passes along the surface of the blood-stream, expanding the arteries as it moves forward, increasing them both in diameter and in length, and producing a sudden increase of blood-pressure. The intermittent expansion of the artery caused at each ventricular systole by the wave of blood passing along the arteries is called the *pulse*. It may be readily felt in any superficial artery, as the radial or temporal artery, where the vessel lies over unyielding bone. The arteries near the heart are more affected by the pressure-wave called the pulse than those more distant, the wave becoming fainter and fainter as it passes along the dividing arteries. For as the expansion travels onwards the wave is spread over the general arterial system, the increase of pressure which produces it is gradually expended on the arterial walls and disappears entirely before it reaches the capillaries. As Professor Foster says :—'The object of the systole is to supply a contribution to the mean pressure, and the pulse is an oscillation above and below that mean pressure ; an oscillation which diminishes from the heart onwards, being damped by the elastic

walls of the arteries, and so, little by little, converted into mean pressure, until in the capillaries the mean pressure alone remains, the oscillations having disappeared.' There is, there-fore, under normal condition·no pulse in the capillaries or in the veins.

Since the pulse-wave proceeds from the root of the aorta through the whole system of arteries it is felt sooner in parts nearer the heart than in those parts remote from the heart. From the difference in time ($\frac{1}{6}$ second) between the pulse felt in the external maxillary·artery and the dorsal artery, and estimating the difference in distance from the heart of these vessels at 1·5 metre, it has been calculated that the pulse-wave travels at the rate of about 9 metres (30 feet) per second. The rate no doubt varies, being greater in less extensile arteries than in those with soft yielding walls. As already noted, the progress of the pulse-wave must not be confounded with the actual current of the blood. This does not travel at a greater rate than half a metre ($1\frac{1}{2}$ foot) per second, even in the larger arteries. The pulse-wave may be compared to the wave pro-duced by a rapid wind on the surface of a slowly moving stream.

The average frequency of the pulse (*i.e.* of the heart's beat) is about 70 per minute in the adult male, 80 per minute in the female ; higher in children, slower in old people (par. 71). It increases with exertion, by taking stimulants, in debilitating disease, and in times of pleasurable excitement. It diminishes during rest, during sleep, and with painful and depressing emo-tion. High temperature, as in a Turkish bath, increases it ; low temperature, as in a cold bath, lowers it.

Various instruments have been devised in order to obtain a graphic record of the expansion of the arterial wall at the time the pressure-wave is passing over a particular spot. Such instruments are called sphygmographs (Gk. *sphygmos*, a pulsa-tion ; *grapho*, I write), and the curve of pressure they register is called a sphygmogram. A figure showing Marey's sphygmo-graph applied to·the radial artery is given below. A small button presses lightly on the artery, and its movements up and down are transmitted to a lever, by which the movements are magnified and recorded on a blackened surface moving towards the hand at a definite rate. A series of curves is thus registered, each of which corresponds with one beat of the heart. A single pulse-curve as recorded by the sphygmograph shows variations with the artery experimented upon, and also with the same

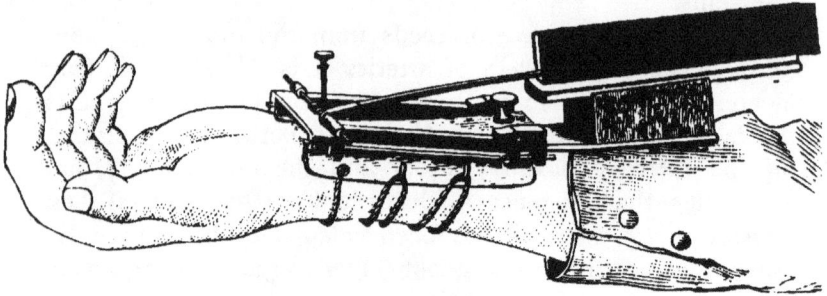

FIG. 86.—Marey's Sphygmograph applied to the Radial Artery. (Waller.)

artery according as the arterial tension or blood-pressure is normal, below normal, or above normal. In a normal pulse-tracing of the radial artery (fig. 87) we see (1) a nearly vertical line of ascent caused by the pulse-wave distending the artery and suddenly raising the lever of the sphygmograph ; (2) a more gradual line of descent, during which the lever falls and the elastic recoil of the artery aids the vessel in recovering its normal calibre. This line of descent, however, is marked by two or three secondary waves, the largest of which, near the middle, is known as the dicrotic wave. The dicrotic wave is most probably due to the recoil from the closed semilunar valves of the aorta as they resist any regurgitation when the elas-

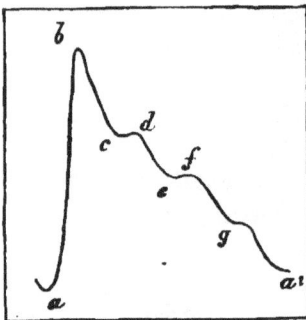

FIG. 87.—A normal pulse-trace magnified. *a b c*, primary wave ; *c d e*, predicrotic wave ; *e f g*, dicrotic wave; *e*, aortic notch ; *a—e*, systole ; *f a'*, diastole of the ventricles.

ticity of the arteries comes into action. The dicrotic wave is best marked when arterial tension is low and the pulse soft. In a hard pulse, where tension or pressure is high, a wave called the tidal or predicrotic wave is well marked.

80. **Circulation in the Capillaries.**— In the capillaries, the structure and distribution of which have already been described, the blood flows with a constant and continuous motion, all pulsation having been extinguished by the elasticity of the arteries and by the far greater sectional area of these vessels as compared with the arteries. The motion of the blood may be seen under the microscope in the web of the frog's foot, in its tongue, or in the mesentery of such an animal as the rabbit. In the larger capillaries two or three rows of corpuscles may be seen moving abreast, the heavier red corpuscles being in the middle of the stream and the white ones rolling along more slowly near the walls, which are bathed with a layer of plasma. In the smaller capillaries the corpuscles move in single file, sometimes being squeezed out of shape for a time, and making their way with difficulty through a channel whose diameter is no larger than that of the corpuscles. Though the walls of the capillaries contain no muscular fibres, these vessels undergo contraction and expansion, probably owing to the activity of the cells composing their walls. These changes are chiefly due to alternate contraction and expansion of the muscular walls of the arterioles, which thus

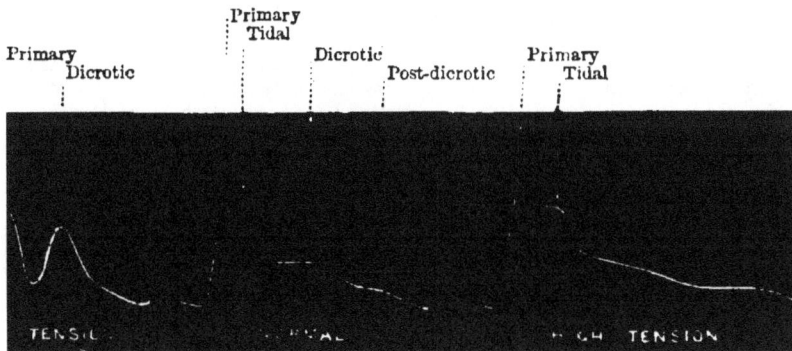

FIG. 88. Pulse Waves. (Waller.)

admit a smaller or larger supply of blood. Through the thin capillary walls a portion of the blood plasma oozes to nourish the surrounding tissue, and in these vessels also occurs a kind of internal respiration, or interchange of gases ; for the oxygen brought by the red corpuscles diffuses from the capillaries into the lymph and tissues, while carbon dioxide diffuses from the lymph of the tissues into the blood stream in order to be carried to the lung, from which it passes into the air. Besides the transudation of a portion of the fluid part of the blood from the capillaries to the surrounding tissues, and the chemical interchange between the blood and the surrounding tissues, corpuscles pass out, or migrate, from the blood by the process known as the *diapedesis* (see fig. 58). The escape of white corpuscles appears to occur normally ; their power of amœboid movement enables them to pass through the thin capillary wall. Under certain conditions of increased blood-pressure in the capillaries, or in inflammation, the migration of the leucocytes increases, and red corpuscles also leave the vessels.

81. **Circulation in the Veins.**—Three causes combine to propel the blood along the veins, in which it moves in the normal condition

I

with a continuous pulseless flow towards the heart. These are : (1) The *vis a tergo*, or power behind—that is, the contraction of the ventricle ; (2) the action of the muscles, which are capable of compressing those veins having valves ; (3) the aspiration, or suction power, of the thorax. The first of these causes, the driving force of the heart, is the main cause of the onward current of blood in the veins, but the other causes are valuable aids. In the various movements of the muscles of the body pressure is brought to bear on many of the veins, and as the valves of these vessels prevent backward flow, a part of the pressure urges the blood onwards in its proper course. Even the part of the pressure acting backwards towards the valves often assists the forward movement of the blood, for the stream may pass along some anastomosing branch of the vein, and thus proceed on its way by another channel. The suction power of the thorax is due to the fact that the pressure in the chest is negative—that is, below that of the atmosphere. This is more marked during inspiration, and it is then that the blood is drawn towards the heart, especially on the right side, where the effect on the thin-walled veins is most marked. In ordinary conditions there is no pulse in the veins, for reasons already given, but in certain exceptional states the pulse may pass through the capillaries into the veins. Thus stimulation of the chorda tympani nerve leads to dilatation of the small arteries of the submaxillary gland ; the peripheral resistance they normally offer is lessened, and the elasticity of their walls is not brought into action to a degree sufficient to change the intermittent flow into a continuous one. A venous pulse is thus found in the veins passing from the gland. Another kind of venous pulse is sometimes observed in the great veins of the neck, when there is some regurgitation through an imperfect tricuspid valve.

82. Velocity of the Blood-current.—The rate at which the blood flows diminishes progressively from the aorta to the capillaries, since the sectional area increases progressively from aorta to capillaries. The united sectional area of the capillaries is said to be five hundred times that of the aorta. Hence, whilst the blood in the aorta moves at about 1 foot per second, it will move in the capillaries at only $\frac{1}{30}$ inch per second. In the veins, whose sectional area diminishes as they proceed towards the heart, some of the velocity is regained. But the calibre of the venæ cavæ is about twice that of the aorta ; so that, although the velocity in the veins exceeds that in the capillaries, yet the rate in the veins where it is highest (in the venæ cavæ) is only about half that in the aorta. By injecting a particular salt into the jugular vein of one side of the neck, and noting the time required for it to appear in the other side of the neck, it has been estimated that the time required for the blood to make a complete circuit is about 25 heart-beats. This is not inconsistent with the small velocity in the capillaries, for it must be remembered that any particular portion of blood may not require to pass through a capillary of more than $\frac{1}{30}$ inch in length.

83. Regulation of the Frequency and Force of the Heart-beats by the Nervous System.

—The nerves of the heart include (*a*) inhibitory fibres to the heart, (*b*) accelerator fibres to the heart. The inhibitory fibres are derived from the medulla by the internal branch of the spinal accessory or eleventh cranial nerve, but run down in the vagus or tenth cranial nerve, reaching the heart by the cardiac branches of the vagus. The accelerator fibres leave the spinal cord by the anterior roots of the second and third dorsal nerves, pass through sympathetic ganglia, and from the inferior cervical ganglion reach the heart along with the cardiac fibres of the vagus. The cardiac inhibitory fibres of the vagus are fine medullated fibres until they reach the minute ganglia or collections of nerve-cells in the heart itself. From these intra-cardiac ganglia, which are placed for the most part superficially, and are more numerous towards the base than the apex, they proceed as fine non-medullated fibres to the muscular tissue of the heart. The accelerator or augmenting fibres from the sympathetic system are medullated from the spinal cord to the cervical ganglia, after which they are non-medullated.

If the heart be quickly removed from the body it goes on beating for a few minutes in the case of a mammal, and in the case of a cold-blooded animal for some hours. Its power of rhythmic contractility is thus seen to be within itself, or automatic. This power of rhythmic contractility was for long thought to be due to nervous impulses sent out from the small ganglia in the substance of the heart ; but recent researches appear to prove that the rhythmical beat of the heart, and the maintenance of orderly sequence from auricle to ventricle, is a property of the peculiar muscular tissue of the heart, and does not depend on motor impulses from the intra-cardiac ganglia. The cardiac muscular tissue has already been shown (par. 30) to differ in structure from voluntary muscle on the one hand and unstriped muscle on the other. Its physiological properties are also different from ordinary muscular tissue in some respects. With a skeletal muscle we get with a strong stimulus a strong contraction, with a weak stimulus a weak contraction, if at all, the height of contraction being, within certain limits, proportional to the strength of the stimulus. With the heart the case is different, for the least stimulus that has any effect has a maximum effect. Further, the heart, unlike voluntary muscle, cannot be thrown into a state of tetanus or continuous contraction by stimuli repeated very rapidly. These histological differences and peculiarities of construction probably indicate a peculiar nature. But the important fact in support of the view that the muscular tissue of the heart possesses the power of inherent rhythmical contractility is that certain parts of the heart which contain no ganglia, as the lower part of the apex, can be induced to carry on a rhythmic beat. It should be noted that the heart's contraction is a single contraction, beginning at the end of the great veins and passing as a continuous wave to the apex, though broken into auricular and ventricular contraction by the lower conductivity of the muscular ring at the junction of auricles and ventricles.

Though the muscular tissue of the heart possesses the quality of automatic rhythmic movement, yet in the living body it is no doubt modified and controlled by the nervous supply received from the vagus and sympathetic in order to adapt itself to varying circumstances and needs. It is thus under the influence of the central nervous system, with which both sets of nerves are connected. This regulating influence is of two kinds—firstly, an influence through the *vagus*, slowing, checking, or inhibiting its beats altogether; secondly, an influence transmitted through the sympathetic supply, quickening or augmenting its beats.

Section of one vagus has little influence, but section of both vagi leads to increased frequency of beat. This shows that the vagi are channels for transmitting a constant restraining or inhibitory influence, so that when the restraint is abolished the heart beats more rapidly. *Stimulation* of the vagus by induction currents, at the cut peripheral end of the vagus, leads to a slowing or weakening of the beat, or, if the stimulation be strong enough, to total stoppage in diastole. From the heart thus arrested no blood is sent to the arteries, and the elastic recoil of the arterial walls sending the blood onwards, a great fall of arterial pressure results. As the influence passes off the heart commences to beat again, and the blood-pressure rises to the normal, or even beyond, in leaps corresponding to the beats. These facts show that the vagus is the inhibitory nerve of the heart, its cardiac fibres transmitting impulses from the centre in the

Heart }
signal }

FIG. 89.

Effect of vagus excitation upon contractions of frog's heart ; the time of excitation is indicated by the rise and fall of the signal line.

medulla that lead to restraint or arrest of the heart-beats. Experiments on the localisation of this centre in the medulla show that it is situated at the lower end of the floor of the cavity of the fourth ventricle in the grey matter where the vagus arises. This lower part of the fourth ventricle, where the central canal of the spinal cord opens out, is called, from its resemblance to the nib of a pen, the *calamus scriptorius*. The *cardio-inhibitory centre* is situated, then, in the grey matter just above the calamus scriptorius. Stimulation of this centre with an electric current stops the heart at once. Near this spot, as we shall afterwards see, is the respiratory centre, and underneath it pass fibres to the respiratory muscles, and hence this region has been called the 'nœud vital,' or vital knot, as its injury causes instant death by simultaneous arrest of the heart and respiration. The restraining influence of the vagus may be increased not only by artificially stimulating the nerve-bed in a reflex manner by afferent nerves carrying impressions to the centre in the medulla, but a blow on the stomach may lead to sensations that so act on the centre that inhibitory influences are thus sent reflexly along the vagus, and lead to such a slowing or stoppage of the heart's action that the person loses consciousness and faints. Fainting from strong emotion or disordered stomach may arise from vagus inhibition.

Stimulation of the sympathetic does not inhibit but augments the action of the heart, increasing either the frequency or the force or both. These fibres are in reality connected with the central nervous system, as they can be traced back through the sympathetic ganglia and rami communicantes to the spinal cord. It is thought that they are then connected with an *augmentor centre* in the medulla. These augmentor fibres carry impulses from the central system that have originated in various ways. Fright and other mental emotions may produce palpitation through these nerves, though quickening of the heart-beat may arise through the reflex removal of the normal inhibiting action of the vagus.

Depressor nerve fibres also pass from the heart (par. 84).

84. Regulation of the Vascular System—Vaso-motor Nerves.

—It has been pointed out that the arteries contain in their middle coat unstriped muscular fibres, for the most part arranged circularly, that these fibres become relatively more abundant as the arteries become smaller, and that the arterial walls contain nerve-fibres in close connection with the muscular tissue. These are the *vaso-motor* nerves. The amount of resistance to the passage of the blood into the capillaries varies with the state of contraction of the walls of the small arteries, and this peripheral resistance is placed under the control of the nervous system, the vaso-motor nerves carrying the impulses that regulate the degree of contraction of the muscular fibres of the arterioles. Veins also contain muscular and nervous elements, but less is known about the variations of calibre in these vessels than is known regarding arterial contraction. Confining our attention to the arteries, therefore, we may say that these vessels contain nerve-fibres of two kinds : (*a*) vaso-constrictor fibres ; (*b*) vaso-dilatator fibres. The vaso-constrictor fibres leave the spinal cord by the anterior roots of the spinal nerves of the middle portion of the cord, enter the sympathetic ganglia of the thorax and abdomen, where they lose their medullary sheath, and thence pass, accompanied by branches from the ganglia, in various ways to their distribution in the blood-vessels. The vaso-dilatator fibres spring from the central system by cranial or spinal nerves, retain their medulla until near their termination, but their distribution has only been made out in a few cases.

The presence and function of vaso-constrictor nerves, by which the calibre of the vessels may be diminished, have been demonstrated in several ways. When the ear of a rabbit is held up to the light, arteries and veins may be distinctly seen. If now the sympathetic nerve in the neck be divided, the ear on that side becomes warmer and redder, the vessels dilate, and more are visible than before. On stimulating the peripheral cut end of the cervical sympathetic, the vessels contract and the ear becomes pale. These experiments show not only that this nerve may on excitation lead to contraction of the arteries of the ear with diminution of blood-supply, but that in ordinary circumstances vaso-constrictor nerves are in moderate action, and keep the arteries in a state of moderate contraction known as *arterial tone*. Similar changes of calibre and blood-flow have been observed in the small arteries of a frog's web under the microscope on stimulation of the sciatic nerve in the leg, and there is evidence to show that vaso-constrictor nerves are widely distributed in all parts, and that the vascular supply of every region of the body is thus under the control of the nervous system. Experiment shows that the tonic contraction of the arteries, and the maintenance of a peripheral resistance to the blood-flow, are dependent on the integrity of the vaso-motor nerves with a particular part of the medulla called the *vaso-motor centre*. Section of the spinal cord below the medulla, if the action of the heart be kept up by artificial respiration, is followed by relaxation of the arterioles and a sudden fall of blood-pressure. Section above the medulla does not lessen the blood-pressure. Stimulation of the vaso-motor centre, which is situated in the grey matter of the medulla about ¼ inch above the *calamus scriptorius*, causes powerful contraction of all the blood-vessels, and a sudden rise of blood-pressure owing to increased peripheral resistance. The normal tonic contraction of the arteries must therefore be due to the persistent impulses conveyed by nerve-fibres under the control of the vaso-motor centre in the medulla, these fibres passing from the centre down the spinal cord and leaving by the anterior roots of the spinal nerves to pass to the vessels in the way already described. Besides the chief vaso-motor centre in the medulla, other subordinate centres are supposed to exist in the spinal cord ; for, although division of the cord in the back cuts off the arteries of the limbs from the chief centre, and leads to these vessels becoming dilated, yet they recover their tone after some time if the animal be kept alive.

The almost constant moderate action of the chief vaso-motor centre may be increased or diminished by reflex action consequent on various afferent stimuli. As a rule, stimulation of afferent nerves produces an increase of arterial contraction and blood-pressure, a *pressor* effect, as it is called, either locally or generally. The circulation of venous blood stimulates the vaso-motor centre. As the blood becomes venous at death, the energetic stimulation of the vaso-motor centre at this time contracts the arteries and drives the blood towards the capillaries and veins. Hence the 'emptiness of the arteries after death' is due to this arterial contraction after the heart has ceased to beat. The centre may be also affected by impulses proceeding from the brain. *Blushing* is an instance of local vaso-motor action, due to an emotion so acting on a part of the vaso motor centre that the muscular walls of the arteries of the neck and face relax on withdrawal of the nervous impulses that kept up their tone, and the superficial vessels become flushed with blood. *Pallor*, due to fright or other emotion, is also an instance of vaso-motor action, resulting in increased constriction of blood-vessels in the face and neck, with temporary diminution of blood-supply. (Pallor may in some instances be due to a diminution of heart-action.)

The vaso-motor centre is, however, directly connected with the heart by an afferent nerve, the *depressor*, which has its origin in that organ. This nerve appears to influence the centre according to the work required from the heart, for it carries impulses to the vaso-motor centre when the heart is over-distended with blood that lead to restraint of the centre, dilatation of the blood-vessels, and a lowering of blood-pressure that bring the heart relief by allowing the blood to flow onwards more easily. The restraining or inhibitory function of the depressor nerve of the heart is shown by dividing it and stimulating the end connected with the medulla, when a marked fall of blood-pressure occurs.

But although the dilatation of the arteries may occur through inhibition of the activity of the vaso-motor centre, the presence of dilatator nerves in the blood-vessels whose action is antagonistic to the first kind has been distinctly proved.   Thus the submaxillary salivary glands receive a nervous supply not only from the cervical sympathetics, but a nerve passes to each gland from the seventh cranial nerve, called the *chorda tympani*, from its crossing the tympanum (see par. 114).   If the chorda tympani be divided no special result is noticed ; but if the cut end that is connected with the gland be stimulated, the small arteries of the gland dilate, become flushed with blood, which flows more quickly, and the secretion of saliva increases.   So rapidly does the blood pass through the dilated vessels that it issues from the veins of the gland still bright red in spite of having passed through the capillaries, though during normal flow the issuing blood is dark and venous.   It is obvious, therefore, that the chorda tympani nerve contains vaso-dilatator fibres only, as the cervical sympathetic appears to contain only vaso-constrictor fibres.   Other vaso-dilatator fibres are found in the nerves going to muscles.   Stimulation of some nerves appears to result sometimes in dilatation of the arteries, sometimes in contraction, and hence they are thought to contain both kinds of fibres.   No special vaso-dilatator centre has yet been found, although one is thought to exist in the medulla.   It does not appear, however, to be in constant action, like the vaso-motor (or vaso-constrictor) centre.

CHAPTER VI

RESPIRATION

85. Nature and Object of Respiration.—Respiration in man is that function of the body by which there takes place an absorption of oxygen from the air into the system, with an accompanying elimination of carbon dioxide.   It may be divided into two stages : *external* respiration, or the introduction of oxygen into and the excretion of carbon dioxide from the blood in the pulmonary capillaries ; *internal* respiration, or the gaseous interchange between the blood and the tissues that occurs in all the general systemic capillaries. Dr. Waller

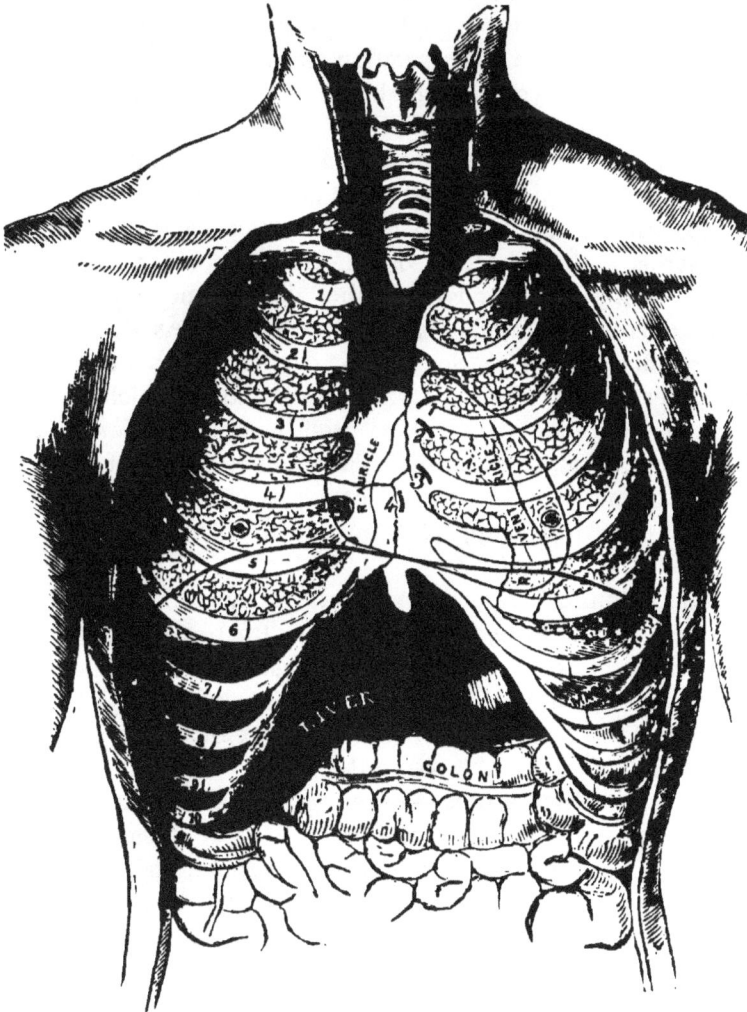

FIG. 90.—Front View of the Thorax. The Ribs and Sternum are represented in relation to the Lungs, Heart, and other Interna Organs.

1, pulmonary orifice ; 2, aortic orifice ; 3, left auriculo-ventricular orifice ; 4, right auriculo-ventricular orifice.

thus summarises the whole process : ' Oxygen, introduced into
the lung by muscular movement, diffuses into the pulmonary
blood and is conveyed to the systemic capillaries, whence it
diffuses into the lymph and tissues ; here it enters and forms
part of some complex compound, which subsequently yields
carbon dioxide as a disintegration product. Carbon dioxide
diffuses from the lymph to the blood, is therein carried to the
lung, whence it diffuses into the air.' The whole object of the
process is to supply oxygen to the various tissues and to remove
the products of oxidation that there takes place, and the essential apparatus is an organ having a thin membrane, on one side
of which is a thin sheet of blood, while the other side is in
contact with the atmosphere or other aërating medium, so that
an interchange of gas can readily take place. As the human
lung offers a surface of blood in the capillaries in an extremely
thin layer of more than 150 square yards to air which is constantly being renewed, the blood itself being also renewed in a
very short period, the condition for an interchange of gases is
highly favourable. To fully understand the process, however,
we must describe the apparatus and mechanism of respiration
in detail.

The pulmonary respiratory apparatus includes (1) the air-
passages, viz. the nose, pharynx, larynx, trachea, bronchi, and
bronchioles, leading to (2) an immense number of minute
cavities in the lung, called air-cells or alveoli ; (3) certain
muscles in the boundary walls of the thorax by means of which
movements of air are effected in the lungs.

86. **The Air-passages.**—Entering at the nostrils, the air
passes along two irregular cavities, the nasal fossæ, into the
pharynx.   In its passage it becomes warmer, more charged
with water vapour, and to a certain extent freed from dust
particles. These passages are thus better adapted for the
admission of air than the mouth, through which, nevertheless,
air is often admitted to the pharynx.   From the pharynx the
air passes downwards into the larynx, the modified and enlarged
upper part of the tube called the trachea, formed mainly of four
cartilages.   In the larynx are two ligamentous bands called the
vocal cords, by the vibration of which voice is chiefly produced.

The chink or opening between the vocal cords is spoken of
as the glottis, or *rima glottidis*. Attached to the upper part of

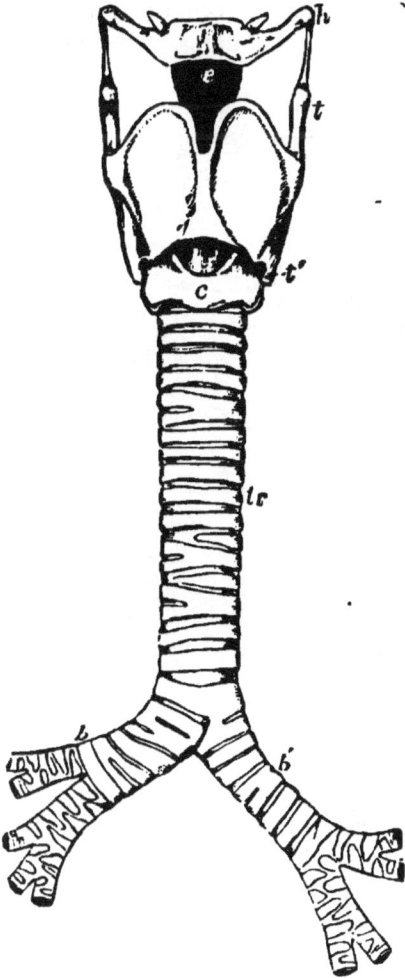

FIG. 91.—The Trachea. Front.

h, hyoid bone; *t*, *t'*, thyroid cartilage; *c*, cricoid; *e*, epiglottis; *tr*, trachea; *b* and *b'*, bronchi.

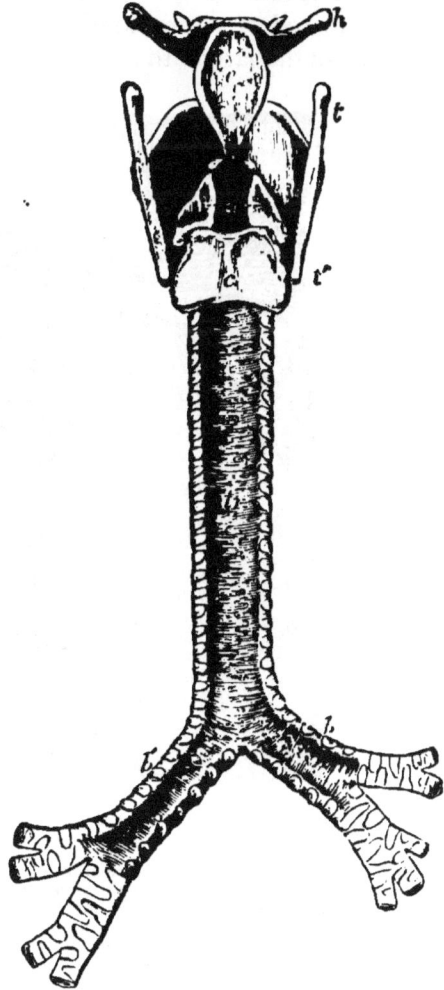

FIG. 92.—The Trachea. Back.

a, arytenoid cartilages; *h*, hyoid bone; *t*, *t'*, thyroid cartilage; *c*, cricoid; *e*, epiglottis; *tr*, trachea; *b* and *b'*, bronchi.

the larynx is a leaf-shaped cartilaginous structure called the
epiglottis, its broad end being free. Below, the larynx (a fuller

account of which will be given in connection witn Voice) joins the trachea.

The trachea is a cylindrical tube, flattened posteriorly composed of several kinds of tissue, extending from the lower part of the larynx to a point opposite the third dorsal vertebra,

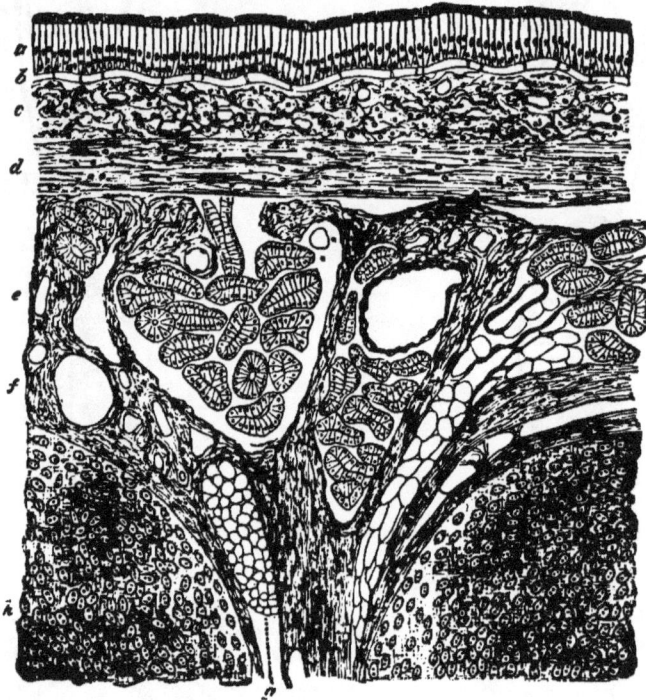

FIG. 93.—Longitudinal Section of the Human Trachea, including portions of two Cartilaginous Rings. (Klein.) Moderately magnified.

a, ciliated epithelium ; *b*, basement-membrane ; *c*, superficial part of the mucous membrane, containing the sections of numerous capillary blood-vessels and much lymphoid tissue ; *d*, deeper part of the mucous membrane, consisting mainly of elastic fibres ; *e*, submucous areolar tissue, containing the larger blood-vessels, small mucous glands (their ducts and alveoli are seen in section), fat, &c. ; *f*, fibrous tissue investing and uniting the cartilages ; *g*, a small mass of adipose tissue in the fibrous layer ; *h*, cartilage.

where it divides into the right and left bronchi. Its average length is about 4½ inches, and its average diameter three-quarters of an inch. The walls of the trachea consist of the following layers :—(1) An external coat of connective tissue containing white fibres and elastic fibres. Embedded in this fibrous coat are 16–18 **C**-shaped rings of hyaline cartilage, which serve to keep the tube open. Some of these cartilaginous rings bifurcate or unite. The outer fibrous membrane completes the tube behind, where the rings are incomplete. (2) A muscular layer within the fibrous external coat, consisting of unstriped muscular fibres. These are for the most part arranged transversely, and are found joining the tips of the cartilages together, and also in the intervals between the rings. These fibres form the *trachealis muscle*, and serve to some extent to regulate the diameter of the tube. A few longitu-

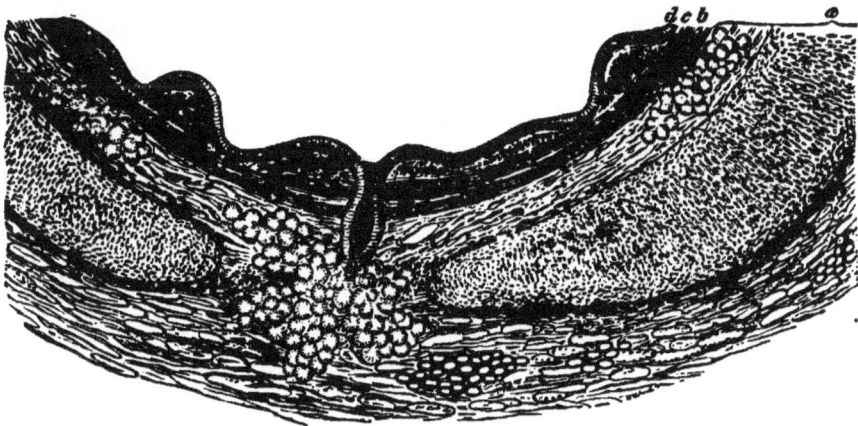

FIG. 94.— Portion of a Transverse Section of a Bronchial Tube (Human), 6 mm. in diameter. (F. E. Schultze.) Magnified 30 diameters.

a, cartilage and fibrous layer with mucous glands, and, in the outer part, a little fat : in the middle, the duct of a gland opens on the inner surface of the tube ; *b*, annular layer of involuntary muscular fibres : *c*, elastic layer, the elastic fibres in bundles which are seen cut across ; *d*, columnar ciliated epithelium.

dinal muscular fibres may be seen outside the transverse fibres. (3) A submucous coat consisting of loose connective tissue (areolar tissue), and containing mucous glands, the ducts of which open into the canal of the trachea. These glands secrete mucus, which, with entangled particles, is carried towards the mouth by ciliary action. (4) A mucous membrane consisting of, starting within : (*a*) an epithelium formed of a layer of columnar ciliated cells, with a few goblet-cells, and two or three layers of immature cells ; (*b*) a transparent basement-membrane ; (*c*) a layer of areolar tissue with longitudinal elastic fibres. These layers should be studied in figs. 93 and 94. Ciliated epithelium cells and 'goblet-cells' are described in par 9.

The right and left bronchi, which diverge at an angle of 100°, are similar in structure to the trachea, the right containing 6–8 imperfect rings of cartilage, the left 9–12. As the branches of these bronchi become smaller the cartilages cease to form regular hoops, and are found as scattered plates on all sides of the tube. When the tubes reach less than 1 mm. in diameter the cartilages disappear, but the circular muscular fibres

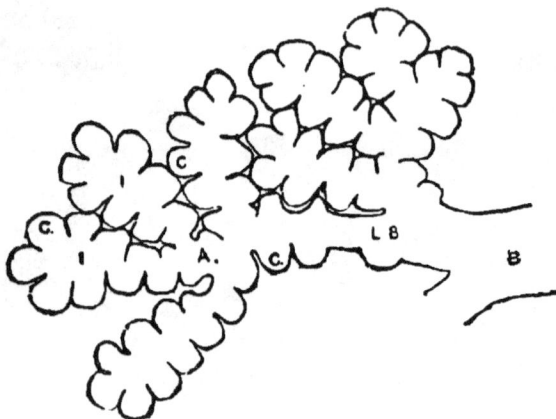

FIG. 95.— Diagrammatic representation of the Ending of a Bronchial Tube in Sacculated Infundibula.

B, terminal bronchus ; L B, lobular bronchiole ; A, atrium ; I, infundibulum ; C, air-cells or alveoli.

are more pronounced. When one of these small branches enters a lobule of the lung, the outer layer of cells has lost its cilia, the muscular fibres have become more scanty, and the mucous glands have disappeared. As the terminal bronchus passes into a lobule it gives off branches called *bronchioles*, each bronchiole terminating in one or two funnel-shaped expansions called *infundibula*, the walls of which are beset with blind cup-shaped pouchès called *air-cells* or *alveoli*. The thin walls of the bronchioles consist of a layer of flattened epithelial cells, with little or no muscular coat, and but a few scattered elastic fibres in the outer connective tissue. Finally, the walls of the infundibula and alveoli consist of a thin membrane of areolar and elastic tissue lined by thin transparent flat cells.

87. The Lungs.—The two lungs are the greyish, spongy, elastic organs which fill during life the whole free cavity of the chest on each side of the mediastinum.[1] Each lung has an apex extending into the root of the neck just above the clavicle, and a concave base resting on the diaphragm. The outer surface is convex, and corresponds in form to the cavity of the chest; the inner surface is concave, and at a depression (the root of the lung) the bronchi and blood-vessels enter. The posterior border of each lung is rounded and broad, fitting into the deep concavity on either side of the spinal column. The anterior edge is thin, and overlaps the front of the pericardium.

The right lung is a little thicker and heavier than the left, but it is about an inch shorter, owing to the diaphragm rising higher on the right side to accommodate the liver. The lungs form about $\frac{1}{37}$th of the weight of the body in the male and $\frac{1}{42}$nd in the female. Each lung is divided into an upper and lower lobe by a deep fissure, the upper lobe on the right side being subdivided by a secondary fissure. The surface of each lung is marked out into polygonal spaces, which are the bases of the lobules. The substance of the lungs is made up of lobules united by connective tissue, called interlobular septa. The interlobular septa are continuous with the subpleural connective tissue and with the peribronchial connective tissue entering the lung at its root, so that the fibrous framework of the lung is continuous throughout, as in other organs. Each lobule may be regarded as a miniature lung, as it consists of (1) a small bronchial branch giving off bronchioles which end in infundi-

[1] The *mediastinum* is the name given by anatomists to the space in the middle of the thorax between the two inner pleural walls of the lungs. It extends from the sternum in front to the spinal column behind, and contains all the organs of the thorax except the lungs. There is an upper portion above the top of the pericardium, called the superior mediastinum, and a lower portion below the upper part of the pericardium. This lower portion is divided into the anterior mediastinum, between the sternum and the pericardium, the middle mediastinum, containing the heart, great vessels, and the roots of the lungs, and the posterior mediastinum, between the pericardium and the eight lower dorsal vertebræ (see fig. 98).

bula and air-cells ; (2) a small artery breaking up into a plexus
of capillaries in the thin connective tissue of the air-cells, these
capillaries uniting to form a small bronchial vein ; (3) nerves ;
(4) lymphatics.   The aggregate of these air-tubes and air-cavi-

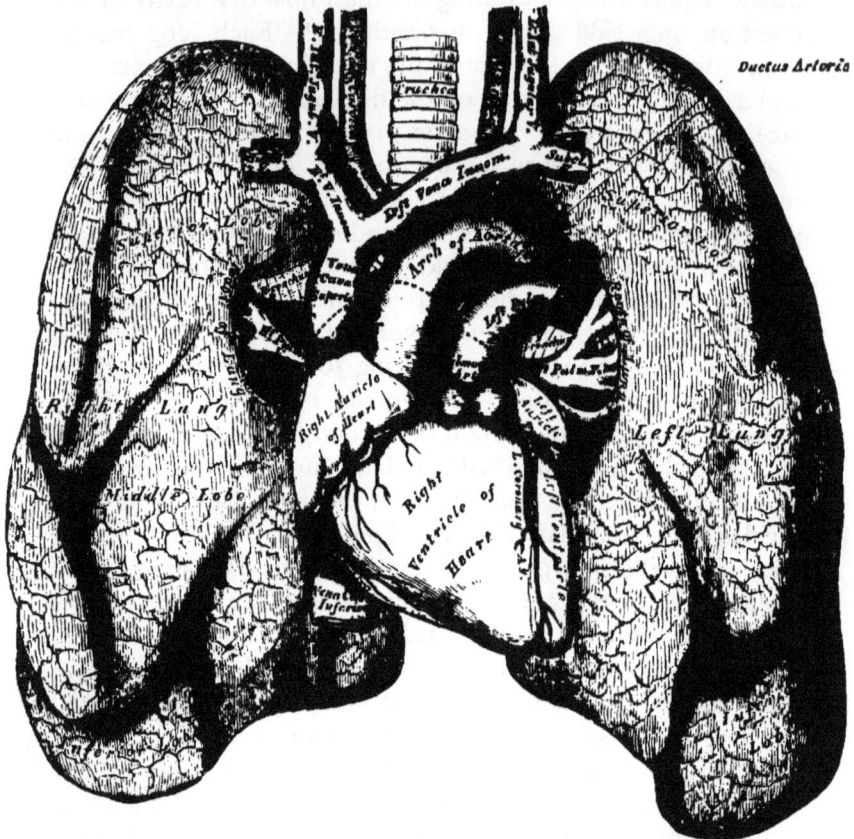

FIG. 96.—Front view of the Heart and Lungs.
(Arteries are red and veins are blue.)

ties, with the vessels distributed to them, together with the
connective tissue, nerves, and lymphatics, make up the substance of the lung. The air-cells or alveoli, the structure of
whose wall has already been described, are about $\frac{7}{10}$ inch in
diameter, and are said to number 725 million, and to present

a surface one hundred times greater than that of the whole
body. It is from the air in these alveoli that the blood obtains
a fresh supply of oxygen, and to the air in the same cavities the
blood gives up carbon dioxide. For supported in the delicate
framework of connective tissue of the alveoli is a close network
of capillaries, the blood of which is only separated from the air

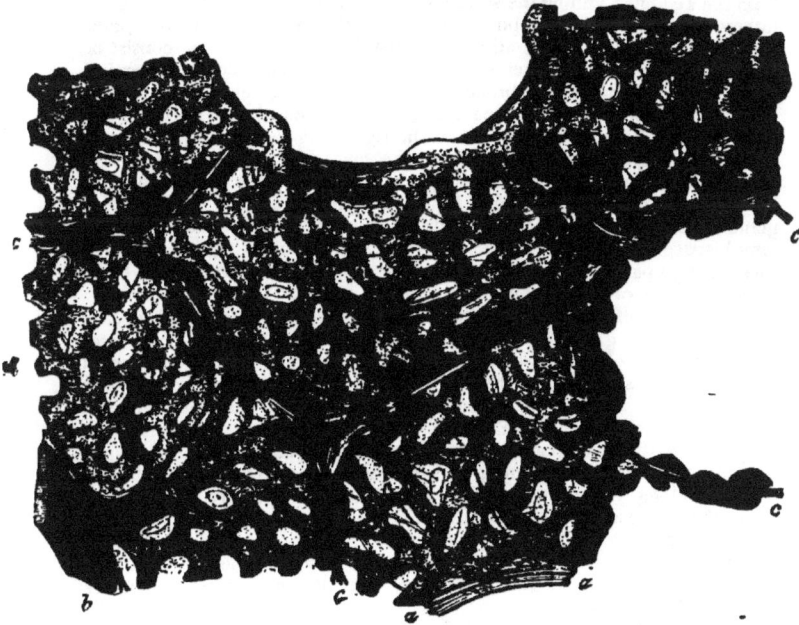

FIG. 97.—Section of Injected Lung, including several contiguous Alveoli. (F. E. Schultze.)
Highly magnified.

a, a, free edges of alveoli : *c, c,* partitions between neighbouring alveoli, seen in section :
b, small arterial branch giving off capillaries to the alveoli. The looping of the vessels
to either side of the partitions is well exhibited. Between the capillaries is seen the
homogeneous alveolar wall with nuclei of connective-tissue corpuscles and elastic fibres.

by the delicate wall of the vessel and the thin epithelial cells
lining the minute air-cavity. Between adjacent alveoli there is
only one layer of capillaries, the vessels twisting first to one
side and then to the other of the septa which separate them.

88. Blood-vessels and Lymphatics of the Lungs.—At the root of the lungs
the bronchi and pulmonary arteries enter, the subdivision of the pulmonary
arteries accompanying that of the bronchial tubes. The terminal arterial

K

branches are about $\frac{1}{1000}$ in. in diameter, and from them arise the closely set capillaries, $\frac{1}{3000}$ in. in diameter. · Their meshes are so close that the interspaces are even narrower than the vessels. By the union of the capillaries are formed the pulmonary veins, which also leave the root, and return the purified blood to the left auricle of the heart. This constitutes the pulmonary or lesser circulation. But besides these blood-vessels there are bronchial arteries, two or three in number, springing from the aorta, and destined to nourish the bronchi, lymphatic glands, and connective tissue of the lungs. The right bronchial vein enters the vena azygos major (a vein from the abdomen joining the superior vena cava), and the left the superior intercostal vein. The lymphatic vessels of the lungs are numerous, and arise from lymph-spaces in all parts of the lung tissue. They consist of a superficial set and a deep set, accompanying the blood-vessels and forming small glands on the smaller bronchi. Both sets emerge at the root of the lungs, where they enter the bronchial glands, passing thence from the left lung into the thoracic duct and from the right lung to the right lymphatic trunk. Foreign particles arrested in the mucus of the air-passages and alveoli often find their way through the epithelium into the lymphatics below, and may then be carried by the lymph-stream into the bronchial glands. Leucocytes that have wandered from the blood-vessels into the lymph spaces and thence into the alveoli are the chief agents by which these foreign particles are removed. A section of coal-miner's lung often shows particles of carbon thus embedded in lymphatic glands.

89. Nerves of the Lung.—The nerves of the lung consist of branches from the pneumogastric or vagus and from the sympathetic. They enter the lungs and follow the distribution of the vessels and bronchi, small ganglia being situated in the walls of the latter. Their exact mode of termination is not clear, though some fibres are distributed to the bronchial muscle and large blood-vessels. Impulses pass from the lungs along the pneumogastric and along other nerves to a respiratory nervous centre in the medulla, and from this centre are reflected efferent impulses along various nerves that bring about respiratory movements. Thus the intercostal muscles are supplied by branches from the spinal cord termed intercostal nerves, and the diaphragm is mainly innervated by branches of the third and fourth spinal nerves termed the phrenic nerves. The action of the respiratory nervous mechanism is further described in par. 100.

90. The Pleuræ.—Each lung is invested on its external surface by a delicate serous membrane enclosing the organ as far as its root, and being then reflected upon the inner surface of the thorax. The pleura thus forms a closed sac, as do the other serous membranes, and consists of two layers, the layer covering the lung being called the *visceral* layer, and the layer reflected on the inner surface of the thorax the *parietal* layer. In health the two layers are always in contact, and contain just so much fluid as will ensure their easy gliding upon each other during life. The continuity of the layers will be understood from the transverse section of the thorax, where a space between them is only shown for the sake of clearness. The right and left pleuræ are distinct, the membranes only touching each other behind the upper third of the sternum. The space between them, called the *mediastinum*, contains all the viscera of the thorax except the lungs. The inner opposing surfaces of the pleura (of the pericardium and of other serous membranes) are smooth and moist, being lined by a continuous layer of pavement epithelium cells. In some places there are apertures between the cells, called *stomata*, leading into subjacent lymphatic vessels, by which the serous cavity is put into communication with the lymphatics of the lungs and diaphragm, so that during the movements of the chest the fluid is constantly being drawn off. New fluid is constantly being secreted to replace tha removed. 'The serous cavities therefore are to be

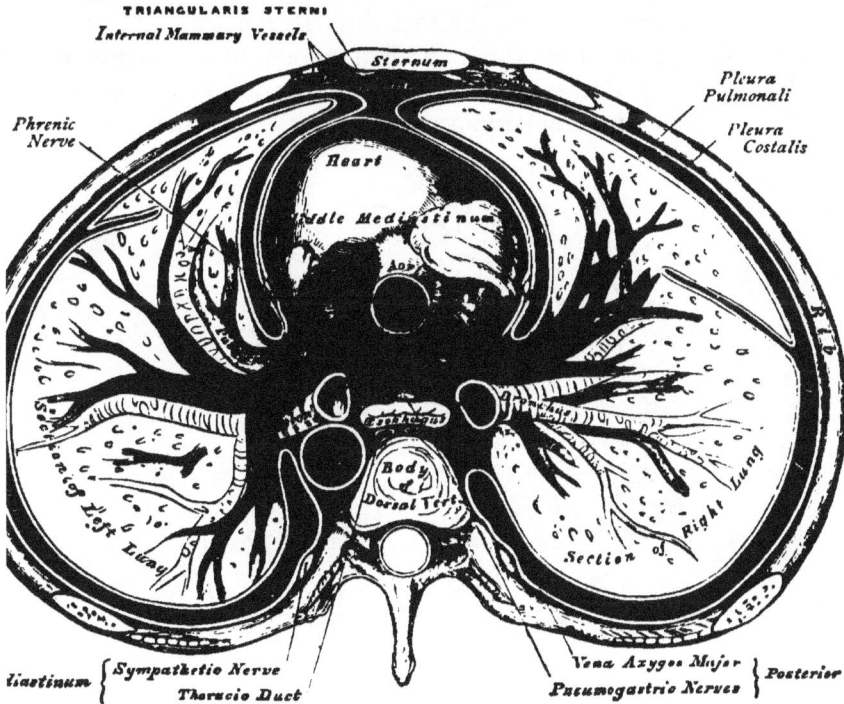

FIG. 98.--A Transverse Section of the Thorax, showing the relative Position of the Viscera and the Reflections of the Pleuræ.

regarded as expanded initial reservoirs from which, as well as from the lymph-capillaries and lymph-spaces of the tissues, the lymph-stream is continually being fed ' (fig. 99).

91. General View of the Respiratory Movements.—We have seen that the lungs may be regarded as a many-chambered elastic bag placed in the air-tight thorax, and having a communication with the exterior only by the trachea. The pressure of the atmosphere passing down the trachea keeps the

lungs so far stretched that the two pleural layers are always in apposition, and together with the heart and great vessels completely fill the thorax. If the thorax is opened the distended lungs collapse, owing to the atmospheric pressure on their external surface counteracting that through the trachea. The internal capacity of the thorax undergoes rhythmical variations, the movements of the elastic lungs following those of the thorax. By the contraction of certain muscles, in a mode shortly to be

FIG. 99.—Small portion of Peritoneal Surface of Diaphragm of Rabbit.
(Klein.) Magnified.

l, lymph-channel below the surface, lying between tendon bundles, *t*, *t*, and over which the surface-cells are seen to be relatively smaller, and to exhibit five stomata, *s*, *s*, leading into the lymphatic. The epithelium of the lymphatic channel is not represented.

described, the thorax is enlarged at intervals, the lungs expand to occupy the increased space, and the pressure in the lungs becoming less than that outside in the atmosphere, a rush of air from the atmosphere through the trachea takes place to establish equilibrium of pressure. This act constitutes *inspiration*. When the muscles that produced inspiration relax, the thorax is brought to its former size, mainly by the elastic recoil of the lungs and chest walls (aided at times by other muscles), and this decrease of chest capacity causing the pressure inside to be greater than that outside leads to an outrush from the air-passages to the external atmosphere. This constitutes *expiration*. Inspiration plus expiration constitutes *respiration*, the respiratory act taking place in an adult about sixteen times per minute. The first part of a respiratory act, inspiration, is essentially a muscular act, enlarging the chest capacity ; the second part of a respiration, expiration, is essentially an elastic recoil, aided by the weight of the chest walls. These movements set up differences

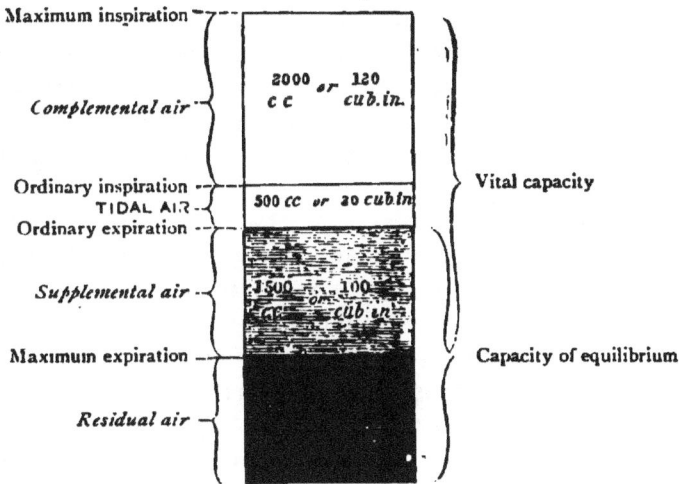

FIG. 100. (Waller.)

Amounts of Air contained by the Lungs in various phases of Ordinary and of Forced Respiration.

of pressure between the inside and outside air, and in both cases the movement of air follows the general law 'that air passes from a region of higher to a region of lower pressure.' The amount of air, 30 cubic inches, taken in at each inspiration and given out at the succeeding expiration in ordinary easy breathing is but a fraction, about $\frac{1}{6}$, of that which the lungs contain. Besides this *tidal* air, as it is called, the lungs contain about 100 cubic inches that may be expelled by a forcible expiration, termed *supplemental* air, and another 100 cubic inches that cannot be driven out by any effort, *residual* air. Moreover, by a great effort of inspiration nearly 120 cubic inches of air, additional to the tidal and called *complemental* air, may be caused to enter the lungs.

The total quantity of air that can be expelled from the lungs after an extraordinary inspiration by an extraordinary expiratory effort is thus equal to 250 cubic inches, and this quantity is spoken of as the 'vital capacity.' Vital capacity is sometimes estimated by blowing as long as possible into a spirometer after the deepest inspiration possible. It usually increases with increase of height, 8 cubic inches for every inch above the average height of 5 feet 8 inches. It is less in women than in men with the same circumference of chest and height in the ratio of 7 to 10. The volume of air taken in during inspiration is rather more than that expired, measured at the same temperature and pressure, some of the oxygen forming other combinations than CO_2.

The changes in the blood of the pulmonary capillaries have been already described in pars. 56 and 58. *Muscular exercise* increases the number of respirations and thereby the quantity of air passing in and out of the lungs. There is, therefore, increased absorption of oxygen by the hæmoglobin of the red corpuscles and increased elimination of carbonic acid from the plasma. Walking at the rate of five miles an hour is said to increase the quantity of air respired five times, the increased consumption of oxygen and formation of carbonic acid probably taking place in the muscles employed (see paragraphs 36 and 37 on the changes occurring in muscle during its contraction).

Since the lungs, therefore, are but partially emptied of air during expiration by the removal of the contents of the nose, trachea, and larger bronchi, it is necessary to inquire how the fresh air passes to the air-cells, which are the parts of the lung where the interchange of gases takes place between the atmosphere and the blood. The process of *diffusion* effects this. It is by the rapid diffusion of the gases in the tidal air and the stationary air that oxygen is supplied to the alveoli, and carbon dioxide given up to the tidal air. Before describing the passage of the air into the blood, and the effect of the blood on the air, we will give a more detailed account of the respiratory movements, inspiratory and expiratory.

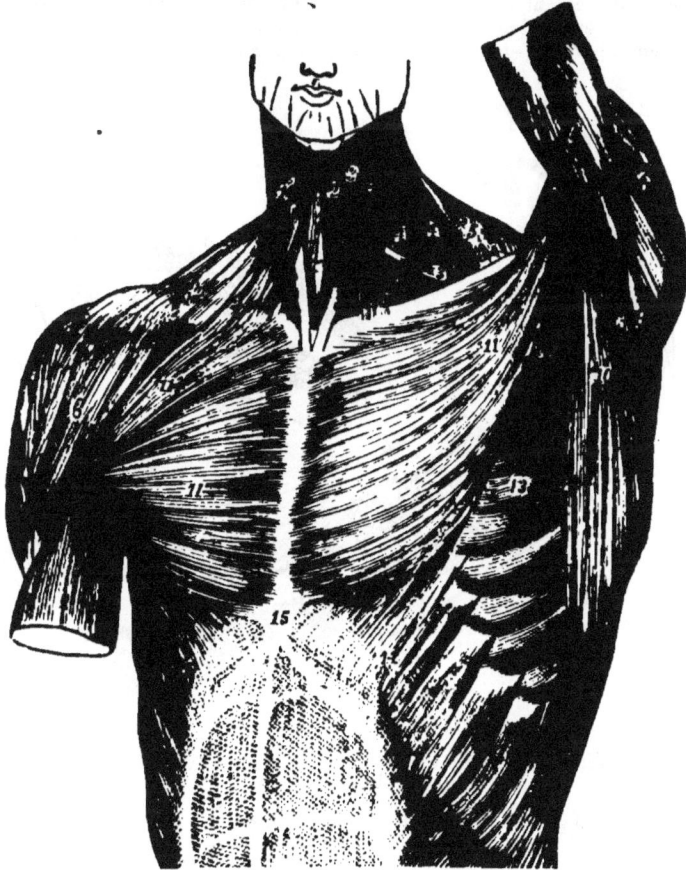

FIG. 101.—Superficial view of the Muscles of the upper part of the Trunk, from before. (Allen Thomson.)

1, sterno-mastoid of the left side ; 1′, 1′, platysma myoides of the right side ; 2, sterno-byoid ; 3, anterior, 3′, posterior belly of the omo-hyoid ; 4, levator anguli scapulæ ; 4′, 4″, scaleni muscles ; 5, trapezius ; 6, deltoid ; 7, upper part of triceps in the left arm ; 8, teres minor ; 9, teres major ; 10, latissimus dorsi ; 11, pectoralis major ; 11′, on the right side, its clavicular portion ; 12, part of pectoralis minor ; 13, serratus magnus ; 14, external oblique muscle of the abdomen ; 15, placed on the ensiform process at the upper end of the linea alba.

92. **Inspiration.**—In inspiration the internal capacity of the thorax is increased, and air enters the lungs through the larynx and trachea to equalise the pressure within the lungs and outside. The enlargement of

the chest is effected by the action of certain muscles, and the increase takes place (1) in the vertical diameter, (2) in the antero-posterior and lateral diameters, *i.e.* an increase in depth, an increase from back to front and from side to side. The vertical diameter of the chest is increased by the contraction and consequent descent of the diaphragm. This is the arched musculo-tendinous sheet separating the cavity of the chest from the abdomen. It is attached all round—to the sternum and ribs in front to the ribs at the side

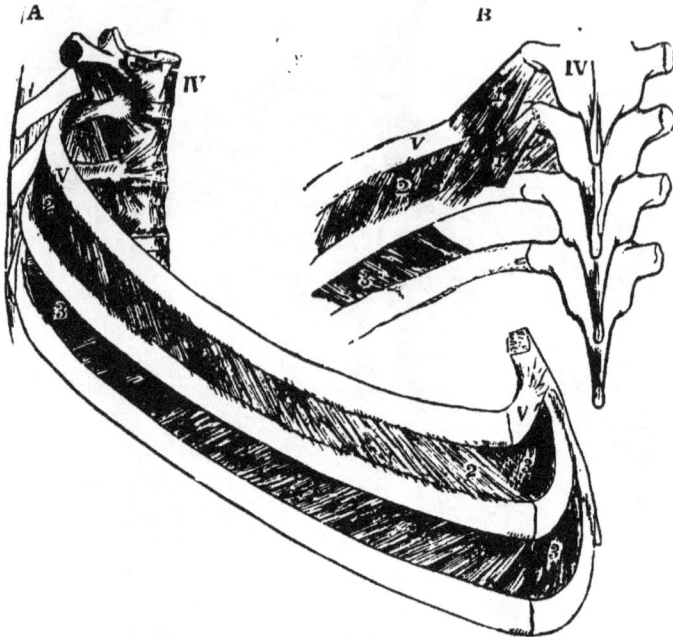

FIG. 102. –Intercostal Muscles of the Fifth and Sixth Spaces. (Allen Thomson, after Cloquet.)

A, from the side ; B, from behind.

IV, fourth dorsal vertebra ; V, V, fifth rib and cartilage ; 1, 1, levatores costarum muscles, short and long ; 2, 2, external intercostal muscle ; 3, 3, internal intercostal layer, shown in the lower space by the removal of the external layer, and seen in A in the upper space, in front of the external layer : the deficiency of the internal layer towards the vertebral column is shown in B.

and to the ribs and spinal column behind. To its upper surface are attached the investing membranes of the lungs (pleuræ) and heart (pericardium). During its contraction the diaphragm becomes flatter to the thorax, the sides descending most. The thorax thus enlarges in depth and the front walls of the abdomen bulge out owing to the pressure on the viscera in that cavity. From its attachment to the sternum and false ribs it tends to pull these downwards and inwards, but its action in this respect is counteracted partly by the vertical direction of the fibres attached to the ribs and partly by the elevation of the ribs that accompanies descent of diaphragm. The increase of the thorax from back to front and from side to side is brought about by muscles that elevate the ribs. These muscles are : (1) **The scaleni,** (2) the intercostal muscles, (3) the levatores costarum or elevators of the ribs.

The scaleni muscles pass from processes of the cervical vertebræ to the first two ribs, and by their action raise or at least fix these ribs. The external intercostals whose fibres run downwards and forwards in the spaces between the ribs so act, when the two first pairs of ribs are fixed by the scaleni, that the ribs are elevated both in front and at the sides, moving on their articulations with the vertebræ. As they slant downwards, the ribs when raised must thrust the sternum forward and enlarge the antero-posterior diameter of the chest ; and since they form arches which increase in sweep, at least from the first to the seventh, the elevation of one into the place of another causes the chest to become wider from side to side. Further as the ribs are raised there is some stretching of the costal cartilages, and a certain amount of rotation of the ribs which brings their outer surfaces more directly outwards, these effects plainly aiding the enlargement of the thoracic cavity. The *levatores costarum* arise from the tips of the transverse processes of the seventh cervical and the upper eleven dorsal vertebræ, and pass obliquely downwards and outwards, being inserted into the outer surface of the rib belonging to the vertebra below that from which they spring. They are regarded as muscles of ordinary inspiration, inasmuch as they assist in elevating the ribs.

FIG. 103.—Diagram of First and Seventh Ribs, in connection with the Spine and the Sternum, showing how the latter is carried upwards and forwards in inspiration. (G.D.T.)

The expiratory position is indicated by continuous lines, the inspiratory by broken lines.

In extraordinary and forced inspiration other muscles are brought into play to enlarge the chest. The *quadratus lumborum,* placed between the last rib and the pelvis, aids the diaphragm by fixing one of its attachments and with the help of other abdominal muscles draws down the lower part of the thorax. The serratus posticus superior, a muscle of the back arising from the spines of the vertebræ in the upper dorsal region, aids in raising the 2nd, 3rd, 4th and 5th ribs ; the sterno-mastoid raises the clavicle ; and the serratus magnus, pectoralis major and minor, all serve to lift the ribs when the arms and shoulders are fixed (see fig. 101).

Associated with the respiratory movements of the thorax are movements of the nostrils and of the glottis. At each inspiratory movement the nostrils expand when breathing through the nose, returning to their previous condition in expiration. During inspiration the glottis is wide

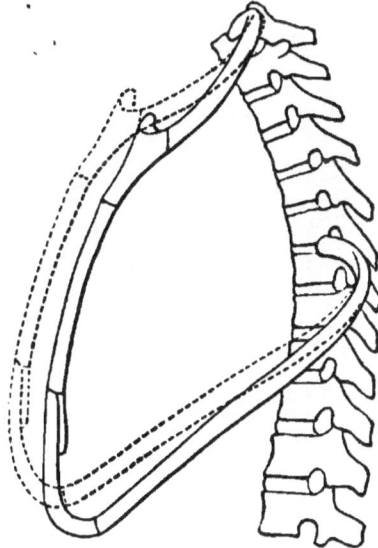

open, but during expiration it is narrowed by the action of the muscles that move the arytenoid cartilages of the larynx In laboured inspiration these facial and laryngeal actions are exaggerated.

93. **Expiration.**—In ordinary easy breathing, expiration or the diminution of the chest cavity and expulsion of air is effected mainly by the elastic recoil of the lung tissue and costal cartilages which had been put on the stretch in inspiration. The diaphragm relaxes and ascends, and the external intercostal and other inspiratory muscles ceasing to act, the ribs fall down and the walls of the chest return to a condition of rest as a result of the elastic reaction. The abdominal muscles may also assist, and possibly part of the internal intercostals. Much discussion has taken place about the action of the internal intercostals, and diverse views are held regarding their action. Some authorities hold that the only office of the internal intercostals is to render the intercostal spaces firm and the whole thoracic cage rigid, but the general view is that the parts of the internal intercostals between the costal cartilages elevate the ribs and assist in inspiration, and that the lateral portions of these muscles between the bony ribs depress the ribs and assist in expiration.

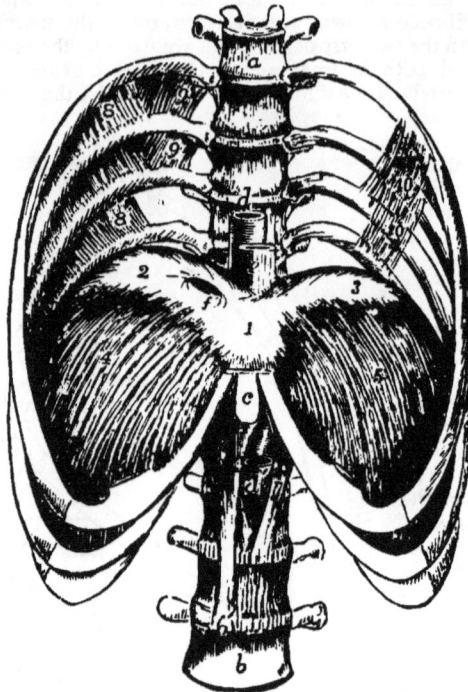

FIG. 104.—The Lower Half of the Thorax, with four Lumbar Vertebræ, showing the Diaphragm from before. (Allen Thomson, after Luschka.)

a, sixth dorsal vertebra ; *b*, fourth lumbar vertebra ; *c*, ensiform process ; *d, d'*, aorta, passing through its opening in the diaphragm ; *e*, œsophagus ; *f*, opening in the tendon of the diaphragm for the inferior vena cava ; 1, central, 2, right, and 3, left division of the trefoil tendon of the diaphragm; 4, right, and 5, left costal part, ascending from the ribs to the margins of the tendon ; 6, right, and 7, left crus ; 8 to 8, on the right side, the sixth, seventh, and eighth internal intercostal muscles, deficient towards the vertebral column, where in the two upper spaces the levatores costarum and the external intercostal muscles 9, 9, are seen ; 10, 10, on the left side, subcostal muscles.

In forced expiration the abdominal muscles play a more prominent part, their contraction forcing the abdominal viscera and the diaphragm upwards and pulling down the sternum and lower ribs. Their action is aided by that of the *triangularis sterni* which depresses the costal cartilages, by the serratus posticus inferior, and as the expiration becomes more forced by every muscle that can depress the ribs or press the abdominal viscera.

Even at the end of the most forced expiration the lungs contain the residual air and are in a more or less stretched condition with the two layers of the pleura in contact. This is proved by the fact that if in an animal or in a corpse the thoracic wall be perforated, the lungs collapse driving out air, the collapse separating the two layers of the pleuræ and air passing into the pleural cavity (pnemothorax). Collapse of one lung is dangerous, but collapse of both lungs through openings into both pleural cavities is fatal, since the effect of the respiratory movements is merely to drive air in and out of those cavities instead of renewing the air in the lungs.

Before birth the lungs contain no air and exercise no elastic force ; but after birth the alveoli and bronchioles are opened out as the animal begins to breathe, a further expansion taking place through the subsequent growth of the thorax.

Expiration follows inspiration immediately, and the two acts take up nearly the same time, though in women and children the expiration is slightly longer. At the end of expiration there is normally a very slight pause. In an adult the number of respirations is 16 to 18 a minute. In a new-born child the number is about 44, and in a child of five, 25 a minute. Muscular effort increases the number, so that in standing even it is more than when lying at rest. In health the number of respirations is to the number of heart beats in the proportion of 1 to 4 or 1 to 5.

94. Types of Breathing.—There are three more or less distinct types of respiration, marked by the mode in which the increase of chest capacity is mainly produced. In young children and some men the diaphragm is the chief muscle employed in tranquil inspiration, and as this causes a pronounced rise and fall of the walls of the abdomen, it is called the *abdominal type*. In boys and many adult males, the action of the diaphragm is not so great, and the movement of the ribs is most distinct from the seventh downwards, so that this is spoken of as the *inferior costal type*. In girls and women the upper part of the sternum and the upper ribs take the chief share in respiration, the walls of the abdomen showing but little motion. This is the *superior costal type*.

95. Respiratory Sounds.—If the ear be placed against the uncovered chest, or if a stethoscope or other conductor of sound be applied, a soft blowing murmur, the *vesicular murmur*, is heard over the regions where the lungs are situated during inspiration. This respiratory murmur is most distinct in children, and may in them be clearly heard during expiration. It is probably caused by the friction and oscillation of the air in the small bronchioles and alveoli. Over the trachea in front, or the larger bronchi at the back, between the shoulder blades, a louder and rougher sound called the *bronchial murmur* is heard both in inspiration and expiration.

By lightly tapping or percussing the walls of the chest a resonant or hollow sound is heard at those parts beneath which the lungs are in contact with the thoracic wall, the part over the heart yielding a dull sound. In inspiration the front part of the lungs moves downwards and inwards, so that their inner borders in front approach. The resonant area is therefore found to increase with each inspiration and to diminish with each expiration, the dull area increasing and diminishing at opposite times.

96. Differences between Atmospheric Air and Respired Air.—After its entrance into the bronchial passages in inspiration the tidal air effects exchanges with the stationary air of the lungs, so that at the following expiration differences of composition are found between the air expired and the atmospheric air. In round numbers dry air contains one fifth its volume of oxygen and four fifths its volume of nitrogen, a trace of carbonic acid with a still smaller trace of ammonia. More accurately the percentage composition of dry atmospheric air is :—

	Oxygen	Nitrogen	Carbon Dioxide
By volume	20·95	79·02 [1]	0·03 to 0·04

Slightly more oxygen is found in the air of the country compared with that of a town. Ordinary atmospheric air always contains an amount of watery vapour, the amount varying with the temperature. As the temperature increases the amount of moisture that the air will hold before it is saturated increases. The air in summer usually contains three times as much water vapour as in winter, but relative to its temperature it is drier than the air in winter. Expired air contains nearly 5 per cent. less oxygen, and over 4 per cent. more carbon dioxide, than the air inspired, the nitrogen being often the same or slightly greater. The average composition of expired air is :—

	Oxygen	Nitrogen	Carbon Dioxide
By volume	16·033	79·58	4·38

There is less CO_2 in the tidal expired air than in the air of the deeper parts of the lung. It may be noticed that the quantity of oxygen lost is greater than the quantity of CO_2 gained, although equal volumes of these gases contain the same volume of oxygen at the same temperature and pressure. Hence there is a slight diminution in the expired air as compared with the inspired air, $\frac{1}{40}$ to $\frac{1}{30}$ when measured under like conditions. The fraction $\dfrac{\text{Vol. } CO_2 \text{ given off}}{\text{Vol. of oxygen absorbed}}$ is called the respiratory quotient. It will be found to be about $\frac{9}{10}$ or ·9 normally, but it increases with muscular

[1] This includes the small amount of the newly discovered gas, argon.

activity, a carbo-hydrate diet, &c., and diminishes with sleep, fasting, etc. The loss of oxygen in respiration is due to the fact that all the oxygen taken into the system does not combine with carbon to form CO_2, for a small part of it is used up to oxidise the hydrogen of fats and thus produce water, and some to partially oxidise the nitrogenous parts of food and thus produce urea.

Expired air further differs from inspired air in being saturated with water vapour, and as the vapour of the atmosphere varies the lungs must give off different quantities of water from the body at different times. The temperature of the expired air is nearly that of the body, being 97° to 98° F. As the air in most countries is below the temperature of the body, expired air is usually warmer than inspired air, heat being acquired as soon as it enters the air passages. In hot climates the expired air may have a lower temperature than the inspired air. Expired air is said to contain traces of ammonia. Expired air also contains slight quantities of organic matter, as is shown by the brown coloration produced on passing it through strong sulphuric acid. This expired organic matter is an important factor in rendering all expired air foul and injurious.

The quantity of oxygen taken in by the lungs of an adult man has been found to amount in 24 hours to 700 grams (11,000 grains), and the quantity of carbon dioxide expired 800 grams (12,300 grains), containing 218 grams or 3,300 grains of carbon. But the quantities vary with exercise, food, etc. The amount of aqueous vapour varies between 350 and 500 grams, the chief cause of this variation being the varying quantity in the inspired air.

97. How Changes are effected in the Blood Gases in the Lungs.—We have seen that expired air mainly differs from atmospheric air in containing about 5 per cent. less oxygen and 4 per cent. more carbon dioxide, and we must now ascertain how this change is brought about. Let us consider the intake of oxygen by the pulmonary blood first. The inspiratory movements already described introduce about 30 c. ins. of tidal air into the upper parts of the bronchial passage, and this by diffusion spreads through the stationary air to the alveoli of the lungs. Here it is separated from the thin sheet of blood in the pulmonary capillaries by the delicate walls of these vessels and by the thin moist membrane forming the alveolar walls. Why should the oxygen pass from the alveoli into the blood? To answer this question we must first state the laws of absorption of gases by fluids. The law established by Dalton and Henry shows that the *weight* of a gas taken up by any fluid at a given temperature is directly proportional to the pressure of the gas, i.e. as the pressure of a gas increases the weight (volume multiplied by density) *dissolved* increases, and as the pressure diminishes the weight *dissolved* diminishes in like proportion. If the atmosphere above a fluid consists of two or more gases, the pressure on its surface is equal to the sum of the pressures of the several gases, but each constituent of the mixture only exerts its own pressure as if it alone existed above the fluid, the pressure of each being known as the *partial pressure* or *tension* of the gas. The normal atmospheric pressure is 760 mm. of mercury, and as air contains 21 vols. per cent. of oxygen and 79 vols. per cent. of nitrogen, the partial pressure of the oxygen in atmospheric air is $\frac{21}{100}$ of $760 = 159 \cdot 6$ mm. of mercury, and the partial pressure of nitrogen is $\frac{79}{100}$ of $760 = 600 \cdot 4$ mm. of mercury. The pressure of the CO_2 in the atmosphere of which it forms only ·04 per cent. is almost zero.

Water therefore absorbs oxygen from the air, the amount dissolved varying directly as the partial pressure of the gas. The dissolved oxygen exerts a tension or pressure, and when this becomes equal to that in the atmosphere above equilibrium is established, and no gas is then taken up or given off by the liquid. Different gases are not absorbed in equal quantities by the same liquid at the same pressure and temperature, for some are more soluble than others. But for the same gas, the same liquid, and the same temperature, the weight of the gas absorbed is proportional to its pressure. Further, when a liquid is exposed to a gas, the gas is absorbed or passes out of the liquid until equilibrium is established between the pressure of the gas above and the pressure of the gas in the liquid. We thus see why CO_2 escapes from a bottle of soda water with such force. The gas was forced into the liquid in the bottle by great pressure, and the bottle tightly corked. When the cork is removed, the gas which had been exerting pressure in its efforts to escape, rushes out until the pressure of CO_2 in the liquid is equal to the small pressure of CO_2 in the atmosphere above.

The statement and explanation of the Henry-Dalton law of the absorption of gases by such liquids as water has been given that we may see that it does not afford a sufficient explanation of the diffusion of gases to and from the blood. For the case is different with such a fluid as blood, where the gases absorbed are not in a state of simple physical solution. Here chemical forces come into action. As was explained in a previous chapter, the oxygen of the blood is for the most part in a state of loose chemical combination with the hæmoglobin of the red corpuscles, less than 1 per cent. being dissolved in the blood plasma. Now venous blood contains 8 to 12 per cent. of oxygen by volume and arterial blood 20 per cent., while such a liquid as water would hardly contain 1 per cent. at the pressure to which the blood in the pulmonary capillaries is subjected. This is proof that the oxygen is not simply dissolved, but that it is chemically united to some substance in the blood. Moreover the absorbed gas uniting with the hæmoglobin by chemical affinity does not exert the outward pressure that a gas simply dissolved would. By exposing venous and arterial blood to atmospheres more or less rich in oxygen, an atmosphere has been found to which arterial blood neither gives off nor takes in oxygen, and the same for venous blood. The partial pressure or tension of oxygen in venous blood, which is the sum of its dissociation tension plus the physical tension of the small amount dissolved, is thus determined, and found to be less than the partial pressure of oxygen in the alveoli. Although, therefore, the oxygen thus begins to pass into the blood because the oxygen-pressure in the alveoli is greater than that in the venous blood of the capillaries, yet the quantity absorbed does not vary directly and regularly according to pressure. If venous blood, or a solution of reduced hæmoglobin of the same strength, which behaves in a similar manner, be exposed to gradually increasing pressures of oxygen, very little is absorbed at first, the greatest absorption taking place between 40 mm. and 60 mm. pressure of mercury, after which the absorption again becomes small. On the other hand, arterial blood, or a solution of oxyhæmoglobin of the same concentration, on being subjected to gradually diminishing pressures gives off but little oxygen until the pressure is reduced to 60 mm., when a rapid evolution of gas begins through dissociation of the oxyhæmoglobin. Knowing these facts, we are now able to understand how venous blood can absorb the 8 to 10 volumes of oxygen required to saturate it. The absorption of O by the venous blood is dependent on the partial pressure of O in the atmosphere to which it is exposed, not in direct proportion to the pressure as in the case of water, but in the way described above, viz., the greatest absorption takes place at the low pressures between 40 mm. and 60 mm. ; and as the oxygen-pressure in the alveolar air has been found to be nearly 100 mm., and that of venous blood but 40 mm., the venous blood takes up oxygen from this air to near saturation.

It may be said that the oxygen does not come into direct contact with the blood-corpuscles in the pulmonary capillaries, being separated from them by the conjoined alveolar capillary membrane and the layer of plasma in which the corpuscles float. This is true, but this thin moist film appears to let oxygen pass into the blood as freely as if it did not exist. At any rate, there is no satisfactory evidence that this living film exerts any influence in absorbing oxygen, though it should be said that some authorities, whose experiments show a greater oxygen-tension in arterial blood than in alveolar air, believe that the living alveolar wall must exercise a secretory activity upon the oxygen.

The exit of carbon dioxide from the blood in respiration is not quite so easily accounted for as the intake of oxygen. From 100 volumes of venous blood 46 volumes of CO_2 can be extracted by the mercurial pump ; arterial blood contains 40 volumes of CO_2 per cent. Experiments have shown that nearly the whole of the CO_2 is contained in the plasma, and as plasma would only take up a small fraction of this quantity at its partial pressure and the temperature of the body if it merely dissolved CO_2 like water, the CO_2 must be in chemical union. It exists, in fact, in loose combination with sodium, forming sodium carbonate chiefly. If the excretion of CO_2 from the blood in the lungs is a mere physical process due to simple diffusion, the only condition necessary will be that the tension of the CO_2 in the air-cells shall be less than the tension of the CO_2 in the blood. This is said by some to be the case, though the difference is but small, and the escape of the gas is then described as simply due to physical diffusion. But many recent authorities maintain that the discharge, though begun as diffusion, goes on beyond equilibrium until the tension in the arterial blood of the pulmonary vein falls below that in the alveolar air, and that, therefore, other forces must operate. Certain experiments seem to show that the absorption of oxygen leads to the simultaneous expulsion of CO_2, the red corpuscles then acting the part of a weak acid. The sudden strok of the heart has also been said to mechanically assist the liberation of the CO_2 from the venous blood, just as a tap on the side of a glass containing aërated water will cause bubbles of CO_2 to pass off. The cells of the alveolar epithelium are also said to exercise an excretory activity upon the CO_2.

98. Internal or Tissue Respiration.— In as sing through the capillaries of the various tissues arterial blood has its oxygen removed, carbon dioxide passes into it, and the blood becomes venous again. These changes are spoken of as internal respiration. It ws at one time thought that the oxygen was used and the carbon dioxide formed in the lungs only. This was disproved when these gases were extracted from the blood of all parts, and when it was found that there is more oxygen and less carbon dioxide in the blood passing away from the lungs than in that entering them. It was afterwards thought that oxidation took place in the blood itself. But the use of oxygen by the substances of the blood itself can only be very small, as the amount of oxidisable substances in the blood is small. Besides, it is possible to keep a frog alive in an atmosphere of oxygen when the whole of its blood is replaced by a saline solution. The metabolic processes of its body, consisting of using up oxygen and excreting carbon dioxide, go on, and must, therefore, have their seat in the tissues. This is further proved by experiment with muscle. A muscle removed from the body contains no free oxygen, for none can be obtained under the vacuum of a mercury pump; yet the muscular substance while still living and irritable continually uses up oxygen and gives off CO_2. This excretion of CO_2 by the muscle goes on in pure nitrogen or any atmosphere free from oxygen, the amount of CO_2 given off increasing with the activity of the muscle. The oxidation by which the CO_2 is produced must have gone on in the muscular tissue, and as there was no free oxygen for the muscle to obtain, it must have come from some substance in the muscle in which it was stored or held in reserve. In other words, the oxygen taken by the muscular substance from the arterial blood does not immediately proceed to form CO_2, but enters into some complex combination more stable than hæmoglobin, the CO_2 given off during muscular life and activity being the final result of a series of changes not fully understood. Possibly the oxygen brought to the tissue forms a complex substance in the muscle-cells, the decomposition of which sets free energy as heat, mechanical work, &c., and leads to the production of CO_2 and lactic acid. This stored-up oxygen explains how it is that a muscle can go on contracting for a little time without receiving oxygen, and why in vigorous muscular action the CO_2 is in excess of the oxygen absorbed. But if muscular irritability is to continue a fresh supply of oxygen is soon needed (pars. 36, 37).

All our knowledge of what goes on in other tissues points to the same conclusion as muscular activity; and that the tissues themselves, and not the blood, are the seat of the oxidation and other metabolic processes is the view most in harmony with what is known of the vital activity of one-celled animals (as the amœba) and of plants.

The cause of the interchange of gases between the blood and the tissues is the same as already explained in describing the interchange of gases in the lungs, viz. diffusion from a place of high partial pressure to a place of low partial pressure. The oxygen-pressure of muscular tissue has been seen to be zero, and hence oxygen will be constantly streaming from the red corpuscles of the systemic capillaries, where the pressure is comparatively high, through the plasma, through the capillary wall, and through the lymph, and through the sarcolemma to the muscular fibres. The carbon dioxide produced in the muscle by its activity exercises a greater tension than the carbon dioxide of the blood, and therefore passes in reverse order, from the muscle to the blood. We may call this stream of O from the blood to the muscular substance, with the stream of CO_2 from the muscle to the blood, muscle respiration. The respiratory changes in the other tissues are doubtless of a similar character, the lungs and the blood only taking part in so far as they are tissues themselves.

One caution is needed. It must not be supposed that CO_2 is the only product of oxidation in the tissues. A slight quantity of the oxygen finally appears with hydrogen to form water, and some of it occurs in urea, uric acid, &c., bodies in which the oxidation is not complete.

99. **Rationale of Ventilation.**—The object of ventilation is to secure for each person such a supply of fresh air that there may always be abundance of oxygen without any injurious accumulation of carbon dioxide and organic matter. A dwelling-room becomes foul and offensive when the percentage of respired CO_2 reaches 1 per cent., but with *half* this quantity it begins to feel stuffy. The offensive and injurious quality of respired air is largely due to the organic exhalations from the lungs and skin, some of these having a poisonous effect. Air containing a high proportion of CO_2 derived from action on a carbonate is not nearly so injurious as air containing the same proportion of CO_2 from breathing. As we cannot measure the organic exhalations in a dwelling-room, the percentage of respired CO_2 is taken as their measure, and the problem of ventilation is to keep the CO_2 from exceeding about ·06 per cent. An adult gives off ·6 cubic feet of CO_2 per hour. Ordinary air contains ·04 per cent. of CO_2, therefore 3,000 cubic feet of air would contain 1·2 cubic feet of CO_2. Adding ·6 cubic feet to this, we get 1·8 cubic feet of CO_2 in a room of 3,000 cubic feet where an adult has been breathing an hour, *i.e.* ·06 per cent. Hence 3,000 cubic feet of fresh air per hour per person is required so that the respired air may never contain above ·06 per cent. of CO_2, the standard of impurity. A room having a capacity of 1,000 cubic feet of space per person, and supplied with an additional 2,000 cubic feet of fresh air per person per hour, the air admitted being of suitable temperature and not giving rise to draughts, satisfies the requirements of good ventilation.

100. **Influence of the Nervous System on Respiration.**—The normal co-ordinated movements of respiration are brought about and regulated by a respiratory nerve-centre acting automatically and rhythmically, a deficiency of oxygen in the blood and the extent of distension of the lungs being its normal stimuli. These normal movements, however, may be modified by the action of various stimuli on certain afferent nerves or by an effort of the will. The nervous mechanism includes : (1) a chief respiratory centre in the medulla oblongata, connected with higher centres in the cerebrum ; (2) subordinate centres in the spinal cord ; (3) afferent nerves passing to the centres ; (4) efferent nerves passing to the muscles of respiration. The

L

chief respiratory centre is in the medulla, just above the vaso-
motor centre, for injury to this part of the brain stops respiration at once, while the upper part of the brain may be unde-
veloped or removed without causing breathing to cease. It is
said to be automatic or reflex, because in normal respiration
regular rhythmical discharges of nervous impulses pass out
without our knowledge from its nerve-cells at the rate of 16
to 18 times a minute, along nerves to the muscles that effect
the movements of the thorax, *i.e.* along branches of spinal
nerves to the intercostal muscles, along the phrenic nerves to the
diaphragm, and along the inferior laryngeal branch of the vagi
to the glottis. These *efferent* discharges are doubtless due to
impulses reaching the centre along afferent nerves, especially
along branches of the vagi, the impulses varying according
to the amount of gases in the blood of the lungs and to the
distension of these organs. Thus normal respiration may be
regarded as a series of involuntary or reflex acts. Numerous observations have shown that the vagi nerves are most
intimately connected with the respiratory acts. Branches of
these nerves, accompanied by branches from the sympathetic
system, end in the lungs, and when these organs are inflated
with air carry impressions to the respiratory centre that lead to
expiratory movements ; while the relaxation of the lungs leads
to the transmission of impulses from the centre that call into
action the inspiratory muscles. Section of one vagus nerve
has little effect on respiration, but section of both vagi completely alters the character of the breathing. The rate
becomes much slower—less than one-half—and each breath
becomes deeper and more prolonged. Stimulation of the
central cut end of the vagus quickens the breathing again, and
if the stimulus be very strong may lead to respiratory spasm.
From these experiments on the vagi it is concluded that the
vagi are in constant action, transmitting afferent impulses that
keep up the respiratory movements.

Besides the vagus, other *afferent* nerves convey impressions
to the respiratory centre that lead to modifications of breathing.
Stimulation of the central cut end of the superior laryngeal
branch of the vagus that passes to the larynx slows the respi-

ratory movements ; excitation of the nasal branch of the fifth nerve in the nostrils has a similar effect. The sensory nerves in the skin may so affect the respiratory centre as to arouse special respiratory movements by the reflex impulses sent out from the centre to the nerves regulating the respiratory muscles. This explains the deep gasps that follow the sudden application of cold water to the surface of the body, the sensory impulses passing from the skin to the respiratory centre, causing it to send out efferent impulses that lead to a great inspiratory movement. The effect of stimuli on a sensory nerve cannot always be foretold, as respiration is thus in some cases accelerated, in others retarded. Lastly, voluntary efforts may affect the breathing, while mental emotions sometimes quicken and sometimes lessen the rate. But though we can alter the respiratory movements by an effort of the will, we cannot altogether arrest them. After a minute or two the reflex power of the centre overcomes the effort to inhibit its action, and we are forced to breathe. This is due to the blood becoming more and more venous while the breathing is suspended, the activity of the centre increasing as the blood becomes deficient in oxygen and charged with carbon dioxide. In ordinary easy breathing, *eupnœa*, the excitation to respiratory activity is probably brought about by stimulation of the sensory nerves in the lungs, the reflex action of the centre producing the necessary muscular movements, but in many other cases the respiratory centre appears to be mainly influenced by the quality of the blood sent to it. If air be rapidly forced into the lungs, or if several powerful inspiratory efforts be made, it is easy to hold the breath for a longer time than usual, as divers well know. When no effort is thus made to breathe, the condition is known as *apnœa*. Whether this state is owing to the blood saturated with O failing to stimulate the centre, or to inhibitory impulses generated by rapid inflation of the lungs, is doubtful. Difficulty of breathing, or *dyspnœa*, however, is undoubtedly caused by deficiency of oxygen and excess of CO_2 in the blood, such deficiency exciting the centre to send out more powerful stimuli to the respiratory muscles. Obstruction to the entrance of air into the lung, as by immersion in water or choking produced by closure of the trachea, increases the venous condition of the blood, quickly produces laboured and painful efforts to respire (dyspnœa), and if not relieved leads to

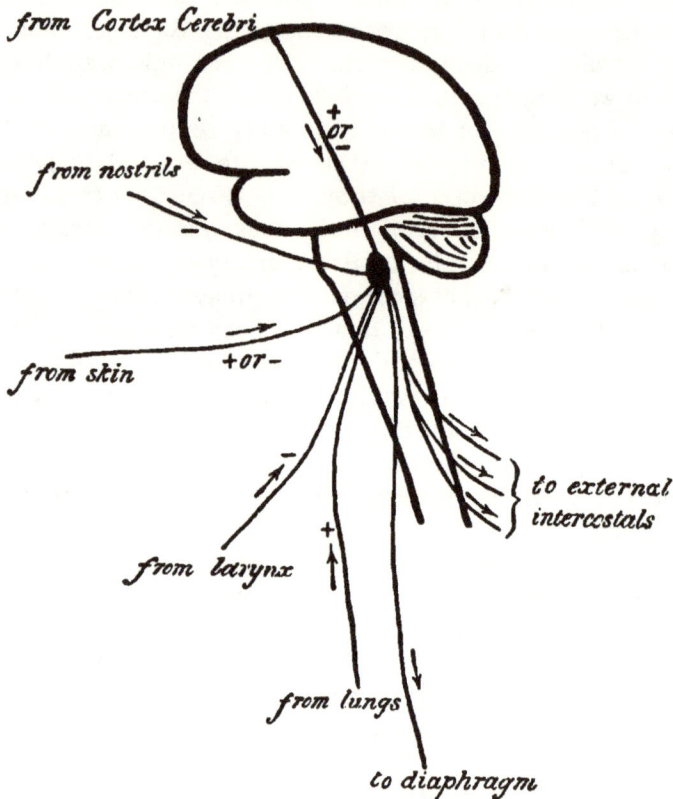

FIG. 105.

Diagram to illustrate the chief Nervous Connections of the Respiratory Centre. (Waller.) Impulses may reach it—

    (1) From the cortex of the brain (*voluntary control, emotional modifications*).
    (2) From the surface of the body (*increase or diminution by cutaneous stimuli*).
    (3) From the lung by way of the vagus (*increase*).
    (4) From the larynx by way of the superior laryngeal (*diminution*).
    (5) From the nostrils by way of the fifth nerve (*diminution*).

Impulses pass from the centre to the diaphragm along the phrenic nerve, to the intercostal muscles along intercostal nerves, and in laboured respiration along many other nerves to many other muscles.

suffocation, or *asphyxia.* Asphyxia is often divided into three stages :—(1) Increasing dyspnœa, during which the extraordinary muscles of inspiration and expiration are brought into action, a state that usually lasts about one minute.   The violent respiratory efforts are due to the venous blood circulating in the medulla strongly stimulating the respiratory centre.   (2) In the second stage the inspiratory muscles become less active, the expiratory muscles contract more energetically, and almost every muscle of the body passes into a state of spasm or convulsion ; the heart makes increased efforts, blood-pressure rises, and the lips become purple.   This convulsive stage is over in another minute and is succeeded by (3) the third stage, or stage of exhaustion, which lasts two or three minutes. The respiratory centre becomes paralysed, consciousness is lost, the pupils dilate, the lips become pale, and all motion soon ceases, the muscles being flaccid.   A few deep inspiratory efforts are made at long intervals, finally ceasing with a gasp. At this time the pulse can be hardly felt, though the heart may beat feebly for a few seconds longer.   A *post-mortem* examination of an asphyxiated person shows the right side of the heart, the large veins, and the lungs to be gorged with dark venous blood, and the left side of the heart to be empty and contracted.   Recovery from asphyxia may often be brought about by resorting to artificial respiration, which may cause the heart to resume its action.   Cutting a distended jugular vein to relieve the engorged right auricle may also assist the heart to recommence.

101. **Special Respiratory Movements.**---Besides the voluntary modifications of the respiratory acts which are made in speaking and singing, there is a series of movements, mostly of an involuntary reflex nature, due to a variety of stimuli.

Coughing consists of a deep inspiration, closure of the glottis, and a sudden violent expiratory effort by which a blast of air is sent through the upper air-passages and mouth, the blast of air often carrying with it a substance that was in contact with the respiratory mucous membrane. The sound is produced by the sudden bursting open of the glottis.   It may be produced voluntarily, though it is usually a reflex act caused by a stimulus applied to certain parts of the air-passages, more particularly the larynx, the afferent impulse travelling along the superior laryngeal branch of the vagus from the larynx, or along other branches of the vagi from the bronchi or lungs.

Sneezing consists of a sudden violent expiratory blast driven through the nose after a deep inspiration, the glottis being open and the mouth shut off from the pharynx by the soft palate and the anterior pillars of the fauces.   It is usually caused reflexly by stimulation of the sensory fibres of the nasal mucous membrane (the nasal branch of the fifth nerve).

Snoring occurs during respiration through the open mouth, when the stream of air causes the uvula and soft palate to vibrate.

Laughing consists of short rapid expiratory blasts, which cause the

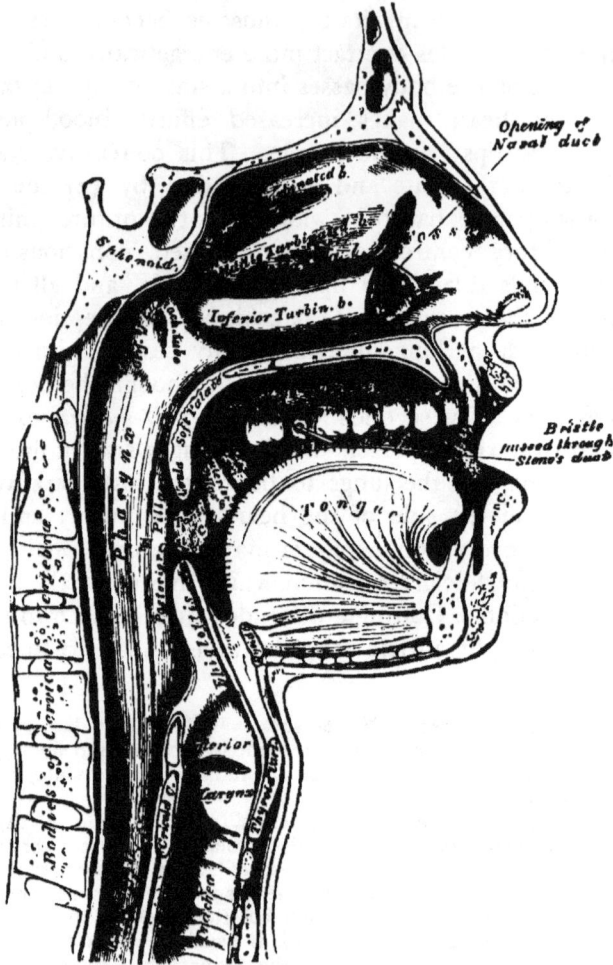

FIG. 106.—Sectional View of the Nose, Mouth, Pharynx, &c.

peculiar sound and lead to peculiar movements of the facial muscles. It is generally involuntary.

Crying, caused by emotional states, consists of short sudden expirations accompanied by long inspirations, facial contortions, and lachrymal secretion.

Sobbing often follows long-continued crying, and is made up of a rapid

series of convulsive inspirations accompanied by sudden involuntary contractions of the diaphragm.

Sighing is a prolonged inspiration accompanied by a plaintive sound,
and often caused by sad recollections.

Yawning is a prolonged deep inspiration attended by a stretching of
the muscles of the lower jaw, the mouth, fauces and glottis being opened
and a characteristic sound emitted. It is involuntary, and is usually
excited by drowsiness or weariness.

Hiccough is an inspiratory gasp caused by a sudden involuntary
contraction of the diaphragm, the inspiration being arrested by a sudden
closure of the glottis which produces the characteristic sound. It is often
due to irritation of the gastric mucous membrane.

102. **Influence of Respiration on Circulation.**—The heart,
the venæ cavæ, the aorta, and the pulmonary vessels situated in
the air-tight thorax are exposed to certain alterations of pressure
during the respiratory movements, and these alterations have a
direct mechanical effect on the pressure of the blood. As the
chest expands during inspiration, air rushes in from the outside,
through atmospheric pressure, until the pressure in the lungs
and outside is equalised. Part of this atmospheric pressure,
however, is spent in overcoming the elasticity of the air-tubes
and alveoli. This elastic force, which increases the more the
lungs are distended, acting in opposition to the pressure of the
atmosphere, supports a part of the full atmospheric pressure
which would otherwise fall on the heart and great vessels
within the thorax but outside the lungs, and hence these
organs are then under a less pressure than the other blood-
vessels of the body. The amount of this pressure thus withdrawn from the heart and intrathoracic vessels by the elasticity
of the lung in normal inspiration is 10 or 12 mm. of mercury,
and may rise in a deep inspiration to 30 mm. The thick-walled
ventricles and the stout aorta, where we know that the blood-
pressure is high, will be but little influenced relatively by the
difference of pressure existing on the vessels inside and outside
the thorax, while the thin-walled venæ cavæ and auricles, where
the blood-pressure is almost zero, will be sensibly affected. An
increased flow of blood from the veins outside the thorax that
are under the full atmospheric pressure will pass along the
venæ cavæ to the right side of the heart ; an increased quantity
will be sent from the right ventricle through the lungs to the

left ventricle, the diminished pressure in the pulmonary vessels assisting the onward flow ; more blood will therefore be sent into the aorta to increase arterial pressure. Inspiration thus aids the circulation, and produces an increase of blood-tension in the arteries.

During expiration the pressure on the vessels within the thorax returns towards that of the atmosphere, but does not

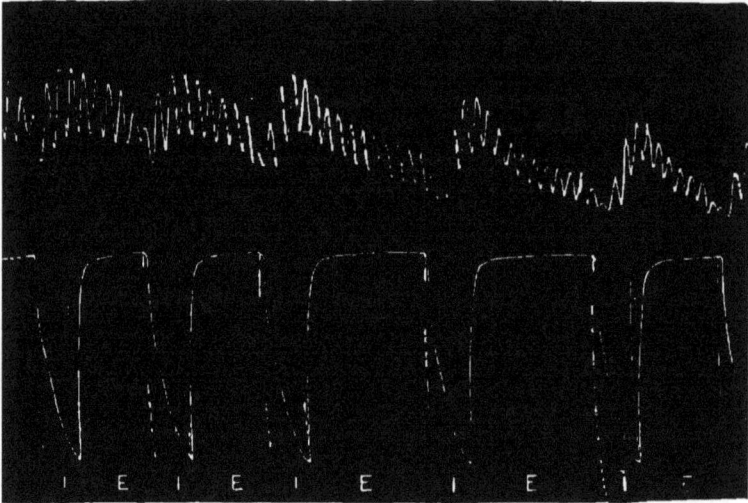

Fig. 107. — Rabbit.   Influence of Respiratory Movements upon Arterial Blood-pressure.   (Waller.)

Showing increased blood-pressure and quickening of heart-beat during inspiration, and fall of blood-pressure with slowing of heart-beat during expiration.

quite reach it.   The blood-flow to the right side diminishes, and the pressure in the arteries sinks down.

This variation of blood-pressure is shown in the tracing furnished by an artery, and if we examine a simultaneous tracing of the thoracic respiratory movements taken by a pneumograph, by which a lever is caused to descend during inspiration and ascend during expiration, we find that the arterial pressure does rise during inspiration and fall during expiration.   But it will be noticed, on comparing the two curves, that the inspiratory rise begins a little later than the commencement of inspiration, and the expiratory fall a little later than the commencement of expiration. This delay may be partly due to the effect of the abdominal movements in inspiration and expiration on the arteries and veins in the abdomen and legs, but more to the effect of the varying pressure within the thorax on the pulmonary vessels. During inspiration there will be at first a widening of these vessels, which will delay the onward flow, though the increased room soon leads to more rapid and easy flow. In expiration a narrowing of the pulmonary vessels leads to opposite effects.

Respiration also produces effects on the circulation in the blood-vessels through the nervous system. In the first place the cardio-inhibitory centre is stimulated by the vagi nerves during a fall of blood-pressure in expiration, and produces a slower rate of heart-beat (see fig. 107). Section of the vagi, by which the centre is cut off from the heart, destroys this effect. In the second place, when the blood is not sufficiently oxygenated the vaso-motor centre and vaso-constrictor nerves are excited, and the blood-pressure rises for a short time. If the blood remains non-arterialised, the excitability of the centre is soon exhausted, the arterioles relax, and pressure falls by peculiar undulations below the normal.

CHAPTER VII

FOOD

103. **Food** is the material required for the nutrition of the body, whose tissues are continually undergoing chemical changes that produce energy in the form of mechanical work and heat, and lead to the excretion of certain waste products that are of no further use in the economy. There is a daily loss of matter amounting to nearly 4,000 grams in a man of average weight, a loss of comparatively simple bodies resulting from the various chemical decompositions :—

By the *lungs*, of carbon dioxide and water vapour.

By the *kidneys*, of water, urea, uric acid, and salts.

By the *skin*, of water, a small quantity of CO_2, and traces of urea.

By the *bowel*, of water, indigestible portions of food, &c.

To replace this waste, if the body is to retain its vigour and health, substances must be taken containing the same elements as those that are thus daily lost, and the food-stuffs must be of such a nature that they may ultimately replace the tissue used up. Now we know that the body consists of certain compounds, or proximate principles as they are termed, of two kinds, inorganic and organic, and it is found that animals require as food, materials closely allied in chemical composition to their own tissues.[1] Plants are able, by the help of the energy of the sun's rays, to form complex unstable compounds out of simple inorganic bodies derived from the soil and air, but animals require in order to renew their tissue complex combustible materials, the oxidation of which sets free the energy required for the performance of their active functions. The substances taken in as food, however, have in most cases to undergo many changes before they can become constituent parts of the body. These changes begin in the alimentary canal, and are there partly of a physical and partly of a chemical nature called digestion. Digestion converts insoluble food into soluble food, indiffusible substances into diffusible substances, through the action of certain juices poured out from the secretory cells of the digestive canal. The food thus digested then undergoes absorption through the walls of the alimentary canal into the blood, either directly into the capillary blood-vessels that lead to the portal vein, or indirectly by passing into the lacteals or lymphatics of the intestine. The blood-stream carries the absorbed products to the tissues, which incorporate them by a process termed assimilation. But the blood not only serves as the medium for distributing food to the tissues, it also carries the oxygen taken in at the lungs to the tissues, and there occurs that process of internal respiration already described. The removal of one product of tissue change, carbon dioxide, has been treated of in a previous chapter, and the removal

[1] See Appendix—Chemistry of the Body.

of the nitrogenous waste by the kidneys will be described in a later one.

Hunger is the peculiar indefinite sensation which is specially referred to the stomach, but arises from the general need of the system. It is relieved as soon as nutrient matter is absorbed into the blood. *Thirst* is a peculiar sensation referred to the palate and pharynx, and is relieved by the passage of water into the blood.

104. **Classification of Food-stuffs.**— Foods are divided into classes called food-stuffs or alimentary principles, which correspond closely to the chief proximate principles or substances of which the body consists. These classes may be thus arranged:—

I. **Proteid, or nitrogenous** substances, containing the elements carbon, hydrogen, oxygen, nitrogen, and sulphur. Examples are *myosin* in muscle, *casein* in milk, *gluten* in bread, *albumen* in eggs, *legumen* in peas and beans. *Gelatine* is a nitrogenous body closely allied to the proteids. They are called nitrogenous food-stuffs because they are the only class containing nitrogen, but it must be remembered that carbon is the most abundant element by weight in them. (See percentage composition of proteids, par. 106, and Appendix, page 413.)

II. **Non-nitrogenous substances,** containing carbon, hydrogen, and oxygen, without any nitrogen, and comprising:—

(*a*) *Amyloids or carbohydrates*, substances in which the oxygen and hydrogen exist combined in the same proportion as in water. Examples are the various kinds of *starch* and *sugar*. (See Appendix, page 416.)

(*b*) *Fats and oils*, substances in which the oxygen is not sufficient in amount to combine with the hydrogen. Examples are *palmitin* and *stearin* in butter, suet, &c., *olive oil*.

Fats possess a greater store of potential energy than carbohydrates, as both their carbon and some of their hydrogen require to be oxidised by oxygen taken in at the lungs.

III **Mineral matter**—*e.g.* such salts as the chlorides, phosphates and carbonates of sodium, potassium, calcium, &c. These are chiefly useful as exercising a beneficial influence on the various chemical processes occurring in the body.

IV. **Water.**—Water is chiefly of use for solvent and mechanical purposes, and is not strictly a food, though necessary to life. About five pints are needed daily.

The first two classes, being derived from animals and plants, are often spoken of as *organic* ; salts and water, belonging to the mineral world, are called *inorganic*.

The chemical food-stuffs or compounds above enumerated are not, except in the case of water, cane-sugar, and sodium chloride, used as articles of diet in a separate and distinct state. We do not have a meal composed of myosin, casein, starch, &c. ; our food materials are meat, bread, milk, &c., which contain these chemical compounds in various proportions. To ascertain the amount of such food materials required we must ascertain the amount of daily loss from the body and the amount of those chemical food-stuffs or alimentary principles in the articles of common consumption.

The old classification of the organic foods by Liebig into plastic or tissue-formers (proteids) and respiratory or heat-producers (the carbohydrates and fats) is inaccurate, for we shall see that all kinds of organic foods are both combustible with generation of energy and assimilable to form tissue. Proteids not only form flesh and yield force, but they may form fat and produce heat, and carbohydrates (fat and starch) yield muscular energy as well as heat (par. 109).

105. Daily Loss of Material.—The amounts of the chief elements discharged from the body in twenty-four hours are set forth in the following table :—

	Water	Carbon	Hydrogen	Nitrogen	Oxygen
By the lungs . .	330	248·8	—	—	651·15
By the kidneys .	1700	9·8	3·3	15·8	11·1
By the skin . .	660	2·6	—	—	7·2
By the bowel (fæces)	128	20·0	3·0	3·0	12·0
Grams . .	2818	281·2	6·3	18·8	681·45

Besides the above, there is a small amount of the following elements in the excreta : sulphur, phosphorus, chlorine, potassium, sodium, &c., these elements forming compounds with one or other of those in the table. These salts amount in all to 32 grams, 26 grams being contained in the urine and 6 grams in the fæces.

In order to balance this expenditure of material our food must contain the requisite amounts of water and of the elements eliminated. It should be noted, however, that a small amount of water is produced in the body by the oxidation of hydrogen in the food. That the amount of water given off in a day is actually more than that taken in has been proved in several cases. In one instance the total amount taken in in all forms was 2,016 grams, and the amount eliminated 2,190 grams, a result which shows that 174 grams were derived from oxidation of hydrogen. This chemical production of water in the body explains why the respiratory quotient, $\dfrac{\text{volume of } CO_2 \text{ given off}}{\text{volume of O taken in}}$ is less than unity (par. 96).

The water required to repair the daily loss, less the small

amount formed in the body, will be replaced by the water taken as drink and in the articles of food ; the oxygen required will be obtained from the air breathed ; and the hydrogen will come with the other food. We may, therefore, confine our atttention to the carbon and nitrogen. The daily loss of carbon is seen to be 281·2 grams (about 4,500 grains, or 10 oz.), and of nitrogen 18·8 grams (about 300 grains, or $\frac{3}{4}$ oz.), the carbon being excreted chiefly as carbon dioxide gas by the lungs, and the nitrogen as urea ·dissolved in the urine. It is evident from the above that some nitrogenous food is essential to life. An animal fed solely on non-nitrogenous food and water soon dies, as it continues to excrete urea when no nitrogen is being supplied. But to feed an animal on proteid alone is injurious, for proteid contains only $3\frac{1}{2}$ of carbon to 1 of nitrogen, and the daily loss (4,500 grains of carbon and 300 grains of nitrogen) shows that the proportion of carbon to nitrogen required is as 15 to 1. To live on albumen, or any other proteid, alone would require, to get the requisite 4,500 grains of carbon, such a quantity that the amount of nitrogen would be in great excess. To get rid of this excess much injurious labour would be cast on the body. If only sufficient proteid was used to supply the body with nitrogen, it would then be starved for carbon. Our daily supply of food must therefore be partly nitrogenous and partly non-nitrogenous, because the latter is richer in carbon. As already remarked the common articles of diet contain the various kinds of chemical food-stuffs in different proportions, and to ascertain which of these may be taken singly or in suitable mixtures to repair the daily loss we must study their composition.

106. **Composition of Common Articles of Food.**—The table on p. 158, from Waller's ' Human Physiology,' sets forth in an instructive way the composition of certain articles of food.

We learn from this that milk contains all classes of food-stuffs, and that the nitrogen and carbon are in the desirable proportion $\dfrac{1N}{15C}$ by weight. Hence milk is regarded as a perfect food. The proteid in milk is in the form of casein, the fat exists in small globules which rise to the surface on standing and form cream, the carbohydrate is lactose or milk-sugar, and the salts are chiefly phosphates and chlorides of potassium and calcium. **Cheese** is an important product of milk, in which the element nitrogen exists in proportion to the carbon in far greater quantity than in milk, as its chief constituent is the curded casein.

Eggs approach the character of a perfect food, but the carbon element is deficient. The whole of the fat is contained in the yolk.

Of the nitrogenous articles of diet, meat is the chief, and of these beef contains the greatest proportion of proteid (myosin and globulin). Lean beef contains over 20 per cent. of proteid material, mutton 18 per cent., and pork 11 per cent. Meat also contains fatty matters (as lecithin), extractives (as kreatin and sarcolactic acid), and salts (chiefly of potassium and calcium). The flesh of poultry and fish (except eels) contains very little fat, and is therefore often eaten with bacon or butter. All these nitrogenous meats contain a quantity of carbon, for it will be remembered that proteid itself contains about 50 parts of carbon to 15 of nitrogen.

Such articles as bread and vegetables, though often spoken of as non-

Approximate Composition of Some Common Articles of Diet
(compiled chiefly from Parke's tables).

Food-stuffs	Proximate principles				Elements	
	Water per 100	Proteid per 100	Fat per 100	Carbo-hydrate per 100	Carbon per 100	Nitro-gen per 100
Milk	86	4	4	4	7	0·6
Butter . . .	7	1	92	—	70	0·15
Eggs . . .	75	14	10	—	15	2
Beefsteak . .	70	22	5	—	15	3·3
Bread . . .	40	8	1·5	50	28	1·25
Potatoes . .	75	2	—	21	10	0·3
Oatmeal . .	15	12	5	65	40	2
Dried peas .	15	22	2	60	40	3·3
Rice . . .	10	5	1	83	40	0·75
Cocoa powder .	10	15	50	25	55	2·2
Cheese . . .	40	35	25	—	35	5·2
Beer . . .	89	1	—	10	5	0·15

nitrogenous, do contain some nitrogen. Only butter, the fat of milk, and the fat of meat are in reality non-nitrogenous. Peas, indeed, contain as large a proportion of nitrogen as meat, but their nitrogen is not as easily assimilated, and they are not, therefore, as nutritive as meat. The proteid of bread is chiefly gluten, and the carbohydrate is in the form of starch. Starch, indeed, is the chief carbohydrate of all cereals.

Green vegetables contain about 90 per cent. of water, and their chief value is due to the presence of organic acids and salts.

Nearly all food-stuffs contain a small quantity of salts, and though they appear to supply little or no energy by oxidation, salts are both necessary and beneficial to the body. They are daily excreted in certain amounts, and must daily be replaced. They appear to exercise important functions in directing the metabolism of the body. Sodium and chlorides are especially prominent in blood-serum ; potassium, iron, and phosphates in red corpuscles ; and calcium salts, especially the carbonate and phosphate, are necessary for bone and teeth. The most important is sodium chloride, which is taken partly in articles of diet and partly as a condiment. One of its functions is to supply the chlorine for the hydrochloric acid of gastric juice. The greater part of the sodium chloride is, however, discharged in solution in the urine. All food-stuffs also contain a certain quantity of water (from fifty to sixty per cent.), and an additional quantity is taken in various liquids, as water is necessary to aid in dissolving the food materials and carrying them to the tissues, as well as being required to assist in removing the secretions and excretions.

107. **Requisites of a Suitable Diet.**—A knowledge of the amount of daily loss from the body and a study of the composition of the various articles of diet enable us to estimate the relative amounts of food-stuffs required to make good the daily waste. But a suitable diet must contain not only the proper amount and proportion of the proximate principles, it must be adapted to the individual and the amount of work he performs, and be supplied in a digestible form. In constructing a suitable diet we first find the quantity of a suitable proteid which will replace the amount of nitrogen lost, and supplement this with a carbohydrate, or carbohydrate and fat, which will bring the quantity of carbon up to the required amount. The salts, about 460 grains, or 30 grams, are supposed to be contained in these food-stuffs, while the water required is contained partly in the food and is partly taken as water, with or without tea, coffee, &c.

It will be seen that the staple articles of diet, meat and bread, may be so adjusted as to supply all that is needed—19 grams, or 300 grains, of nitrogen, and 281 grams, or 4,500 grains, of carbon.

	N grams	C grams
¾ lb., or 12 oz., of lean meat (340 grains) contain .	10	37
2 lb., or 32 oz., of bread (906 grains) contain .	9	252
	19	289

But this is not a customary diet, as a certain proportion of fat is taken with the meat, and part of the bread is usually replaced by potatoes, rice, &c. Various combinations from the common articles may be made, of which the following is given as an example of a liberal diet :—

	N grams	C grams
16 oz. bread, containing	5·0	125
8 ,, meat ,, ...	7·5	34
4 ,, fat ,, ...	—	84
16 ,, potatoes ,, ...	1·3	45
½ pint milk ,, ...	1·7	20
8 oz. eggs ,, ...	2·0	15
4 ,, cheese ,, ...	3·0	20
	20·5	343

Moleschott's experiments with ordinary articles of diet, the percentage composition of which was known, led him to assign the following quantities of proximate principles as suitable amounts for a man doing ordinary work :—

			N grams	C grams
120 grams of proteid (4·232 oz.), containing			18·88	64·18
90 ,,	fat (3·174 oz.)	,,	—	70·20
330 ,,	carbohydrate (11·64 oz.) containing		—	146·82
			18·88	281·20

with salts 30 grams and water 2,500 grams.

Another experimenter found that he could do work and maintain his weight (the nitrogen of the body remaining the same) on a diet of 100 grams of proteid, 100 grams of fat, and 240 grams of carbohydrate.

The *energy* derived from such normal diets may be ascertained by burning the food in a calorimeter and measuring the amount of heat given off by its combustion. Its oxidation into carbonic acid, urea, and water yields 2,310,000 calories of heat. (A calorie is the unit of heat, or the amount of heat required to raise 1 gram of water 1° C.) This amount of heat represents 1,000,000 kilogram-metres of mechanical work. About one-sixth (150,000) of this energy would be expended in a good day's work, and the remaining five-sixths are needed to keep up the temperature of the body (see par. 166).

Of course the diet will vary with the weight, work, and age of the individual. Dr. Waller states that a man in full work requires daily 1 per cent. of his body weight in solid food and 3 per cent. in water, the 1 per cent. of solid food being made up of ·2 per cent. of proteid, ·15 per cent. of fat, ·6 per cent. of carbohydrate, and ·05 per cent. of salts. The diet for idleness would be less, as the waste of tissue and the output of energy in work and heat would be more easily balanced. In children a larger proportion of food to body weight is required, as they require food to put on flesh and fat so as to increase in size, as well as food to repair waste of tissue and supply energy.

108. Food as a Source of Energy.—The body is not only losing matter continually, but also energy in the shape of motion and heat, and food is required not only to make up the loss of material, but also to balance the energy set free. Energy is the power of doing work against resistance, or the capacity for producing physical change. *Kinetic energy* is the power of doing work which a body possesses in virtue of its motion. A moving bullet possesses visible kinetic energy. If it strike a target its visible energy is transformed into heat, and the heated bullet possesses energy in the motion of its molecules, for it can then do work by expanding bodies colder than itself. One form of energy can often be transformed into another. A current of electricity can be converted into heat. Chemical actions often give rise to heat, light, electrical and other phenomena. *Potential energy* is the power of doing work or producing physical change in virtue of a body's position.

A bent spring possesses potential energy, which becomes kinetic as it changes its shape. Carbon and oxygen when separate possess a store of potential energy, and during their combination the energy becomes kinetic, producing heat and mechanical motion, as in the steam engine. The animal body is a machine for converting the potential or positional energy of the food into actual or kinetic energy, for the chemical changes undergone by the food in the body, as it gets oxidised, convert the potential energy into the actual energy of heat and mechanical motion. This loss of energy in heat and motion is replaced by the combustion of materials which ultimately come from the food digested, *i.e.* by the oxidation of proteids, carbohydrates, and fat into urea, carbon dioxide, and water.

Energy is measured either in terms of mechanical work done, or in equivalent heat-units. In English measures the *unit of heat* is the quantity of heat necessary to raise one pound of water through 1° F., and as it is found by experiment that one pound of water falling freely under the influence of gravity must fall through 772 feet in order to have its temperature raised 1° F., 772 foot-pounds is the mechanical equivalent for 1° F. In the metric system the heat-unit or *calorie* is the amount of heat necessary to raise one gram of water 1° C., and the mechanical equivalent of this heat-unit is 425 gram-metres.

The heat value of the various proximate principles of food has been determined by completely burning known quantities of these substances in a calorimeter. The amount of heat thus produced is the same whether the oxidation be quick or slow, in one stage or several stages. One gram of dry proteid yielded on combustion 5,000 calories or heat-units ($= 2,124$ kilogram-metres of work). Now one gram of proteid yields one-third gram of urea, and this is not a completely oxidised product. This one-third gram of urea would yield 735 calories on being completely oxidised, and this number must be deducted from 5,000 to give the energy derivable by the body from one gram of proteid. The physiological heat value of one gram of proteid is therefore $5,000 - 735 = 4,265$ calories

M

which is equal to 1,812 kilogram-metres in terms of work. One gram of fat on oxidation yields 9,070 calories (= 3,841 kilogram-metres of work); and one gram of carbohydrate yields 3,910 calories (= 1,657 kilogram-metres of work).

If, now, we take Moleschott's diet as a basis, we can calculate the amount of energy in it that is available for the body:—

```
Proteid        120 grams × 1,812 = 217,440 kilogram-metres
Fat             90 grams × 3,841 = 345,000      ,,
Carbohydrate   330 grams × 1,657 = 546,810      ,,
                         Total   1,109,250      ,,
```

If we knew the amount of energy lost by the body in a day, we should be in a position to say whether the diet is sufficient for the supply of this energy. The energy of the body is expended (*a*) in maintaining its temperature, (*b*) in internal muscular work and physiological processes, *e.g.* circulation, respiration, and secretion, (*c*) in external mechanical work, as in carrying weights, locomotion, &c. Professor McKendrick has made the following estimate of the daily output of energy in terms of mechanical work :—

```
Equivalent of heat produced  .  .  620,000 kilogram-metres
Work of the heart  .  .  .  .  .  .  50,400      ,,
Work of the respiratory muscles  .  11,700      ,,
Eight hours' active labour  .  .  . 125,000      ,,
                         Total   807,100      ,,
```

As the energy available in the daily diet exceeded one million kilogram-metres, it is clear that the potential energy of the food is quite sufficient to balance the daily expenditure of energy by the body.

109. **Effects of Various Diets.**—(*a*) *Effect of a Proteid Diet.* It is found that feeding a man on a diet of proteids alone leads to digestive disturbances and diarrhœa, but experiments have been made on dogs which teach important facts. After a period without food, during which the nitrogen excreted in urea and derived from the tissue proteids of the body falls to a certain amount, an amount of proteid is given just sufficient to replace the nitrogen lost in urea. This does not stop the loss of nitrogen in the body, but leads at once to excretion of double

the quantity of urea, so that the output of nitrogen still exceeds the income. Only when the proteid of the food contains three times the amount of nitrogen excreted during the starvation period does the amount of nitrogen given out correspond with that taken in. The animal is then said to be in ' nitrogenous equilibrium,' nitrogenous equilibrium signifying equality between the nitrogen taken in in food and that excreted in urine. One effect, therefore, of a purely proteid diet is to increase the metabolism of the tissues and the excretion of urea. It is found that the body weight may increase during nitrogenous equilibrium, owing to the laying on of fat. This suggests the idea that the proteid food is split up into a urea portion and a fatty portion—a decomposition which is quite possible when we consider the chemical composition of proteid, urea, and fat. Moreover, the fact that to maintain the body in a condition of nitrogenous equilibrium requires three times as much proteid as is necessary to replace the proteid of the disintegrated nitrogenous tissues shows that all the proteid cannot go to repair tissue waste. Some of it is used to carry on the activities of the body, passing after digestion into the lymph, and being oxidised there without actual building up into protoplasm, while another part is built into living protoplasm to supply the tissue waste. Hence the distinction into ' fixed or tissue proteids ' and ' circulating proteids.' Ultimately all are reduced to urea, CO_2, and water.

If more nitrogenous food is taken than is required to supply the loss of nitrogen, a certain amount of this remains in the body to form flesh, though the amount of urea excreted increases with increase of body weight.

All nitrogenous foods are not of the same value as articles of diet. *Gelatine*, though allied to proteids and containing the same elements, will not supply the place of such proteids as albumen. Animals cannot live on gelatine alone, nor on gelatine together with fats or carbohydrates. Still it causes nitrogenous equilibrium to be maintained with less true proteid, so that it would appear to be able to take the place of the circulating proteid that splits up into urea and fat, but not the place of the tissue proteid.

(b) Effects of Fat and Carbohydrates as Food.—Every animal requires a certain amount of nitrogen in its food, and neither fats nor carbohydrates, which are non-nitrogenous foods, can long sustain life without proteids. An animal fed on non-nitrogenous food takes nitrogen from its own tissues for a time, but its food soon ceases to be digested, and death from starvation follows. But fats and carbohydrates taken in moderation diminish the amount of proteid necessary to keep up nitrogenous equilibrium. With a small quantity of fat or carbohydrate there is no storage of carbonaceous substances in the body, but when taken in quantity, carbon is stored in the tissue as fat, or as glycogen in the liver, from carbohydrates. Glycogen does not appear to be formed from fat. It will be remembered that fat may be derived from proteid, and it also appears that carbohydrates may give rise to fat, for it has been found that cattle put on an amount of fat more than four times that gained from the food. The ultimate fate of fat is to be oxidised into CO_2 and water.

Fat is utilised in the body to produce heat and muscular work. As it contains more unoxidised carbon and hydrogen than carbohydrates (starches and sugar) it has a greater capacity for force production. The amount of heat developed by the combustion of certain alimentary substances, as well as the mechanical work obtainable by their oxidation, is shown in the following table :—

	Calories	Kilogram-metres
1 gram of beef fat yields	9,069	3,841
1 ,, ,, butter yields	7,264	3,077
1 ,, ,, beef muscle yields . . .	5,103	2,161
1 ,, ,, arrowroot (starch) yields . .	3,912	1,657

Whether muscular tissue derives its energy under a normal diet from the oxidation of the proteids or the carbohydrates of the food is a question not yet definitely settled. But the old idea that the proteids were exclusively tissue-forming and work-producing foods, and that the carbohydrates gave rise to heat only, has been overthrown by the results of experiments. Thus Fick and Wislicenus in 1865 ascended the Faulhorn, 1,956 metres high, on a diet from which nitrogen was excluded. The amount of nitrogen excreted in the urea of the urine was carefully ascertained in each case, and from this the amount of proteid used up was calculated. Knowing the amount of energy required to raise their bodies the height of the mountain, and calculating the amount of energy set free from the proteid disintegrated, it was found that the energy from the proteid was far short of accounting for the work done in climbing. Besides, there was much muscular energy expended within the body in cardiac and respiratory movements, etc. Hence it is evident that energy was set free from other than proteid substances. Other observations on soldiers and pedestrians confirm this conclusion. Muscular energy is not derived from the metabolism of proteids alone, nor does it increase the excretion of urea proportionately, for the excretion of urea is either not altered at all, or but slightly augmented. On the other hand work does increase the consumption of oxygen and the output of CO_2, and the examination of muscle itself (par. 37) shows that its contraction leads to the discharge of CO_2 without any nitrogenous products being evident. It would thus appear that normally the sources of muscular energy are to be looked for in the fats and carbohydrates, and that proteids are mainly used to repair tissue waste, though it must be remembered that energy can be derived from proteids when needful, as is evident from the case of the dog fed on an exclusively proteid diet.

110. **Starvation or Inanition.**—When an animal is deprived of nourishment it continues to live for some time, but instead of using food to supply its heat and other forms of energy, it uses its own tissues. Excretions continue to pass from kidney, skin, and lungs, and the animal diminishes in weight by loss of fat and flesh from its various organs, the total loss amounting to nearly one-half the body weight in some cases. This loss is accompanied by great thirst, weakness, and pallor. The temperature shows a large daily fluctuation, and gradually sinks to about 70° F. A torpid condition comes on with mental weakness or delirium until death intervenes. A dog may so live for more than 20 days, but in the case of man death occurs with entire absence of food and drink in from 8 to 10 days. A small quantity of water lengthens the time, and adults last longer than the young.

The excretion of nitrogen in the form of urea sometimes falls quickly at first, then reaches a minimum where it remains several days; it rises again when the fat has been used up, and again falls quickly as death approaches. The fæces become gradually less and the secretion of bile greatly diminishes. The intake of oxygen and the excretion of carbonic acid also fall.

In the loss of weight the different tissues and organs take a very different part. The fat suffers most, as may be imagined, seeing that it is mainly a reserve or storehouse which the body utilises in the absence of sufficient food from without. Next to the fat, the spleen and liver dimi-

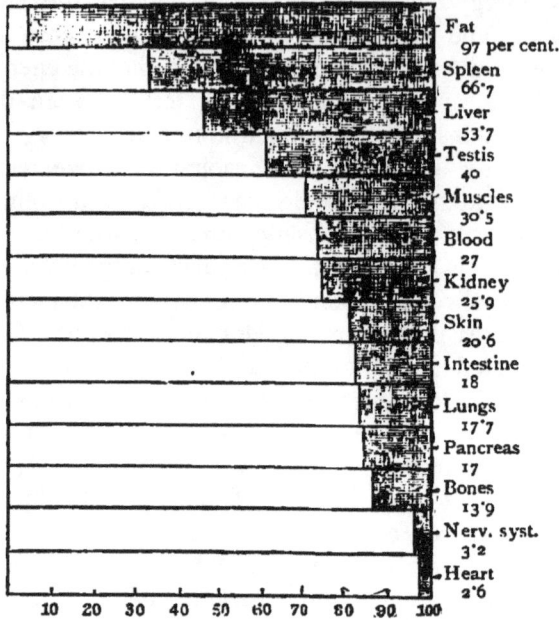

Fat
97 per cent.
Spleen
66·7
Liver
53·7
Testis
40
Muscles
30·5
Blood
27
Kidney
25·9
Skin
20·6
Intestine
18
Lungs
17·7
Pancreas
17
Bones
13·9
Nerv. syst.
3·2
Heart
2·6

10 20 30 40 50 60 70 80 90 100

Fig. 108.—Graphic representation of the percentage of different tissues lost during starvation ; the shaded areæ represent loss, the unshaded areæ amounts remaining at death. (According to Voit's analyses.)

nish in weight (glycogen ceasing to be stored in the liver). The muscles also lose heavily, while the brain and heart are but little affected, being probably nourished by materials drawn from less important organs. The annexed diagram represents graphically the percentage of loss of weight in various tissues or organs at the close of starvation.

CHAPTER VIII

DIGESTION

111. **Mastication.**—The first stage in the digestive process consists in the mastication or chewing of the food in the mouth. This operation is of great importance, as the thorough breaking up of the food enables the digestive juices to gain

free access, and thus more easily to exert their action upon it. The act is accomplished by means of the movements of the lower jaw, tongue, and cheeks. The articulation of the lower jaw by means of its two condyles moving in the glenoid fossæ of the temporal bones is of such a nature that it is movable not only up and down against the fixed upper jaw, but from side to side, so that the food may be cut by the sharp edges of the incisor and canine teeth, or, crushed and ground between

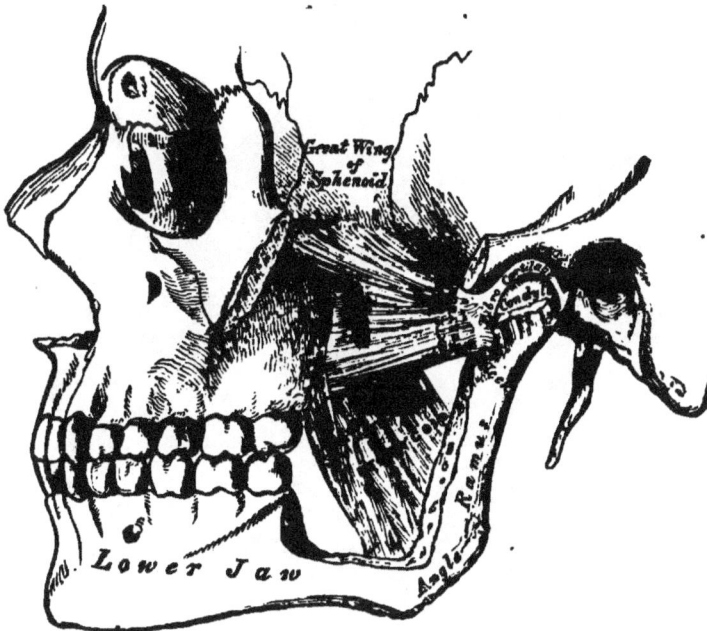

FIG. 109.—The Pterygoid Muscles; the Superficial Muscles, the Zygomatic Arch and a portion of the Ramus of the Jaw having been removed.

the surfaces of the bicuspids and molars. The mouth is opened by the digastric and other muscles passing from the lower jaw to the hyoid bone situated at the base of the tongue ; it is closed and the biting movement effected by the temporal muscles, which pass from the temporal bone to the upper end of the lower jaw, by the masseter muscles of the cheeks which arise from the malar bones and adjoining parts of the zygo- matic arch and are inserted into the outer surface of the ramus of the lower maxillary, and by the internal pterygoid muscles. The external pterygoid muscles draw forwards the condyles and thrust the lower jaw forward, but when acting alternately lead to the grinding movement. Mastication is partly a voluntary and partly a reflex act. The afferent branches of the fifth

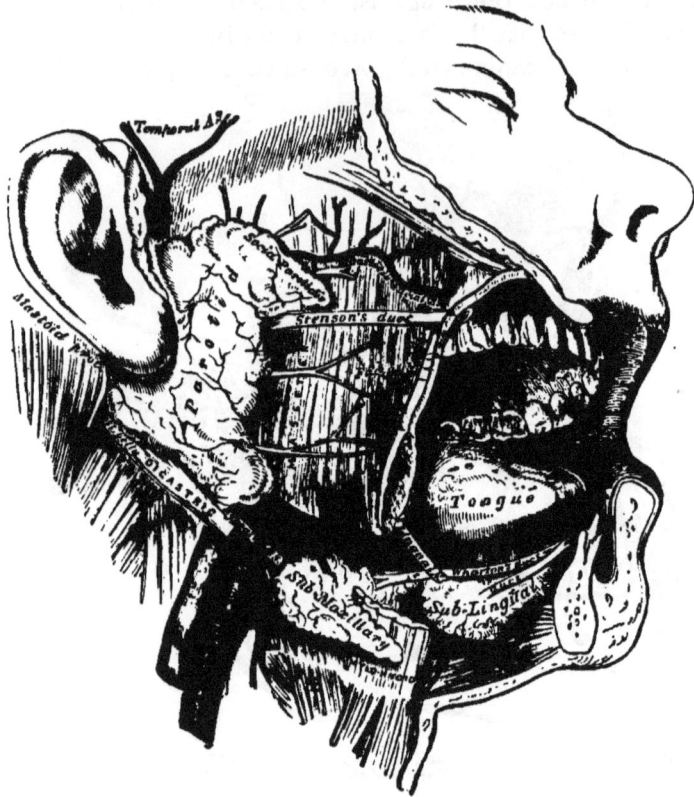

FIG. 110 —The Salivary Glands.

and ninth cranial nerves convey to the brain the tactile and other sensations produced by food in the mouth, and the efferent impulses pass to the muscles by the motor-fibres of the same nerves. The nerve-centre for mastication through which the reflex action occurs, and by which the various movements are co-ordinated, is situated in the medulla.

112. **Insalivation.**—During mastication the food becomes mixed with the first of the digestive juices. This is the saliva secreted by the three pairs of salivary glands, the parotid, the submaxillary, and the sublingual, and the secretion passes into the mouth by the ducts leading from these glands (fig. 110). Small buccal glands in the mucous membrane of the mouth also contribute their secretion to the mixed saliva. The general

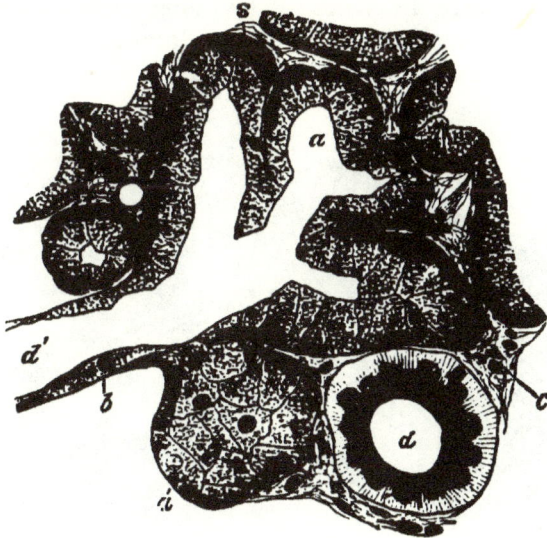

Fig. 111.—Section of a Racemose Gland, showing the commencement of a Duct in the Alveoli. Magnified 425 diameters. (E. A. S.)

a, one of the alveoli, several of which are in the section shown grouped around the commencement of the duct, *d′*; *a′*, an alveolus, not opened by the section; *b*, basement-membrane in section; *c*, interstitial connective tissue of the gland; *d*, section of a duct which has passed away from its alveoli, and is now lined with characteristically striated columnar cells; *s*, semilunar group of darkly stained cells at the periphery of an alveolus.

position and size of the salivary glands, parotid, submaxillary and sublingual, may be again learnt from fig. 110, and we confine ourselves now to their minute structure. Tracing one of the main ducts backwards, it is found to divide and subdivide in the body of the gland into smaller and smaller ducts, the ultimate ducts terminating in spaces, tubular *acini* or *alveoli* lined by secreting epithelium cells. A group of alveoli with the small ducts arising from blind ends are bound together by connective tissue carrying blood-vessels, lymphatics, and nerves to form a lobule, and the whole gland consists of lobules of various sizes bound together and encased by connective tissue.

The interior of the ducts are canals lined with a layer of columnar epithelial cells resting on a basement membrane of connective tissue, with a small quantity of unstriped muscular tissue in the larger branches. The lining epithelial cells of the ducts shows a nucleus near the centre, and the half away from the lumen is seen to be faintly striated. If we examine sections

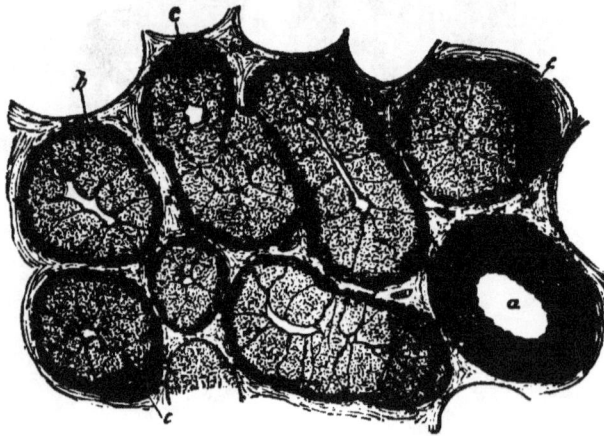

FIG. 112.—Section of Dog's Submaxillary, stained. (Kölliker.)
a, duct ; *b*, alveolus ; *c*, crescent.

of the glands under the microscope, the alveoli present different appearances according as they are cut longitudinally, transversely, or obliquely, while the real salivary cells lining the alveoli are soon found to be of two kinds, with difference of structure according to the substances they secrete. This leads to a division of the salivary glands into two classes, (a) *mucous glands* and (b) *serous or albuminous glands*.

Mucous glands have acini or alveoli lined by large conical cells resting on a basement membrane, and leaving only a small central lumen or opening. These cells do not stain well, secreting a highly refracting substance called mucigen, which is discharged as mucin. Mucin is a complex slimy body consisting of a proteid and an animal gum.

The thick and viscid secretion of mucous salivary glands is called mucous saliva. Between the large clear cells that form the lining of the alveoli of the mucous glands and the basement membrane there may often be seen small cells called from their shape *crescents* or *demilunes*. The large cells of mucous glands differ according to the condition of activity of the gland. In a resting or loaded condition the cells are larger than during activity, and appear crowded with granules of mucigen (fig. 113). In a discharged gland the cells become less and the lumen of the acinus larger, a few granules only being found in the part of the cell near the opening. The sublingual are mucous glands in man.

113. **Serous or Albuminous Glands** are lined by polyhedral granular cells, which readily stain in

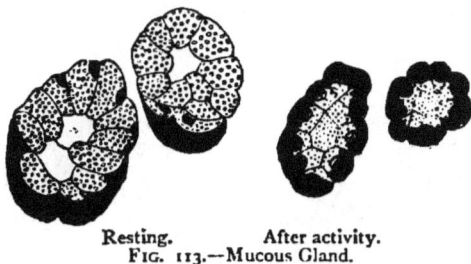

Resting.   After activity.
FIG. 113.—Mucous Gland.

all parts. No cells corresponding to the crescents or demilunes of mucous glands are present. The secretion is thin and clear, consisting chiefly of water with minute quantities of salts and the ferment ptyalin. This limpid saliva contains a proteid called serumalbumin, but no mucin. During the resting or loaded state the cells lining the alveoli of the gland almost close up the lumen and appear densely granular, the fine granules being supposed to be the precursors of the ptyalin ferment. After activity the granules become less numerous (fig. 114). The fine, thin, watery saliva secreted by the parotid contains no mucin.

The parotid gland of man is an entirely albuminous gland. The submaxillary glands are mixed, being mucous in some parts and serous in others. Some of the small buccal glands in the mucous membrane of the mouth are mucous, and some serous or albuminous. The sublingual glands are almost entirely mucous.

The blood-vessels of the salivary glands form a rich capillary network in the fibrous tissue around the tubular alveoli, and the lymphatics begin as lacunar spaces in the same tissue. The nerve fibres, which are derived both from the cerebrospinal nerves and from the sympathetic, appear to pass both to the blood-vessels and to the secretory cells.

114. The Nervous Mechanism of Salivary Secretion.—Experiments on the submaxillary gland of the dog and other animals have taught us the chief facts regarding the effects of nerves on salivary secretion. It has already been shown that this gland receives efferent or motor nerve fibres,(1) from the *chorda tympani*, a branch of the seventh cranial nerve, and (2) from the cervical sympathetic; and we have further shown that, as regards the blood-vessels, the chorda tympani dilates them, *i.e.* is vaso-inhibitory, but that the cervical sympathetic constricts them, *i.e.* is vaso-motor (vasoconstrictor (par. 84). Experiments also prove that fibres of both these nerves act directly on the secreting cells of the gland, and that secretion is not the

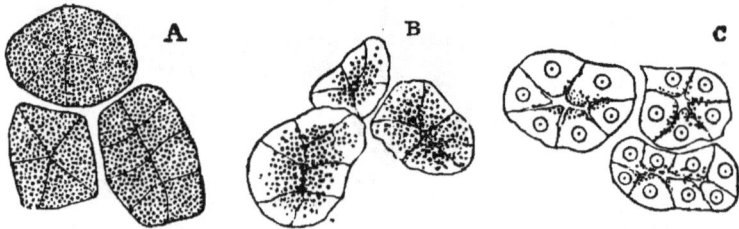

FIG. 114.—Cells from Alveoli of a Serous Gland. A, at rest ; B, after a short period of activity ; C, after a prolonged period of activity. (Langley.)

In A and B the nuclei are obscured by the granules of zymogen.

passive filtration of a fluid through the cell, but is due to the vital activity of the cell, for secretion continues some time after the blood-vessels are ligatured, and drugs are known that stop secretion without affecting the blood supply. Stimulation of the chorda tympani not only produces increased flow of blood through the gland through dilatation of the blood-vessels, but an increased flow of watery saliva. That the increase of secretion is not the mere consequence of increased blood flow, though normally associated with it, is proved thus : (a) The pressure in the duct of the gland is often much higher than the pressure in the arteries, and the greater pressure cannot be entirely due to the less ; (b) Stimulation of the chorda tympani in a decapitated rabbit leads to a secretion of saliva, and cannot then be due to variations of blood pressure ; (c) The drug atropin paralyses secretory fibres, but has no effect on vaso-dilatator action. Stimulation of the chorda tympani in an animal poisoned with atropin causes dilatation of the blood-vessels of the gland but no secretion, and hence secretion cannot be a necessary consequence of increased blood supply. Hence the chorda tympani is both vaso-dilatator and secretory.

Experiments have also shown that stimulation of the sympathetic not only leads to constriction of the blood-vessels, but to a slight flow of thick saliva rich in corpuscles, and to a microscopic change in the structure of the cells indicative of the production by protoplasm of the material of discharge. Hence the cervical sympathetic is both vaso-constrictor and secretory.

The natural and normal secretion of saliva is a reflex act. The afferent or sensory nerves are in particular the nerves of taste (lingual branch of the fifth) and branches of the ninth pair called glosso-pharyngeal. Stimulation of these nerves by the food causes afferent impulses to pass to the nerve-centre of salivary secretion in the medulla, and from this centre efferent impulses are sent out along the vaso-dilatator and secretory fibres of the chorda tympani and along the secretory fibres of the sympathetic, so that the gland becomes flushed with blood and saliva is freely discharged. Afferent impulses that lead to increased salivary secretions may also pass to the centre through the nerves of sight or of smell on sight or odour of savoury food. Mental emotions may also pass from the cerebrum *via* the centre, some of which, as the thought of a pleasing taste, lead to secretion, and some, as fear, inhibit secretion and produce dryness of the mouth.

115. Composition of Saliva. — The quantity of saliva secreted daily averages about a quart. Its rate of flow is greatest when the afferent nerves are stimulated by food and mastication is going on ; at other times only enough is secreted to keep the mouth moist. By inserting a small metallic tube or cannula into the duct, saliva may be obtained separately from each

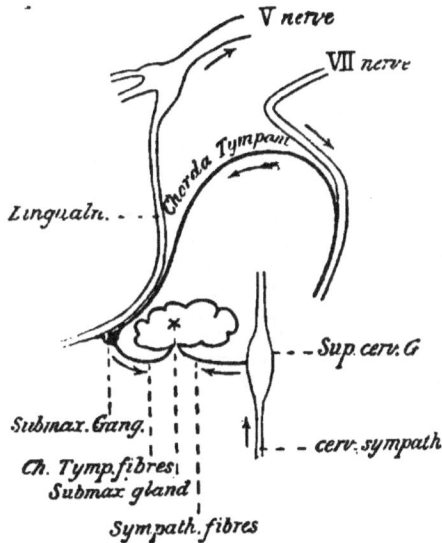

FIG. 115.—Diagram to illustrate the nervous channels to the submaxillary gland of the right side. Chorda tympani fibres pass to it through the submaxillary ganglion. Sympathetic fibres reach it by branches from the superior cervical ganglion which accompany the arteries of the gland. (Waller.)

kind of salivary gland. Parotid saliva is a clear watery fluid free from mucin and containing serum-albumen, inorganic salts, and the important organic ferment ptyalin. Sublingual saliva is a viscid fluid distinctly alkaline in reaction, contains much mucin (which is precipitated by acetic acid), inorganic salts, and numerous escaped leucocytes called 'salivary corpuscles.' The saliva from the 'mixed' submaxillary gland is somewhat viscid, owing to mucin, and contains ptyalin, though in less proportion than in the parotid. Ordinary saliva is a mixture of these secretions with that from the tubular glands of the mouth Its chemical composition is as follows :—

Chemical Composition of Mixed Saliva (Frerichs)

Water	994·10
Solids	
Ptyalin	1·41
Proteids and epithelium . . .	2·13
Inorganic salts in solution . . .	2·29
	5·9
	1,000·0

The chief inorganic salts are sodium and potassium chlorides and calcium carbonate and phosphate, the two latter of which sometimes form concretions on the teeth called ' tartar.' Carbon dioxide is also dissolved in appreciable quantity.

Examined under the microscope, saliva shows protoplasmic globular cells termed salivary corpuscles, epithelial scales from the mucous membrane of the cheeks and tongue, and various micro-organisms. Most of the saliva is re-absorbed in the alimentary canal.

116. **Uses and Properties of Saliva.**—(a) *Mechanical.*—It serves to keep the mouth moist, and thus facilitates the movements of the tongue ; it dissolves sapid substances, and thus renders them capable of exciting the nerves of taste ; and it moistens and softens the food, thus aiding deglutition. These mechanical functions are important.

(b) *Chemical.*—Its chemical or physiological action is the transformation of insoluble starch into dextrin and a soluble form of sugar called maltose. This is due to the ptyalin, which is a hydrolytic ferment that causes starch to take up water and become a soluble sugar, the ferment itself undergoing no change in the process. Starch grains consist of starch enclosed in an envelope of cellulose. The saliva acts but very slowly on raw starch, but if it be boiled the cellulose envelopes are burst. The chemical action of the ferment appears to consist of several stages, several varieties of dextrin being produced, but the final result has been represented by the following equation :--

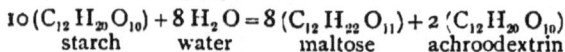

$$10\,(C_{12}H_{20}O_{10}) + 8\,H_2O = 8\,(C_{12}H_{22}O_{11}) + 2\,(C_{12}H_{20}O_{10})$$
$$\text{starch} \qquad \text{water} \qquad \text{maltose} \qquad \text{achroodextrin}$$

The action of ptyalin on starch has led it to be called an amylolytic or starch-digesting ferment, and the process is seen to be one of hydration or taking up of water. The diastase of germinating barley effects a similar transformation of starch into the sugar termed maltose.

Saliva has no action on cane sugar, gum, proteids, or fat. Nearly 50 per cent. of the starch of the food is usually acted on in the mouth if properly masticated. The ferment ptyalin, like other ferments, acts best at a certain medium temperature ($35°$ to $40°$ C.), is delayed by cold and destroyed by heat, and has almost unlimited power if the product of its action (maltose) is not allowed to accumulate.

117 **Deglutition.**—The complicated act of swallowing, by means of which the food is passed from the mouth into the œsophagus, may be divided into three stages. (a) The food, sufficiently ground and moistened with saliva, is gathered

together by the muscles of the cheeks and tongue into a mass or bolus and carried to the back of the mouth ; (b) it is next passed down the slope of the tongue and epiglottis through the fauces or back part of the mouth into the pharynx, the soft palate having been raised so as to shut off the posterior nares or cavity of the nose. The act of raising and drawing back the tongue causes the epiglottis to fall down and close the entrance to the larynx, the vocal cords at the same time coming together and the whole larynx being drawn forward and upward by the thyro-hyoid muscle. (c) The constrictor muscles of the pharynx grasp the bolus and move it on into the œsophagus, where, by a successive wave-like or peristaltic contraction of the circular fibres, with a simultaneous shortening of the longitudinal fibres, the mass is squeezed along to the cardiac orifice of the stomach. The first stage of the deglutition is voluntary, but the complex set of co-ordinated movements that follow are due to reflex action. The afferent nerves of the pharynx and œsophagus are mechanically stimulated by the food, and the impulses are reflected from the deglutition centre in the medulla along the efferent nerves to the muscles of these structures. In traversing the œsophagus the food is carried forward not by gravity but by the special kind of action (peristaltic action) in which the tube contracts from above downwards upon its contents in a ring-like wave.

118. **The Palate and Tonsils.**—The palate forms the roof of the mouth and floor of the nose, and consists of the hard palate in front and the soft palate behind. The hard palate has a bony basis consisting of the maxillary and palate bones, and is covered by a dense structure formed of the periosteum and mucous membrane. The soft palate is the musculo-membranous movable fold attached to the posterior part of the hard palate and having its lower edge free. Hanging from the middle of its lower border is the conical-shaped process called the *uvula*, also composed of muscles and covered with mucous membrane (fig. 106). From the base of the uvula and under surface of the soft palate folds of muscle and mucous membrane, the anterior arches or pillars of the fauces descend on each side to the tongue. Other folds run downward and backwards on each side to the pharynx and form the posterior pillar of the fauces, being projections of the palato-pharyngeal muscles covered by mucous membrane. In the triangular recess between the anterior and posterior pillars of the fauces are lodged two oval bodies called the tonsils. A tonsil is a glandular mass of lymphoid tissue. The free surface is covered with stratified epithelium which presents from twelve to fifteen orifices that lead into small recesses or crypts from

which follicles branch out into the substance of the gland. In these follicles or nodules active multiplication of lymph-cells occurs, and these pass through the epithelium into the crypt, mingling with the saliva as salivary corpuscles (fig. 117).

119. The Pharynx and Œsophagus.—The pharynx is the dilated upper part of the alimentary tube, extending from the base of the skull to the œsophagus (fig. 106). It is composed of an outer fibrous membrane, striated muscles arranged in two layers, an outer layer of constrictors and an inner layer of elevators, and an internal lining of mucous membrane. The upper part of the mucous membrane is coated with ciliated epithelium continuous with that of the nostrils and Eustachian tubes, but below the level of the soft palate the epithelium is stratified like that of the gullet. Numerous mucous glands are found opening on its surface. The œsophagus or gullet is the tube extending from the pharynx to the stomach. This tube has an outer fibrous covering, a muscular coat consisting of an outer layer of longitudinal fibres and an inner layer of transverse fibres running circularly, a submucous layer of connective or areolar tissue containing some mucous glands, and an internal pale mucous membrane.

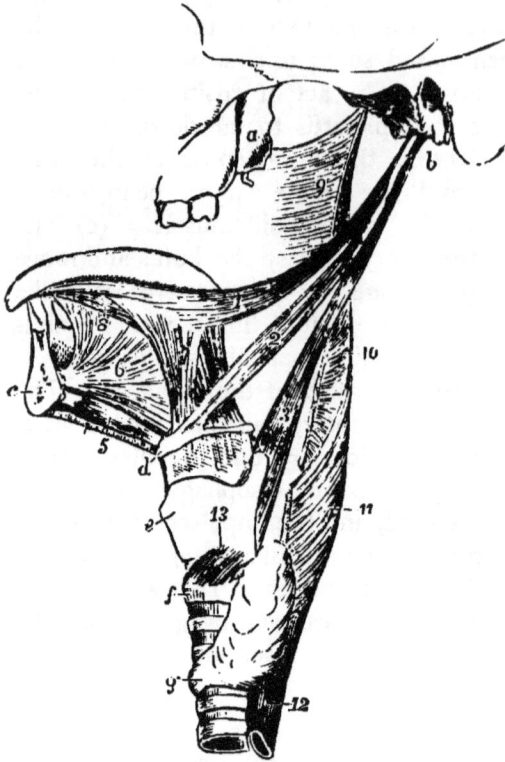

FIG. 116.—Muscles of the Tongue, Pharynx, &c., of the Left Side. (Allen Thomson.) ⅔

a, external pterygoid plate ; *b*, styloid process ; *c*, section of lower jaw ; *d*, hyoid bone ; *e*, thyroid cartilage ; *f*, cricoid cartilage ; between *d* and *e*, the thyro-hyoid membrane ; *g*, isthmus of thyroid body ; 1, stylo-glossus muscle ; 2, stylo-hyoid ; 3, stylo-pharyngeus ; 4, cut edge of mylo-hyoid ; 5, genio-hyoid ; 6, genio-glossus ; 7, hyo-glossus ; 8, lingualis inferior ; 9, part of superior constrictor of pharynx ; 10, back part of middle constrictor ; 11, inferior constrictor ; 12, upper part of œsophagus ; 13, crico-thyroid muscle.

A narrow layer of plain longitudinal muscular fibres, the *muscularis mucosæ*, is found between the mucous membrane and areolar tissue (fig. 118).

120. The Stomach and its Structure.—The bolus of food passes from the œsophagus through the left or cardiac opening

into the stomach, where a juice secreted by the gastric glands
acts on certain constituents of the food during the churning
movements of its contents effected by the muscular walls, the
partially digested food being ᷱat length expelled through the
right or pyloric orifice into the duodenum. The stomach
when moderately distended is in the adult 10 to 12 inches in
length and 4 to 5 inches broad in the widest part, and it is
capable of holding about six pints.

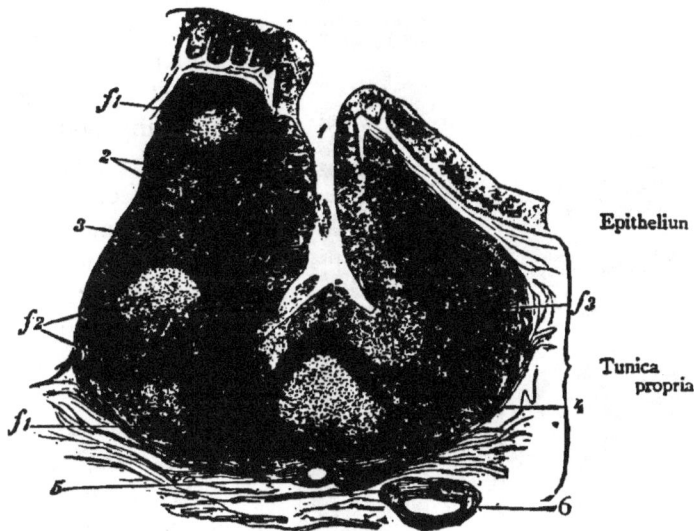

FIG. 117.—Vertical Section of a Crypt of a Human Tonsil × 20. (Landois and Stirling.)

1, crypt; epithelium infiltrated with leucocytes below and on the left, but free on the
right; 3, adenoid tissue with sections f_1, f_2, f_3, of follicles of it; 4, fibrous sheath;
5, section of gland duct; 6, blood-vessel.

The walls of the stomach consist of four coats, which are
from without inwards.

(1) The *serous coat*, derived from the peritoneum.

(2) The *muscular coat*, composed of three layers of non-
striped muscular fibres, an outer longitudinal layer continuous
with that of the œsophagus, a circular layer most abundant at
the middle and at the pyloric portion, and an oblique layer
found at the cardiac end continuous with the circular fibres of
the œsophagus.

N

(3) The *submucous coat*, of loose connective tissue with the larger blood-vessels, lymphatics, and nerves that have passed inwards through the muscular coat.

(4) The *mucous coat*, containing the tubular secreting glands and a network of capillary blood-vessels between the glands.

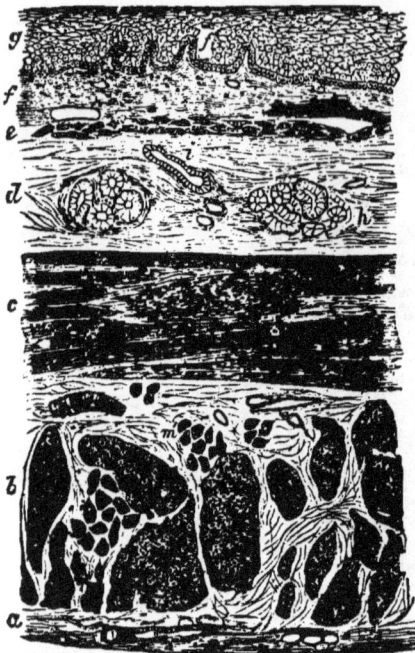

FIG. 118.—Section of the Human Œsophagus. (From a sketch by V. Horsley.)

The section is transverse, and from near the middle of the gullet. *a*, fibrous covering; *b*, divided fibres of the longitudinal muscular coat; *c*, transverse muscular fibres; *d*, submucous or areolar layer; *e*, muscularis mucosæ; *f*, papillæ of mucous membrane; *g*, laminated epithelial lining; *h*, mucous gland; *i*, gland-duct; *m'*, striated muscular fibres in section.

The chief uses of the muscular fibres in the walls of the stomach appear to be the adaptation of its walls to the quantity of food in the organ, the closure of the pyloric orifice until the food is acted upon by the gastric juice, and the effecting of peristaltic movements which mix the contents with the gastric juice until the unabsorbed portions are ready to be driven through the pylorus.

The mucous membrane of the stomach consists of an epithelial layer and a corium of fine connective tissue with a basement-membrane between, the corium of connective tissue being separated from the submucous tissue by a thin layer of unstriped muscular fibres, the *muscularis mucosœ*

The mucous membrane is smooth and soft, of a pale pink colour during life, and in the empty state thrown into folds or rugæ. When examined with a hand-lens the internal or free surface of the mucous membrane of the stomach presents a peculiar honeycombed appearance owing to small shallow pits or depressions $\frac{1}{100}$ to $\frac{1}{200}$ of an inch in diameter, separated by slightly elevated ridges. These pits are the mouths of the ducts of the gastric glands, and the thickness of the mucous membrane of the stomach is due to the fact that it is closely studded with tubular glands set vertically side by side (fig. 119), and bound together by a small quantity of connective tissue which contains the blood-vessels and lymphatics. At the bottom of the pits or ducts are seen minute orifices, which are the openings of the gastric follicles or secreting tubules that branch from the duct. Examined with the microscope the ducts leading into the walls of the stomach are in all cases seen to be lined by a columnar epithelium of mucous secreting cells similar to that which forms the free inner surface of the mucous membrane, but the epithelium of the glands proper or secreting follicles differs from

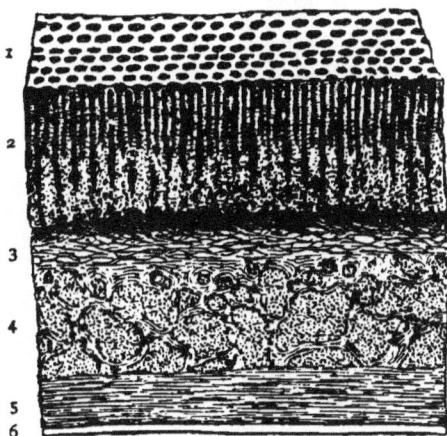

FIG. 119.—A Section through the Walls of the Stomach. Magnified 15 diameters.

1, surface of the mucous membrane, showing the openings of the gastric glands ; 2, mucous membrane, composed almost entirely of glands ; 3, sub-mucous or areolar tissue ; 4, transverse muscular fibres ; 5, longitudinal muscular fibres ; 6, peritoneal coat.

this, the difference being of a different kind in the glands of the cardiac and pyloric ends.

We have therefore two chief varieties of gastric glands, (*a*) cardiac or peptic glands and (*b*) pyloric glands.

The *cardiac* or *peptic glands* are tubular glands with a short duct from which branch two or three long tubules, though some are simple or unbranched. The epithelium of the tubules of the cardiac glands differs from the columnar epithelium of the duct, being composed of two different kinds of cells—cubical granular cells, called *central* or *chief* cells, forming an almost continuous layer in the tubule; and ovoid opaque granular cells with clear nucleus, called *parietal* or *oxyntic* cells and found scattered among the chief cells down to the blind end of the tubule. The chief cells are believed to be the source of the

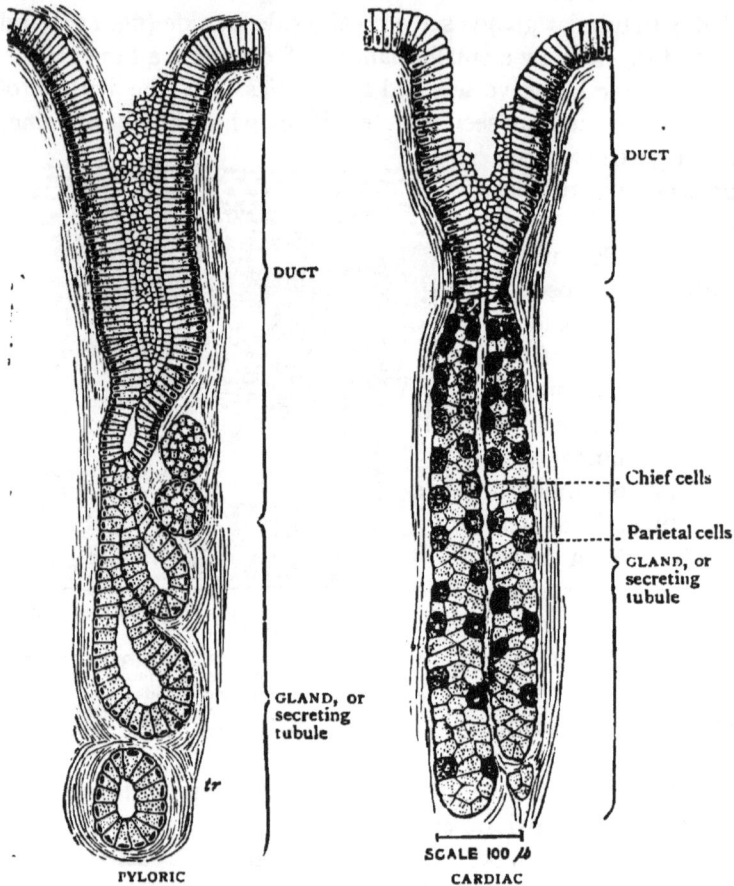

FIG. 120.—Gastric Glands opening by Ducts lined with Columnar Epithelial Cells on the free surface of the Stomach. The Cardiac Gland shows two kinds of gland cells, and the Pyloric Gland one kind.

tr, a deep portion of a gland tubule cut transversely.

pepsin in the gastric juice, and the oxyntic or parietal cells the source of the hydrochloric acid. (Gk. *oxys*, acid, sharp.)

The *pyloric glands* occur only in the region of the pylorus, have longer ducts with well-defined lumen from which branch short wavy secreting tubules with narrow necks. The duct is lined with columnar mucous epithelium cells as in the cardiac gland, but the cells of the secretory part are all of one kind, corresponding to the chief cells of the cardiac glands. The pyloric glands secrete pepsin but no acid. As in the case of the salivary glands the appearance of the cells in the gastric glands differs according to their physiological activity. During rest the central cells in both cardiac and pyloric glands are loaded with granules which are discharged into the lumen during activity. After activity each cell is seen to have a clear outer zone, the size of which increases with prolongation of activity. The ovoid parietal cells of the cardiac gland show shrinkage only during digestion. The granules of the central cells furnish another instance of a zymogen or ferment precursor, in this case *pepsinogen.* During discharge into the lumen of gland pepsinogen becomes transformed into pepsin.

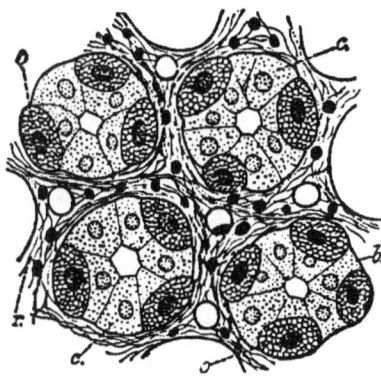

FIG. 121.—Section of the Gastric Mucous Membrane taken across the direction of the Glands (cardiac part).

b, basement membrane; *c,* central cells; *o,* oxyntic cells; *r,* retiform tissue (with sections of blood-capillaries) between the glands.

The blood-vessels of the stomach are very numerous, and pass to the organ along its curvatures. The arteries, after giving off branches to form a capillary network in the mucous coat, run through the submucous coat to form a longitudinal capillary network between the tubular glands, terminating at the surface in larger horizontal capillaries around the orifices of the ducts. From this superficial horizontal network the veins run downwards.

Lymphatics arise in the mucous membrane and pass into larger vessels in the submucous coat. A gangliated plexus of nerve-fibres with many ganglia is found in the submucous and muscular coats. These nerves of the stomach are derived from the pneumogastric and sympathetic system.

121. Composition and Properties of Gastric Juice.—Gastric juice is a clear colourless acid fluid with a specific gravity of 1·003, and containing only ⅓ per cent. of solids in solution. Its chief constituents are two ferments, pepsin and rennet ; free hydrochloric acid ; mineral salts, and 99 per cent. of water. That there is free hydrochloric acid in gastric juice has been proved by estimating the total quantity of chlorine and assigning to the metals present the due proportion. A quantity of chlorine remains over, which must therefore be combined with hydrogen to form free hydrochloric acid. Moreover, chemical tests show the presence of free hydrochloric acid.

Analysis of gastric juice gives the following results in parts per 1,000 :--

	Human.	Dog
Water	994·4	973·06
Organic substances, chiefly pepsin . .	3·19	17·13
Free hydrochloric	·20	3·05
Calcium, sodium, and potassium chlorides	2·08	4·36
Calcium, magnesium, and iron phosphates	·12	2·00

The total quantity of gastric juice secreted daily has been estimated to range from 10 to 20 pints and to be nearly one-tenth weight of the body. It is mostly reabsorbed in the small intestine with the food.

Gastric juice has been obtained through an accidental fistula or permanent opening from the outside into the stomach. The mucous membrane was excited to action by introducing hard substances, and the liquid secreted was then drawn off for examination. It can also be obtained from the stomach of a recently killed animal by treatment of the mucous membrane with glycerin and hydrochloric acid. Pepsin can also be prepared from the gastric mucous membrane, though probably not entirely isolated from other bodies. It is a proteid-like substance, only acting in an acid medium. The functions of gastric juice have been ascertained by observing the changes undergone by the food in the stomach and by experimenting with the gastric juice obtained from a fistula. It has thus been shown that it acts only on proteids, which it dissolves and converts into peptones. Peptones are distinguished from other proteids by the property of diffusing readily through animal membranes. They are not coagulated by boiling. But no peptone is found in the blood of the portal vein, as during its passage through the vascular walls of the stomach it is converted into serum-albumen.

If a piece of lean meat or some shreds of fibrin be placed in gastric juice and the mixture kept at a temperature of about 40° C. for some time, artificial digestion will go on. In about an hour the fibrin will be in great part dissolved, and by appropriate tests three other forms of proteid will be found. These are acid-albumen or syntonin, albumoses, and peptone. If a solution of pepsin alone were added to the fibrin, no change would appear. If a weak solution (·2 per cent.) of hydrochloric acid alone were used, the fibrin would become swollen and transparent merely. Putting the solution of pepsin and the weak solution of hydrochloric acid together would lead to the fibrin being dissolved. Boiling the solution of pepsin, however, would render it inactive. It thus appears that both pepsin and free hydrochloric acid are necessary for the conversion of the proteid of food into peptone, and that the change into peptone takes place in stages, the action being of the nature of hydration, or taking up the elements of water. We may put the matter thus. Proteid food stuffs acted on by the pepsin and hydrochloric acid of gastric juice are converted in the end into a hydrated kind of proteid called peptone ; intermediate substances termed proteoses being formed in the process. The need for this change is obvious when we remember that food proteids are not diffusible through animal membranes, while peptone is readily diffusible. Among the chief proteids are the two important classes 'albumins' and 'globulins,' their intermediate substances or proteoses being termed 'albumoses' and 'globuloses' respectively. Taking albumin as a type of proteid food, the action of gastric juice may be thus illustrated :

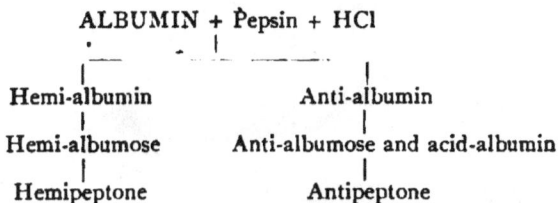

ALBUMIN + Pepsin + HCl

Hemi-albumin	Anti-albumin
Hemi-albumose	Anti-albumose and acid-albumin
Hemipeptone	Antipeptone

The hemipeptone differs from the antipeptone in the way in which it is affected by the ferment trypsin of pancreatic juice : hemipeptone yields leucine and tyrosine ; antipeptone is unaffected by the trypsine.

All the insoluble coagulable proteids of the food are thus acted on and finally converted into diffusible peptone. In the case of milk, a curd or clot of casein is first produced by the action of a ferment called rennet or rennin (which is distinct from fibrin), and this is then acted upon by the pepsin. The collagen of connective tissue and the gelatin into which collagen is changed by boiling are also dissolved by gastric juice and changed into peptone, but mucin and the horny tissues are unaffected. Gastric juice has no action on carbohydrates and fats, though the connective tissue and the proteid cell-walls of the fat cells are dissolved so that the fat is set free to mingle with the other substance in the stomach. Certain saline constituents of the food are dissolved in the stomach, and it is said to change cane-sugar into dextrose. Lastly, gastric juice is antiseptic, possessing the property of preventing and checking putrefaction.

Why the stomach does not digest itself during life by its own gastric juice is not very clear. An animal killed during digestion and examined some hours later is found to have its stomach partially digested. In most deaths secretion ceases before circulation stops. The leg of a living frog and the ear of a living rabbit have been passed through a gastric fistula into the stomach of a dog and digested, but it is probable that death of the tissues took place before digestion. In some way, however, the living tissue of the stomach resists digestion during life.

The flow of gastric juice is not continuous, for during fasting the gastric glands are inactive. The natural stimulus is the advent of food into the stomach. This causes a copious outflow, and is accompanied by an increased flow of blood, which causes the pale mucous membrane to become red. Even the introduction of indigestible substances, as pebbles or mechanical stimulation with a feather, induces a small amount of secretion. Dilute alkalis have powerful stimulating effects. The stomach is connected with the central nervous system by nerve-fibres from the two vagi. It also receives fibres from the so-called sympathetic system, by branches from the solar plexus of the splanchnic nerves. But there is no proof that any of these nerve-fibres regulate secretion, for it goes on when all nervous connections are severed. There is no nerve passing to the stomach whose stimulation excites a secretion of gastric juice as the chorda tympani does in the submaxillary gland. Possibly there are local centres in the walls of the stomach. Emotional states are known to interfere with gastric secretion, and this shows that the central nervous system must be connected in some way, direct or indirect, with the gastric glands.

The acid mixture of finely divided food and gastric juice is called *chyme.*

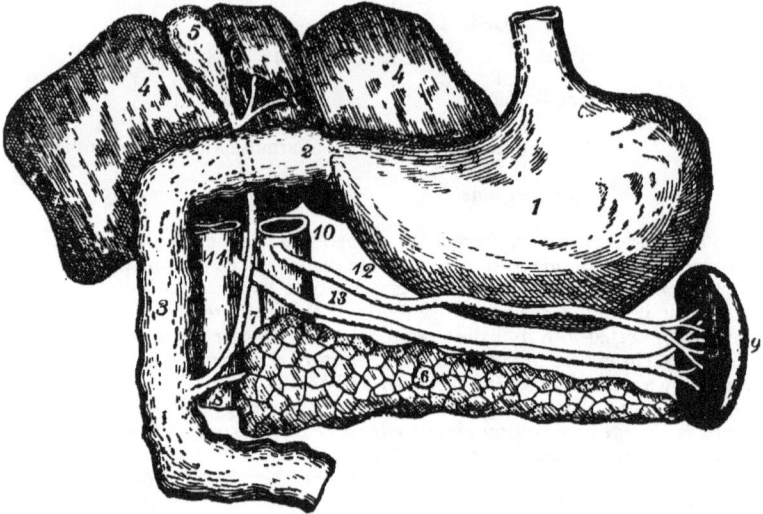

Fig. 122.—The Stomach, Duodenum, Liver, Spleen, and Pancreas.

1, stomach; 2, pylorus; 3, duodenum; 4, liver (under surface); 5, gall-bladder; 6, pancreas; 7, bile-duct; 8, pancreatic duct; 9, spleen; 10, aorta; 11, portal vein; 12, splenic artery; 13, splenic vein.

122. **Structure of the Pancreas.**—The pancreas is a long narrow compound tubular gland, of a light colour and soft texture, lying transversely across the abdomen opposite the first lumbar vertebra. Its broad end is embraced by the duodenum, and the narrow end is in contact with the spleen. Its structure closely resembles that of the salivary glands, being, like them, divided into lobes, lobules, ducts, and acini. The gland has a thin connective-tissue capsule which sends divisions or septa between the lobules, and these septa carry blood-vessels and nerves.[1] The main duct commences in the narrow end by the junction of the ducts from the lobules of this part, and,

¹ The secretory nerves of the pancreas are derived from the vagus.

running along the whole length of the gland, receives nearly at right angles contributory ducts from the other lobules. It passes into the duodenum very obliquely, terminating in this portion of the intestine by an orifice common to it and the bile-duct. A small accessory duct, the duct of Santorini, opens into the duodenum independently in many instances.

Each lobule, as in the salivary glands, contains one of the ultimate branches of the chief duct, terminating in a number of blind alveoli or acini, with narrow lumen. The acini are more tubular in character and more numerous relative to the ducts than in the salivary glands, and the connective tissue of the gland is somewhat looser. The small ducts in a lobule that lead to the secreting tubes or alveoli have a distinct lumen lined with a single layer of flattened cells, resting on a membrana propria ; but the convoluted tubular acini are lined with cubical granular cells, the lumen between the cell-layers being scarcely visible. The granular contents

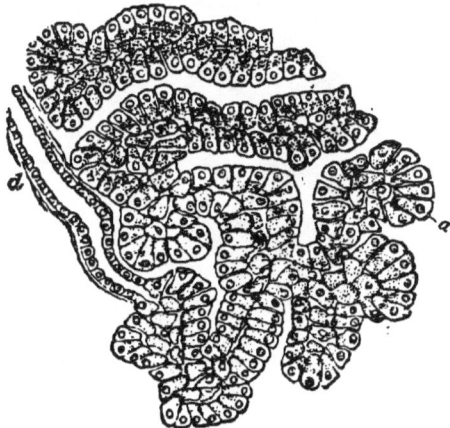

FIG. 123. – Section of the Pancreas of the Dog. (Klein.)

d, end portion of a duct leading to an alveolus *a*, lined with cells.

of the cells are considered to be products of the activity of the cell-protoplasm (zymogen) which is about to be transformed into ferment in active secretion. The increase in the size of the outer clear zone of the pancreatic cells, as the granular contents are discharged transformed into ferment, has actually been observed during life in the pancreas of the rabbit, which is scattered between the layer of the mesentery, so that individual lobules may there be microscopically examined.

123. Composition and Properties of Pancreatic Juice. – Pancreatic juice is a clear, viscid alkaline fluid, having a specific gravity of 1·030, coagulating with heat owing to the presence of albumen. It contains when obtained from a newly opened duct—

Water about 90 per cent., solids about 10 per cent. The solids include proteid substances, inorganic salts, especially sodium carbonate, and four ferments, viz. :—*trypsin*, a proteolytic or proteid-digesting ferment ; *amylopsin*, an amylolytic or starch-digesting ferment ; *steapsin*, a fat-splitting ferment ; and *rennin*, a milk-curdling ferment. If obtained from a permanent fistula the secretion in time becomes more watery. Secretion is normally excited by the chyme, and is probably a reflex act, but the nervous channels are not known. The presence of the four ferments makes the pancreatic juice the most important digestive juice of the body.

Action on Proteids.—Proteids are converted by pancreatic juice into albumoses and peptones as in the case of gastric juice. But trypsin acts only in an alkaline medium, and pepsin only in an acid medium. Hence the first stage of the change into peptone is alkali-albumin and not acid-

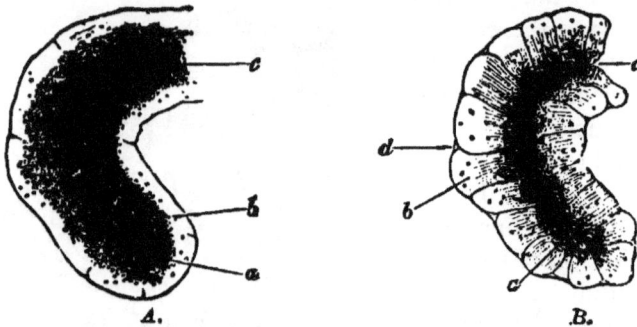

FIG. 124.—Part of an Alveolus of the Rabbit's Pancreas. *A*, at rest ; *B*, after active secretion. (From Foster, after Kühne and Lea.)

a, the inner granular zone, which in *A* is larger and more closely studded with fine granules than in *B*, in which the granules are fewer and coarser ; *b*, the outer transparent zone, small in *A*, larger in *B*, and in the latter marked with faint striæ ; *c*, the lumen, very obvious in *B*, but indistinct in *A* ; *d*, an indentation at the junction of two cells, only seen in *B*.

albumin. Trypsin also acts more powerfully than pepsin, the proteid not being first swollen and then dissolved as in the case of gastric juice, but gradually eaten away and dissolved at the edges. Further, trypsin carries the process of digestion further, decomposing the hemi-peptone into the simpler products leucine and tyrosine.

Action on Carbohydrates.—Carbohydrates as starch are converted by the amylolytic ferment of pancreatic juice into the same products, dextrin and maltose, as those produced by the ptyalin of the saliva. But the action is much more rapid and more powerful. Glycogen, or animal starch, is affected in the same way.

Action on Fats.—Fresh pancreatic juice contains a ferment that causes fats to take up water and split up into glycerin and a free fatty acid. The fatty acid then combines with an alkali present to form what is called a soap, the whole chemical change being called saponification. The presence of a soap in the contents of the duodenum assists in breaking up the rest of the fat into fine particles, as it forms a thin layer outside each liquid fat globule, and thus prevents them running together again. This mechanical division of oils and fats into finely suspended particles is called emulsification. Pancreatic juice, therefore, emulsifies and saponifies fats.

Action on Milk.—Pancreatic juice also contains a ferment similar to rennet, which has the power of forming a curd in milk.

124. **Bile.**—The bile is the fluid secreted by the cells of the liver from the blood brought to it by the portal vein. As this gland has several other functions, its minute structure will be considered separately, and we shall now treat only of the composition and properties of the bile. This fluid passes into the duodenum by the hepatic duct during digestion, but is carried by the cystic duct to be stored in the gall-bladder when digestion is not proceeding, as it is then easier for the bile to pass in that direction, for the orifice into the duodenum is narrower than the duct and requires some pressure to open it. This required pressure is only great enough when the stimulus of the acid chyme, passing into the duodenum, sets up contraction in the gall-bladder and the gall-ducts.

Bile is of a brown or greenish colour, has a bitter taste, is slightly alkaline and slimy from the presence of mucin. Its specific gravity varies from 1·020 to 1·040, and its average chemical composition is as follows :—

Chemical Composition of Bile.

100 parts contain :—

Water	86
Bile salts	9
Fat, lecithin, and cholesterin	1
Mucus and colouring-matter	3
Inorganic salts	1
	100

The bile salts are glycocholate and taurocholate of sodium, the former being by far the more abundant. The colour of the bile is due to the two pigments, *bilirubin* (yellow) and *biliverdin* (green), the former of which is the more abundant. When the flow of bile is obstructed the colouring matter is absorbed unchanged, and produces the yellow colour of the tissues seen in the disease *jaundice*. These bile-pigments are made by the liver from the hæmoglobin of the red corpuscles destroyed in the liver, the bile-pigment differing from blood-pigment in not containing iron. Powerful reducing agents remove some of the oxygen from the bile-pigments, and produce hydrobilirubin. This is identical with *urobilin*, which is found in urine, and with *stercobilin*, found in the fæces. Hence the bile appears to be the agent by which the effete colouring matter of the blood is excreted.

Cholesterin belongs to the class of organic bodies called alcohols, and is found in bile and the brain. It crystallises in rhombic plates, and though containing no nitrogen is probably produced from a proteid. *Lecithin* is the compound of bile containing phosphorus.

125. **Functions of Bile.**—Bile is a fluid secreted mainly for aid in digestion, though it is excreted in part also. Most of the bile secreted by an adult (about one quart daily) is reabsorbed in the intestine after it has done its work. A small part is excreted in the fæces, especially in the colouring matter, and a small part of that reabsorbed portion is removed as urobilin by the kidneys. As a digestive agent, the bile appears to act in concert with the pancreatic juice.

(1) Bile aids in emulsifying the fats so as to render them ready for absorption by the lacteals. Bile and oil shaken together form a fine emulsion, and the pancreatic juice still further assists the process by splitting up some of the fatty particles into glycerine and a fatty acid, the fatty acid uniting with sodium to form a soap. A soap favours emulsion and diffusion through an animal membrane. Experiment and observation point to the fact that both bile and pancreatic juice working together lead to the absorption of fat, as where either is excluded the absorption of fatty particles is incomplete.

(2) Bile precipitates the albumoses and peptones of gastric digestion, stopping the action of the pepsin in the acid chyme in the duodenum. It thus prepares chyme for the action of the pancreatic juice by neutralising the gastric juice.

(3) Bile excites contraction of the muscular walls of the intestine, thus stimulating the intestines to propel forwards their contents.

(4) Bile also acts as an antiseptic.

126. **The Peritoneum and Mesentery.**—The peritoneum is the serous membrane of the abdomen, and like other serous membranes forms a closed sac (except in the female), one layer of which (the *parietal* layer) lines the boundary walls of the cavity, and the other layer of which (the *visceral* layer) is reflected more or less over the contained organs. The membrane is very thin, and its free surface is smooth and moist, being covered by a layer of flattened endothelial cells ; its attached surface is rougher, being connected to the abdominal walls and viscera by areolar tissue. The peritoneum consists of a main cavity or *great sac*, connected by a narrow neck with a pouch or small sac. A general idea of the peritoneum and its com· plex reflections may be obtained by supposing all the viscera of the abdomen in position and a closed sac of extreme thinness to be placed over them. Wherever there is a cleft between viscera, a process of the peritoneum derived from the part of the sac in contact with the viscera is tucked in between them, ' so as to cover the adjacent surfaces of the viscera and separate them from each other, and at the same time, by becoming adherent to the viscera, form an investment for them.' The reflections of the peritoneum from the abdominal walls thus serve to invest and keep in position the various viscera, and the processes or folds thus formed are of various kinds. Some of these folds, as those passing to the liver and bladder, are termed *ligaments*. Others connected with the stomach are called *omenta*, the great omentum consisting of four folds of peritoneum stretching from the greater curvature of the stomach to the transverse colon, from which it hangs down in front of the small intestines as a great protecting flap or apron. The great omentum always contains some adipose tissue which in fat persons accumulates largely. The *mesenteries* are the folds connecting the intestines to the posterior abdominal walls. The mesentery proper is the broad double fold of peritoneum retaining the small intestines in position and connecting them with the posterior abdominal wall. Between its layers, which are in apposition where it does not invest the intestine, are blood-vessels and nerves, lacteal vessels and the mesenteric lymphatic glands (figs. 131, 132).

127. **The Structure of the Intestines.**—The intestinal canal from the pylorus to the rectum is divided into small intestine and large intestine, the two being continuous at the ileocæcal valve. The small intestine, whose length is about 20 feet in the adult, is divided into *duodenum* (10 inches), *jejunum* (8 feet), and *ileum* (11 feet). The small intestine is supported by the double fold of peritoneum termed the mesentery.

Its walls are constructed of the four coats already referred to, the serous from the peritoneum, the muscular with its internal circular and external longitudinal layer of fibres, the submucous or connective tissue in which the blood-vessels and lym-

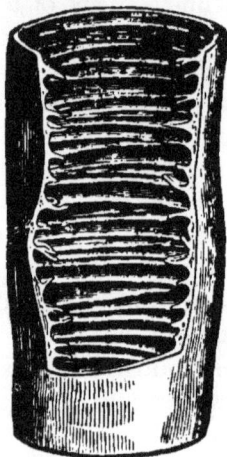

FIG. 125.—A Portion of the Small Intestine laid open to show the Folds of the Mucous Membrane (Valvulæ Conniventes).

FIG. 126.—A Small Portion of the Mucous Membrane of the Small Intestine. Magnified 12 diameters.

a, Peyer's glands, surrounded by tubular glands; *b*, villi; *c*, openings of the tubular glands.

phatics ramify, and the mucous membrane lining the interior. A gangliated plexus of non-medullated nerve fibres lies between the two muscular coats, and another in the submucous coat.[1] In regard to the function of digestion, we require to note carefully the following structures in the mucous membrane: (a) *valvulæ conniventes* ; (b) three varieties of *glands* ; (c) *villi*. The valvulæ conniventes are crescentic folds of mucous membrane arranged transversely in the small intestine from a little below the pylorus to the middle of the ileum. They do not disappear when the tube is distended, but form vertical ridges

[1] The former is known as Auerbach's nerve plexus; the latter as Meissner's plexus.

on the interior surface. They serve to increase the surface for secretion and absorption, to check the rate of transit of the liquid products along the canal, and to assist in mingling the particles of food.

The glands of the intestine are of three kinds. 1. The glands or crypts of Lieberkühn are simple tubular glands lined by columnar epithelium distributed all over the surface of the small and large intestines, though increasing in size in the large intestine as they approach the anus. In the small intestine the crypts of Lieberkühn open between the villi. (2) Brunner's glands are convoluted tubules similar in structure to the pyloric glands of the stomach, found only in the duodenum, and their epithelium undergoes similar changes during secretion. Their blind ends are found in the submucous coat, and the duct passing through the muscularis mucosæ opens on the surface of the mucous membrane. (3) Peyer's glands are found in the lymphoid tissue between the tubular glands, and are closed sacs or nodules of the size of millet-seeds, and composed of cells and blood-vessels without any ducts. When occurring singly, they are known as *solitary glands* ; when in groups, they are known as Peyer's patches. These patches occur chiefly in the ileum. Villi are minute thread-like processes found exclusively on the inner surface of the small intestine. They are concerned in the absorption of food, and will presently be described in detail. The large intestine has the usual four coats, the longitudinal fibres of the muscular coat being gathered into longitudinal bands that produce a puckering of the wall. Its mucous membrane is lined by columnar epithelium containing numerous mucous secreting cells, and there are numerous simple tubular glands along nearly the whole length.

128. **Succus Entericus, or Intestinal Juice.**—Intestinal juice is the secretion of the glands of Lieberkühn (and in the duodenum of Brunner's glands). It is a thin alkaline fluid having little or no action on proteids, fats, and starch, but converting *cane sugar* (saccharose) into *invert sugar* in virtue of a ferment called *invertin.* Cane sugar is caused to take up

water, forming a mixture of lævulose and dextrose, termed invert sugar.

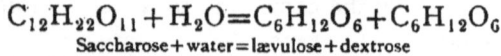

$$C_{12}H_{22}O_{11} + H_2O = C_6H_{12}O_6 + C_6H_{12}O_6$$

Saccharose + water = lævulose + dextrose

129. **Digestive Changes in the Small Intestine.**—The chyme as it leaves the stomach consists of the starchy matters that are in process of conversion into dextrin and sugar and the sugar thus converted and dissolved in the fluids but remaining unabsorbed ; the proteid material con-

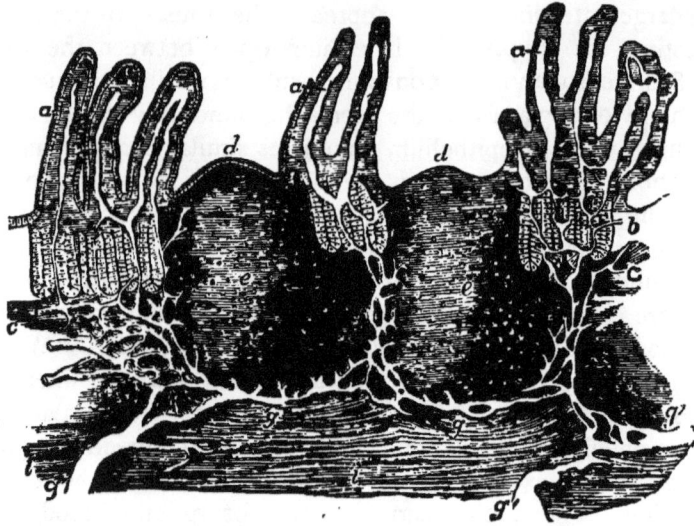

127.—Vertical Section of a Portion of a Patch of Peyer's Glands wf h the Lacteal Vessels injected. (Frey.) 32 diameters.

The specimen is from the lower part of the ileum *a,* villi, with their lacteals left white ; *b,* some of the tubular glands ; *c,* the muscular layer of the mucous membrane ; *d,* cupola or projecting part of the nodule ; *e,* central part ; *f,* the reticulated lacteal vessels occupying the lymphoid tissue between the nodules, joined above by the lacteals from the villi and mucous surface, and passing below into *e,* the sinus-like lacteals under the nodules, which again pass into the large efferent lacteals, *g'* ; *i,* part of the muscular coat.

verted into peptones and partially converted ; fatty matter broken up and melted ; gastric juice, saliva, and fluid that has been taken as drink ; and indigestible parts of food.   In the duodenum the chyme is subjected to the action of the pancreatic juice, the bile, and the succus entericus, and this action continues as the food is driven along the digestive tube by the peristaltic contractions of its walls.   The starchy or amyloid portions of the food, the conversion of which into sugar had been arrested in the stomach by the acid gastric juice, are acted on vigorously by the amylolytic ferment of the alkaline pancreatic juice as soon as the acidity of the chyme is neutralised (the contents of the duodenum become alkaline a few inches from the pylorus), and the sugar in the form of maltose or glucose is dissolved in the intestinal fluids and absorbed by the intestinal blood-vessels. The proteid or albuminous substances which have been only partly dissolved and absorbed in the stomach are brought in the intestines under the influence of the pancreatic ferment trypsin. The action of the pepsin of gastric juice, which is only active in an acid medium, appears to be destroyed by trypsin, for pepsin is precipitated along with the gastric peptones and albumoses on meeting the pancreatic juice and bile in the duodenum. Trypsin then continues the digestion of the proteids by converting them and the allied gelatinous matters into diffusible peptones, which are absorbed as they pass along the intestines by the blood-vessels and lymphatics. Possibly some of the peptone is split up into the crystalline bodies leucine and tyrosine. If so, these substances leave the body as urea without having been built up into tissue material, and only contributing to the energy of the body heat during the chemical changes that convert them into urea. This apparently wasteful expenditure of proteid food has been called ' luxus-consumption,' or wasteful consumption. Another important digestive change in the small intestine is the alteration of fat in such a way as to render it ready for absorption. As already described, this is brought about by the combined action of the pancreatic juice and the bile, and consists partly in producing a fine subdivision of the particles of fat called an emulsion and partly of a chemical decomposition by which a soap is formed. Most of the fat thus changed is absorbed by the lacteals of the villi, as will presently be

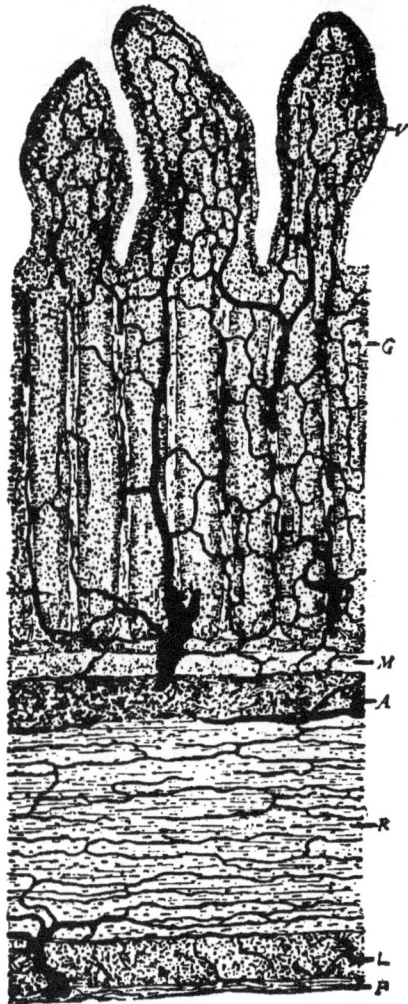

FIG. 128.—Small Intestine, Vertical Transverse Section, with the Blood-vessels injected. (Heitzmann.)

V, a villus ; G, glands of Lieberkühn ; M, muscularis mucosæ ; A, areolar coat ; R, ring-muscle (circular layer of muscular coat); L, longitudinal layer of muscular coat ; p, peritoneal coat.

O

explained ; but a small part of that converted into soap is probably absorbed by the blood-vessels of the intestine.

The cane sugar that reaches the small intestine is mostly converted into invert sugar by the succus, entericus before it is absorbed.[1] The liquids taken as drink, as well as the digestive juices poured into the intestine and containing the dissolved nutrient materials, are also in great part absorbed. The unabsorbed chyme that passes into the large intestine is still half-liquid, but of a light yellow colour and possessed of a distinct fæcal odour.

130. **Digestive Changes in the Large Intestine.**—From the absence of villi in the large intestine we may conclude that little absorption of fatty matter occurs in the large intestine, but other digestive changes may continue. It is clear that absorption of the liquid parts continues, for the contents become firmer and more solid as they approach the rectum. Moreover, nutrient material injected into the large intestine is found to nourish the body when food cannot be taken into the stomach.

Other changes of a putrefactive nature are produced by minute living beings called micro-organisms (bacteria). These bacteria are unicellular beings that multiply by division and produce changes in the media in which they live. Numerous bacteria are found in the intestines, and in the large intestine especially set up putrefactive processes. These changes are probably checked by the antiseptic action of the bile. Some of these changes may be beneficial, but others lead to the production of such gases as carbon dioxide, sulphuretted hydrogen, and marsh gas in the intestines.

131. **The Fæces.**—The fæces expelled from the body consist of the indigestible residue of the food, substances taken in too large quantity for digestion, certain excrementitious matter from the digestive juices not absorbed, substances produced in the intestinal canal, and inorganic salts. Among these bodies are :—

(a) Cellulose, woody fibre, uncooked starch, horny matter, &c.

(b) Mucin, cholesterin, stereobilin, and other pigments (derived from the bile pigments as they disappear when the bile is cut off), excretin, skatol, C_8H_9N (a nitrogenous body produced by bacteria and the chief cause of the fæcal odour), fatty acids.

(c) Lime and magnesia soaps, earthy phosphates, &c.

The fæces on a mixed diet are about one-eighth the weight of the food.

132. **Absorption of Food.**—By the secretory activity of the cells in the various glands of the alimentary canal we have seen that ferments are prepared that so act on the proteids, carbohydrates, and fats that a fluid is produced in the alimentary canal that contains these alimentary principles of the food in a diffusible condition. How, then, does this fluid pass through the living epithelial wall of the mucous membrane into the underlying capillary blood-vessels and lacteals? In the first place, we know of a physical process called *filtration*, by which is meant the passage of fluids through the pores of a membrane under pressure. Substances that may be obtained in the form

[1] See Appendix, ' Carbohydrates.'

of crystals, or crystalloids, as they are termed, filter easily when in solution. Glue-like substances, or colloids, as they are termed, filter with difficulty. The greater the pressure the greater the amount that will filter through in a given time. Filtration may possibly occur to a small extent under the pressure exerted on the digested food by the contraction of the intestinal walls. (The condition known as *œdema*, or dropsy, in which the lymph-spaces of the connective tissues and other structures become charged with abnormal accumulations of lymph, appears to be due mainly to excessive filtration or transudation of this fluid consequent on increased pressure in the capillaries and veins.) In the second place, we can partly explain the passage of the diffusible and dissolved food materials into the blood and lacteals by the physical process called *osmosis*. This is a kind of diffusion that takes place between two different solutions separated by a membrane. If a strong solution of sugar or salt be separated by a thin membrane, as dead bladder or parchment paper, from a weak solution, sugar or salt passes from the stronger solution to the weak solution, and water from the weaker solution to the stronger, in consequence of the osmotic current set up. The exchange of fluid particles in osmosis takes place independently of pressure, and the rate at which osmosis takes place varies as well as the total amount of exchange varies with the differences of concentration in the solution, the qualities of the dissolved substances, and the nature of the separating membrane. Crystalloids pass through a membrane, or dialyse, as it is termed, much more readily than colloids. Sugar passes more easily than peptone. The conditions for osmosis clearly exist in the alimentary canal. On one side of the mucous membrane of the intestine there are relatively concentrated solutions of diffusible salts, peptones, sugar, and soap, and on the other side are the blood-vessels containing proteids scarcely diffusible. Moreover, the circulation of the blood is constantly removing the fluid that has imbibed material from the intestine and bringing fresh fluid capable of absorption. Hence we may say that some of the digested food passes into the capillary vessels in the mucous membrane of the intestine by osmotic diffusion, or at any rate that osmotic diffusion is one factor in absorption of digested food.   But the absorptive epithelium of the intestine is unlike the dead bladder or parchment in being a living membrane, and there are certain facts that make it probable that the vitality of the membrane exerts an influence in the process of food-absorption.   Thus a loop of intestine can be isolated in a living animal, and when various substances in solutions of various strength are introduced and their disappearance studied, it is found that there is a difference between the ordinary diffusion of these substances through a membrane and their diffusion through the mucous membrane of the living intestine.   It thus appears likely that the secretory activity of the living epithelial cells in the intestinal mucous membrane plays an important part in the absorption of food, and that the phenomena do not agree entirely with the physical processes of filtration and osmotic diffusion.   Why, then, does the process of digestion essentially consist in rendering foodstuffs soluble and diffusible?

Dr. Foster answers the question thus : 'Because though the cell is not an apparatus for diffusion, diffusion is an instrument of which the cell makes use.   When we say that peptone does not enter the blood by ordinary diffusion, we do not mean that diffusion has nothing to do with the matter. The activity of a living cell is an activity built up upon and making use of various chemical and physical processes ; in it the processes of ordinary diffusion play their part as to the ordinary processes of chemical decomposition ; but the cell uses and modifies them for its own ends.   If, as we have every reason to believe, the cell of a villus passes the sugar unchanged from the intestine into the blood-capillary, it makes use of diffusion to effect that passage ; and if it does change the proteid into something else before it passes it on, it receives it into itself in the first instance by the help of diffusion. When we say that substances do not enter the blood by ordinary diffusion, we mean that the diffusion which takes place in a living cell is something so different in the results from ordinary diffusion through a dead membrane that it is undesirable to speak of it by the name.'

What has been said above applies chiefly to the proteids and carbohydrates of the food, and may be taken to refer to the passage of these digested food-principles both through the epithelium cells of the mucous membrane and through the flat endothelial cells of the capillary vessels situated beneath the basement-membrane. The sugar formed by the ferment of the salivary glands and the amylopsin of pancreatic juice is maltose. Before absorption it is changed by the succus entericus into glucose, and as such is found in blood. Lactose is also changed into glucose, as is also most of the cane sugar. Most of the proteid is absorbed as peptone or peptone and albumose, but, as already noted, neither peptone nor albumose is found in the blood, and we therefore conclude that through the activity of the epithelium-cells or lymph-cells, or both, peptone is changed into blood-proteids during the actual process of absorption. As to the fats, it has been shown that they undergo either a chemical change (saponification) or a physical change (emulsification) ; a small part is saponified, and the soaps so formed are absorbed like the other soluble materials. But the greater part is reduced to a fine state of subdivision or emulsified, and these minute particles are absorbed by a special mechanism in the villi.

133. **Structure of a Villus.**—The absorbing surface of the mucous membrane of the small intestine is increased by the transverse folds called valvulæ conniventes, and by the numerous small processes projecting into the interior along its whole length. They are largest and most numerous in the duodenum and jejunum, and become smaller and fewer in the ileum. Some are conical in shape, some cylindrical, and some triangular, and the average size is about $\frac{1}{12}$ inch. Their total number has been estimated to exceed four millions.

Each villus consists of a small projection of mucous membrane coated with columnar epithelium continuous with that of the other parts of the intestine, a basement-membrane supporting the columnar epithelium, an interior framework or stroma of adenoid tissue in which is a network of capillary vessels, and a central lymphatic or lacteal with unstriped muscular fibres around it. The central lacteal begins in a

blind end near the summit of the villus, and is continued below
into a plexus of lymphatics in the submucosa, which joins
together to form larger vessels that proceed to lymphatic glands.
The endothelial cells that form the walls of the lacteal appear
to have openings between the cell-plates that are in connection
with the spaces or meshes of the adenoid tissue of the villus,

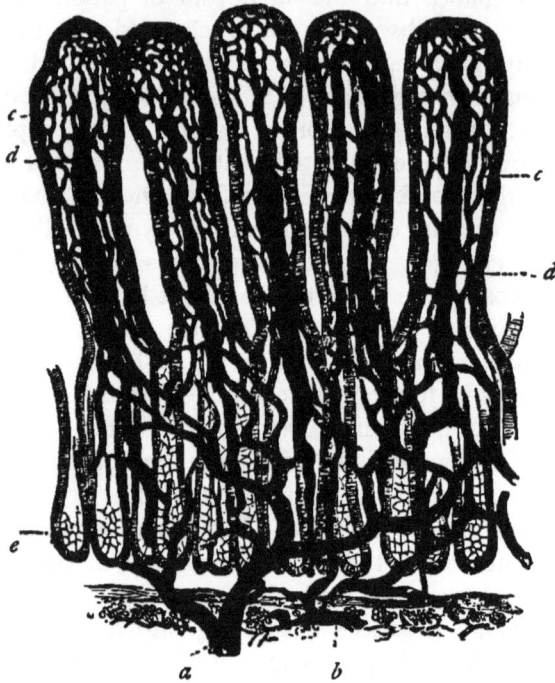

FIG. 129.—Villi of Small Intestine. (Cadiat.)

a, small artery; *b*, subjacent lymphatic plexus; *c*, blood-capillaries traversing villi; *d*,
  lacteal beginning blindly near the free end of the villi; *e*, Lieberkühn's glands.

and these spaces are, as usual in this kind of tissue, occupied by
leucocytes. The muscular fibres of the villi are derived from the
muscularis, and lie for the most part outside the lacteal. Their
contraction shortens the villus and tends to empty the lacteal.
The network of blood capillaries is supplied by a small artery
that enters from the submucosa, and the blood is carried out of
the villus by one or two small veins, the veins of the intestines,
the inferior mesenteric and superior mesenteric, uniting with

those from the spleen, pancreas, and stomach to form the **portal vein** which passes to the liver (fig. 131).

The single epithelial layer that forms the free surface of the villus consists for the most ~part of columnar or conical epithelial cells having the outer broader end marked by fine striations. Mixed with the columnar cells that possess this refractive striated border are mucous cells of a goblet shape which discharge mucin from their mouths into the intestine.

134. **Absorption of Fat.**—After a meal containing fat the lacteals contain a fluid having a milky appearance owing to the presence of fat in a finely divided condition, and this fluid is termed *chyle*. During the intervals between digestion the lacteals contain ordinary lymph. Chyle is ordinary lymph plus fatty particles absorbed from the intestines, and under the microscope lymph-corpuscles and minute fat globules may be seen, the fat of chyle forming about 5 per cent. Like blood and lymph, chyle will coagulate spontaneously, yielding a soft white clot (par. 64). The richer the meal in fat the greater the proportion of fat in the lacteals, though for every 100 parts of fat absorbed from the alimentary canal only about sixty parts find their way into the thoracic duct. Where the missing quantity goes has not been ascertained. It cannot be found in the blood of the portal vein, nor does it pass into the blood-vessels of the villi, as none can be seen there when their capillaries are examined under the microscope after treatment with osmic acid (an acid which blackens particles of fat). But if sections be made of the villus of an animal killed during digestion and stained with osmic acid, black fatty granules are seen in the epithelial cells and in the lymphoid spaces surrounding the central lacteal. The fat-globules in the lacteal itself are in most cases much more minute than those seen in the emulsified contents of the intestine, and these exceedingly fine granules of fat are termed the 'molecular basis' of the chyle. As to the mode by which the emulsified contents of the intestine reach the lacteals much discussion has taken place. It is probably as follows : (1) Absorption of fat into the columnar epithelium-cells of the surface of the villus ; (2) passage of the globules thus absorbed through the spaces of the

lymphoid tissue of the villus to the lacteal, or carriage of the fat globules by the lymph-corpuscles (leucocytes) through the lymphoid tissue to the lacteal ; (3) passage into the lacteal, possibly through openings between the endothelial cells forming the wall of the lacteal ; (4) a breaking up of the fat-globules thus brought into the lacteals to form the 'molecular basis' of the chyle.

The peristaltic contractions of the intestine exert a pressure that assists the onward flow of the chyle into the valved lymphatic vessels below, while the muscular fibres of the

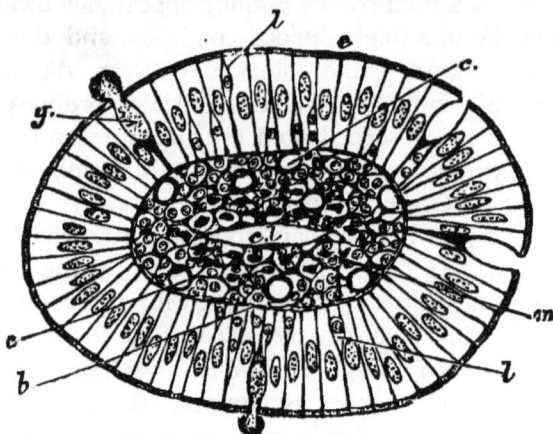

FIG. 130.—Cross-section of an Intestinal Villus.

e, columnar epithelium ; *g*, goblet-cell, its mucus is seen partly extruded ; *l*, lymph-corpuscles between the epithelium-cells ; *b*, basement-membrane ; *c*, blood-capillaries *m*, section of plain muscular fibres ; *c.l*, central lacteal.

villi derived from the muscularis mucosæ exert a kind of pumping action as they contract, squeezing the contents of the tissue-spaces into the villi, and emptying the lacteals and blood-capillaries of the villus into the underlying vessels.

As chyle contains a larger proportion of proteids than lymph, it appears to take some share in the absorption of this kind of food, though it is certain that the proteids after conversion into peptones are almost wholly absorbed into the capillary blood-vessels of the villi and other parts of the intestinal mucous membrane. But it must be noted that the peptones formed from the nitrogenous foods by the action of the gastric, pancreatic, and intestinal juice, although absorbed into the capillary vessels that form rootlets of the portal vein in the stomach and intestines, cannot be found as such in any part of the portal blood. They must therefore be reconverted into albumins as they pass through the alimentary or capillary wall.

Absorption goes on to some extent in all parts of the alimentary canal, its various sections having been arranged as regards activity of absorption in the following order :—Small intestine, large intestine, stomach, mouth, pharynx, œsophagus. The most active part is undoubtedly the small intestine, especially the upper part, owing to the presence of the valvulæ conniventes and the villi. Watery solutions of salt and sugar, with some peptone, are absorbed into the blood-vessels of the stomach. A little water with very soluble matter may be absorbed in the mouth. After the digestive juices by the action of their special ferments have rendered the food-materials soluble and diffusible, absorption takes place partly by osmotic diffusion, and partly through the vital activity of the epithelium of the digestive tract. The cells lining the tract ingest the nutrient food-materials that have been digested, and then discharge the matter into the tissue-spaces. From the tissue-spaces the more diffusible substances pass for the most part into the capillary blood-vessels of the portal system, while the less diffusible fatty particles pass into the central lacteal, and only reach the blood after traversing the lymphatic system. In other words, ' the greater part of the water, salts, sugar, and proteid absorbed passes into the blood-vessels, and by the portal vein to the liver ; the greater part of the fats absorbed passes into the chyle-vessels, and by the thoracic duct into the venous system.'

Into the rootlets of the portal vein are absorbed
- Peptones (major part)
- Sugar ,,
- Salt ,,
- Soaps ,,
- Water ,,
- Fat a trace.

Into the lacteal vessels are absorbed
- Fats (major part)
- Soaps (small part)
- Peptones ,,
- Water ,,

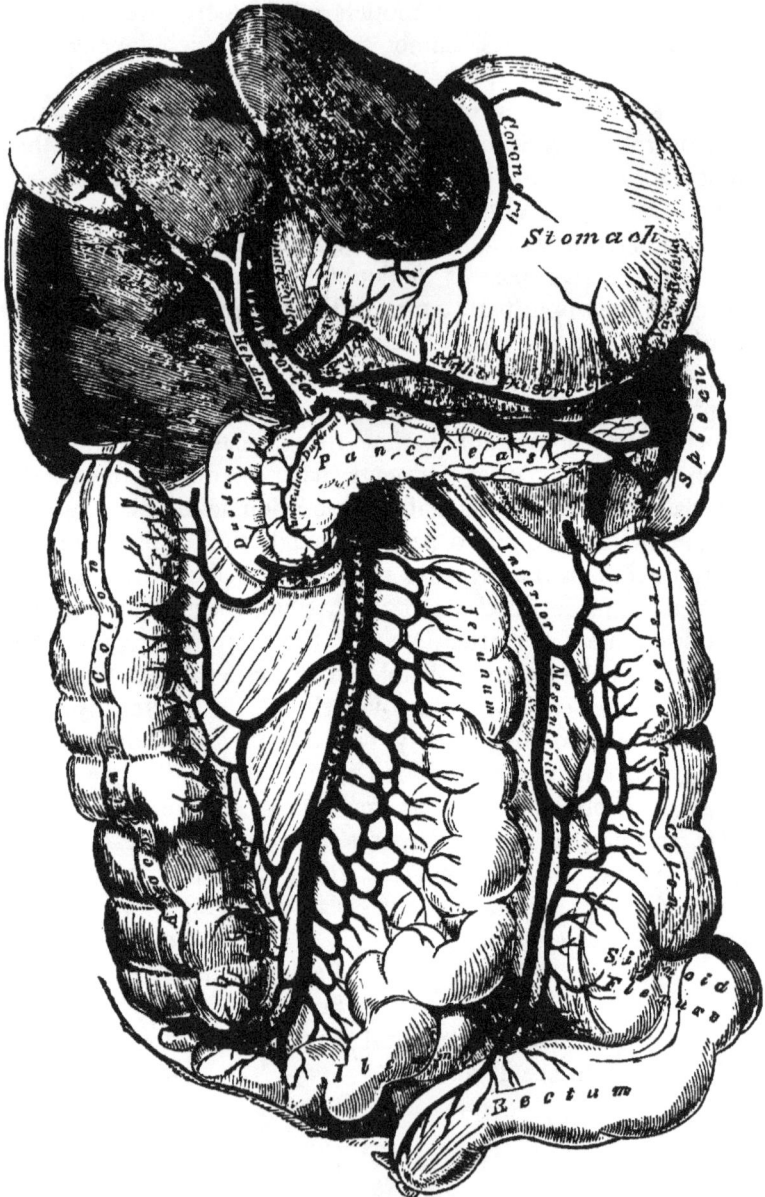

FIG. 131. - Portal Vein and its Branches.

In this figure the liver and stomach are tilted upwards; part of the duodenum and the whole of the transverse colon are cut away.

135. Movements of the Stomach and Intestines.—It has already been mentioned that the walls of the stomach contain non-striped muscular fibres arranged in three layers. The circular fibres are thickened and abundant at the pylorus, and there form the sphincter muscle termed the *pyloric valve*. When the stomach is empty its walls are usually at rest though contracted, and the mucous membrane is pale ; but when food enters, the membrane becomes red owing to increased blood-supply, and movements of its walls begin. These movements serve to bring the mucous membrane in contact with the food, as well as to propel the food forwards. There is a

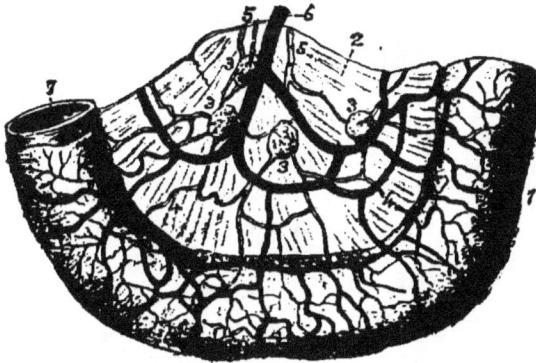

FIG. 132.—Lymphatics of the Intestine.

1, Portion of small intestine ; 2, with layer of mesentery ; 3, mesenteric lymphatic glands with lacteals (4) passing in, and (5) passing out ; 6, branch of portal vein formed by smaller branches from intestine.

rotatory or churning movement in which parts of the wall of the stomach in contact with the food glide to and fro with a rubbing movement, the food and gastric juice being rolled about and mixed together. A periodic peristaltic or wave-like movement also begins soon after the food has entered the stomach, setting up currents in its contents, and after a time this motion becomes so marked towards the pyloric end that the food, now reduced to the condition of chyme, is propelled through the relaxed pyloric valve into the duodenum.

Vomiting is a reflex act by which the contents of the

stomach are expelled. It may be produced by afferent impulses of various kinds, as tickling the throat, exciting the vagus endings in the stomach by emetics &c. It is generally preceded by a sensation or nausea and a reflex flow of saliva. A deep inspiration is then taken, the glottis is closed and the diaphragm fixed. The abdominal muscles contract and press the stomach against the diaphragm ; the cardiac sphincter of the stomach relaxes, and its muscular walls, with the reversed peristaltic action of the œsophagus, complete the act.

FIG. 133.--The Ileo-cæcal Valve.

a, ileum ; *b*, ascending colon ; *c*, cæcum ; *d*, junction of the cæcum and colon ; *e* and *f*, loose folds of the mucous membrane, forming the ileo-cæcal valve ; *g*, vermiform appendage.

The food mingled with the various secretions, and subjected to the absorbent action of the villi and blood-vessel, is forced along the small intestines by slow, rhythmical, wave-like contractions of the longitudinal and circular muscular fibres in the intestinal wall. Here is the best example of peristaltic action. By the contraction of the longitudinal fibres a portion of the intestine is drawn backwards, and then the circular fibres, contracting from above downwards, progressively narrow the tube and force the contents forwards, just as pressing along a flexible tube containing a liquid moves it onwards.

At the junction of the small intestine with the large one is placed the ileo-cæcal valve to guard against reflux into the ileum. This valve is formed of two semilunar folds of the mucous membrane of the ileum, projecting into the inner and back part of the large intestine, a narrow transverse aperture lying between.

Distension of the cæcum stretches the margin of the folds and brings them together, as does also the pressure exerted by peristaltic movements. These movements in the large intestine are more sluggish than in the small intestine, for it is said that it takes about three hours for food to pass along the small intestine, and about twelve hours to pass through the large intestine.

CHAPTER IX

THE LYMPHATIC SYSTEM

136. Through the thin walls of the blood-capillaries there passes into the spaces part of the blood plasma, which thus comes to bathe the surrounding tissues. This exuded plasma becomes the lymph, and contains in solution absorbed nutritious matters, as well as oxygen set free from the hæmoglobin of the red corpuscles. It is thus both a nutritive and respiratory fluid for the cells and other form-elements of the tissues, each tissue taking from it the substances it needs for life and activity. The surplus of lymph, together with certain waste products arising from the metabolism of the tissues, is carried away from the tissues, and back to the blood-stream, by a set of absorbent vessels called *lymphatics*. These vessels have their origin within the tissues, and are found in nearly all parts—even in some parts, as the cornea, which contain no blood-vessels; but they are, as we shall see, most intimately associated with connective tissue. Under the term *lymphatic system* are included : (1) the vessels specially called lymphatics or absorbents, together with the glands belonging to them ; (2) the lacteals, which are the lymphatics of the intestines, and differ from other lymphatics only in their power of absorbing chyle during digestion, as well as exuded lymph ; (3) the serous membranes which enclose cavities that are in reality large lymph-spaces. The lymph that fills up the interstices, or spaces between the cells and other tissue-elements of the body, together with that contained in the lymphatic system exceeds in quantity the fluid in the blood-vascular system, and forms more than a quarter of the weight of the body. It is the intermediary or middleman between the blood and the tissues.

From it all the tissues obtain their oxygen and nutritive materials ; into it are excreted the waste products of tissue activity. Oxygen taken into the alveolar capillaries of the lungs and nutritive materials absorbed from the alimentary

FIG. 134. Principal Lymphatic Vessels and Glands of the Head and Neck on the Right Side (after Bourgery in part). (Allen Thomson.) ⅓

a, right innominate vein at the place where it is joined by the principal lymphatic trunk ; *a'*, the left vein ; *b*, arch of aorta ; *c*, common carotid artery ; *d*, thyroid body crossed by the anterior jugular vein ; *e*, cut surface of sternum ; *f*, outer part of clavicle ; 1, submaxillary lymphatic glands ; 1', lingual ; 2, parotid ; 3, 3, suboccipital and mastoid ; 4, superior deep cervical ; 5, 5, inferior deep cervical glands ; 6, 6, axillary glands ; 7, on the superior vena cava, some of the anterior mediastinal vessels ; 8, on the innominate artery, some of the superior mediastinal ; to these last are seen descending some of the lymphatics from the thyroid body and lower part of the neck.

canal into the portal capillaries, or into the lacteals of the villi, eventually pass from the blood into the lymph, as the blood is driven under pressure through the general or systemic capillaries ; from the lymph they are supplied to the tissues for building up protoplasm and supplying energy. The waste products of tissue activity, carbon dioxide, urea, enter the lymph and are then for the most part diffused into the blood-vessels of the tissues, to be afterwards excreted, though some of the waste products are carried off in the lymph and enter the blood indirectly by the lymphatic circulation. The chemical and physical properties of lymph are described in par. 63.

137. **Origin and Structure of the Lymphatics.**—The lymphatic vessels are those vessels which, beginning as minute channels in the tissues, collect and return into the circulation the fluid passed out of the capillaries, after it has supplied nutritive matter and oxygen to the tissues and received certain products of tissue waste.   These absorbent vessels thus form a system for draining the tissues, and, uniting in their course to form wider tubes, they at last combine into two large terminal trunks which open on each side of the thorax near the heart at the junction of the jugular and subclavian veins—the trunk on the left side being the largest and most important, and known as the *thoracic duct*.   The fluid lymph after entering the lymphatic vessels always moves in one and the same direction—towards the point of discharge into the venous system—for the larger vessels are provided with small semilunar valves, the free edges of which are turned towards the heart, so that any reflux towards the tissues is impossible. Valves are also placed at the entrance of the lymphatic trunks into the great veins.

Some lymphatics appear to begin in the form of a plexus or network of fine tubes as in membranes, but many of these vessels commence in irregular lacunar spaces in connective tissue, and from these spaces (which are the same as those containing connective-tissue corpuscles and wandering leucocytes), often called lymph-spaces or lymphatic canaliculi, as they are partly occupied by this fluid, the lymph passes into more definite channels termed lymph-capillaries.   It is not clear whether there is direct open communication between the cell-spaces and the lymphatic capillaries or not. The lymph-capillaries form a network in the various organs of the body, and have for their walls a single layer of endothelial cells with sinuous margins.   They are generally larger than blood-capillaries, have numerous anastomosing branches, and soon join to form the lymphatic vessels proper.   As the vessels become larger they resemble in structure the corresponding veins, except that their three coats are thinner and the valves more numerous.   During their course towards the point of discharge the lymphatic vessels become connected with the peculiar structures termed lymphatic glands (par. 138).

Other modes of origin of the lymphatics are found besides those described above.   Thus the lacteals or lymphatics of the intestine begin in blind extremities in the villi, as already described.   Other lymphatics arise by means of free *stomata* on the walls of the larger serous cavities, which are not the absolutely closed spaces they were once thought to be (see par. 90, fig. 99), but lymph-lacunæ from which the fluid is drained away through the minute stomata into the network of lymph capillaries lying in the subserous tissue.   In the central nervous system minute arteries are found, having around them for a small distance a lymph-capillary in the form of a tubular sheath, so that the plasma that leaves the blood-vessels enters at once into this tubular space and is then carried away into the regular lymphatic canals. This mode of origin is known as 'perivascular lymphatic.'

As the lymph in the lymph-spaces is free it gets into direct contact with the tissue-elements and percolates between the cells, thus reaching parts of tissues devoid of blood-vessels. In this way it supplies nutriment to the epithelium of mucous membrane, to the interior of cartilage, the cornea, and the innermost cells and substance of bone. The total quantity of lymph and chyle that passes into the blood in 24 hours is estimated to be in a well-fed animal equal to the whole quantity of the

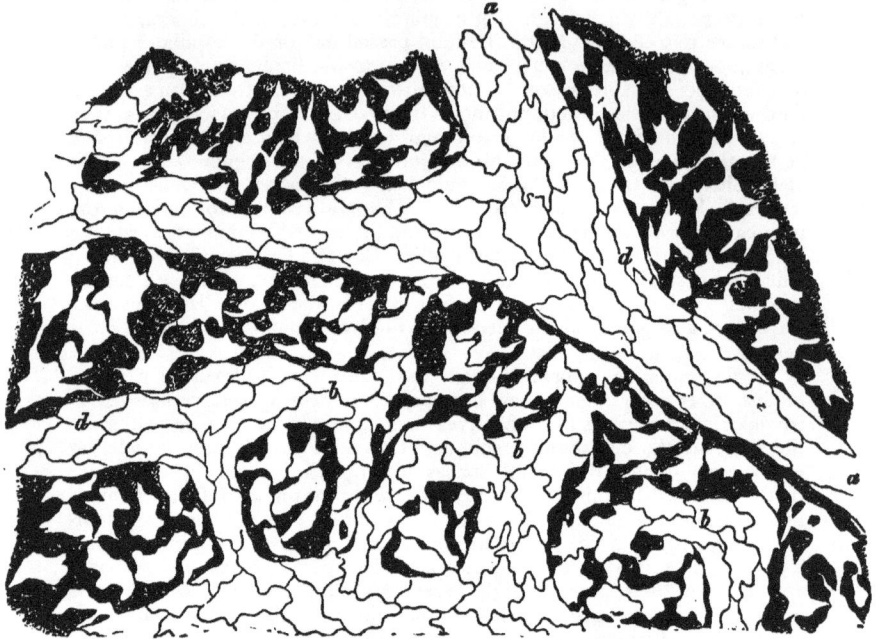

FIG. 135.—Portion of Serous Membrane of Diaphragm (pleural) from the Rabbit, treated with nitrate of silver after removal of superficial epithelial layer. (Recklinghausen.)

c, c, cell-spaces of tissue ; *d, d*, commencing lymphatic vessels connected at *b, b*, with the cell-spaces.

blood in the body, about half of this coming from the lacteals and half from the other lymphatics of the body.

[The watery solution of salts and proteids that leave the blood-plasma to become lymph passes from the blood-vessels into the lymph-spaces by a process or processes not fully understood. It is partly a process of osmotic diffusion through a porous membrane, partly a process of filtration under pressure through a similar partition, and partly a process depending on the activity of the living endothelial cells of the capillary wall. That it is not mere diffusion is shown by the passage of about one-half the indiffusible proteids of blood-plasma into lymph ; that it is not mere filtration under pressure is shown by increase of pressure not always being accompanied by increased passage of fluid as in an ordinary filter, and by the passage of waste products in the opposite direction, from the lymph-space into the blood. That the capillary wall plays an important part will be evident

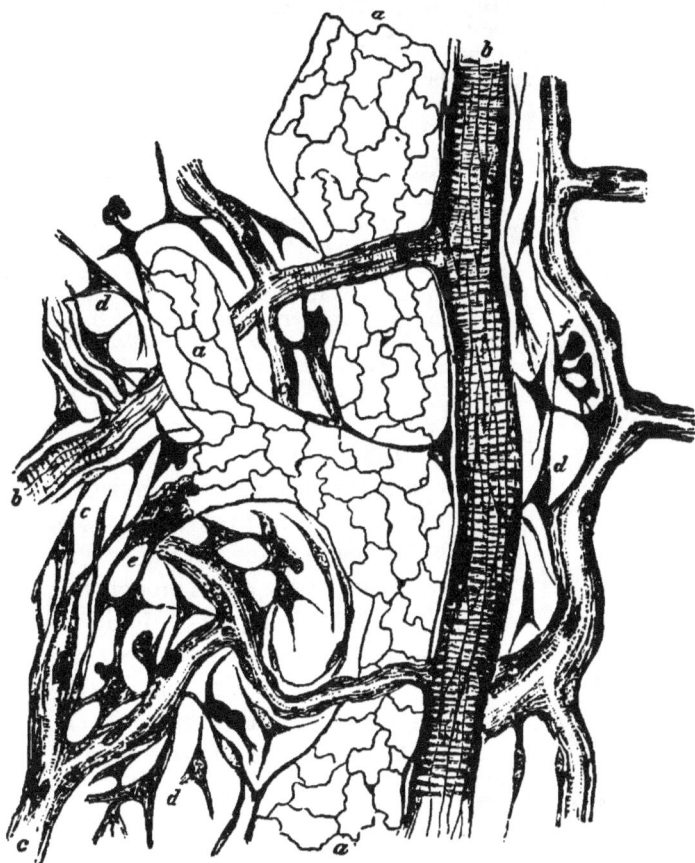

FIG. 136.—Nitrate of Silver Preparation from Rabbit's Omentum. (Klein.) Magnified.

a, lymphatic vessel ; *b*, artery ; *c*, capillaries ; *d*, branched cells of the tissue which are seen to be connected both with the capillary walls and, as at *e*, with the lymphatic. The cells are, in this instance, stained by the nitrate of silver.

when we consider that it is a living, changing thing continually acting upon and being influenced by the blood. When peptone is injected into the circulation of a living animal it leads to an increased flow of thicker lymph, a more concentrated plasma passing through the vessel walls, and the peptone disappearing at first by being transferred from the blood to the lymph. The increased lymph formation and the transference of the

P

peptone to the lymph point to an active process of secretion by the cells of the capillary walls.   Lymph capillaries unite to form vessels having the general structure of veins, but the thinness of the walls of these lymphatic vessels makes them difficult to see when empty.   When the lymphatic vessels are distended they have usually a beaded appearance owing to their numerous semilunar valves.   Lymphatic vessels are arranged in a superficial set near the surface of the body, just beneath the integument, and a deep set in the interior.   The lymphatics of any part or organ exceed the veins in number, but are of smaller size.   They also anastomose or intercommunicate more freely than veins. Both the lacteals that convey the chyle from the intestines to the thoracic duct, and the general lymphatics that take up the lymph from all other parts of the body, become connected in their course with lymphatic glands.   These lymph glands occur in groups around joints, in the recesses of the neck, in the organs of the thorax and abdomen ; a large number placed in the mesentery being connected with the lacteals.   As these glands effect important changes in the fluid flowing through them, their structure must now be described.]

Fig. 137.—A Lymphatic Gland (1) is shown with (2) afferent vessels, and (3) efferent vessels.   The numerous valves cause the beaded appearance of the lymphatic vessels.

138. **Lymphatic Glands.** — Lymphatic glands are small compact bodies, varying in size from a hempseed to a kidney bean, placed in the course of the lymphatic vessels.   Through them the lymph and chyle passes in its onward course.   They exist in great numbers in the mesentery and alongside the great vessels of the abdomen, thorax, and neck.   Some are found in the axilla and groin, but none below the elbow or knee.   At a slight depression on one side, the hilum, blood-vessels enter and leave the gland.   The efferent lymphatic vessels also leave at the same spot, but the afferent lymph-vessels enter at various points of the periphery.   Section of a lymphatic gland shows an external cortical substance of light colour and an internal darker medullary substance coming to the surface at the hilum.

The gland consists of (1) a fibrous envelope or capsule of connective and muscular tissue from which a framework of processes (trabeculæ) passes inwards and divides the gland into

spaces (alveoli) with free communication, having the form of
converging chambers in the cortex, but diminishing in size and
regularity of shape in the medulla ; (2) the proper glandular
substance, a mass of lymphoid tissue (*i.e.* a fine network of
fibres with lymph-corpuscles in the meshes) occupying the
central part of the spaces, and thus leaving a narrow channel,
bridged across by cells and fibres between the lymphoid tissue
and the trabeculæ ; (3) a free supply of blood-vessels, the arterial

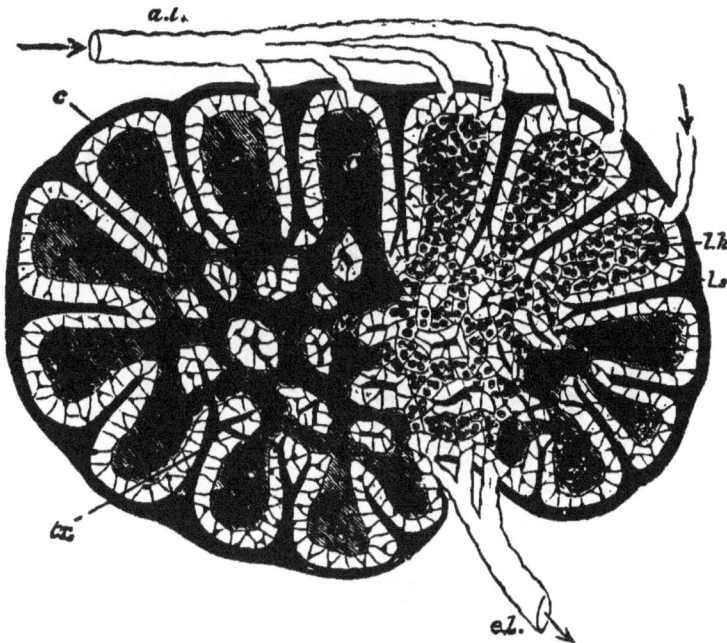

Fig. 138. – Diagrammatic Section of Lymphatic Gland. (Sharpey.)

a.l, afferent ; *e.l*, efferent lymphatics ; C, cortical substance ; M, reticulating cords of medullary substance ; *l.s*, lymph-sinus ; *c*, fibrous coat sending trabeculæ, *tr*, into the substance of the gland.

branches running in the trabecular framework, and breaking
up in part in capillary networks in the glandular substance ;
(4) the afferent and efferent lymph-vessels.

The afferent lymphatic vessels enter the lymph-sinuses of
the cortex after passing through the capsule, and the lymph is
slowly carried along through the channels in the substance of

the gland towards the hilum, taking up in its passage many lymph-corpuscles, probably derived from the actively multiplying leucocytes that occupy the meshes of the adenoid tissue.

The small spaces or alveoli of the medullary part of the lymphatic gland contain adenoid tissue crowded with leucocytes of various sizes, and separated from the fine trabeculæ by channels of open reticular tissue forming the lymph-sinuses or lymph-paths, the whole of the lymph-sinuses forming an intercommunicating labyrinth throughout the gland. Active cell-multiplication by karyokinetic division goes on in the

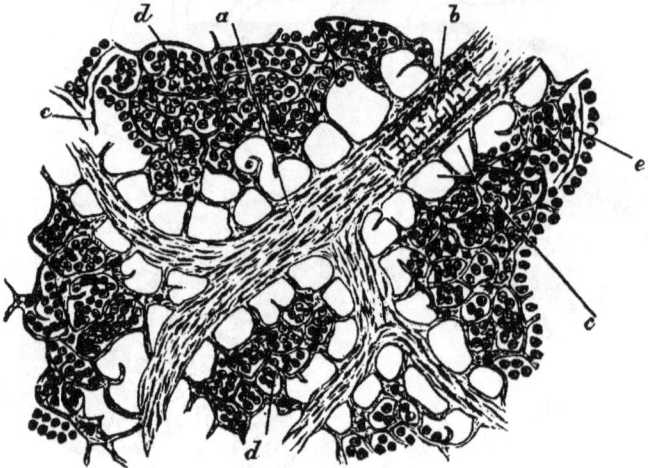

FIG. 139.— Section of Lymphatic Gland Tissue.
a, trabeculæ ; *b*, small artery in substance of same ; *c*, lymph-paths ; *d*, lymph-corpuscles ; *e*, capillary plexus.

small portions of tissue crowded with leucocytes, and some of these cells pass into the lymph-stream of the lymph-sinuses, as it is found that lymph-corpuscles are more numerous in the issuing lymph than before. Leucocytes are also found, and appear to multiply in other localities containing adenoid tissue as in the tonsils, certain parts of the air passages, the solitary glands and Peyer's patches of the intestine, the spleen, &c.

One function of lymphatic glands, therefore, is the production of leucocytes that become the white corpuscles of the blood. Besides the increase in corpuscles in lymph that has passed through the glands, it becomes more coagulable, the lymph in the rootlets of the system having but a feeble power of coagulation. The same changes occur in chyle after it has passed through the lymph-glands of the mesentery. Before entering the mesenteric glands, chyle is but an opaque milky fluid, hardly coagulable at all ; but after passing through the glands there is a diminution of its molecular basis, numerous white corpuscles and the elements of fibrin being also then present. The lymph leaving a lymphatic gland therefore differs from that entering it by being more coagulable, by being freed from accidental matters, and by being enriched with numerous young leucocytes.

The absorbent power of the lymphatics is taken advantage of when ointments are rubbed into the skin, or when a drug is administered by a hollow needle inserted under the skin, and the fluid forced in by a syringe (hypodermic injection). The absorbent vessels gather up the matter and pass it into the blood. That the lymphatic glands act as a sieve and arrest deleterious matters is evidenced by their inflammation after a poisoned wound. Thus a wound in the foot that introduces poisonous matter may inflame the lymphatic glands of the thigh.

139. **The Thoracic Duct.** — The thoracic duct is the common trunk that receives the lymphatics from the lower limbs, from the abdominal viscera (except the upper part of the liver), from the walls of the abdomen, from the left side of the thorax, from the left lung, the left side of the heart, the left arm, and the left side of the neck and head. It is from 15 to 18 inches long, and extends from opposite the second lumbar vertebra to the root of the neck. Its diameter is nearly $\frac{1}{3}$ inch, but at its lower end it is dilated into an expansion called the *receptaculum chyli*. Its course is somewhat tortuous, and it contracts and enlarges at irregular intervals. Throughout its course it is supplied with valves, and these are more numerous in its upper part. Its termination at the junction of the left jugular and subclavian veins is also guarded by a valve of two segments which allows the contents to pass into the veins, but prevents any reflux. Its

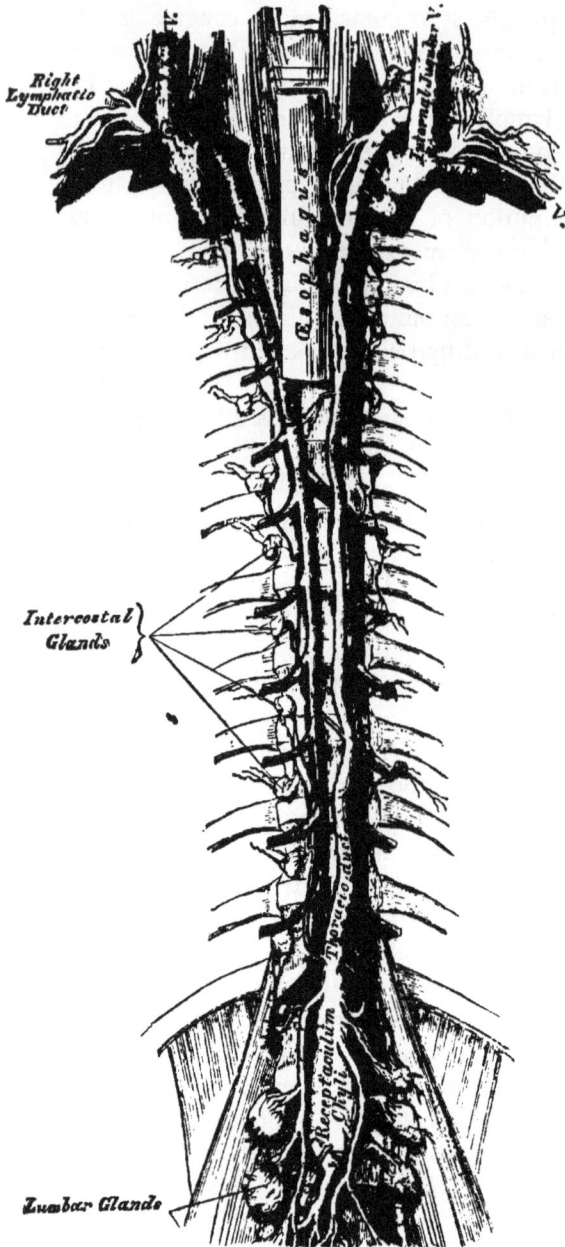

Right Lymphatic Duct

Intercostal Glands

Lumbar Glands

FIG. 140.—The Thoracic and Right Lymphatic Duct.

walls consist of three coats, an epithelial lining of flattened lanceolate cells, outside of which are longitudinal elastic fibres; a middle coat of connective tissue, outside of which are numerous muscular fibres arranged transversely and longitudinally; and an external coat of alveolar tissue with isolated bundles of muscular fibres. The muscular character of the wall of the thoracic duct is therefore well marked.

140. Conditions affecting the Amount of Lymph and Chyle.— Various causes combine to determine the amount of lymph in the lymphatic spaces and vessels of any region of the body. Thus the mere position of a limb, as when the hand is kept hanging down for a time, may lead to an increase of lymph in the spaces and channels of the organ, as is evidenced by the swollen limb and tense skin. A tight bandage causes a similar swelling on the peripheral side of the ligature. This swelling, though partly due to the dilated blood-capillaries consequent on the hindrance to the return flow along the veins, is mainly caused by the unusual fulness of the lymph-spaces and vessels in the skin and subjacent tissues. Active exercise long continued may produce a similar result, for muscular activity not only increases the formation of lymph but promotes its more rapid outflow. The tendons and fasciæ of muscles have numerous small stomata that absorb the lymph from the muscles into lymphatic vessels. Increase of blood-pressure in the capillaries of any area due either to arterial dilatation or more effectively to obstruction of the venous outflow usually increases the transudation of lymph, though the condition and activity of the living cells of the capillary wall is of great importance, and may prevent the increased outflow of lymph.

A great increase in the amount of chyle occurs during digestion after a full meal, the lacteals becoming-white and distended, though collapsed and containing only a small quantity of clear lymph during a period of fasting.

Abnormal conditions of the vascular system may lead to an excessive transudation of lymph, with an undue accumulation of lymph in the lymph-spaces. This lymph congestion is spoken of as *ædema,* or dropsy.

141. **Inflammation.**—Irritation of a part of the tissue of any organ leads to increased flow of blood to that part ; the vessels dilate and become congested, and the white corpuscles cling to the sides and begin to emigrate through the vessel wall. If the irritation is great, a morbid condition termed inflammation is set up. This is characterised by swelling, heat, and redness, accompanied by marked vascular changes and much exudation of plasma and corpuscles into the injured tissue. The phenomena of inflammation have been observed by the aid of the microscope in the transparent parts of animals where it has been set up. A frog's tongue or web spread out for microscopical examination shows the normal circulation through the arterioles, capillaries, and small veins. A severe irritant sets up the following inflammatory changes. A dilatation of the small arteries and a quickening of the blood-stream is first noticed, more vessels also becoming visible. After a time the stream becomes slower, and the dilated vessels are seen to contain two layers of corpuscles—an inner axial layer of red corpuscles and a layer of white corpuscles adhering to the walls. Leucocytes or white corpuscles also crowd the capillaries, and they soon begin to pass in quantity through both veins and capillaries into the surrounding tissues. Small buds or processes first appear on the outer side of the wall, until the whole corpuscle passes outside by amœboid movements (see fig. 58). With the emigration or diapedesis of numerous white corpuscles there is great exudation of coagulable plasma or plastic lymph, and the swelling that marks an inflammation is produced. The redness is due to the unusual quantity of blood in the affected part ; so also is the heat, which is not really greater than the maximum heat of the blood of the interior.

If the inflammation subsides the white corpuscles cease to emigrate, the stream of blood quickens again, and the normal circulation is again set up. The migrated corpuscles and surplus lymph pass away eventually into the lymphatic channels. But if the inflammation continues, complete arrest of blood-flow may occur, some red as well as white corpuscles may pass into the tissue, and the accumulated leucocytes degenerate into pus-cells and form an abscess which must be opened if it fail to burst naturally.

Dr. Metschnikoff of Paris explains the phenomena of inflammation as essentially a reaction of the leucocytes contained in animal bodies to the presence of injured tissue or intrusive particles. The irritation caused by lesion of tissue or the presence of foreign bodies—bacteria and other micro-organisms—leads to active movement of leucocytes to the spot, and these, migrating from the vessels, devitalize or engulf and digest the disintegrated tissue or foreign particles, and so prepare the way for repair of tissue. The white corpuscles that thus combat irritative agents are called *phago-cytes*, and diapedesis is one form of struggle against offending particles. The process has often been watched, and leucocytes have been seen to take up the irritating bodies or found containing the dead and dying microbes. If the phagocytes succeed in devouring and removing the irritating substances, they disappear probably by being carried off into the lymph-stream ; if the invading particles are too powerful, the corpuscles themselves are destroyed and collect in the tissue as pus-cells.[1]

142. **Movements of Lymph.**—Several causes combine to promote the onward flow of lymph towards its place of discharge into the blood. In the first place, there is a *vis a tergo*, or pressure from behind, derived from the blood-pressure, *i.e.* ultimately from the heart. The lymph in the lymph-spaces must be under a higher pressure than the blood in the great veins into which the thoracic duct opens, and where we know that the pressure is very low, and any fluid will flow from a region of high pressure to a region of lower pressure. In the next place, the numerous valves present in the lymphatic vessels are so arranged that every pressure exerted on the tissues in muscular movements must assist in driving forward the lymph. The respiratory movements also assist in drawing the lymph onwards as they assist the flow of the blood towards the heart, every inspiration diminishing the pressure in the vessels within the thorax (par. 102). Further, the muscular fibres in the walls of the lymphatics may probably undergo rhythmical contractions, while the intestinal movements and the contraction of the muscular fibres of the villi appear to assist in driving the chyle from the lacteals into the valvular lymphatics below (par. 134). Lastly, as the lymph-vessels gradually unite to form larger vessels, and finally end for the most part in the thoracic duct, the sectional area of which is less than that of the combining vessels, an increased pressure must result from this union of trunks into one. But the movement is very slow—far slower than the slowest blood movement in the capillaries.

---

[1] The term *Chemiotaxis* is applied to the physical and chemical stimuli that determine the wandering of leucocytes, so that the process of inflammation is said to be a chemiotactic one, since it consists of the gathering together of the migratory cells to parts of the living tissue that have been injured or invaded by foreign particles, as if the disintegrated tissue or disease-producing particles were attractive to them. And the process of destroying the injurious microbes is believed to be a chemical one, the source of the destructive agent being in the cells thus acting under chemical stimuli.

CHAPTER X

THE LIVER AND THE DUCTLESS GLANDS

143. The liver is the largest gland in the body, and weighs from fifty to sixty ounces. It is situated just under the diaphragm, chiefly on the right side, being under cover of the ribs in front and in contact with the posterior wall of the abdomen behind. Its upper surface is convex, and in relation with the under-surface of the diaphragm ; its under-surface is concave and in relation with the stomach and duodenum. The thin anterior border corresponds with the margin of the ribs in the standing position, but the whole organ rises and falls somewhat with the breathing. The liver is retained in position by five ligaments, four of which are folds of perito-neum that attach it to the diaphragm. On the under irregular surface five fissures are seen, which divide the organ into five lobes. They are the longitudinal fissure, the fissure of the ductus venosus (a vein in the fœtus communicating with the inferior vena cava) the transverse fissure, the fissure for the gall-bladder, and the fissure for the inferior vena cava. The two chief lobes are the right and the left, the right being five or six times larger than the left. A serous coat derived from the peritoneum invests the greater part of the liver and is closely adherent to a fibrous coat or capsule beneath, which covers the entire surface.

At the transverse fissure this fibrous coat is continuous with Glisson's capsule, an areolar investment that surrounds the portal vein, hepatic artery, and hepatic duct, and accom-panies these vessels in their branches in the interior of the organ.

The vessels of the liver are : (1) the hepatic artery, which comes off from the cœliac axis, a branch of the abdominal aorta, to supply the organ with nutrient blood ; (2) the portal vein, which conveys to the liver the venous blood of the stomach, spleen, and intestines ; (3) the hepatic veins which arise in the substance of the organ and convey the blood from the liver into the inferior vena cava. (4) The hepatic ducts which arise by minute branches in the organ, and convey away the bile. (5) Numerous lymphatics. (Note five ligaments, five fissures, five lobes, five kinds of vessels.) The nerves of the liver are derived from a plexus of the sympathetic and from the pneumogastric.

The bile is carried out of the liver by two hepatic ducts which leave the right and left lobes at the transverse fissure. These combine, and from the united duct thus formed the cystic

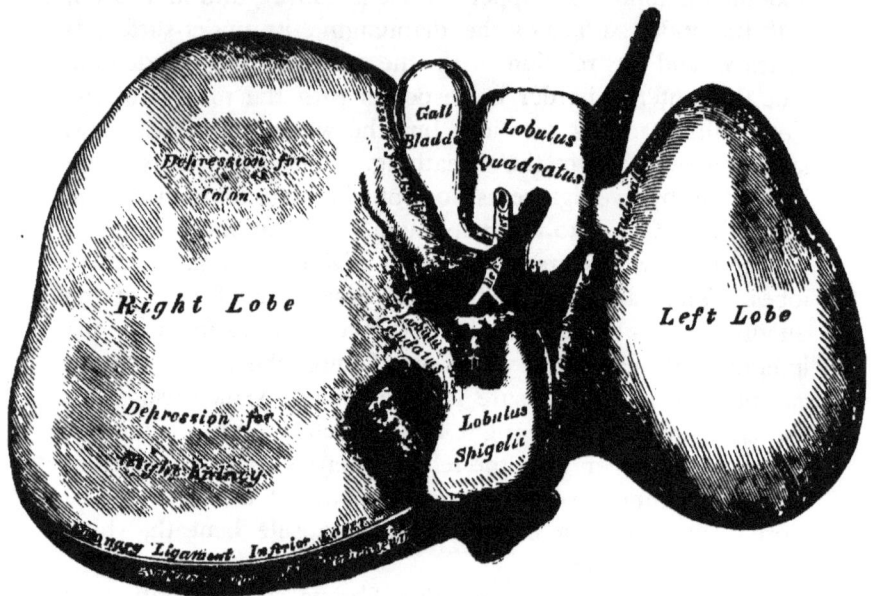

Fig. 141.—The Liver. Under Surface.

duct passes off to the tapering end of the gall-bladder. The united duct now known as the *common bile duct* passes on and enters the duodenum very obliquely along with the pancreatic duct. The gall-bladder is a musculo-membranous pear-shaped bag lodged under the right lobe of the liver, about four inches long and one inch broad in its widest part. Its walls consist of three chief coats, an external serous coat, a middle fibrous or areolar coat containing plain muscular fibres, and an internal mucous membrane with columnar epithelium, that secretes a viscid mucus. Bile is formed continuously by the cells of the liver, and during the intervals of digestion it enters for the most part into the gall-bladder, from which it is discharged shortly after food is taken. The ducts have a fibrous external coat and an internal mucous coat.

144. **Minute Structure of the Liver.**—The surface of the liver shows to the naked eye small spots about the size of a pin's head. These mark the surfaces of the lobules, and the liver is made up of innumerable lobules, with interlobular connective tissue, each of which is about $\frac{1}{20}$ inch in diameter. Microscopic examination shows that each lobule consists of numerous polygonal cells about $\frac{1}{1000}$ inch in diameter, and that among these

FIG. 142.—Diagrammatic Representation of Two Hepatic Lobules.

The left-hand lobule is represented with the intralobular vein cut across : in the right-hand one the section takes the course of the intralobular vein. *p*, interlobular branches of the portal vein ; *h*, intralobular branches of the hepatic veins ; *s*, sublobular vein ; *c*, capillaries of the lobules passing inwards. The arrows indicate the direction of the course of the blood. The liver-cells are only represented in one part of each lobule.

hepatic cells are seen minute branches of the portal vein and the hepatic artery, with the commencement of a hepatic vein. Minute channels for the bile are also seen to begin between the hepatic cells.

The blood-vessels of the liver (portal vein and hepatic artery) enter the liver on its under surfaces, where also the bile duct leaves the organ. The branches of these three vessels are invested with connective tissue (Glisson's capsule), which is continuous with that surrounding the organ, and the three vessels continue together to the outside of the lobules, the connective tissue there terminating in septa dividing the lobules. The parts or spaces enclosing these three sets of vessels are called *portal canals*. We will follow the course of each set of vessels in detail.

The portal vein accompanied by the hepatic artery enters the liver at the transverse fissure, and is eventually distributed between the liver-lobules, forming *inter*-lobular veins. From the interlobular veins surrounding each lobule a close network of capillaries passes into the lobule, the meshes of the network being occupied with rows or columns of cells. These lobular capillaries converge towards the centre of the lobule, to form a central or *intra*-lobular vein.

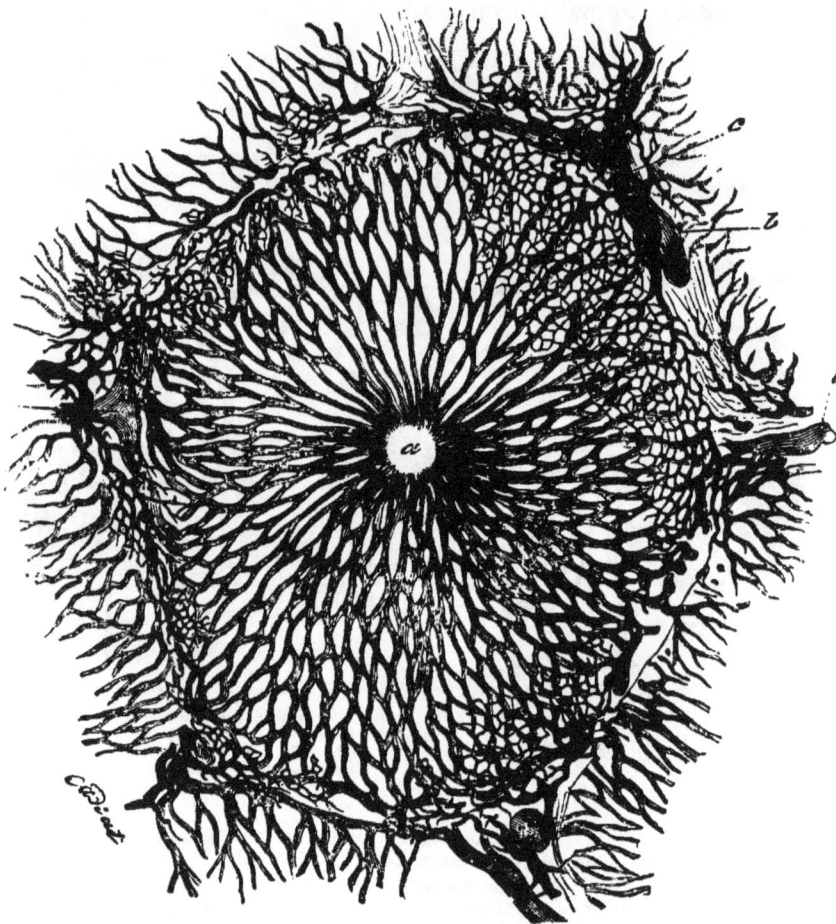

FIG. 143. Lobule of Rabbit's Liver, vessels and bile ducts injected. (Cadiat.)
a, central vein ; *b, b*, peripheral or interlobular veins ; *c*, interlobular bile-duct.

Each intralobular vein passes through the lobule to its surface, and joins similar veins from other lobules to form *sublobular* veins. The sublobular veins unite into larger and larger trunks, and end at last in the *hepatic veins*, three large veins which leave the liver at the back and open into the inferior vena cava.

The portal vein, which thus breaks up into capillaries like an artery, has comparatively thick and muscular walls, and neither it nor its branches, nor the branches of the hepatic veins, contain any valves.

The hepatic artery divides into branches, which accompany the branches of the portal vein and bile duct until it comes between the lobules. These branches are much smaller than those of the portal vein, and smaller than the accompanying bile duct. In their course they give off capillaries to supply the branches of the portal vein, the larger bile ducts, and the capsule, these capillaries terminating in the hepatic veins that enter branches of the portal vein. Other capillaries are said to break off from the interlobular portion of the artery, and entering the lobule unite with the capillaries of the portal vein near the margin of the lobule. There are thus two sets of vessels carrying blood to the lobules, but only one set returning blood.

The hepatic ducts *commence* in minute passages between the cells of a lobule, and are called bile capillaries. These bile capillaries form in the lobule a closer capillary network than that of the blood-capillaries, and as they run between the opposed surfaces of adjacent cells, while the blood-capillaries run along the edges of the cells, the two sets of vessels are kept separate. The bile capillaries radiate to the circumference of the lobule, and form interlobular ducts between the lobules. The interlobular ducts unite, and entering Glisson's capsule accompany the branches of the portal vein, until by union

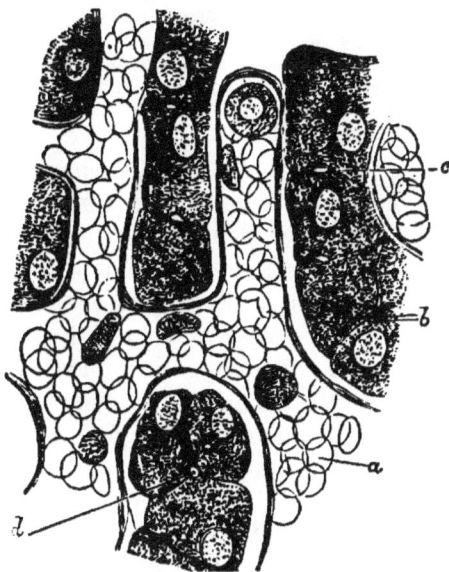

FIG. 144.—Microscopical Section from the Liver of a Child Three Months Old, hardened in Chromic Acid.

The hepatic cells (*b*), with their single nuclei, are separated from the capillary wall by a small intervening space. The capillaries (*a*) contain closely compressed coloured, and a few colourless, blood-corpuscles. A few elongated nuclei belonging to the capillary wall are seen. Within the line of junction (septum), between two hepatic cells, the transverse section of a biliary duct is seen as a small transparent space (*c*). There is also one at the angle where several of these cells come into contact (*d*).

they form the two main ducts which leave the liver at the transverse fissure. The large bile ducts possess a fibrous and elastic coat, with some fibres of unstriped muscular tissue arranged circularly.

The liver cells are polygonal in shape and about $\frac{1}{1000}$th of an inch in diameter. They are arranged in columns between the meshes of the blood-capillaries, and under high powers the protoplasm and nucleus of each cell is seen to be fibrillated. Their appearance differs according as the body is fasting or digesting food. In fasting they are finely granular and cloudy, but after a meal, especially of starchy food, coarse particles of glycogen are visible (fig. 145).

It will now be seen that each lobule of the liver contains all the elements of a gland—protoplasmic cells to form a secretion, blood-capillaries in close relation with the cells to provide material for secretion, ducts to carry away the secreted products. But, as we shall presently see, the cells of the liver perform other functions besides that of secreting the bile. The liver, in fact, discharges at least four functions :—(1) Its cells secrete bile which passes away by the hepatic ducts. (2) Its cells form glycogen or animal starch, which passes in an altered form into the blood-stream. (3) It takes part in the formation of urea and other excretory products. (4) It is concerned in the destruction of the red corpuscles of the blood. Moreover, it is believed to assist in arresting and changing deleterious matters in the blood.

145. **Secretion of Bile.**—The secretion of bile appears to be continuous, but the rate is much influenced by food. The bile does not exist as such in the blood, but is elaborated by the secretory activity of the hepatic cells from constituents in the portal blood. Passing into the bile capillaries and onwards into the bile ducts, it is carried along the common bile duct into the duodenum during digestion, but during periods of fasting it returns along the cystic duct into the gall-bladder, where it accumulates until food is again taken. The rate of secretion rises rapidly after meals, then falls slightly for a time, after which there is a more gradual rise and fall. The discharge of bile in quantity shortly after the ingestion of food is probably a reflex action, the presence of the acid chyme in the duodenum acting as a stimulus and leading to contraction of the muscular walls of the gall-bladder and bile ducts. The pressure of the bile in the bile ducts is very low, though nominally a little higher than in the portal veins—a proof that it is not a mere filtrate from the blood, though the amount is influenced by the quantity of blood sent to the liver and by blood pressure.

If the pressure in the ducts be much increased by any obstruction, the bile is reabsorbed from the distended ducts into the lymphatics of the liver, and this lymph entering the thoracic duct, the bile thus passes into the blood and produces the condition called *jaundice*, when the skin and white of the eye are yellow, while the fæces are light-coloured.

The quantity secreted daily is about sixteen ounces. Its appearance and chemical composition have been described in par. 124. Some of the elements of bile (*e.g.* the colouring matters and cholesterin) are probably merely separated from the blood, but the more special constituents (*e.g.* the bile salts, sodic glycocholate and sodic taurocholate) are formed by the activity of the hepatic cells. Its chief uses in digestion are to assist in emulsifying fat, so as to enable it to pass through the intestinal membrane without alteration, to precipitate the peptones and form a sticky deposit on the valvulæ conniventes, thereby giving time for absorption,.to stimulate the muscular walls of the intestine, and to act as an antiseptic or hindrance to putrefaction (see par. 125).

146. **Glycogenic Function of the Liver.**—On killing a well-fed animal and at once removing the liver, suitable methods of analysis show the presence of a carbohydrate body called glycogen or animal starch. It can also be seen under the microscope in a section of the liver of such an animal, stored up as granules in the liver-cells. Glycogen has a formula $C_6H_{10}O_5$, or some multiple of this, and is distinguished from ordinary starch by giving a wine-red colour with iodine. If the liver be left several hours in a warm place, little or no glycogen will then be found, but abundance of sugar (dextrose). By some means, therefore, the glycogen soon changes into sugar after death, and the same change of glycogen into sugar is believed to occur during life, though more slowly. But where do the liver-cells obtain their glycogen, for the blood contains none? Glycogen is formed in the liver from the sugar and some other absorbed food materials in the portal blood. A process of dehydration or taking away the elements of water from the sugar, $C_6H_{12}O_6$, may be one method by which the liver-cells produce it. During digestion, especially after a meal rich in carbohydrates (starchy and saccharine food), the blood entering the liver by the portal vein contains more sugar than the blood leaving the liver by the hepatic veins, but in the intervals of digestion the hepatic venous blood contains about twice as much as the portal venous blood, the *average* amount in portal blood being 1 per 1,000, and in the blood of the hepatic veins 2 per 1,000. It thus appears that the liver regulates the amount of sugar in the blood, storing it as glycogen during digestion and reconverting this glycogen into sugar to be discharged by the hepatic veins for use by the tissues during the time that no food is ingested. This storage is rendered necessary by the fact that if the sugar in the blood rises above ·3 per cent. the excess is excreted by the urine. As evidence of the above function of the liver, it may be mentioned that if a solution of sugar be slowly injected into a branch of the portal vein, no sugar appears in the urine, but if the sugar be injected into the jugular vein its presence in the urine may be easily detected.

The source of glycogen in the liver is undoubtedly the food, for in the liver of a starved animal glycogen is practically absent. That it is derived from carbohydrates especially is shown by the large quantities that accumulate in the liver-cells in animals fed on such a diet, the liver thus forming a storehouse of carbohydrate material. But it is also derived from proteids, as has been proved by feeding a dog on proteid material freed from all carbohydrate. It is not derived from fats, because the glycogen soon disappears from the liver of an animal fed exclusively on fat. The destination of the glycogen of the liver is to some extent still under discussion. But the most generally accepted view is that it is conveyed from the liver as sugar to undergo combustion in the tissues. Heavy muscular work soon leads to the disappearance of hepatic glycogen, and the amount of sugar in the venous blood of an active muscle is slightly less than in its arterial blood. Thus the sugar produced by the liver-cells from the stored glycogen is consumed by muscle, and the 'sugar-cycle' of a well-fed animal is as follows:—The sugar absorbed from the alimentary canal enters the portal blood, is in great part stored as glycogen in the liver-cells, is gradually reconverted into sugar that passes away by the hepatic veins, is consumed by living muscle and discharged as CO_2 and H_2O.

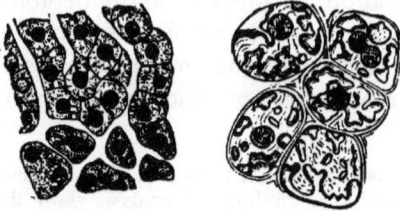

Fasting. FIG. 145. After Food.

Liver-cells of dog after a thirty-six hours' fast, and fourteen hours after a full meal—in the latter case swollen with glycogen. (Heidenhain.)

Besides existing in the liver, glycogen is also found to a small extent in muscular tissue, especially in the skeletal muscles, where it possibly forms a local reserve to supply the energy of muscular contraction. It is especially abundant in the rudimentary muscles of the embryo, where it appears to become transformed during development into the contractile substance of the muscular fibres.

[Glycogen may also be regarded as a source of heat, for it is found to disappear in a few hours from the liver of a rabbit that is kept in a cold bath.]

147. Diabetes.—When the amount of grape-sugar (dextrose) in the blood rises above ·3 per cent., the sugar is excreted by the kidneys and appears in the urine. The presence of grape sugar in the urine is spoken of as *glycosuria*, or *diabetes*. Temporary diabetes may be produced in a well-fed animal by a puncture made in the lower part of the floor of the 4th ventricle. For a day or two the animal passes sugar in its urine, and then sugar ceases to be excreted. In a starved animal the puncture does not lead to sugar in the urine. The glycosuria in this case, therefore, appears to be caused by the conversion of the glycogen of the hepatic cells into sugar so rapidly that the excess is excreted by the kidney. The nature of the nervous influence that thus causes glycosuria is not definitely known. The puncture is effective even after section of the vagi, so that these nerves cannot be the channels along which the influence is conveyed to the liver.

Glycosuria, or temporary diabetes, may also be caused by several drugs, as strychnine, phlorizidin, &c. Phlorizidin will produce glycosuria in starved animals which are free from carbohydrates, and in this case the sugar must be formed directly from proteid.

In the disease known as diabetes, or *diabetes mellitus*, the blood contains an abnormal amount of sugar, and large quantities are passed in the urine. This is probably due to the incapacity of the tissues to consume the sugar, or to the inability of the liver-cells to retain glycogen, for in severe cases sugar continues to be excreted even when carbohydrates are excluded from the food.

It has been recently found that the pancreas has some peculiar relation to the sugar function of the body, for if this organ be removed from a dog the animal becomes affected with severe diabetes that continues until death ensues. It is thus concerned in the metabolism of the carbohydrates.

On the formation of urea in the liver, see par. 155.

148. **The Spleen.**—The spleen, thyroid, thymus, and suprarenal capsules, are often spoken of as *blood-glands*, from the peculiar changes they are believed to bring about in the blood circulating through them. As they have no excretory duct, but pour their products into the blood-stream, they are also spoken of as *ductless glands*. In this respect they are closely related to the lymphatic glands already described. The spleen is the largest of these so-called ductless glands. It is situated at the cardiac end of the stomach, has a deep bluish-red colour, a somewhat oval shape, and measures about five inches in length and three inches in breadth. On the internal concave surface is the hilum, or recess where vessels and nerves enter and leave the organ. The spleen is invested with an external serous coat derived from the peritoneum, while under this is a tough fibrous and muscular capsule which is reflected inwards upon the vessels of the hilum. The general structure of the organ is similar in many respects to that of a lymphatic gland. From the inner surface of the capsule numerous partitions or trabeculæ pass off and join those arising from the covering of the blood-vessels, the whole organ being thus divided into a large number of irregular spaces. In those spaces formed by the interstices of the fibrous framework lies a soft mass of dark red substances, the spleen pulp. This is found to consist chiefly of cells, the branching processes of one kind forming a delicate network of small communicating spaces in the interior of the larger spaces formed by the trabeculæ of the capsule. These small spaces, formed by branching flattened cells, contain blood, granular corpuscles resembling lymph corpuscles, and large amœboid cells containing coloured blood-corpuscles in various stages of transformation.

The splenic artery, a branch of the aorta, divides into

Q

several branches that enter the organ at the hilum and ramify in the interior. As the small arteries leave the trabecular sheaths, their external coat of connective tissue becomes gradually changed into adenoid or lymphoid tissue, the altered coat having here and there spheroidal swellings termed the *Malpighian corpuscles of the spleen*. These small bodies, visible to the naked eye as whitish specks in a section of the spleen, are from $\frac{1}{30}$ to $\frac{1}{80}$ inch in diameter, and in minute structure

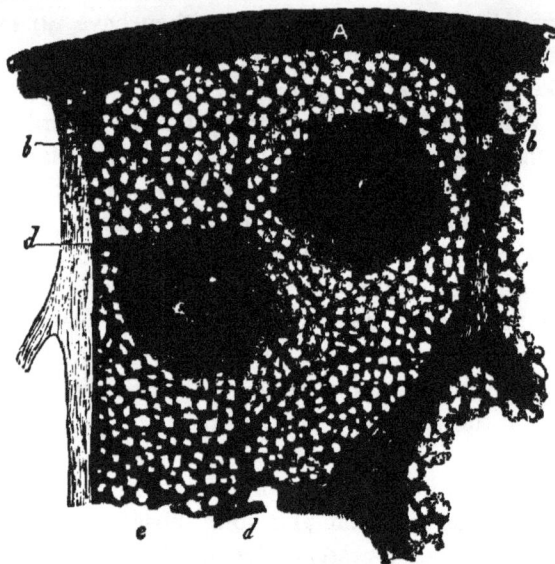

FIG. 146.—Vertical Section of a small Superficial Portion of the Human Spleen, as seen with a low power. (Kölliker.)

A, peritoneal and fibrous covering ; *b*, trabeculæ ; *c, c*, Malpighian corpuscles, in one of which an artery is seen cut transversely, in the other longitudinally ; *d*, injected arterial twigs ; *e*, spleen pulp.

are found to consist of delicate adenoid reticulum traversed by minute capillaries from the small artery, and containing lymph corpuscles in the meshes. The small arteries terminate in capillaries that traverse the pulp in all directions, and it has been shown that in many cases the capillary vessels terminate in the reticulated tissue, so that the blood thus comes into actual contact with the tissue elements. Venous channels arise in the pulp, and convey the blood that has escaped into the interstices formed by the branched capsules into veins wnich run in the fibrous trabeculæ and ultimately leave the organ at the hilum.

149. **Functions of the Spleen.**—(1) The spleen appears to take some share in the formation of blood-corpuscles. The splenic vein certainly contains more colourless corpuscles than the splenic artery, and according to some coloured corpuscles are also added, the new-formed corpuscles entering the blood as it filters through the spongy network of the pulp. That colourless corpuscles are produced within the organ seems to be proved by the enormous increase in their number when the spleen is enlarged in the disease termed leucocythæmia. (2) There is evidence to show that the spleen destroys and removes worn-out red corpuscles and particles of foreign matter, chemical analysis revealing the presence of nitrogenous extractives probably thus derived. The large amœboid splenic cells

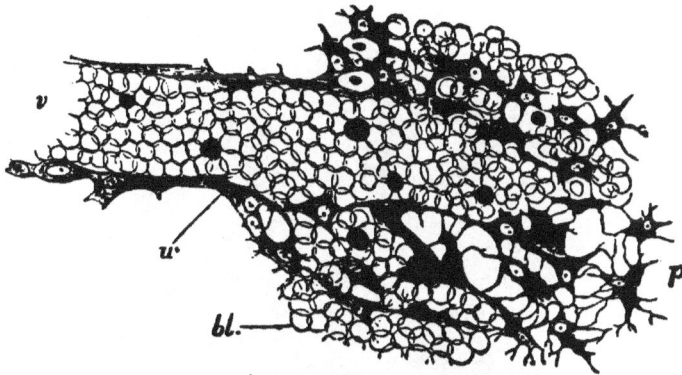

FIG. 147.—Thin Section of Spleen Pulp, highly magnified, showing the mode of origin of a small Vein in the Interstices of the Pulp.

v, the vein, filled with blood-corpuscles, which are in continuity with others, *bl*, filling up the interstices of the retiform tissue of the pulp; *w*, wall of the vein. The shaded bodies amongst the red blood-corpuscles are pale corpuscles.

seen in the splenic pulp often contain pigment granules and fragments of red corpuscles that finally disappear, the spleen thus playing the part of a purifying filter. (3) Towards the end of digestion the spleen increases in size, owing chiefly to an increase of granular albuminous plasma in the organ, a diminution gradually following as this material becomes less. Uric acid is almost constantly found in the organ. It thus appears that the spleen has some relation to the storing and elaborating of nitrogenous food materials absorbed during digestion. It is a remarkable fact, however, that removal of the organ from an animal produces no serious effect, the only apparent result often being an enlargement of some of the lymphatic glands.

To assist the circulation through the network of the splenic pulp, the muscular tissue of the capsule and trabeculæ has the power of rhythmic contraction. When the organ is enclosed in a plethysmograph, or volume recorder, its bulk is found to undergo slow variations, each contraction and expansion occupying about a minute. The muscular activity of the spleen is under the control of the nervous system, a rapid contraction being brought about directly on stimulation of the vagus or splanchnic nerves, as well as by stimulation of the splenic nerves themselves. These latter nerves have their centre in the medulla, as stimulation of the medulla either directly or by asphyxiated blood induces contraction of the organ. The nervous system also in some way regulates the flow of blood through the spleen.

150. **The Thyroid Gland.**—The thyroid consists of two lobes of isolated ductless alveoli, bound together by connective tissue. It is situated beneath the muscles of the neck on each side of the larynx and trachea, a cross-piece in front of the trachea uniting the two lobes (fig. 67). Each lobe is somewhat conical in shape, about two inches in length and one inch in breadth. The gland is covered by a thin layer of areolar tissue, and from this capsule fibrous partitions pass inwards and divide the organ into spaces occupied by spherical or oval vesicles. Each vesicle is closed and lined by a layer of cubical epithelial cells resting on a basement membrane, and is filled either with a clear glairy fluid or a more solid material termed

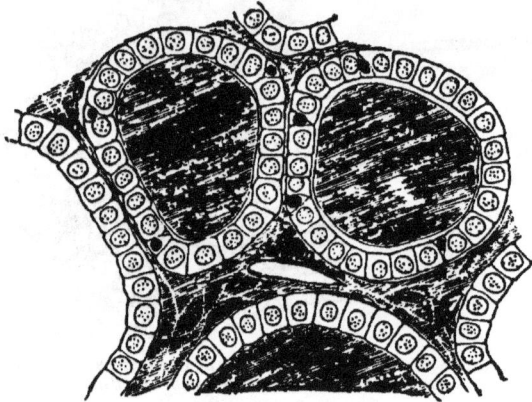

FIG. 148.—Section of the Thyroid Gland of a Child (Quain's Anatomy).
Two complete vesicles are seen. In the middle of one of the interspaces a blood-vessel is seen. Between the epithelial cells there are small cells like lymph corpuscles.

colloid. A rich supply of blood is furnished by the two thyroid arteries, the capillaries forming a dense plexus round the vesicles. Numerous lymphatics arise in the organ, and an abundant supply of nerves pass off from the cervical ganglia of the sympathetic.

Little is known as to the function of the thyroid. It probably forms an internal secretion in the vesicles which is carried off by the lymphatics and poured into the blood. In the disease known as *goitre* there is great enlargement of this organ, often accompanied by a peculiar form of idiocy termed *cretinism.* Disease of the organ or its extirpation also frequently results in changes in the composition of the blood, an accumulation of mucin in certain tissues, and distressing nervous symptoms. The disease *myxœdema*, occurring in adult life, is connected with waste or atrophy of the thyroid, and is marked by excess of mucinoid substance in the subcutaneous tissues and by general stupidity. An extract of sheep's thyroid, or eating the raw organ, has been found to relieve the disease, so that the gland must be in some way connected with the healthy nutrition of the body.

151. The Thymus.—The thymus gland is a temporary organ placed over the trachea behind the upper part of the sternum. At birth it weighs about half an ounce, increases in size during the first two years, and then shrivels away gradually, being represented in the adult by a small portion of fatty substance. The organ is surrounded by a capsule from which fibrous partitions pass inwards, dividing the body into lobes and lobules and carrying the blood-vessels. Each small lobule is further subdivided into follicles by delicate connective tissue, while each follicle consists of a cortical portion containing lymphoid cells and a medullary portion containing granular cells and the concentric cells of Hassall. The thymus is probably concerned in some way in producing changes in the chemical or cellular character of the blood. Coloured corpuscles and hæmoglobin granules are said to be found in the issuing lymph. Various extractives can be obtained from the gland, as well as a peculiar proteid body of the nature of globulin, which causes intravascular clotting when injected into the veins. But the early development and rapid disappearance of the thymus suggest that its chief functions are performed during that period of life when growth and tissue formation are most active.

152. The Suprarenal Bodies.—The suprarenal capsules are two small bodies of a flattened triangular shape resting upon the upper surface of the kidney (fig. 149). They are about $1\frac{1}{2}$ inch long and weigh about 2 drams. A section through a fresh organ shows an external or cortical zone and an internal or medullary zone. Each capsule is invested by a fibrous coat that sends vertical divisions into the cortex. The cortical portion thus consists of a fibrous framework with cells arranged in columns, the blood-vessels running between in the fibrous septa. In the medullary part the framework of connective tissue has more the arrangement of a close network, and the cells of the medulla lying in the interstices are irregular and branching. The arteries supplying these bodies break up into capillaries in the fibrous septa of the organ, forming in the medullary part venous sinuses which converge and ultimately form the large vein which emerges at the centre of the gland. Nerve fibres with ganglion cells in connection, whose mode of termination is unknown, are numerous. Complete removal of the suprarenal capsules leads to an alteration in the composition of the blood, to great muscular weakness and to nervous prostration. Their removal is also fatal to the animal experimented upon after a short time. They thus appear to decompose effete pigment in the blood and secrete something necessary to muscular tone.

CHAPTER XI

EXCRETION.—THE KIDNEYS AND THE SKIN

153. Elimination of Waste Products.—It has already been shown that the various food-materials taken into the alimentary canal are prepared by the digestive processes for being taken into the blood, that by absorption into the rootlets of the portal vein, or into the lymphatics of the intestines, they are carried into the blood-stream, and that by exuding through the thin walls of the systemic capillaries the blood-plasma or lymph carries the absorbed material to the cells and other tissue elements, to serve for their nourishment. Oxygen taken up in the alveoli of the lungs is also being continually carried to the tissues. As a result of tissue activity or metabolism, which is mainly a process of oxidation, certain waste products are formed which enter the venous blood, either directly in the capillaries or indirectly by the lymphatic circulation. Carbo-hydrates, fats and proteids are converted into carbon-dioxide, water, urea, and other allied nitrogenous bodies. Certain inorganic salts, as sulphates and phosphates, are also produced by oxidation of the sulphur and phosphorus in some articles of food. These waste products must be removed from the body in some way, and their elimination is effected by special parts termed excretory organs, the effete substances so removed being called *excretions*. As already shown, the lungs serve as the chief channel for the elimination of carbon-dioxide, as well as for a considerable amount of water in the form of aqueous vapour. The kidneys also remove a large quantity of water in which are dissolved nearly all the urea and allied bodies, with the main portion of the salts, the liquid being known as the urine. A variable amount of water, with a small quantity of salts and carbon-dioxide, is eliminated from the blood by the skin in the form of sweat. The fæces also contain, besides undigested portions of the food, a small amount of waste products mainly derived from the bile ; for the bile, though chiefly a secretion to aid digestion and so reabsorbed by the intestine, furnishes a small amount of excrementitious matter that colours the fæces. Even a small portion of the reabsorbed bile is afterwards excreted in the urine as urobilin (par. 124), or other colouring matter of the urine.

154. **Structure of the Kidneys.**—The two kidneys are situated in the abdomen close to its hinder wall and behind the peritoneum and intestinal canal, one being on each side of the spinal column. They are surrounded by a mass of fat and loose areolar tissue, and capped by the suprarenal bodies. Each kidney, which has a characteristic shape, is about four

inches long, two to two and a half broad, and about one inch thick. The weight varies from four to six ounces. They are invested by a tough fibrous capsule which can be easily removed from the substance of the organ, being only attached by fine processes of connective tissue and minute blood-vessels. The deep longitudinal fissure named the *hilum*, on the internal concave border, allows the passage of the blood-vessels, nerves, and ureter into and out of the organ.

On making a vertical section of a kidney, the naked eye distinguishes a deep red outer *cortical* portion of the kidney sub-stance and a pale red inner *medullary* or pyrami-dal portion, the latter being composed of about twelve papillæ or pyramids, the apices of which project into a wide funnel-shaped sac called the *pelvis*. The pelvis is the dilated ex-tremity of the ureter or excretory duct, and is divided into several trun-cated branches called calyces, or infundibula, around each apex of the pyramids. The cortical matter lying outside also invests the bases of the

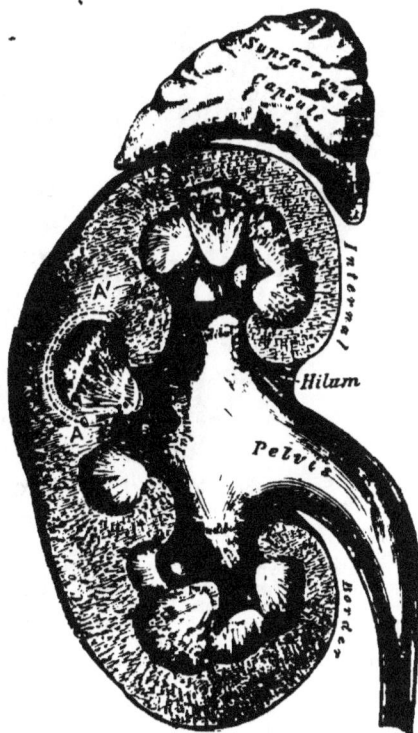

FIG. 149.—Vertical Section of Kidney.

pyramids and sends processes between them known as the columns of Bertin. It is soft and friable and shows small granules due to the presence of Malpighian corpuscles.

The medullary portion, which is divided into a *boundary layer* and *papillary part*, is denser than the cortex, and exhibits striæ or radial streaks passing into the cortex, and there known as *medullary rays*, or pyramids of Ferrein. The portion of the cortex lying between the medullary rays is termed the *labyrinth*, and exhibits the red granules just mentioned. Careful examination with a hand lens shows small openings at the apex of each papilla, and on squeezing the part under examination a drop or two of urine exudes.

Microscopic examination shows that the kidney is made up of tubules, blood-vessels, and a small amount of connective tissue. The organ is in fact a compound tubular gland, the *tubuli uriniferi*, or urine carrying tubules, arising within the cortex, and these, after pursuing a complicated course and uniting with other tubules to form collecting tubules, discharge their fluids into the calyces of the pelvis, through the minute openings at the end of each pyramid. Both cortex and medulla are thus largely composed of these uriniferous tubules. Those parts of the tubules in the medulla have a straight direction, and those in the cortex for the most part a contorted arrangement, with some groups of straight tubules passing from the medulla to form the medullary rays (fig. 151).

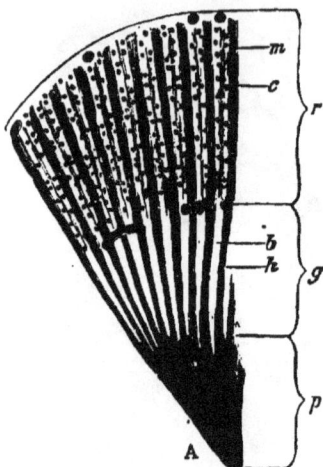

FIG. 150.—Section through part of the Dog's Kidney. (Ludwig.)

p, papillary, and *g*, boundary zones of the medulla ; *r*, cortical layer ; *h*, bundles of tubules in the boundary layer, separated by spaces, *b*, containing bunches of vessels (not here represented), and prolonged into the cortex as the medullary rays, *m* ; *c*, intervals of cortex, composed chiefly of convoluted tubules, with irregular rows of glomeruli, between the medullary rays. At apex of pyramid A, excretory ducts open into a calyx of the pelvis.

Each uriniferous tubule consists of a layer of epithelial cells resting on a basement membrane, the blood being brought into close connection with these cells at certain parts of the tubule in ways presently to be described. The tubules begin in the cortical part of the kidney in a cup-shaped expansion about $\frac{1}{120}$ inch in diameter, termed the *Malpighian capsule* (fig. 153), each capsule enclosing a bunch of convoluted blood-capillaries termed a *glomerulus*. The tubule leaves the capsule by a narrow *neck* (2), becomes twisted on itself to form the *first convoluted tubule* (3), straightens somewhat to form the *spiral tubule* (4), and then passes directly down with diminished width into the medulla to form the *descending loop of Henle* (5). Before reaching the apex of the pyramid the tubule from the loop of Henle (6) by an upward turn runs towards the cortex, parallel to its previous direction, as the *ascending loop of Henle* (7, 8, 9), again enters the cortex in a wavy manner as the *zigzag tubule* (10), and again takes on a contorted course as the *second convoluted tubule* (11). Now narrowing into a *junctional tubule* (12), it joins a straight or *collecting tubule* (13), which passes through the medullary substance (14), and unites with other collecting tubules to form a *duct of Bellini*, or excretory duct, that opens at the apex of the papilla, thus allowing the secreted urine to be discharged into the pelvis, or expanded upper end of the ureter.

The epithelium cells lining the tubules vary in character in

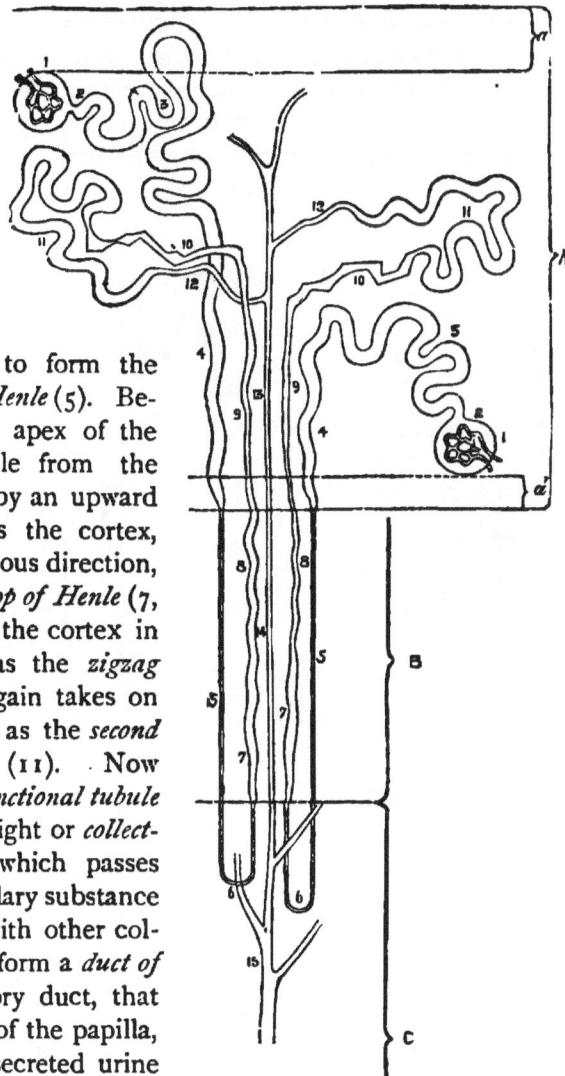

FIG. 151.—Diagram of the Course of Two Uriniferous Tubules. (Klein.)

A, cortex; B, boundary zone; C, papillary zone of the medulla; *a, a'*, superficial and deep layers of cortex, free from glomeruli. For the explanation of the numerals, see the text.

the different parts ; three varieties may be noted. In the capsule the epithelial cells are squamous or flattened, and are reflected over the glomerulus (fig. 152), there being thus two layers, the inner of which is fused with the glomerular loops.

In the first convoluted tubule, in the spiral tubule, in the ascending loop of Henle, in the zigzag tubule, and in the second convoluted tubule, the epithelial cells are granular

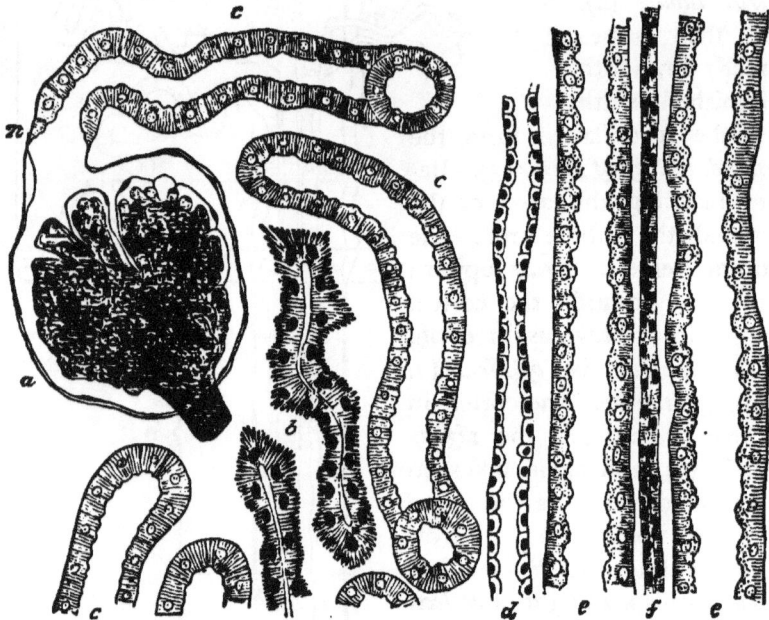

Fig. 152.—Tubules from a Section of the Dog's Kidney. (Klein.)

a, capsule, enclosing the glomerulus ; *n*, neck of the capsule ; *c, c,* convoluted tubules ; *b,* irregular tubules; *d,* collecting tube ; *e, e,* spiral tubes ; *f,* part of the ascending limb of Henle's loop, here (in the medullary ray) narrow.

cubical cells, and show a fibrillated or rodded structure. In the descending loop of Henle and in the straight collecting part of the tubule the cells are clear cubical cells.

Thus the cells of the convoluted and irregular parts of the tubules exhibit the characters of active secreting cells, and their protoplasm is believed to be concerned in extracting from the blood the chief organic substances of the urine, while the cells lining the collecting and discharging parts of the tubule are similar to those seen in the conducting part of a gland.

The kidney is abundantly supplied with blood. The renal artery, a direct branch of the aorta, enters the organ at the hilus, dividing in its passage into several branches that pass round the pelvis into the kidney sub-stance between the pyramids in the columns of Bertin. On reaching the boundary between the cor-tex and medulla these branches divide, and spreading laterally form incomplete anastomosing arches at the bases of the pyra-mids. From these arterial arches vessels proceed on the one hand outwards towards the cortex and inwards to the medullary pyra-mids. The vessels from the arterial arches that run outwards to the cortex between the medul-lary rays are the *interlobular*

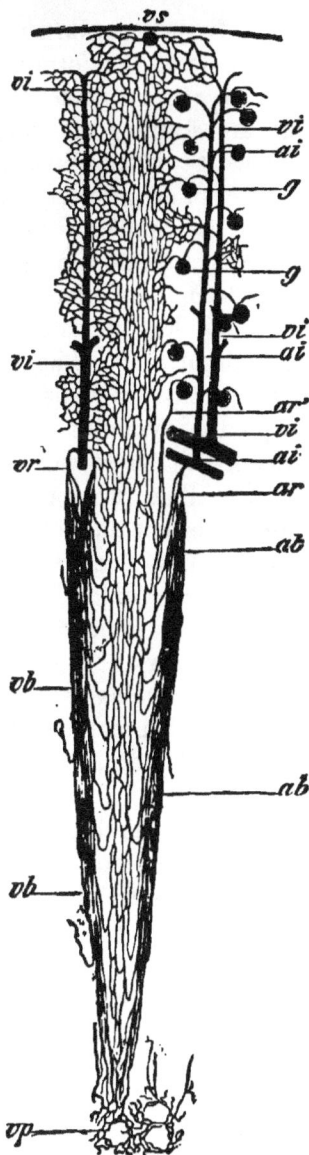

FIG. 154.—Diagram of the distribution of the Blood-vessels in the Kidney. (Ludwig.)

ai, ai, interlobular arteries; *vi, vi,* in-terlobular veins; *g,* glomerulus; *vs,* stellate vein; *ar, vr,* arteriæ et *venæ recta* forming bundles, *ab* and *vb*; *vp,* venous plexus in the papillæ.

FIG. 153.—Minute Structure of Kidney.

arteries. In their outward course the interlobular arteries give off lateral branches that form the *afferent* vessels of the Malpighian bodies, an afferent vessel entering the capsule of a uriniferous tubule and forming the glomerulus. From each glomerular tuft a somewhat smaller *efferent* vessel passes out of the capsule, and again breaks up into a network of capillary vessels over and between the tubules. This double capillary network—a first capillary distribution in the capsules to form the glomeruli and a second capillary distribution to form the vascular network over the tubules—repeats on a small scale the peculiarity of the portal circulation. The branches of the arterial arches that run off to the medullary substance in a straight direction through the pyramids as the *arteriæ rectæ*, break up into a plexus of capillaries with elongated meshes. From these arise *venæ rectæ* that run into venous arches corresponding to the arterial arches at the boundary of the cortex and medulla, these venous arches also receiving the interlobular veins that are formed by the union of the capillaries from the Malpighian bodies, and by some capillaries formed by small branches of the interlobular arteries that do not enter the glomeruli.

The small veins at the surface of the cortex being arranged in a star-shaped manner, are spoken of as the *venæ stellatæ.*

It will be observed that there are at least two ways by which the blood can pass through the kidney without traversing the glomeruli, (1) through the straight vessels of the medullary portion, whose blood supply is distinct from that of the cortex ; (2) through the small branches given off from the interlobular arteries of the cortex, that break up into capillaries without entering the capsules of Malpighi.

The lymphatics of the kidney arise in lymph spaces of the scanty framework of connective tissue in the organ, and the fluid collected into vessels emerges from the organ by lymphatics that make their exit either at the tubes or through the capsule. The nerves derived from the renal plexus and lesser splanchnic nerves form small trunks with ganglia, and for the most part accompany the blood-vessels. Little is known as to their mode of termination. No secretory nerves similar to those found in the submaxillary gland have been discovered.

155. The Urine.—Fresh urine is a clear straw-coloured fluid of peculiar odour and acid reaction. It consists of water holding in solution urea and other solids, and has an average specific gravity of 1·020. The specific gravity, being dependent on the amount of solids relative to water, may vary from 1·002 after drinking much wa'er, to 1·040 after abstinence from liquid, and after copious perspiration. Any cause which draws off a large quantity of fluid from the body through any other channels than that of the kidney, e.g. the skin, or the lungs, or the bowels, lessens the quantity of water in the urine and increases the specific gravity, for the amount of solids discharged daily by the kidney is fairly constant. Thus in cold weather when perspiration is scanty the urine is more abundant and more dilute than in summer, when the amount of water vapour discharged from the skin is great. A healthy adult passes on the average about 50 fluid ounces or 2½ pints of urine daily, the amount of solids in solution being about 2 ounces, or 4 per cent. Urea is the chief solid, forming 2·2 per cent. Sodium chloride comes next, forming 1 per cent. Other solids dissolved in small quantity are acid sodium phosphate, phosphates of calcium and magnesium, alkaline sulphates, uric acid, creatinin, &c. Nearly two-thirds of the solid substances are organic bodies. A quantity of the gas carbon-dioxide, with a small amount of nitrogen, is also held in solution in urine.

The acidity of urine is not due to the presence of free acid, but to acid sodium phosphate. (The sulphuric acid is combined to form sulphates, the phosphoric acid to form phosphates, and the uric acid to form salts termed urates.) After standing for some time in contact with air urine becomes alkaline owing to the conversion of urea into the alkaline body, ammonium carbonate. The conversion of urea into ammonium carbonate, by taking up two molecules of water, is brought about by a micro-organism.

The colour of urine is due to pigments concerning the nature of which much discussion has arisen. According to some authorities the chief colouring matter is the pigment *urobilin*, which is said to be a derivative of bile-pigment.

The average daily quantity of the chief constituents of urine in twenty-four hours is thus stated by Waller :—

	per 1000	per diem	
Water	960	1440	c. cm.
Urea	20	30	grams
Uric acid		0·75	,,
Hippuric acid	2	0·75	,,
Creatinin		1·5	,,
Phosphates		3·0	,,
Chlorides	15	7·5	,,
Sulphates		3·0	,,
Mucus and Extractives	3	—	

Urea, the chief solid constituent of urine, is an almost colourless body, which can be obtained from a solution in water or alcohol in the small silky four-sided prismatic crystals or in delicate white needles. Its chemical composition is represented by the formula $CO(NH_2)_2$. About 33 grams (500 grains) are excreted daily by an adult man, but the amount varies with circumstances, being increased by large quantities of nitrogenous food and diminished by a vegetable diet. Active muscular exercise is also said to produce a slight increase of urea, though the excretion mainly increased by muscular activity is CO_2 from the lungs. There is no doubt that urea represents the chief end-product of the changes undergone by the proteid or nitrogenous food stuffs, just as carbon dioxide represents the chief end-product of the carbon contained in the food, and water, the chief end-product of the chemical change undergone by hydrogen. The production of carbon dioxide and water can be brought about by the simple process of oxidation. But the stages intermediate between proteids and urea are complex and ill understood, while urea differs from CO_2 and H_2O in being only a partially oxidised product. It is thought that creatin [1] may be a waste product of muscular tissue that passes into the blood and forms a precursor of urea. Another source is believed to be such substances as leucin and tyrosin, which are two of the

FIG. 155. —Crystals of Urea.

a, four-sided prisms ; *b*, indefinite crystals, such as are usually formed from alcohol solutions.

products that may arise from pancreatic digestion in the alimentary canal. These substances being absorbed in the intestines are carried by the portal vein to the liver and there converted into urea by the hepatic cells. The urea is then taken up by the hepatic veins, and passing into the general circulation is ultimately eliminated by the kidneys. That some urea is formed by the liver seems evident, because (1) when leucin is introduced into the alimentary canal the amount of urea in the urine is increased, though leucin itself does not appear ; (2) in acute yellow atrophy of the liver, in which the liver-cells lose their function, leucin and tyrosin partially replace the urea of the urine ; (3) on passing blood through a liver from a recently killed animal, the amount of urea in the blood is increased ; (4) when the circulation of an animal is so altered that the blood from the portal vein is made to flow into the inferior vena cava, without passing through the liver, the amount of urea in the urine is considerably lessened. Other tissues, as the spleen and the lymphatic glands, are thought to take part in the formation of urea, but beyond what has just been said little is known as to the stages intermediate between the highly complex proteids absorbed by the blood and the comparatively simple nitrogenous end-product urea. One important point seems well established. Observations on starving animals show that the proteid in food is partly built up from the exuded blood-plasma into living protoplasm to supply the tissue waste (tissue or organ proteid), but is chiefly acted upon by the living tissues and converted into products that give rise to urea without having formed an integral part of the living substance (circulating proteid). The

---

[1] Creatin is $C_4H_9N_3O_2$. The removal of H_2O leaves Creatinin.

increase of urea excretion after the ingestion of proteid food, and the proportion that the urea discharged bears to the proteid food consumed, indicates that the urea excreted is primarily derived from the circulating proteid, that is, from the recently ingested proteid food, and only secondarily from the disintegration of muscle and other organs. As Dr. Waller observes: ' From proteid to urea there are three usual roads : (1) the short cut viâ leucin and tyrosin in the intestine, and urea in the liver ; (2) the high road viâ circulating proteid ; (3) the long, narrow way viâ circulating and organ proteid. The centre of action is the living tissue element, which, while undergoing little change as to its own proteid, effects considerable change in the proteid solution which soaks through and around it.' That urea is not formed but is eliminated by the kidneys is proved by the result that follows extirpation of these organs. In this case urea accumulates in the blood and tissues. Further, the blood normally contains urea—1 in 4000—and the renal vein carrying blood away from the organ has been found to contain less than the renal artery that leads to it. Uric acid ($C_5H_4N_4O_3$) occurs in normal urine in small quantity as a potassium or sodium salt, and is a less completely oxidised product of proteid metabolism than urea.

156. Secretion of Urine.—Two more or less distinct processes may be distinguished in the secretion of urine by the kidney. (1) A process mainly of *filtration*, by which the water and some of the highly soluble salts pass from the blood in the capillaries forming the glomeruli into the capsule at the beginning of the uriniferous tubules. (2) A process of true secretion, by which the urea and other specific constituents of the urine are secreted by the fibrillated epithelium cells of the tubuli from the blood in the second set of capillaries ramifying over the convoluted tubules.

A Malpighian body, as already explained, consists of a glomerulus or tuft of capillaries lying inside the capsule that forms the dilated end of a uriniferous tubule, the efferent or outgoing vessel of the capsule having a smaller calibre than that of the afferent or ingoing vessel. Hence the blood in the capillary vessel will be under high pressure, and fluid will pass through the filter formed by the thin capillary walls and the layer of flattened cells covering them into the cavity of the capsule where pressure is low. This view is supported by the fact that the quantity of urine secreted increases with increase of blood-pressure in the renal arteries, or rather with increased blood-flow through the kidney. But urine differs materially from the fluid that could be obtained as a filtrate from blood, especially in the absence of serum-albumin in which the blood is rich. It is evident, therefore, that the process is not one of physical filtration alone, but one in which the epithelial cells of the tubules take an active part.

Various observations show that the amount of urine depends largely on the blood-pressure in the glomeruli. Increase of blood-pressure in the glomeruli with increase in the amount of fluid excreted may be produced by—

(1) An increase in the general blood-pressure, due to increased action of the heart or to the constriction of the small arteries of the skin or organs other than the kidney.

(2) Dilatation of the renal artery unaccompanied by compensating dilatation elsewhere. Such dilatation may be caused by section of the renal nerves, a result which shows that these nerves normally exercise a vaso-constrictor action. It is probable that these nerves also contain dilating fibres, so that there appears to be a local nervous mechanism in the kidney controlling to some extent the flow of blood.

(As compression of the renal vein arrests the secretion of urine, while at the same time the pressure in the glomeruli may increase, it is more correct to say that the secretion of urine increases or diminishes with increased or diminished *blood-flow* through the kidney.)

Diminution of blood-pressure in the glomeruli with diminution of fluid secretion of urine may be produced by—

(1) A lowering of the general blood-pressure of the system due to diminished force of the heart's action, or to dilatation of the small arteries of parts other than the kidneys.

(2) A constriction of the renal artery by stimulation of the renal nerves.

The second part in the process of the secretion of urine— the separation of the specific organic constituents, urea, uric acid, &c. from the blood—is effected by the epithelium cells of the convoluted tubules. It has been pointed out that the cells lining these portions of the uriniferous tubules resemble in certain respects the active secretory cells of other glands, while those lining the Malpighian capsules and portions of Henle's loops are simple flattened cells ; and further, that these relatively large secretory cells are surrounded by a set of capillaries derived from the breaking up of the efferent vessels from the glomerular tufts. Thus the differences of epithelium in the uriniferous tubules, and the differences in the capillary blood supply at different parts of the tubule, forcibly suggest different modes of action in the different parts. The fluid filtered or secreted from the blood in the glomerular tufts into the capsules passes along the tubules, and washes down the urea, &c. excreted by the cells of the convoluted tubules. This view of the nature of urinary secretion is confirmed by various experiments. When the blue pigment sulphindigotate of sodium is injected into the blood of a mammal it is excreted by the kidneys, rendering the urine blue, and the course of its excretion has been traced, since it colours the cells through which it passes. Sections of the kidney of an animal so treated, and killed at appropriate times, reveal the presence of this pigment in the fibrillated cells of the convoluted tubules, but not at all in the capsules. After a time it is seen in the lumen of the tubules also, being washed down by the watery fluid from the capsule. Moreover, stoppage of the action in the capsule, either by destruction of the glomeruli with cauterisation or by section of the cervical spinal cord, which reduces blood-pressure so low as to stop the flow of urine, leads to the discovery of the pigment in the cells alone, the water to carry it along not having been secreted.

The same experimenter (Heidenhain) has also traced sodium urate into the convoluted tubules, and crystals of uric acid have been observed within the renal epithelial cells of birds.

Not only, therefore, does the structure of the epithelium cells of the convoluted tubules suggest a secretory activity similar to that of the cells lining the alveoli of a salivary gland, but direct experiments support this view. Yet the case of such glands as the salivary and gastric glands, as well as the pancreas, differs from that of the kidney in the fact that their specific constituents—mucin, pepsin, trypsin, &c.—are elaborated in the cells of the alveoli from antecedent substances in the blood, while the urea and some other important constituents of the urine are simply removed from the blood brought to the kidney, having been preformed elsewhere. There is evidence, however, to show that the small quantity of hippuric acid daily excreted in the urine is formed by the kidney-cells, from a combination of the benzoic acid in the blood and the glycin produced by its own cell metabolism.

157. **Variations in Urinary Secretion.**—It has been stated that the average quantity of urine in 24 hours is about 50 ounces of water and 2 ounces of

R

solids. But the proportion of water to solids, and therefore the specific gravity and colour of urine, vary within wide limits. The kidneys and skin are so correlated that, with the normal daily supply of water, an increase in the amount discharged by the one diminishes the amount discharged by the other as already stated (par. 155). It has also been shown that the amount of fluid removed into the capsule depends largely on the amount of blood passing through the organ, and as the kidney has its own vaso-motor nerves, with both vaso-constrictor and vaso-dilator fibres, under the control of the central nervous system, it is evident that nervous influences may increase the flow of clear watery urine by determining a greater flow of blood with increased pressure. Such an effect is produced by fear and other emotions, as well as by the nervous affection termed *hysteria*. Section of the renal plexus, or nerves passing to the kidney around the renal artery, causes an increase of watery urine owing to the great rise of pressure within the glomeruli. The nerve centre for the renal nerves lies in the medulla in the floor of the fourth ventricle, just in front of the origin of the vagus nerves. Puncture of this part also increases the flow of watery urine (*diabetes insipidus*). Close to this centre lies the centre for the vaso-motor nerves of the liver, and injury to this part leads to *diabetes mellitus*, copious urine containing sugar (par. 147).

By means of an instrument termed the *renal oncometer*, the actual volume of the living kidney may be observed to undergo variations, an increase of size resulting from dilatation of the renal vessels and a diminution resulting from their contraction. By registering these variations of volume on a revolving cylinder at the same time that the arterial blood-pressure is recorded, it is found that the kidney curve rises and falls in ordinary circumstances with the blood-pressure curve. This close connection between the volume of the kidney and the supply of blood to it enables the latter to be estimated by means of the oncometer. Increase of volume is followed by increased secretion of urine, and decrease of volume by decreased secretion. It is also found that changes in the composition of the blood soon affect the renal circulation and volume of the kidney. Injection of water into the blood or rapid absorption through the alimentary canal produces local dilatation of the renal vessels.

Certain drugs called *diuretics* are efficacious in promoting the flow of urine. Thus urea itself is a diuretic, and when injected into the blood is excreted along with watery urine by the cells of the convoluted tubules, even when secretion in the glomeruli has been arrested by section of the spinal cord. Other diuretics act differently. Sodium acetate produces an increase in kidney volume and increased secretion of urine without any increase in general blood-pressure, even when the renal nerves are cut—a result which appears to show that changes in the composition of the blood may produce local dilatation and increased secretory activity, by acting upon some vaso-motor mechanism within the kidney itself or directly upon its blood-vessels. Such changes in the blood probably explain the increased secretion after meals and the diminution during fasting and sleep.

Abnormal constituents of urine are sugar, albumin, and bile. In normal urine there is only a mere trace of grape-sugar or dextrose, but under certain conditions the amount becomes greatly increased with great increase of water also. This constitutes the disease *glycosuria*, or *diabetes mellitus*. The physiological conditions giving rise to this affection and its artificial production have been treated of in par. 147. In *diabetes insipidus*, which probably depends on some derangement of the central vaso-motor control of the kidney, there is a copious secretion of watery urine without sugar. The presence of albumin in urine is indicative of the condition termed *albuminuria*. Various causes bring about this affection, and it can be produced experimentally by ligature of the renal vein for a short time, the stoppage of the blood-flow so interfering with the activity of some of the cells that they seem unable to prevent the passage of the albuminous constituents of the blood-plasma. Resistance to the outflow of bile into the intestine leads to its absorption into the lymphatics of the liver, and thence by the thoracic duct into the blood. From thence the bile-pigments and bile-acids pass into the tissues and urine, producing the condition termed *jaundice*.

158. **Passage of the Urine into the Bladder.**—Partly owing to the high pressure under which it is secreted, partly through gravity, and partly owing to the rhythmical peristaltic contraction of the ureters, the urine is driven on through these tubes into the bladder. These tubes are from sixteen to eighteen inches in length, and about the diameter of a goose quill. They are formed of an outer fibrous coat of connective tissue with blood-vessels and nerves, a middle muscular coat with longitudinal and circular unstriped fibres, and an inner mucous coat having several layers of stratified transitional epithelium. Reflux from the bladder into the ureter is prevented by the oblique manner in which these tubes enter, a sort of valvular opening being thus formed.

159. **The Bladder and Urethra.**—The bladder, which is situated in the pelvis, has an average capacity of about twenty ounces. Its structure is very similar to that of the ureters, though the mucous and muscular coats are thicker. From its narrow end, or neck, a canal termed the *urethra* passes to the outside. At the beginning of the urethra the circular non-striped muscular fibres become thicker, but the urethra is kept closed by a transversely striped muscle termed the *sphincter urethræ*, which consists of circular fibres extending along the urethra to its middle, the tonic contraction of this muscle being maintained reflexly by a centre or centres in the lumbar part of the spinal cord. The urine gradually accumulates in the bladder until the tension in the organ leads to the contraction of its muscular walls, and the sphincter relaxing, expulsion of the urine (micturition) follows. In young children the whole act is reflex. With advancing age there is acquired more

R 2

or less control of the act. Thus, when the bladder becomes distended, the afferent impulse passing into the cord may thence reach the brain, but by an effort of the will the sphincter muscle may be assisted and the expulsion arrested for a time. Or the will may inhibit the sphincter, and assist the reflex contraction of the muscular walls of the bladder by producing contraction of the abdominal muscles.

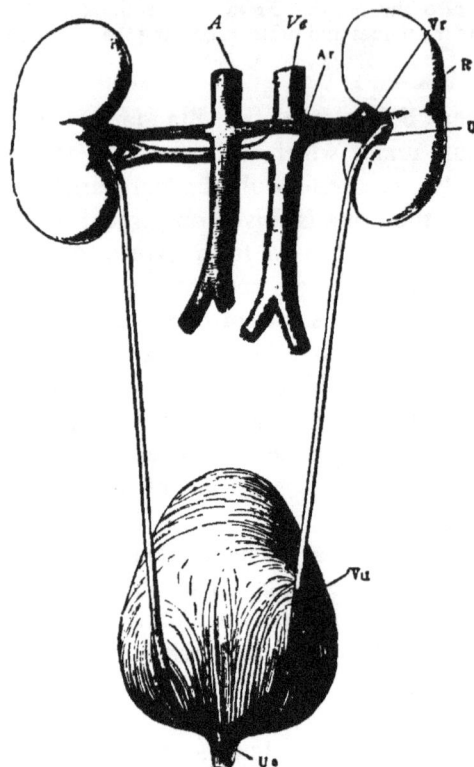

FIG. 156.—The Kidneys, Bladder, and their Vessels. Viewed from behind.

R, right kidney; U, ureter; A, aorta; Ar, right renal artery; Ve, vena cava inferior; Vr, right renal vein; Vu, bladder; Ua, commencement of urethra.

160. **Structure of the Skin.**—The skin forms an external covering for the deeper tissues over the whole body, and consists of a superficial layer termed the *epidermis* or cuticle, and a deep layer termed the *corium, dermis,* or *cutis vera* (true skin). As already described (par. 6), the epidermis consists of layers of stratified epithelial cells united together by a small amount of cement substance. The dermis, or true skin, is made up of an interlacing network of white fibrous connective tissue, with some elastic fibres, numerous blood-vessels, lymphatics, and nerves. Below, the skin gradually becomes blended with the subcutaneous tissue through a layer of areolar tissue of varying thickness containing fat-cells. Certain structures, as the nails, the hairs,

the sebaceous glands, and the sudoriparous glands, may be termed appendages of the skin. At the orifices of the body the skin gradually passes into mucous membrane, soft membrane moistened by their secretion `mucus`, and composed like the

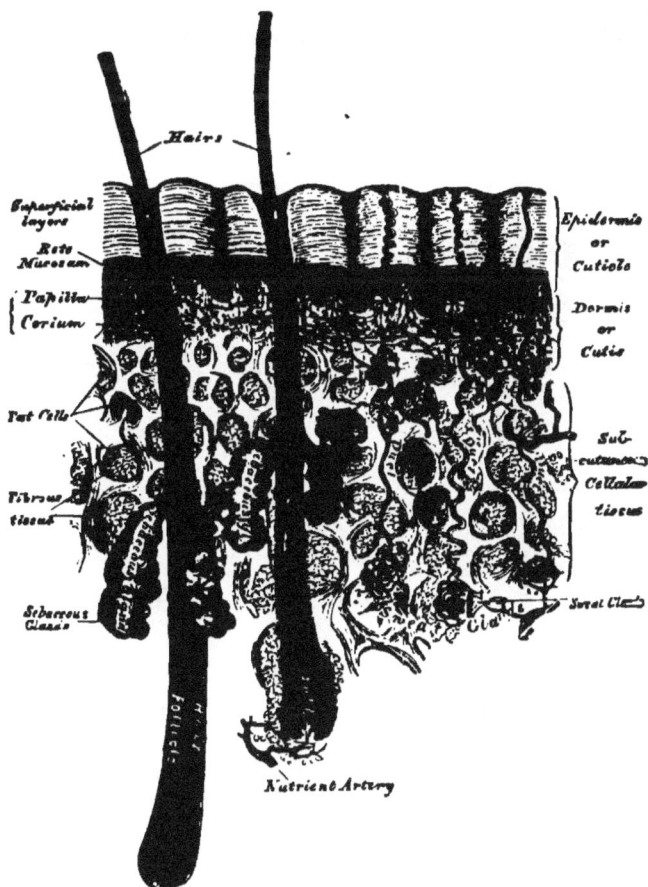

FIG. 157.—A Sectional View of the Skin (magnified).

skin of corium and epidermis, though the epidermis differs greatly in different parts.

The corium, or true skin, whose connective tissue shows a more open texture in its deepest parts, bears on its upper surface numerous small papillæ which project up into the epidermis, the inmost layers of epidermal cells, or *rete Malpighii* (rete mucosum), being moulded over the papillæ and forming processes between them. The papillæ are highly sensitive vascular eminences of a conical or club shape, about $\frac{1}{100}$ inch in length. On the general surface of the body they are comparatively few in number, but in sensitive situations, as the palm of the hand and fingers, they are numerous, and being arranged in parallel curved lines, form the elevated ridges seen on the free surface of the epidermis. The papillæ contain

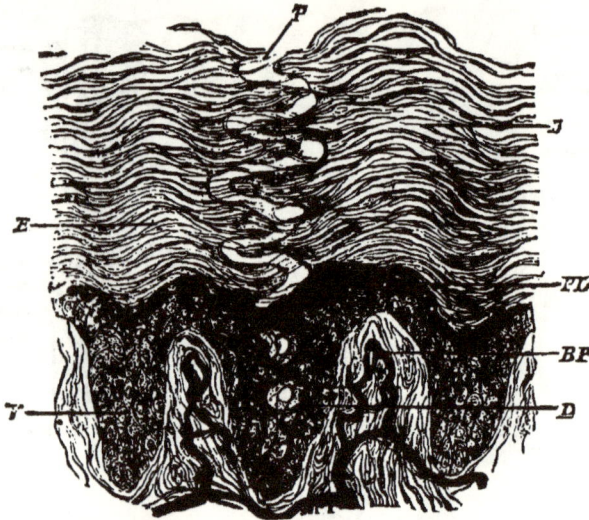

FIG. 158.—Section of Human Epidermis with Two Vascular Papillæ of the Corium. (Heitzmann.)

BP, loop of capillary vessels in papilla ; *V*, rete mucosum ; *PL*, stratum granulosum ; *E*, stratum corneum ; *D* to *p*, duct of sweat gland passing through the epidermis.

capillary loops derived from a small artery in the cutis, and most of them have also one or more terminal nerve fibres, those of the hands and feet terminating as touch corpuscles (par. 49). It will be remembered that fine nerve fibrils pass into the epidermis, where they end either between the cells or in the deep epithelial cells themselves (fig. 6).

161. **Hairs and Nails.**—Hairs are modifications of the epidermis developed in pits—the *hair follicles*—that often extend down into the subcutaneous tissue. The substance of the hair consists of a central pith or medulla, a fibrous horny part, and an external cortex or cuticle. The medulla

formed of angular cells is often wanting. The fibrous part constitutes the chief part. It consists of long tapering cells united to form fusiform fibres. The cuticle consists of thin flat scales overlapping like tiles. The part of the hair lying within the follicle and termed the *root* ends in a *knob* of soft growing cells fitting over a vascular papilla. The pit or follicle is formed of two coats, an outer or *dermic coat* continuous with the corium, and an inner epidermal coat termed the *root sheath*. The hair grows from the bottom of the follicle by the multiplication and elongation of the cells covering the papilla. Connected with the hair follicles are small bundles of involuntary muscular fibres forming the muscles of the hairs, or *arrectores pili* (fig. 160). Passing from the upper part of the corium on the side to which the hair slopes to the bottom part of the follicle, the contraction of this

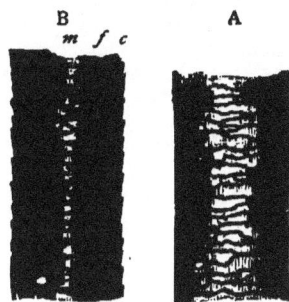

FIG. 159.—Piece of Human Hair (magnified).

A, seen from the surface; B, in optical section. *c*, cuticle; *f*, fibrous substance; *m*, medulla, the air having been expelled by Canada balsam.

FIG. 160.—Hair Follicle in Longitudinal Section. (Biesiadecki.)

a, mouth of follicle; *b*, neck; *c*, bulb; *d*, *e*, dermic coat; *f*, outer root sheath; *g*, inner root sheath; *h*, hair; *k*, its medulla; *l*, hair knob; *m*, adipose tissue; *n*, hair muscle; *o*, papilla of skin; *p*, papilla of hair; *s*, rete mucosum, continuous with outer root sheath; *ep* horny layer; *t*, sebaceous gland.

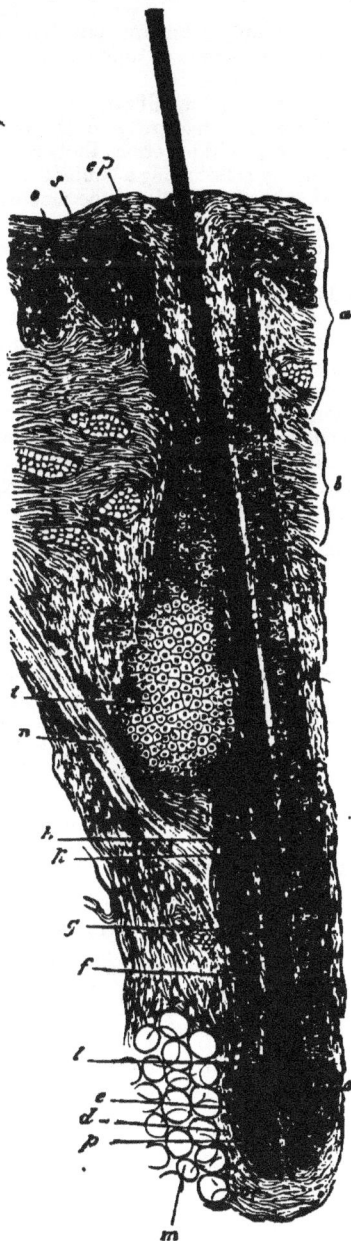

muscle not only erects the hair, but also raises one part of the general skin surface and depresses another, thus producing the roughened condition known as 'goose-skin.'

Nails are also modifications of the epidermis implanted by a portion termed the root in a groove of the skin, and growing from a modified portion of the true skin termed the bed, or *matrix*, of the nail. The nails consist of horny cells having a laminated arrangement, the deepest layer lying in contact with the papillæ of the matrix. The growth of the nail, like that of the hair and the epidermis, is effected by the successive production of new cells at the root and under surface.

162. **Sebaceous Glands.**—The sebaceous glánds are simple saccular glands found all over the skin, except in the palm of the hand and the sole of the foot. The duct usually opens into a hair follicle (fig. 160, *t*), though in a few situations it opens free on the surface. Both the duct and its alveolar expansions are lined by secretory cells, some of which become charged with fatty matter. Excretion appears to take place by the rupture or disintegration of the loaded cells, the transformed cells and their contents being pushed out of the duct through the hair follicle on to the surface of the skin by the newly generated cells behind. The secretion (or excretion) called *sebum* appears, when fresh, to be an oily substance that sets on cooling. Under the microscope it is seen to consist of fatty particles, epithelial cells, and crystals of organic bodies. As to the physiological cha·racter of the material Professor Foster says : ' The secretion of sebum, in fact, is a modification of the particular kind of secretion taking place all over the skin, and spoken of as shedding of the skin. It is chiefly the chemical transformation which is different in the two cases. In the skin generally the protoplasmic cell substance of the Malpighian cells is transformed into keratin, in the sebaceous glands it is transformed into the fatty and other constituents of the sebum. Some, perhaps, may hesitate to apply the word secretion to such a process as this ; but as we shall see later on, the formation of milk, which certainly deserves to be called a secretion, is a process intermediate between the secretion of saliva or gastric juice and the formation of sebum.' The chief purpose of the seba·ceous secretion appears to be to lubricate the hairs and to keep the skin moist and supple. Two or more sebaceous glands may be connected with a single hair follicle.

163. **The Sweat Glands.**—The sweat glands, called also *sudoriparous* or *sudoriferous* glands (Latin *sudor*, sweat), are found in the human skin in nearly all parts of the body, their total number exceeding two millions. They are most numerous on the palm of the hand and sole of the foot, largest in the axillæ (armpits) and groin. In the neck and back there are com-paratively few. The orifices of the ducts or pores are about $\frac{1}{800}$ inch in diameter, and may be seen with a hand lens. Each gland consists of a single tube with a blind end forming a close coil about $\frac{1}{60}$ inch in diameter, situated in the deep part of the true skin or in the subcutaneous fatty tissue. This is the

secreting portion of the gland, and it is surrounded by a network of capillary blood-vessels. From the coil the duct passes
towards the surface as a conducting portion in a somewhat
wavy manner through the čorium, but spirally through the
epidermis (fig. 157). The secreting portion of the coiled tube
consists of a fine basement membrane, a layer of longitudinally
disposed fibres usually regarded as muscular, and a single
layer of columnar epithelium cells lining a central cavity or

lumen. The effe-
rent or conducting
tube, which in-
cludes about one-
fourth of the coiled
part, has a base-
ment membrane,
an epithelium con-
taining two or thre
layers of cells, an
internal delicate
lining, and finally
the central lumen.
It has no muscular
layer and is smaller
than the secreting
tube (see fig. 161).
The part of the
duct passing
through the epi-
dermis is merely

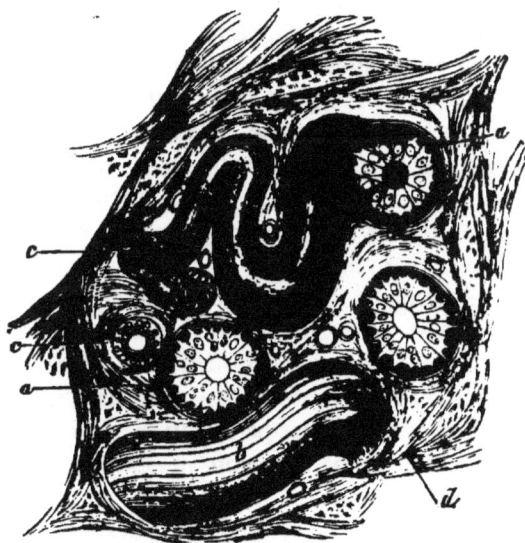

FIG. 161.—Section of a Sweat Gland in the Skin of Man.

a, *a*, secreting tube in section ; *b*, a coil seen from above ;
c, *c*, efferent tube ; *d*, intertubular connective tissue with
blood-vessels. 1, basement membrane ; 2, muscular fibres
cut across ; 3, secreting epithelium of tubule.

a channel through the epithelium cells. The wax formed in
the external passage or meatus of the ear appears to be a mixture
formed from the so-called ‘ceruminous’ glands (which are
modified sweat glands), and from the sebaceous glands of the
passage.

164. **Composition and Amount of Sweat.**—Sweat has been
obtained for examination by enclosing a limb in an air-tight
caoutchouc bag. As thus obtained it is mixed with a few
scales shed from the epidermis and with a small quantity of

fatty matter from the sebaceous glands. It consists of water with only 1·2 per cent. of solids in solution. These are chiefly sodium chloride, various fats and fatty acids, and a small trace of urea. Its reaction is said to be acid when scanty, but alkaline when abundant. A small amount of carbonic acid, about ten grains in 24 hours, passes off from the skin, and an equivalent amount of oxygen is absorbed. Cutaneous respiration in man and other mammals is thus very slight. Hence the death of such an animal as the rabbit, when its skin is covered with impermeable varnish, cannot be due, as was once thought, to arrest of cutaneous respiration. It seems rather to be due in such a case mainly to the rapid loss of heat from the surface, for the animal can be preserved for a considerable time by wrapping it in cotton wadding. In frogs and other amphibia that have a thin naked skin cutaneous respiration is active and important, as they continue to live some time after the lungs have been removed.

As long as the excretion is small in amount it passes off the surface at once, and is called *insensible* perspiration. When the secretion forms drops on the skin, either owing to increase in amount or prevention of sufficiently rapid evaporation, it is spoken of as *sensible* perspiration. The relation between these two kinds of perspiration in similar conditions of the body depends chiefly on the rapidity of evaporation, and this is dependent on the degree of saturation of, and the amount of movement in, the surrounding air. The total amount secreted by a man in 24 hours may be said to average about 1,000 grams, or 2 lbs., but this amount is increased in a hot atmosphere, by active muscular exercise, and generally by any cause that increases the circulation in the capillaries of the skin. That the amount of sweat bears a certain inverse relation to the amount of urine excreted has already been pointed out (par. 155). The regulation of the temperature of the body by the evaporation of sweat is explained in the chapter on Animal Heat.

165. **The Nervous Mechanism of Perspiration.**—As already stated, dilatation of the cutaneous blood-capillaries leads to increased secretory activity of the sweat glands, and constriction of these vessels leads to diminished activity. Such changes

may be brought about not only by external heat relaxing the vessels of the skin and external cold contracting them, but by the vaso-motor nerves that regulate the blood supply of the skin. But apart from variations in the vascular supply, it appears certain that there are special secretory nerve fibres passing to the sweat-glands, stimulation of which causes an increase of sweat. These fibres appear to be contained in the same nerve trunks as fibres having vaso-motor. and other functions. Thus, if the peripheral end of the divided sciatic nerve in the cat be stimulated, drops of sweat appear on the hairless sole of the foot, even when there is constriction of the blood-vessels, or even in the amputated limb. Other secretory fibres have been found in other nerves, as the ulnar. These special sweat nerves appear to spring from nerve-centres in the spinal cord, a dominant centre being said to exist in the medulla. A venous state of the blood, as in dyspnœa, appears to excite the sweat centres. So also do certain drugs called sudorifics, as pilocarpin and strychnine. Sweating may also be produced reflexly by stimulating certain sensory nerves. Thus the pungency of mustard in the mouth often causes perspiration on the face. Certain emotional states, as *fear*, produce sweating through the agency of the central nervous system.

Some observers have described structural changes as occurring in the secreting cells of the coiled part of the tubule, but these are not so well marked as in the case of the salivary glands, as they discharge neither mucus nor proteid material.

166. **The Mammary Gland.**—The rounded eminences situated on the breast form the lacteal or mammary glands, by which milk is formed during the process of lactation. The gland tissue is composed of large divisions or lobes subdivided into lobules, and connected together by areolar tissue and blood-vessels. From fifteen to twenty small channels termed lactiferous ducts open separately at the nipple. Each of these main ducts has a small dilatation or reservoir at a short distance from its orifice, and as it passes inwards it is found to arise by the union of smaller branches from the various lobes and lobules. During lactation the finest branches are seen to terminate in microscopic saccular pouches or acini. Each acinus of the gland consists of a basement membrane lined by a simple layer of epithelial cells. When the gland is inactive, the cells are flattened and show a single nucleus. But during the activity of the gland the secretory cells undergo peculiar and characteristic changes. The cells swell and become cylindrical, the nucleus often divides, and fatty granules appear in the cell protoplasm at the edge near the lumen. This portion of the cell becoming more promi-

nent is at last detached, the fatty granules being shed into the fluid in the lumen to become the microscopic milk-globules, and the discharged proto-plasm dissolving to form the proteid of the milk. The water, salts, and milk-sugar are also secreted by the cells from the blood or lymph bathing the basement membrane, and discharged into the lumen to be carried off by the ducts.

It is thus seen that in the case of milk secretion the superficial part of the cell is cast off and forms part of the secretory product, while the basal part is left to grow and become an active cell once more.

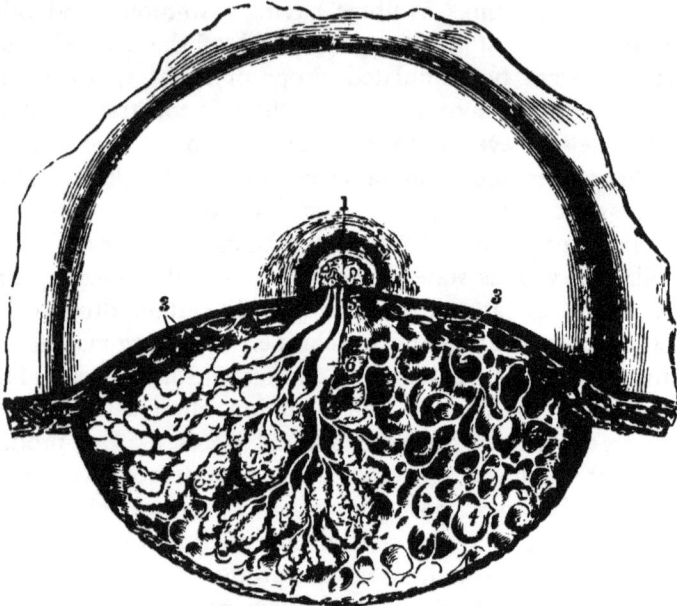

FIG. 162.—Dissection of the lower half of the Female Mamma during the period of Lactation. (From Luschka.)

a, a, a, undissected part of the mamma ; 1, the mammilla ; 2, areola ; 3, subcutaneous masses of fat ; 4, loculi in the connective tissue which supports the glandular substance ; 5, three lactiferous ducts passing towards the mammilla where they open ; 6, one of the sinuses or ampullæ ; 7, some of the glandular lobules which have been unravelled ; 7', others massed together.

The regular secretion of milk begins three or four days after birth of the young. For the first two or three days the fluid, termed *colostrum*, is of a yellowish colour and coagulates on boiling. It contains a number of particles, ' colostrum corpuscles,' which consist of cells filled with minute fat-globules that are larger than the milk-globules. A gradual transition into true milk takes place. As already stated, milk is a perfect food for infants, since it contains all the three kinds of food-stuffs in proper pro-portions. For adults there is too large a proportion of water, as about 11 pints per day would be required to supply the food-stuffs of a normal diet (see pars. 106 and 107).

there, and it also prevented me from writing what I did observe. But casting back the reminiscent eye—with my foot moving forward toward the Pyrenees—my first glance rests upon thy stately castle and sacred towers, O Toledo!

Toledo is but three hours by rail from Madrid, but it is switched off the usual track of travel. It is situated in a mountainous district. The Sierras, which divide the waters of the Tagus from the Guadiana, are seen from its walls. Toledo stands upon a rocky eminence of the Tagus, which here bursts through mountains of granite, encircles the city, and turns many old Moorish mills as it girdles the walls. My sketch gives a fair idea of its position. It was ever a strong place. There is but one approach to it. It is on the land side. The Moorish towers remain to show how well that part was defended. Toledo is a city set on a hill; it cannot be hid. Nay, like Rome, it is set on seven hills; or, rather, it sleeps on seven hills, and with the somnolence of the seven sleepers. It has no longer its two hundred thousand souls. Its population is less than its far off Ohio namesake's by many thousands. It is now the simulacrum, the ghost, or, rather, the skeleton, of a city. Its spirit has departed, but its substance remains. Its bones are perfect. The Goths came like a tempest; the Moors poured down like the rains, and they have gone; but Toledo, in substance, is still found, because founded on a rock. It is called Imperial. That is a memory. It is called the Crown of Spain. That is a fly in the amber of history. It is called the 'light of the whole world.' In the dark ages, when the other portions of Europe were shrouded, Toledo, like Salamanca, was the home of learning, the capital of empire, and the seat of chivalry. She was a Pharos in the world of letters, arts, and theology.

I do not mean to convey the impression that the
Tagus, whose valley we follow from Aranjuez to
Toledo, is all the way rocky and rushing. It is not.
It ruminates—or eats its way wearily — over level,
grassy plains, after it leaves its fretting, working, irri-
gating duties, through the palatial pleasure grounds at
Aranjuez. It is thus like the Guadalquivir. It is so
nearly on a level with the banks and fields, that we see
numerous donkeys and mules at work along the banks
pumping up the water for the fields of potatoes and
wheat. You see that potatoes, Toledan citizen of the
Western world, do grow in the vicinity of your namesake. What their size is I do not know. I hope
they merit a more gigantic tribute than that conveyed in the verse about your Ohio stream :—

'Potatoes they grow small,
And they eat them skins and all,
On Maumee! on Maumee!'

The fields alternate along the Tagus with potatoes
and wheat. Here and there is a violent dash of red
poppies. As we approach Toledo the city towers shine
aloft in the sun's glare as if they were polished, like her
blades. The depot is not in the city, but outside the
walls. To reach the city requires an effort. With
three mules abreast, and hitched to a rickety omnibus,
the high bridge is crossed; then around and up the
hill, we wind with a rush and halloo (for every one in
Spain drives mule and horse with all speed *up-hill*);
then enter on a gentle gallop through an ancient gate-
way, into the narrow, steep, tortuous, and badly paved
streets. Very ancient looks the city. The houses
have a serious, solid, sacerdotal, and comatose look.
They remind us of the Moorish mansions. If within
these walls once lived two hundred thousand souls—
and now there is less than one-tenth of that number
—either there are many unoccupied tenements or rent

is cheap, or both. The city is as still as death. It is
like a cloister, so cool and still. It is a pity to awake
it by the irreverent, reverberating rumble of our
vehicle. It seems as if it ought not be disturbed by
the clatter of the mulish hoof. The inhabitants gaze
out of their shops and houses as if amazed at our
intrusion. We feel conscience-stricken, as if we had
aroused a weary person who had done much and needed
repose. Besides, the Toledans *have* done much, and
have much pretension. They have a right to sleep.
Are they not the *élite* of Spain? Is not theirs the
true Spanish tongue? Is not here the slow, deep, and
guttural enunciation, and here the perfect grammar?
Is not all else *patois?* And is not Toledo still crowned,
if not on earth, in heaven, as the proud metropolis of
the Spanish ecclesiastical world?

It is very hard to find one hotel. I am writing all
the time with Toledo, Ohio, in my eye. The contrast
will force itself. Indeed, I should not have been here
but for a sort of foolish Buckeye pride to make some
felicitous comparisons. One hotel! and that so poor
that I am reminded by it of our Corsican accommoda-
tions amidst the mountains of Evisa! And only one
public carriage. Think of that. Tagus, cease to brawl
your praises beneath these historic walls! Maumee,
raise your Ebenezer from the turrets of your grain
elevators! We could not complain, for we got the
only carriage—a sort of half omnibus. It opened at
the rear. Every part was creaky. Its voice was
cracked, and so were its panels. The harness was
made of ropes. One carriage. What a fall is here!
Once Toledo, besides her immense cathedral, had
twenty-six churches, nine chapels, three colleges, fourteen convents, twenty-three nunneries, and ten hos-
pitals. Now—only one carriage. No one comes here
for business. There is no trade; it would be sacrilege

to traffic. The painter and tourist, with palette and
pen, may come. Old Mortality may come with his
chisel. He may distinguish the Gothic from the
Roman ruin ; decipher the Moorish from the Hebraic,
and both from the Christian inscription and monu-
ment ; wander amidst palaces whose halls have but an
empty echo; gaze at fragmentary statues dust-laden,
and pictures time-tarnished ; and see great rooms and
much room for worship, but no worshippers !

Did I not say that the hotel was very poor and very
small ? It was also very old. How the omnibus ever
worked up and round in the angular streets, with its
three mules, and these half blinded by the red head-
trappings, and dashed us into the little court of the
hotel, where we were suddenly spilled, I do not know.
Toledo has a good driver. He was the only wakeful
person I met. The hotel people gave us a room, with
a dungeon door and a monstrous lock. It had a
ceiling about seven feet high, though the room was
nine feet by seven in dimensions. It had a few prints
on the walls, very old and dusty. One was the
portrait of the 'Illustrious D. George Juan, profundo
matematico,' who died at Madrid in 1773 ; another
was of a great medicine man, Dr. Valles, who doctored
Philip II. I pity him. He must have had a hard
time of it with his distinguished patient. The picture
represents a melancholy man. I also pity the patient.
Then there was a picture of D. Pedro Calderon de la
Barcas, poet ! His face I know. It makes my room
glorious ! Our feast at the hotel was not Olympian.
It was a feast of the imagination. I soon perceived
that we must make our supper here with the 'convo-
cation of public worms' who are supposed to be
preying on the dead past in the old crypts. So,
without more ado, we enter our carriage and proceed
first to the Alcazar, or palace.

CHAPTER XII

ANIMAL HEAT

167. General Statements.—The food that we take contains a store of potential energy which is converted into the kinetic form of heat and motion by the chemical changes that go on in the tissues. These chemical changes, included under the term *metabolism*, are of various kinds, but the great source of heat is the katabolic process termed oxidation. As already remarked, two kinds of processes are continually going on in the body. Protoplasm is being formed out of the food materials, a process called constructive metabolism or *anabolism*, and unaccompanied by any production of heat ; and protoplasmic materials are constantly being oxidised, a process known as *katabolism*, and resulting in the evolution of heat and mus-cular movement with production of carbonic acid, water, and urea. The oxidation process, as stated in treating of muscular tissue, is not direct and immediate, but complex, and occurring in several stages. Yet the result is the same, for we know that the oxidation of a given weight of any substance liberates the same amount of energy, whether the process takes place in one stage or in several. A gram of carbon, whether free or combined, on being oxidised sets free a definite and known amount of heat. A calorie or heat-unit is the amount of heat required to raise the temperature of one gram of water 1° C. The number of gram-calories or heat-units liberated by one gram of various materials has been estimated. In round numbers we get :

Hydrogen	3,450	Carbohydrate	4,000
Carbon	8,100	Proteid	5,000
Fat	9,000	Urea	2,205

These numbers supply the *physical* heat-value of 1 gram of the above substances, *i.e.* the amount of heat produced when complete combustion takes place in a calorimeter. But the *physiological* heat value of proteid is not the same as its physical heat value, as it is not completely oxidised in the

Human Physiology

body. Each gram of proteid yields one-third gram of urea, and hence we must deduct the heat value of this from that of each gram of proteid (from 5,000) to get the physiological heat value of a gram of proteid : 5,000 − 735 = 4,265.

The estimates that have been made of the heat required to maintain the temperature of the body, and replace the loss by evaporation and radiation in a given time, are found to correspond closely with that derived from the quantity of the food supplied. Production of heat and discharge of heat, or income of energy and expenditure of energy, balance. Add to this the physical truths known about the production of heat during the combustion of various substances, and there can be no doubt that the source of heat in the body is the chemical processes there taking place, especially the process of oxidation. Were there no loss, the amount of heat produced in the body in an hour would raise its temperature 2°. Helmholtz has estimated that about 7 per cent. of the heat produced in the body is converted into external mechanical work, that nearly 75 per cent. is lost by evaporation and radiation from the skin, and about 18 per cent. passes off by the lungs and excreta (see par. 170).

168. **Temperature of the Body.**—As regards temperature animals are divided into two great classes :

1. Warm-blooded or homoiothermal animals (Greek *homoios*, like), those which have a *constant* temperature. This class includes mammals and birds, birds having a higher temperature than mammals.

2. Cold-blooded or poikilothermal animals (Greek *poikilos*, variable), those whose temperature *varies* with the medium in which they are living, being only about a degree above the varying medium. This class includes reptiles, amphibians, fish, and invertebrates.

The average *surface* temperature of a healthy human body taken in the axilla is 98·6° F., 37° C. (in the blood of the interior about 2° F. higher), an approximate uniformity being brought about by the circulation of the blood, which carries heat from the parts where it is produced and distributes it to the parts where heat is lost. In the interior of the body, especially where chemical changes are most active, the temperature is somewhat higher than at the surface, where heat is being constantly lost. The warmest organs of the body are the liver, the brain, and the muscles ; the coolest parts are the skin and the extremities, the difference amounting to more than 1° C., or about 2° F. Children have a temperature about $\frac{1}{2}$° C. above that of an adult, and in all persons of regular habit there is a slight diurnal variation from a maximum at 3 p.m. (37·5° C.) to a minimum at 3 a.m. (36·8° C.). Active muscular exercise may raise the temperature of the body 1° C., a feeling of great warmth being produced by the increased blood supply to the skin consequent on the dilatation of its vessels. In fact, our *feeling* hot or cold is due to the state of the cutaneous capillaries. When these vessels are full of warm blood the sensory nerves terminating in the skin are affected with the sensation of heat, but when these vessels are contracted and comparatively empty, we feel the cold of the external air

FIG. 163.—A Clinical Thermometer.

affecting the nerves of the skin. Thus our feeling is no real guide to the body temperature as a whole, which in health is always about the same, however hot or cold we *feel*.

169. **Where Heat is Produced.**—Though heat is produced wherever chemical changes go on, the chief tissues in which heat-production occurs are the muscles, the glands, and the nervous centres. Every manifestation of muscular energy has been found to be accompanied by evolution of heat and carbon dioxide, and, as the carbon dioxide is given off though in a less degree even by resting muscles, it is evident that active chemical change with production of heat must also go on in muscles at rest. The secreting glands, especially the liver, are also the seat of metabolic processes that result in heat, and the temperature of an organ is greater during activity than during rest. Blood leaving the liver is hotter than that entering this organ, probably the hottest in the body during active digestion and absorption, while stimulation of the chorda tympani so increases the activity of the submaxillary gland that the saliva in its duct is found to have a temperature one degree higher than the blood in the carotid artery. The brain is also a source of heat, as the temperature of the blood leaving this organ when mental effort is going on is distinctly higher than the temperature of the blood entering it. Certain physical processes, as friction of the blood against the vascular walls, friction of muscles and tendons, and electrical currents, contribute a small amount to the heat production of the body.

As, therefore, the production of heat depends on chemical action rising and falling as that rises and falls, and as it has been ascertained that in a cold atmosphere more carbonic acid is given off than in a hot one, it is evident that the supply of heat is increased in a cold medium and diminished in a hot medium.

170. **Where Heat is Lost.**—As the temperature of the surrounding air is below that of the body in all temperate climates, heat will be lost from the surface of the body by conduction and radiation, as well as by evaporation of the perspiration. By conduction heat passes to the air in contact with the body and from one air particle to another ; by radiation heat passes from the body into the surrounding medium by producing in it progressive waves of energy derived from the particles of the body. When the external temperature is below that of the body, the loss of heat by conduction and radiation, together with the heat consumed in warming the respired air, may be almost sufficient to remove the heat not required for maintaining the body at its normal temperature, except during times of vigorous activity. Active muscular exercise leads to increased chemical changes, with increased activity of respiration and increased perspiration. Where the external temperature is higher than 98·6° F., as in the tropics or in a Turkish bath, no heat will be lost from the body by conduction and radiation, but a gain from the surrounding medium will occur. In these circumstances a large quantity of heat is lost in evaporation of water from the skin and lungs. The conversion of water into water-vapour uses up a large quantity of heat. In hot air the capillaries of the skin dilate, the vessels becoming flushed with blood, the sweat glands increase in activity, and evaporation of the sweat goes on rapidly. When the secretion of sweat is but small, it evaporates as fast as it reaches the surface, and the skin remains dry—*insensible perspiration.* When the secretion is abundant, sweat may be formed faster than it is evaporated, and it then appears on the skin in drops—*sensible perspiration.* Even in cold weather, after severe muscular exertion and consequent production of heat, sweat is produced in abundance, so that evaporation leads to the removal of this excess of heat. It should be noted, too, that the degree of moisture in the air is of great importance. Air almost saturated with water-vapour interferes with the evaporation from the skin, and thus hinders the loss of heat in this way. But in dry air a high temperature may be borne for some time, especially if liquid be taken freely, as perspiration is then copious and evaporation takes place rapidly.

Clothing is used in temperate and cold climates to keep in the heat of the body by protecting the skin from the chilling influences of the cold air, which, if allowed to come freely into contact with the body, would carry off its heat. The clothing checks conduction and radiation. The low conducting power of atmospheric air renders a number of layers of clothing with sheets of air between them more efficacious than one thick layer.

' Heat is dissipated by radiation and evaporation from the surface of the body ; in hot weather the skin is flushed with blood and moist with perspiration, and superfluous clothing is put off ; in cold weather the skin is pale and dry, and extra clothing is put on ; in the first case the dissipation of heat from the body is accelerated, in the second case it is retarded.'— *Waller.*

It has been estimated that the loss of heat by the skin in the two ways already mentioned—*i.e.* (1) by radiation and conduction in an atmosphere below that of the body temperature, and (2) by the conversion of sweat into water-vapour, amounts to 77 per cent. of the total loss. By respiration, in which the inspired air when cooler than 98·6° F. is warmed and in which much water-vapour evaporates from the lungs, 20 per cent. of the total loss occurs. About 3 per cent. of the total heat loss passes off in the urine and fæces. A small quantity of heat is expended in warming the food and drink ingested.

171. **Regulation of the Temperature of the Body.**— As the bodily temperature of man is nearly constant, notwithstanding the variations of temperature and moisture in the surrounding medium and variations of bodily activity, there must be some regulating mechanism that brings about this uniformity. It is mainly effected by variations in the amount of heat lost from the surface of the body under varying conditions, though some small variation in the amount of heat generated may occur. On the mechanism by which the temperature is kept nearly uniform in all latitudes, Dr. Hale White thus writes in *Nature* :—

‘ The temperature of a man at the equator is within a degree Centigrade the same as that in the Arctic regions. This is because, in the first place, in the Arctic regions the loss of heat from the body is very slight, and in the tropics it is very great, for (*a*) in the tropics more perspiration is secreted by the skin, and this, in consequence of the high temperature of the air, evaporates very quickly, and hence the body is kept cool. It is true that in the tropics people may not be observed to perspire freely, but that is simply because as fast as the perspiration is secreted it is evaporated. It is what is called insensible perspiration. (*b*) More water is secreted by the bronchial mucous membranes in the tropics, and in consequence of the higher temperature of the air it, like the perspiration, evaporates very quickly. The excessive secretion of moisture by the body when the temperature of the air is high is shown in a Turkish bath, and leads, in a bath of about two hours’ duration, to a loss of weight amounting with some persons to three pounds, and to a great diminution in the quantity of urine secreted. (*c*) In the tropics the vessels of the skin are more widely dilated than in the Arctic regions ; hence there is more blood in it, and therefore heat is more readily radiated and conducted from the skin to the external atmosphere. (*d*) The specific heat of the body is very high, and so it cools very slowly in the Arctic regions. It is very highly probable that in the Arctic regions the quantity of heat produced by the body is much greater than in the tropics. The human body in the tropics must often be the coolest of surrounding objects ; in this case it cannot lose anything by radiation or conduction, but it is kept cool by the rapid evaporation of perspiration (usually insensible) and fluid secreted by the bronchial mucous membrane. Whether or not a man in the tropics produces any heat under such circumstances has not been demonstrated, but probably, although the production of heat falls very low, it does not entirely cease.’

S

There are thus variations in loss of heat and variations in production.

172. Variations of Heat-loss and Heat-production.—The skin is the chief organ by which the loss or expenditure of heat is regulated, the regulation being effected (*a*) by varying the quantity of blood exposed to cooling influences at the surface of the body, (*b*) by varying the quantity of sweat produced for evaporation. Through the vaso-motor nerves (dilatators and constrictors) its blood-supply may be increased, and result in greater loss of heat by conduction, radiation, and evaporation ; or the supply of blood to it may be diminished and result in diminished loss of heat. It is probable also that the nerves of the sweat glands assist in increasing or checking the production of the amount of moisture to be evaporated. Moreover, increased temperature of the body as a result of exercise, fever, &c., causes an increase in the number of heart-beats, so that in a given time the whole volume of the blood is more frequently exposed to the cooling influence of the skin. At a temperature of 41° C. (106° F.) the pulse rate reaches 110. The other organs by which variations of loss are produced are the lungs. An increase in the number of respirations leads to a larger volume of air becoming heated to the temperature of the body, as well as to an increase of water vapour which has abstracted heat in being evaporated. Animals like the dog, which perspire but little by the skin, pant freely on a hot day, in order that the increased respiratory activity may carry off their surplus heat, the protruded tongue assisting the loss by evaporation from its surface.

It has been mentioned previously that cooling of the surrounding medium increases the amount of carbon dioxide excreted, and consequently the amount of heat *produced*. Heating the surrounding medium has the opposite effects. Cold excites the action of muscles (shivering, &c.) that promote oxidation processes, and also the appetite for food, especially those substances whose physiological heat value is high. The ingestion of all foods leads to increased tissue metabolism and its accompanying heat.

173. Nervous Control of Heat Production.—The influence of the nervous system on the dissipation and production of animal heat is undoubted. The regulation of the heat mechanism of the body is probably due to reflex actions of various kinds, and possibly to the action of a special nerve centre. Afferent impulses received at the vaso-motor centre lead to efferent impulses by which the calibre of the blood-vessels of the skin or of the internal organs is altered, with variation of blood supply and variation of heat dissipation. The secretory activity of the sweat glands is also under the control of the central system by means of their nerves. Vaso-motor nerves and the nerves of the sweat glands thus affect the regulation of temperature on the side of loss or dissipation. But besides these influences ' physiologists have long suspected that afferent impulses arising in the skin or elsewhere may, through the central nervous system, originate efferent impulses, the effect of which would be to increase or diminish the metabolism of the muscles and other organs, and by that means increase or diminish respectively the amount of heat thus generated.' The evidence adduced in support of a thermogenic (heat-producing) centre that may be reflexly stimulated is as follows :—Warm-blooded animals are so affected by the temperature of the surrounding medium that cold induces increased tissue activity with increased production of heat, while warmth diminishes these processes. In cold-blooded animals it is the reverse ; tissue activity rises and falls with that of the surrounding medium. But if a warm-blooded animal be curarised, not only is there paralysis of the motor nerves, but the temperature rises and falls with that of the surrounding medium, the reflex arc being broken at the muscular end. Injury to certain deep-seated parts of the brain is also said to be followed by a rise of temperature and increased heat production, without the manifestation of any special motor movements. It thus appears that nervous impulses may lead to chemical changes that produce heat without producing muscular action. It is maintained, however, by some authorities that all these heat variations may be explained by the admitted nervous mechanisms that control vascular and glandular changes, and thus indirectly the production and distribution of heat. If this be so, there is no need to suppose that there is a special set of nerves directly influencing heat production, or a special heat centre in the central nervous system.

174. Fever.—Pyrexia, or fever, is a morbid condition characterised by an increase of the bodily temperature above the normal, with increased tissue waste, owing to disturbance of the mechanism regulating heat formation and expenditure. An increase to 99° F. indicates a feverish condition, 100° F. to 102° F. moderate fever, 104° F. to 106° F. high fever. Various facts go to show that this increased temperature is mainly due to an increased production of heat without compensating loss of heat. Diminished loss may also be a contributing cause, as the skin although hot is often dry at a certain stage of the disturbance. But the greatly increased output of CO_2 and urea shows that increased tissue metabolism is going on, and this excessive oxidation must lead to increased production of heat. The wasting character of fever and the incapacity for mechanical work also indicate the transformation of energy into heat. Further, an increased heat production has been observed by direct calorimetric measurements.

A temperature of 7° F. or 8° F. above the normal, if continued, soon causes death, the heated blood increasing the number of heart-beats and respiratory movements until exhaustion ensues. The nervous system is also greatly affected, and fails to regulate the bodily functions properly.

CHAPTER XIII

THE LARYNX AND VOICE

175. The Larynx is the chief part of the organ of voice, and is situated at the upper and fore part of the neck, between the large vessels of the neck, and below the tongue and hyoid bone. It consists of 'a framework of cartilages articulated together, and connected by elastic membranes or ligaments, two of which, projecting into the interior of the cavity, are named the *true vocal cords*, being more immediately concerned in the production of voice. It possesses special muscles which move the cartilages one upon another, and modify its form and the tension of its ligaments, and it is lined by a mucous membrane continuous above with the mucous membrane of the pharynx, and below with that of the trachea.'

176. **Structure of the Larynx.**—The cartilages of the larynx are nine in number, three single ones and three pairs : —

Thyroid	Two Arytenoid
Cricoid	Two Cornicula Laryngis
Epiglottis	Two Cuneiform

The last two pairs are very small. Only the thyroid and cricoid are visible in front and at the sides of the larynx ; at the back the arytenoid cartilages may be seen surmounting the cricoid cartilages, the epiglottis being placed in front of the upper opening of the larynx. The epiglottis is a leaf-shaped plate of yellow fibro-cartilage, attached by a foot-stalk to the interior of the thyroid cartilage at the front. The epiglottis is of little service in voice production. During respiration it

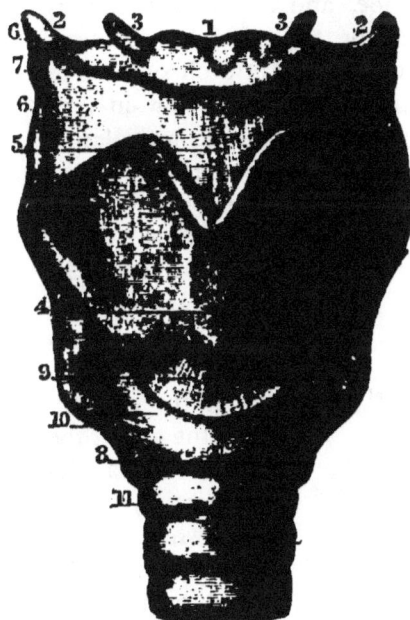

FIG. 164.—Front View of the Laryngeal Cartilages and Ligaments. (Sappey.)

1, hyoid bone ; 2, its large cornua ; 3, its small cornua ; 4, thyroid cartilage ; 5, thyro-hyoid membrane ; 6, lateral thyro-hyoid ligament, containing the *cartilago triticea*, 7 ; 8, cricoid cartilage ; 9, crico-thyroid membrane ; 10, lateral crico-thyroid ligaments.

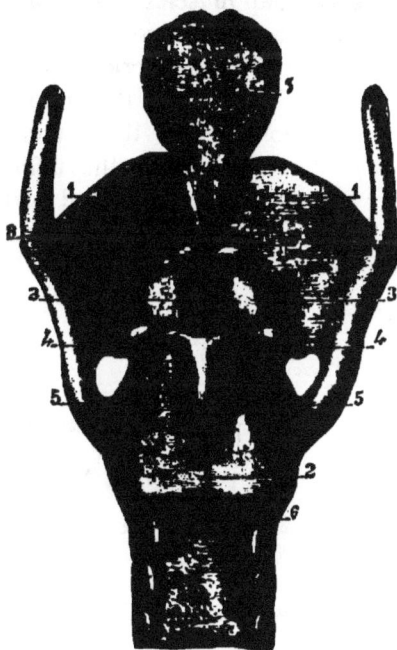

FIG. 165.—Back View of the Laryngeal Cartilages and Ligaments. (Sappey.)

1, thyroid cartilage ; 2, cricoid cartilage ; 3, arytenoid cartilages ; 4, their muscular processes ; 5, a ligament better marked than usual, connecting the lower cornu of the thyroid with the back of the cricoid cartilage ; 6, upper ring of the trachea ; 7, epiglottis ; 8, ligament connecting it to the angle of the thyroid cartilage. The cornicula are seen surmounting the arytenoid cartilages.

stands upwards, but in swallowing it falls downwards and backwards, to close the entrance to the larynx.

The thyroid is a large ridge-shaped cartilage formed of two lateral halves or wings, open behind, but forming a ridge at an acute angle in front, the prominent point of the ridge forming Adam's apple (pomum Adami). Each lateral piece, or ala, is of quadrilateral shape, and to the outer surface are attached the sterno-thyroid and thyro-hyoid muscles. To the inner surface of the thyroid at the angle is attached the epiglottis, and lower, the true and false vocal cords and the thyro-arytenoid muscles. The upper edge of the thyroid is united by a membrane to the hyoid bone, and its lower border is connected to the cricoid cartilage by the crico-thyroid membrane in front, and the crico-thyroid muscle at each side. The posterior edge of the thyroid ends above and below in projections or cornua, the inferior norns or cornua articulating with the posterior lateral portion of the lowest cartilage of the larynx, the cricoid. The cricoid cartilage is named from its

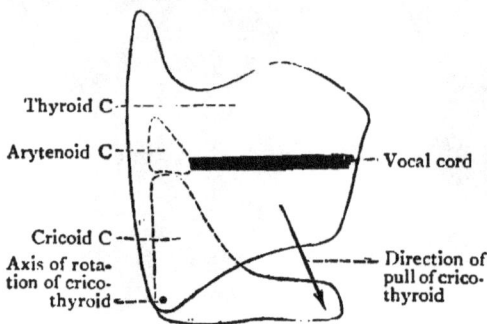

FIG. 166.—Diagrammatic Vertical Section of Larynx.

resemblance to a signet ring (Gr. *krikos*). It is situated below the thyroid, its broad part lying behind in the gap between the alæ of the thyroid, and its narrow part lying in front, where it is united above to the thyroid cartilage by the crico-thyroid membrane. Its lower border is attached to the first ring of the trachea ; its upper border, at the highest part behind, presents on each side a smooth oval surface for articulation with the triangular arytenoid cartilages.

The two arytenoid cartilages are like irregular three-sided pyramids in shape, and they rest by their bases on the posterior and highest part of the cricoid cartilage, while their pointed apices turn backwards and inwards, and are surmounted by the two small cartilaginous nodules that form the *cornicula laryngis*. To the anterior angles of the arytenoid cartilages are attached the posterior ends of the true vocal cords, while the anterior ends are attached in front to the thyroid, in the angle between the two lateral alæ or wings (fig. 167).

These true vocal cords are bands of yellow elastic tissue covered over with mucous membrane, whose epithelium is of the squamous or pavement variety. They serve as the vibrating membranes in the production of voice.

The false vocal cords are two folds of mucous membrane enclosing some fibrous tissue, and situated above the true vocal cords. Between the true and false vocal cords are the passages termed ventricles of the larynx, which lead into recesses between the false vocal cords and the thyroid cartilages, termed the pouches of the larynx.

The glottis, or rima glottidis, is the chink or opening between the true vocal cords, which, as before mentioned, pass horizontally from the arytenoid cartilages behind to the angle of the thyroid in front. Its length in the male measures rather less than an inch, and its breadth when dilated a little

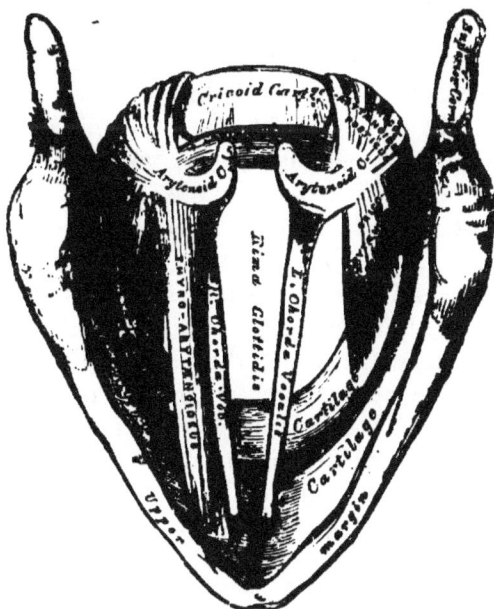

FIG. 167.—Interior of the Larynx, seen from above (enlarged).

more than a third of an inch. Its form varies. In quiet breathing it is widely open and somewhat triangular, but when voice is produced the fissure becomes narrowed, the inner margins of the arytenoid cartilages meet, and the free edges of the vocal cords approximate and are made parallel, the degree of approximation and tension of the cords corresponding to the height of the note given forth.

The chief intrinsic muscles of the larynx are in pairs, and named from their connections and positions as follows : Crico-thyroids, lateral crico-arytenoids, posterior crico-arytenoids, the thyro-arytenoids, and the single arytenoid. The crico-thyroids pass from the external surface of the thyroid forwards and downwards to the outer surface of the cricoid. This contraction, bringing the two cartilages together, will pull up the front of the cricoid cartilage and so depress its back, to which the arytenoid cartilages are articulated, so that the tension of the vocal cords is increased. Certain fibres of the thyro-arytenoid muscles, which lie parallel to the vocal cords, aid in relaxing the cords. The movements of adduction and abduction of the vocal cords, by which the opening of the glottis is regulated, are effected mainly by the muscles connecting the cricoid cartilage with the arytenoid cartilages. The lateral crico-arytenoids adduct the vocal cords, and produce closure of the glottis by rotating the arytenoid cartilages inwards ; the posterior crico-arytenoids abduct the cords, and widen the glottis by rotating the arytenoid cartilages outwards. The single arytenoid muscle that passes between the two arytenoid carti-lages behind brings these cartilages together by its contraction, and assists in closing the glottis.

The nerve supply of the larynx consists of the superior laryngeal and inferior laryngeal branches of the vagus, the former supplying the mucous membrane and the crico-thyroid muscle, the latter the remaining muscles.

177. **Voice and Speech.**—That the sound of the human voice is the result of the vibration of the true vocal cords which bound the glottis, such vibrations being caused by upward blasts of air and being in turn communicated to the column of air in the passages above, has been proved both by observation of the living subject and by experiment on a dead body. By means of an instrument termed the laryngoscope the interior of the larynx and the movements of the cords may be rendered visible. During ordinary respiration the glottis is about half open and widens slightly with each inspiration ; with a forced inspiration it becomes widely dilated, but when vocalisation takes place and the cords are set in vibration, the glottis becomes a mere chink with parallel sides.

In physiology *voice* signifies the sound produced in the larynx when an expiratory blast of air sets the vocal cords in vibration. In order to pro-duce voice, the chink of the glottis, which is about half open in ordinary respiration, must be rendered narrow by the free edges of the vocal cords being brought close together and rendered parallel. The cords must also be rendered more or less tense. Both the width of the glottis and the tension of the cords are regulated by the action of muscles as just explained—the lateral crico-arytenoids bringing the cords together and making them parallel, while the posterior crico-arytenoids separate them, and the crico-thyroids increasing the tension of the cords while the thyro-arytenoids relax them. The *loudness* of the voice depends on the force of the expiratory blast, that is, on the amplitude of vibration of the cords. The *pitch* of the voice depends on'—(1) The length of the cords, for the longer a cord is the lower is the note produced, the pitch of a stretched string

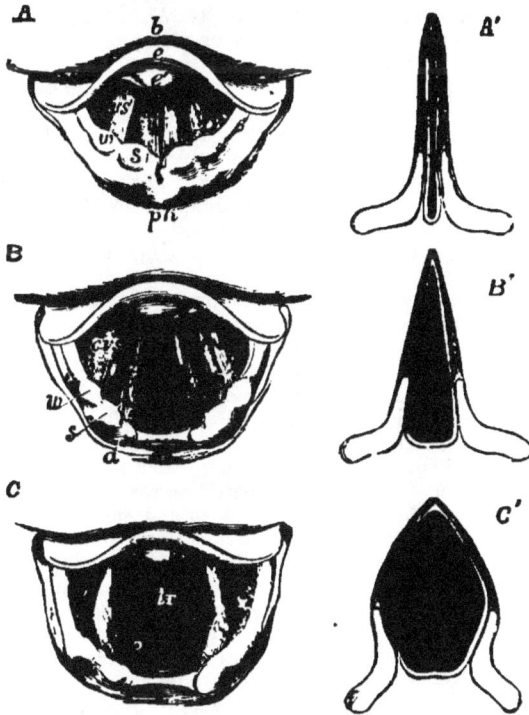

Fig. 168.—Three Laryngoscopic Views of the Superior Aperture of the Larynx and Surrounding Parts in Different States of the Glottis during Life. (Czermak.)

A, the glottis during the emission of a high note in singing ; B, in easy inhalation ; C, in taking a deep breath ; *b*, base of tongue ; *e*, the upper free part of the glottis ; *e'*, the tubercle or cushion of the epiglottis ; *ph*, part of the anterior wall of the pharynx behind the larynx ; in the margin of the aryteno-epiglottidean fold *w*, the swelling of the membrane caused by the cuneiform cartilage ; *s*, corniculum ; *a*, tip of arytenoid ; *cv*, true vocal cords ; *cvs*, false vocal cords ; *tr*, trachea ; *b*, bronchi in C.

A', B', C , diagrams of the glottis and positions of the arytenoid cartilages in the three states.

varying inversely as its length. The vocal cords of women are about one-third shorter than those of men, and hence their voice is of higher pitch. Tenor singers have shorter cords than basses, and sopranos than contraltos in most cases. In males at the age of puberty the larynx enlarges and the vocal cords become longer. Hence the voice becomes deeper and is said to 'break.' (2) The tension of the cords, for the tighter the cords are the higher the pitch, and *vice versâ*. Variations in the pitch of the voice can be made by producing changes of tension in the cords as the actions of the muscles are brought under the control of the will. *Falsetto*, or head notes, are believed to be produced either by the vibration of the edges of the cords only, or by the vibrating portion of the cords being shortened. In ordinary vocal sounds the vibrations of the cords are communicated to the column of air both in the resonating tubes above the larynx and in the trachea below, and the vibration of this column affecting the chest wall, such sounds are spoken of as *chest* notes. The *quality* or *timbre* (par. 234) of the voice depends on the prominence given to particular overtones or harmonics which accompany the fundamental tone. A voice sound, apparently simple, is in reality composed of a fundamental tone and certain accessory tones or harmonics, and the audible sound produced in the larynx, or primary sounding apparatus, may be so affected in its passage outwards through the adjustable resonance cavities of the pharynx and mouth as to become speech. Different voices uttering the same note differ in quality owing to the different set of overtones that predominate, these overtones being determined not only by the length and physical condition of the cords, but by the structure and form of the throat and mouth.

Speech is voice modified by alterations and additions made in the pharynx, mouth, and nose. The sounds formed in the larynx by the vibrations of the vocal cords being communicated to the air may undergo modifications from the varying size of the cavities above and the varying position of tongue and lips, and by these changes laryngeal sounds may be transformed into articulate speech consisting of syllables jointed together to form words. Speech may indeed exist without voice (*i.e.* sounds produced by the vocal cords), as in whispering, when the sounds are produced in the mouth alone. In ordinary speech, however, the sounds or air waves produced in the larynx are moulded and modified in the resonant cavities above. Speech sounds are divided into vowels and consonants. Vowels are the most open and continuable sounds uttered in the process of speech, and they differ from one another not in the pitch of the note produced by the larynx, but in the quality given to the sound by the overtones that are reinforced and modified in the cavity of the mouth. Each vowel sound acquires its special character from the reinforcement or addition of particular overtones as the oral cavity changes shape, and Helmholtz, by means of resonators, has not only been able to analyse the vowel sounds into their component vibrations, but he has succeeded in reproducing them synthetically. When the mouth is adjusted to produce the most open sound, the broad *a* in *far*, it has a sort of funnel shape with the wide part outward ; for *o*, as in *more*, the vowel chamber is like a bottle with a wide neck ; and for *u*, as in *poor*, the chamber is large with a narrow opening at the mouth. With *e* and *i*, sounded *eh* and *ēē*, the mouth has the form of a bottle with a long and narrow neck, formed by raising the tongue towards the hard palate.

In pronouncing *diphthongs* the mouth cavity changes its form as it passes from one vowel sound to the other. *Consonants* are the closer and less continuable sounds of speech, but there is no sharp line of distinction between vowels and consonants, some of the most open of the consonants having at times the value of vowels. Most of the consonant sounds are produced by irregular vibrations, and are of the nature of noises rather than musical sounds. They are accompanied by a narrowing of some part of the pharynx or mouth, which interrupts and modifies the air vibrations from the larynx, or sets up other vibrations in particular parts. According to the parts used in producing them, they are divided into *labials*, or lip letters, p, b, f, v, m ; *dentals*, or teeth letters (made with the tip of the tongue near the teeth), t, d, th ; *gutturals*, or throat letters (made at the top of the throat with the back of the tongue), k, g. According to their degree of closeness or mode of production, consonants are divided into *mutes* (*checks* or *explosives*) as b and p, which involve a complete shutting off of the passage of the breath ; *fricatives* (*spirants* and *sibilants*), as th, f, v, s, z, in which there is a rustling or friction of the breath through nearly closed parts ; *nasals*, as n, m, ng, in which the breath is admitted through the nasal passages and made to acquire a peculiar resonance.

The tongue, it will be seen, plays only a subordinate part in speech, and its loss only affects the pronunciation of those consonants in which it takes part. *Stammering* is due to irregular and spasmodic contraction of the diaphragm, which interferes with the expiratory blast of air on which speech depends. Mental disturbances or emotional excitement produce it in some people. *Stuttering* is defective speech due to inability to manage the larynx and other parts so as to form the proper sounds.

CHAPTER XIV

THE SPINAL CORD

178. **General View of the Cord.**—The spinal cord is the cylindrical mass of nervous matter contained in the spinal canal, which is formed by the superposed rings of the vertebræ. It is 17 to 18 inches long, about ¾ inch in diameter, and reaches from the margin of the foramen magnum of the occipital bone to the first lumbar vertebra, where it terminates in a slender thread, the *filum terminale*, which lies among a mass of nerve roots, termed the *cauda equina*. Above it is continued into the medulla oblongata, or bulb, which lies within the cavity of the cranium. Passing out from the cord at intervals, on each side, are the thirty-one pairs of spinal nerves, the first pair coming off between the skull and the atlas vertebra, the next pair between the atlas and the axis, and so on. The lower pairs, however, come off close together from the lower end of the cord, and pass downwards in the *cauda equina* to be distributed to the lower limbs. The spinal nerves leave the bony

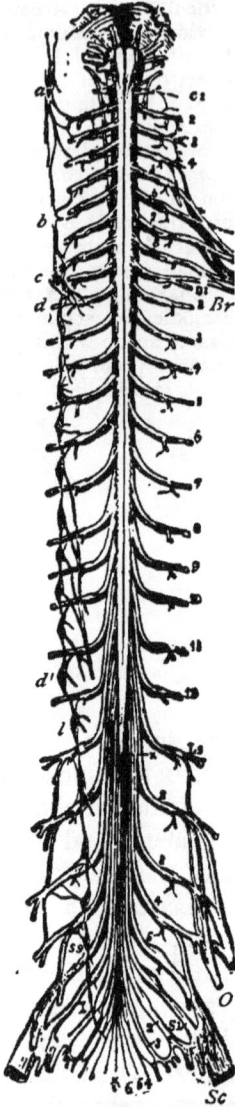

• canal by apertures between the vertebræ, termed intervertebral foramina. The diameter of the cord is not uniform throughout, being marked by two enlargements, the upper or cervical, and the lower or lumbar. Two fissures run along the length of the cord and separate it into a right and left half. The anterior median fissure is wider but not so deep as the posterior median fissure, and it contains a fold of the closely investing membrane of the cord, the *pia mater*, from which blood-vessels pass in to nourish the cord. At the bottom of this fissure is a transverse connecting portion of white substance, termed the *anterior* or *white commissure*. The posterior fissure is not an actual fissure, but a septum of connective tissue and blood-vessels, passing nearly to the centre of the cord, having at its base the *posterior grey commissure*. Besides the two median fissures, there is a lateral furrow at the line of attachment of the posterior roots of the spinal nerves, named the *postero-lateral groove*, and the attachment of the anterior roots, though not marked by any furrow and spread over some space,

Fig. 169.—Diagrammatic View from before of the Spinal Cord and Medulla Oblongata, including the roots of the Spinal and some of the Cranial Nerves, and on one side, the Gangliated Chain of the Sympathetic. (Allen Thomson.) ¼

The spinal nerves are enumerated in order on the right side of the figure. *Br*, brachial plexus; *Cr*, anterior crural, *O*, obturator, and *Sc*, great sciatic nerves, coming off from lumbo-sacral plexus; ×, ×, filum terminale.

a, *b*, *c*, superior, middle, and inferior cervical ganglia of the sympathetic, the last united with the first thoracic; *d*; *d'*, the eleventh thoracic ganglion; *l*, the twelfth thoracic (or first lumbar); below *s s*, the chain of sacral ganglia.

may be regarded as indicating another division in the cord. Further in the upper part of the cord there is a slight longitudinal furrow (s, fig. 170), the *postero-lateral furrow*, a little distance from the posterior median fissure. Each half of the cord is thus divided into four columns : an anterior column between the anterior median fissure and the anterior roots, a lateral column between the line of origin of the anterior roots and posterior roots, a posterior median column between the line of origin of the posterior roots and the postero-lateral fur-

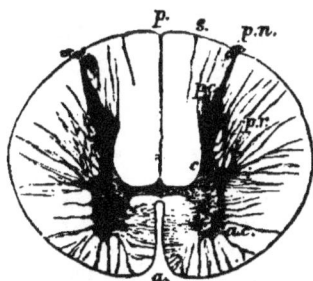

CERVICAL

DORSAL

FIG. 170.—Section of the Spinal Cord in the Upper Part of the Dorsal Region. (E. A. S.) ‡

a, anterior median fissure ; *p*, posterior median fissure; *p*, *n*, posterior nerve roots entering at the postero-lateral groove ; *a*, *c*, anterior cornu of grey matter ; *p*, *c*, posterior cornu ; *i*, intermedio-lateral tract (lateral cornu) ; *p*, *r*, processus reticularis ; *c*, posterior vesicular column of Clarke ; *s*, pia-matral septum forming the lateral boundary of the postero-mesial column.

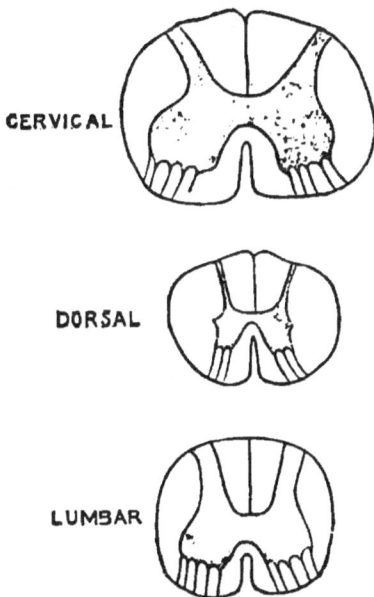

LUMBAR

FIG. 171. — Outline Sketch of Three Sections (× 3). (Waller.)

Taken from the cervical, thoracic, and lumbar regions of spinal cord (human).

row, and a posterior column between the postero-lateral and posterior median fissure.

A transverse section of the cord shows that it consists of white and grey matter. The white matter is on the outside and gives the cord its white opaque appearance ; the grey matter is arranged internally in the form of a crescent in each half, the two crescents being joined in the middle line by a grey commissure. In the centre of the grey commissure is the central canal of the cord, which is lined in early life by cylindrical columnar epithelium. At the outer side of each crescent the grey matter forms a sort of network, termed the *processus reticularis*. The two horns of the crescents of grey matter are named, from their position, anterior and posterior, and a triangular projection of grey matter in some parts of the cord about the middle of the crescent is called the *intermedio-lateral tract*, or lateral horn. The grey crescents vary in form in different parts of the cord. In the cervical region the anterior cornua are large and broad, and the posterior cornua narrow ; in the dorsal and thoracic regions, both anterior and posterior cornua are narrow ; in the lumbar regions the anterior and posterior cornua are broad (fig. 171). The grey matter relative to white is least in the thoracic and dorsal regions, and greatest in the lumbar regions.

179. **Membranes of the Cord.**—The spinal cord does not completely fill the spinal canal, for it is invested by three membranes with a space separating the outermost and the middle membrane, and another space between the middle one and the innermost. The outermost membrane is termed the *dura mater*. It is continuous with that which invests the brain, and is composed of tough connective tissue with a small amount of elastic fibres. In the cranium the dura mater is adherent to the bones, but it is separated from the bony walls of the spinal canal by loose alveolar tissue and a plexus of veins. It sends a tubular sheath along the spinal nerves for a short distance. By slitting up the dura mater and folding it back there is seen a thin delicate membrane around the cord, called the *arachnoid membrane*. It is continuous with the cerebral arachnoid above, passes for some distance as a sheath over the issuing nerves, and is separated in part from the dura mater by a narrow space termed the *sub-dural space*. This space contains a small quantity of fluid. On removing the arachnoid there is exposed the *pia mater*, a vascular membrane closely attached to the surface of the brain and spinal cord, dipping down into the fissures, and sending investments along the nerves. The pia mater is loosely connected with the arachnoid by strands of connective tissue forming a spongy network, but the space between the two is considerable, and is known as the *sub-arachnoid space*. This space contains a fluid called the cerebro-spinal fluid, and may be regarded as a serous or lymphatic space in communication with the perivascular lymphatics of the small arteries that pass into the nervous tissue of the brain and spinal cord from the pia mater, as well as with the lymph spaces in the nervous matter itself. It is also in communication with the central canal of the cord and the ventricles of the brain by a small aperture through the pia mater in the roof of the fourth ventricle, the *foramen of Magendie*. Cerebro-spinal fluid differs from ordinary lymph in the very small percentage of proteids that it contains, in the absence of fibrin ferment, and in the absence of cells ; it probably gets back into the circulation by escaping into the lymphatics of the nerves in

the sub-arachnoid space around their roots, and from them into the general lymphatics of the body.

A narrow fibrous band of connective tissue, attached to the pia mater along its whole length on each side between the anterior and posterior roots of the spinal nerves, is joined at intervals by tooth-like projections to the dura mater. This band is called the *ligamentum denticulatum.*

180. **Minute Structure of the Cord.**—The white matter of the cord is composed of medullated nerve fibres without the external sheath of Schwann, and running longitudinally. They vary in size in different parts of the cord, those near the surface being usually larger than those near the grey matter. These fibres are supported by a peculiar fibro-cellular tissue termed *neuroglia,* made of very small cells with many branching fibrils. The neuroglia is abundant at the surface of the cord, and extends into

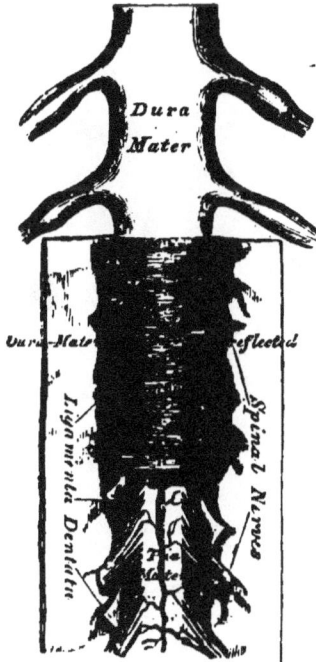

FIG. 172.—The Spinal Cord and its Membranes.

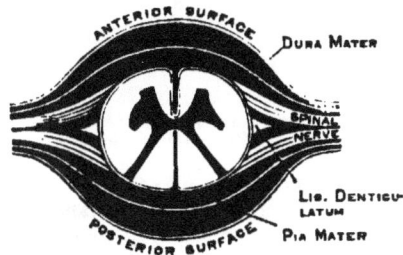

FIG. 173.—Transverse Section of the Spinal Cord and its Membranes.

the grey matter forming the *substantia gelatinosa* around the central canal and at the tip of the posterior cornu. The grey matter of the cord consists of nerve cells with branching processes, an interlacement of fine medullated fibres and neuroglia, the cells being arranged in columnar tracts in the crescents. The cells of the anterior cornu, termed the *vesicular column of cells* of the anterior cornu, are large multipolar cells, each of which has an axis-cylinder process continuous with a fibre of the anterior root of a spinal nerve, the other processes breaking up into a fine meshwork of fibrils. These cells are particularly numerous in the cervical and lumbar enlargements of the cord. The cells of the posterior cornu are much smaller, and any axis-cylinder process they may have does not appear to be connected with the fibres of the posterior root, but passes towards the anterior horn. A group or columnar tract of medium-sized cells, situated at the inner angle of the base of the posterior horn, is known as *Clarke's column.* Another group is found in the lateral cornu of grey matter.

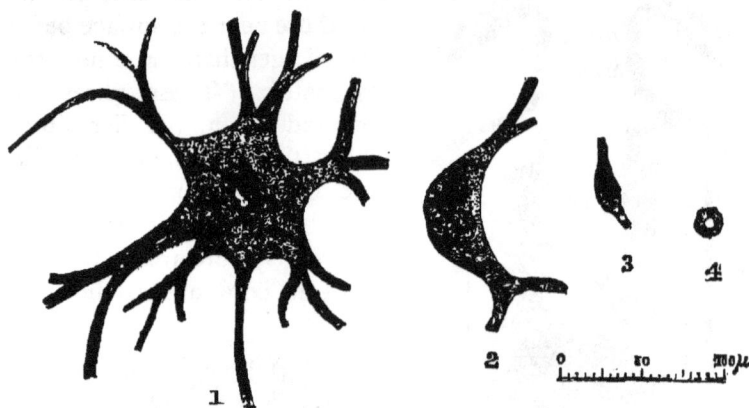

FIG. 174.

1, cell of anterior cornu ; 2, cell of Clarke's column ; 3, 'solitary' cell of posterior cornu ; 4, large nerve fibre. (Drawn to the same scale, viz. × 200 diameters.) (Waller.)

181. **Arrangement of the Nerve Fibres of the White Column in Tracts.**—In addition to the arrangement of the white matter into columns marked out by the superficial fissures and furrows, it has been found possible to arrange it into nerve tracts. If a nerve fibre be separated from its cell, it wastes or degenerates, and the degenerated fibres can be distinguished by their appearance under the microscope, especially after treatment with special staining reagents (par. 47). It has also been found that in the development of the spinal cord the nerve fibres of different tracts acquire their medullary sheath at different intervals, and different groups are thus recognised. Moreover, Gotch and Horsley have been able to distinguish the different paths of nervous impulses by the electrical changes set up. By a combination of these methods various tracts of fibres have been made out, as indicated in fig. 175. If the fibres are found degenerated below a lesion in the cord, the tract is said to be of *descending degeneration* ; if the degeneration is above the lesion, the tract is said to be of *ascending degeneration.* Tracts of descending degeneration are : —

(1) The *crossed pyramidal tract* is a *descending* tract of fibres found in the lateral column at the outer part of the posterior horn, of grey matter

throughout the length of the cord. This tract contains rather large fibres mingled with smaller ones, and is known to descend from the opposite side of the cortex of the brain, the fibres crossing at the pyramids of the medulla.

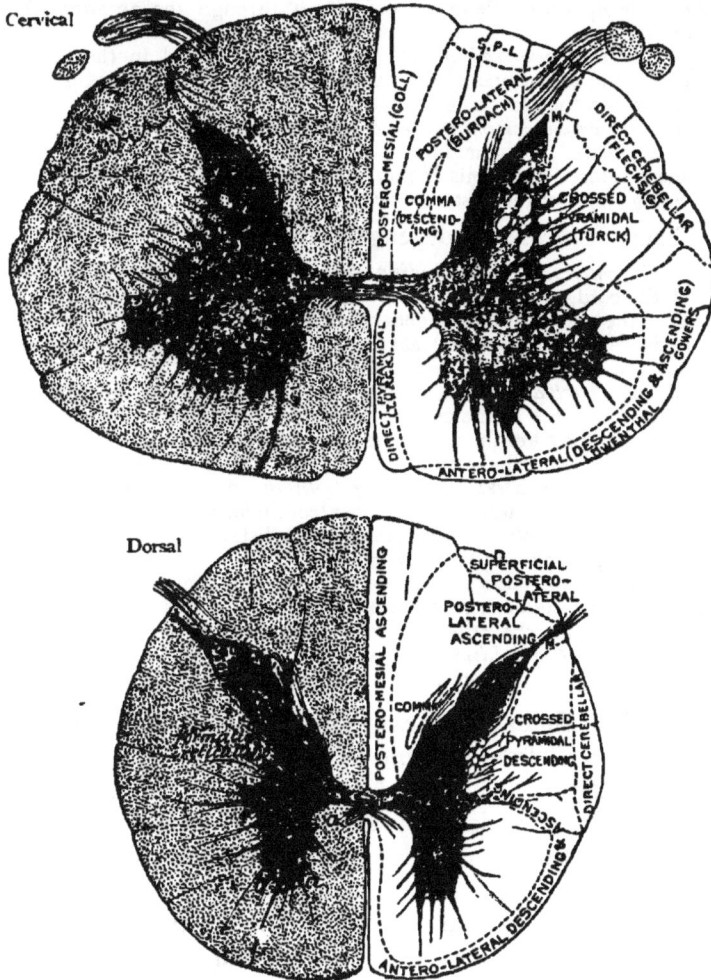

FIG. 175.—Section of the Human Spinal Cord from Lower Cervical (A) and Mid-Dorsal (B); showing the principal groups of the nerve cells, and on the right side of each section the conducting tracts as they occur in the two regions. (Magnified about 7 diameters.)

a, *b*, *c*, groups of cells of anterior horn; *d*, cells of the lateral horn; *e*, middle group of cells; *f*, cells of Clarke's column; *g*, cells of posterior horn; *c.c.*, central canal; *a.c.*, anterior commissure.

(2) The *direct pyramidal tract* is a *descending* tract of large fibres situated in the anterior column by the side of the anterior median fissure. It gradually diminishes in size on passing downwards, ending about the middle of the dorsal region. It belongs to a portion of the same tract as (1), the fibres not undergoing decussation in the medulla, but passing probably to the opposite side in the cord itself through the anterior white commissure.

(3) The *antero-lateral descending tract* is situated in the antero-lateral column, and contains fibres which are connected with cells in the brain cortex on the same side.

(4) The *comma descending tract* is a small tract in the upper part of the cord, situated in the middle of the postero-lateral column. Whether its fibres originate from cells higher up the cord or are derived from the descending fibres of the posterior roots is uncertain.

Tracts of ascending degeneration are :—

(1) The *antero-lateral ascending tract of Gowers*, situated at the outer part of the cord, mingles with the corresponding descending tract. On section of the cord its fibres degenerate above the lesion, and it is thought that its fibres, which can be traced upwards to the cerebellum, may arise from cells in the posterior cornu of the cord.

(2) The *direct ascending cerebellar tract* is situated at the outer part of the cord, external to the crossed pyramidal tract in dorsal and cervical regions. Its fibres are thought to be derived from the axis-cylinder processes of the cells in Clarke's column.

(3) The *postero-lateral ascending column*, or tract of Burdach, is mainly composed of large fibres that are continuous with the fibres of the entering posterior roots. After passing some distance in the cord numerous filaments pass off into the grey matter.

(4) The *postero-mesial column*, or tract of Goll, contains fine fibres also derived from the posterior root fibres and passing up this column into the medulla oblongata, where they terminate among cells of the nucleus gracilis.

(5) The *tract of Lissauer* is a small tract of fine ascending fibres (marked M in fig. 175), close to the posterior roots and derived from them.

182. The Spinal Nerves.—The thirty-one pairs of spinal nerves come off from the cord by two roots at intervals along its length, the portion of cord to which each pair of roots is attached being regarded as a 'segment,' or ganglionic mass of nerve cells, with which fibres are connected. But the segments are fused together into a continuous mass without any mark of separation. Each spinal nerve is attached to the cord by two roots, an anterior or ventral, and a posterior or dorsal, but the two roots after passing through separate openings in the dura mater unite to form a mixed nerve from which branches are given off to the ventral and dorsal parts of the body, as well as a ramus communicans or visceral branch to the sympathetic system (see par. 38 and p. 421). Before the two roots of a spinal nerve coalesce, a ganglion is found on the posterior root, where the roots lie in the intervertebral foramen.   As already stated, experimental excitation shows that the anterior root contains efferent fibres, and the posterior afferent fibres.   As will be seen shortly, however, the anterior root contains a few afferent fibres, though the posterior root is entirely afferent or sensory.

The anterior root arises by several converging bundles of fibres from the anterior column of white matter, and on following the fibres into the grey matter of the cord many of them are seen to be continuations of the axis-cylinder processes of the large cells in the anterior horn.   The fibres that can be thus traced are motor fibres for the skeletal muscles.   Other fibres pass by these cells and do not appear to be connected with them.   What their nature is cannot yet be stated, though we know that the anterior root contains, besides motor fibres to skeletal muscles, vaso-motor fibres to blood-vessels, and secretory fibres to glands.

The posterior root fibres enter the cord in a more compact mass than the anterior root fibres, and after their entrance into the cord separate into two sets, taking different courses.   The smallest fibres enter at the tip of the posterior horn, and join the ascending fibres in the tract of Lissauer ; the other set of somewhat larger fibres passes into the postero-lateral white column, proceeding for the most part upwards, and entering either the posterior median tract or the adjacent grey matter. As they enter the cord the fibres of the posterior roots bifurcate into two principal branches, running upwards and downwards in the posterior white column or adjacent posterior cornu, and give off collateral branches which run inwards towards the grey matter, and end in a plexus of fibrils around the nerve cells.   The fibres of the posterior root originate in the cells of the posterior root ganglia.   A section carried through the ganglion of the posterior root in the direction of the nervous cord shows a number of nerve cells with nerve fibres passing between.   The cells are unipolar, and a cell process joins a traversing nerve fibre by a T-shaped junction (fig. 48, par. 42), though it is possible that some fibres may pass through without any connection with a cell.   The function of

* T 2

the ganglia on posterior roots is to act as centres for the nutrition of the nerve fibres of the root. This is proved by severing
the connection of the fibres with the ganglion, when the fibres cut
off from connection with the ganglion degenerate, while those
that remain in connection do not (fig. 177). Thus section of the
posterior root between the ganglion and the cord leads to degeneration of the central part connected with the cord, while the
peripheral part connected with the ganglion remains unaffected.
Section below the ganglion leads to waste in the peripheral
part, but not in the central part. On the other hand, section

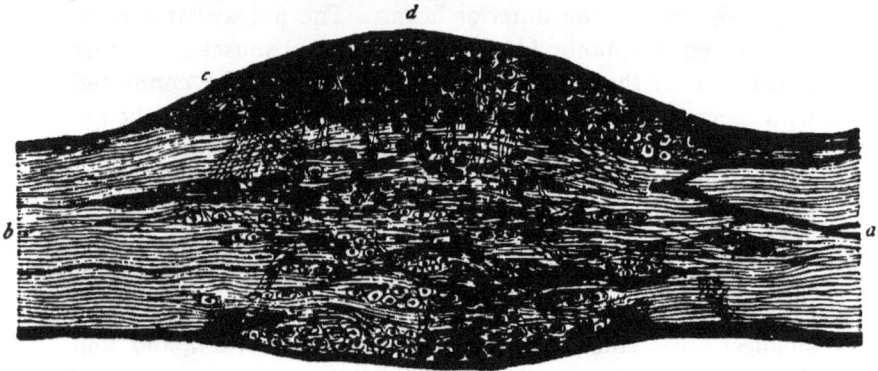

FIG. 176.—Longitudinal Section through the Middle of a Ganglion on the Posterior
Root of one of the Sacral Nerves of the Dog, as seen under a low magnifying power.
(E. A. S.)

a, nerve root entering the ganglion; *b*, fibres leaving the ganglion to join the mixed
spinal nerve; *c*, connective-tissue coat of the ganglion; *d*, principal group of nerve
cells, with fibres passing down from amongst the cells, probably to unite with the
longitudinally coursing nerve fibres by T-shaped junctions.

of the anterior root leads to degeneration in the peripheral
part, but not in the central, as its nutrition centre is in the cells
forming the grey matter of the anterior horn of the cord. Thus
the *trophic* (Greek *trophe*, nourishment) centre for sensory
fibres is in the ganglion of the posterior roots, and the trophic
centre for the motor or efferent fibres is in the grey matter of
the spinal cord. A few fibres, however, remain unaffected in
the peripheral end of a cut anterior root, and the central end
contains a few degenerated fibres among the mass of unaffected
ones. These are the *recurrent sensory fibres*. We have
already stated that the anterior root is motor or efferent,

and the posterior root sensory or afferent. But if the anterior root of a spinal nerve be divided, and the peripheral end be stimulated, there is not only movement of the muscles supplied by the nerve, but, in some cases, there is evidence of pain. This is spoken of as *recurrent sensibility*, and this is due to a few recurrent sensory fibres which leave the cord by the

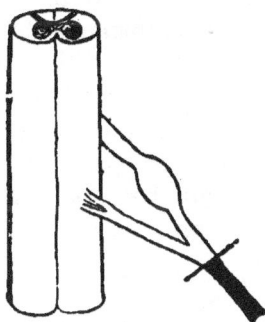

Degeneration of efferent and of afferent fibres below a section of entire nerve.

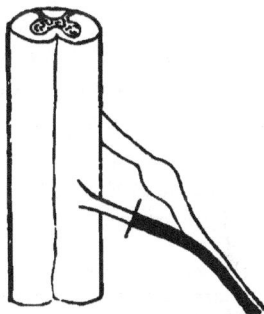

Degeneration of efferent fibres below a section of anterior root.

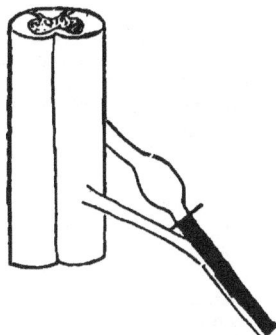

Degeneration of afferent fibres below a section of posterior root beyond the ganglion.

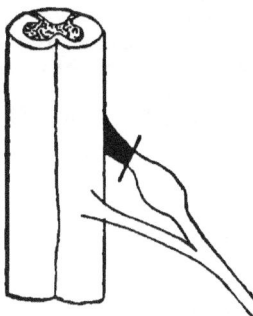

Degeneration of afferent fibres above a section of posterior root above the ganglion.

FIG. 177. Diagrams to illustrate Wallerian Degeneration of Nerve Roots. Degeneration black.

posterior roots, but turn back into the anterior root. On dividing the posterior root, recurrent sensibility disappears, as we should expect.

183. **Functions of the Spinal Cord.**—The spinal cord is both a *conductor* of nerve impulses and a *centre* for reflex action. From its structural connections with various parts of

the body, it is clear that sensory or other afferent nerve impulses from the periphery in the trunk and limbs can only be conducted to the brain and there perceived when the cord is uninjured and intact. The mandates of the will, also originating in the brain, can only pass to their destination along the efferent nerves when the communication through the cord is uninterrupted by disease or injury of the cord. Any disease or injury that affects the whole thickness of the cord leads to *paraplegia*, or complete paralysis, with loss of sensation and voluntary motion in the parts of the body that receive their nerve supply from the part of the cord below such injury. By experiment and observation the paths of the sensory and motor impulses along the cord have been ascertained to a certain extent, although much remains to be learnt.

(1) *Paths of Sensory Impulses in the Cord.*—Affe ent impulses con veyed to the cord by the posterior roots pass up some or all of the ascending tracts to the brain, for degenerations occur above the injury in the postero-external column and the postero-median column on the same side. But the tract taken by different impulses is not definitely determined in many cases. Recent experiments have, however, definitely proved that all sensory impulses do not decussate or cross over in the cord, as was supposed until lately, but that they pass up the same side for the most part, and cross chiefly in the medulla to reach the opposite side of the brain. Mott concludes that sensations of touch, pressure, and impressions of the muscular sense pass up the same side, but that painful sensations pass up both sides.

(2) *Paths of Motor Impulses in the Cord.*—Efferent impulses pass from the brain along the two pyramidal tracts, viz. in the crossed pyramidal tract chiefly, and in the direct pyramidal tract to a less degree, these tracts being undoubtedly the channels of voluntary impulses. For the most part they originate in the cortex of the cerebrum at one side, and cross to the opposite side in the pyramidal decussation in the medulla. A few fibres do not cross in the medulla, but pass down the direct pyramidal tract to decussate in the cord by the anterior white commissure. The vaso-motor impulses to the limbs travel along the lateral columns of the cord on the same side. Hemisection or transverse division of one-half of the cord produces the following results :—Motor paralysis of the muscles of the same side, but not complete in bilateral muscles that act together ; impaired sensation on the same side ; temporary vaso-motor paralysis with vaso-dilatation ; degenerations above and below the lesion on the same side.

184. **Reflex Action of the Spinal Cord.**—We have already (par. 44) explained the nature of reflex actions, and have pointed out how they may differ in nature and character. The reflex power of the spinal cord is the property which the cells

of its grey axis have of transforming afferent into efferent impulses. In such animals as the frog it is found that the cord alone can carry out numerous reflex acts both simple and complex. A frog from which the brain has been removed recovers from the shock in about an hour, and if protected from all stimulating influences remains still until death occurs. But if a gentle stimulus be applied to the skin of one foot, that foot is drawn up. This is an example of a *simple* reflex, and illustrates what is often called the law of *unilateral reflection*, for in general a slight excitation of a sensitive region causes a reflex movement in the neighbouring muscles. A strong stimulus not only leads to movements of the same side, but in a less degree to movements in the corresponding muscles of the opposite side, and if the stimulus be very strong, or if the cord be in an excitable condition, the sensory impulse may be reflected along most of the motor nerves of the body, so as to produce a reflex spasm. This spreading has been termed the *law of radiation* or *diffusion*. Strychnine so affects the grey matter of the spinal cord that the slightest touch on the skin sends all the muscles into this state of spasmodic contraction.

If one of the toes of the brainless frog be dipped into dilute sulphuric acid, the leg is only drawn up after some time. This shows that a weak impulse may not in itself be capable of discharging a reflex act, but that a succession of such impulses sent to the cord may combine their influence until a movement is caused. This phenomenon is known as *summation of stimuli*.

Complex co-ordinated reflex actions are exhibited by a brainless frog in which stimulation of afferent nerves leads to the discharge of complicated movements involving whole groups of different muscles, the movement being of a protective or purposive character. Thus, stimulating the flank of a frog with acid causes the leg of the same side to be swept over the spot, and if this leg be held, the other leg tries to remove the irritation. So well adapted are these and similar movements to secure the removal of the offending irritant, that they seem at first sight to indicate intelligence as existing in the cord. But the absence of all *spontaneous* movements which are a distinguishing mark of intelligence, will forbid this view, and the *mechanical* nature of reflex acts is well illustrated in the case of a decapitated snake, in which complex movements may be stimulated that indicate an effort to twine round a red-hot iron as readily as round a stick in contact with its body.

It thus appears that the disturbance set up when afferent impulses enter the cord spreads in the first place among the cells and network of fibres near the point of entrance on the same side, that it may then cross over at the same altitude to the opposite side, and that in some cases it may even spread along the length of the cord. Excitation of the cord in any part thus tends to spread in various directions, but with a preference for certain paths marked out by the structure and habits of the cord.

Reflex actions occur more readily in a brainless frog than in one whose brain has not been removed, and from this it appears that the brain exerts an inhibitory influence on the cord. This inhibition of reflex action that normally follows a slight impulse may also be brought about by placing crystals of sodium chloride on the optic lobes of a frog, or by strongly stimulating any sensory nerve. The influence of the brain in checking the reflex activity of the cord is well illustrated by the greater ease with which reflex actions occur in sleep, and by the absence of any sense of pain for a time that may occur in a soldier wounded in battle, when his mental energies are all centred on the fight.

Reflex time, or the time taken to transmute afferent into efferent impulses in the cord, varies somewhat with the nature of the reflex act and with the amount of resistance, and has been estimated at ·01 of a second to ·06 (par. 46).

In mammals, as the dog and man, the spinal cord appears to act habitually under the more direct influence of the brain, and the reflex actions of the cord severed from the brain appear to be of a more simple character than in lower animals. Further, the cord takes a longer time to recover from the shock of the operation or injury by which its connection with the brain is broken. The vital powers are much depressed, and for some time stimulation of a sensory surface below the lesion fails to evoke the simplest reflex act. But after some weeks simple reflex actions are called forth. A dog, whose spinal cord has been severed in the back, performs the reflex movements required for micturition and defæcation, and if its paralysed hind limbs be raised, a gentle push causes it to move forwards a few paces before it again sinks into a sitting posture. A man whose cord has been crushed in the dorsal region draws up his legs unconsciously when the soles are tickled, but the cord appears to have no power to carry out co-ordinated movements. In certain diseases of the cord more extensive movements are sometimes witnessed, but they are not of the purposeful co-ordinated character such as those described in the frog.

A special form of reaction, spoken of as ' tendon-reflex,' is the sudden contraction of a muscle when its tendon is struck. If the knee be half flexed, a sharp tap on the patellar tendon leads to the contraction of the rectus muscle of the thigh with raising of the leg. This ' knee-jerk,' as it is called, is present in health, but absent or exaggerated in certain diseases. Hence its importance to the physician. But it is doubtful whether it is a true reflex act, as the time between the blow and the contraction is as short as in muscular contraction, though the severance of its nerves from connection with the cord abolishes it.

185. **Reflex Centres of the Spinal Cord.**—As already intimated, the spinal cord should be regarded as a collection of nervous centres, formed by the union of its segments into a continuous column. These centres subserve various functions, and carry out numerous complicated movements, but it acts for the most part under the controlling and modifying influence of the brain. A *nerve centre* is to be regarded as a collection of nerve cells that have the power of controlling or modifying some action or function of the body, either automatically as the result of intrinsic changes in the nerve centre, or reflexly as the result of an immediate afferent impulse. The chief special centres of the spinal cord are :—

(*a*) The *musculo-tonic centre.*—The spinal cord exercises an influence over the muscular system that keeps the muscles of the body in a continual state of slight contraction, a state known as muscular tone. In injury or removal of the brain, the tone of the muscles still remains, but if the spinal cord be destroyed the muscles become flabby and loose. Section of the sciatic nerve in one leg of a frog causes the muscles of that limb to become relaxed—a proof that muscular tone is a continuous reflex action due to continuous afferent impulses.

(*b*) The *defæcation centre.*—The tonic contraction of the sphincter muscle of the rectum is a reflex action of the spinal cord, the centre being in the lumbar part of the cord. Afferent impulses pass in certain conditions of the lower bowel up to this part, through nerves in the mesenteric plexus, and are reflected by efferent fibres to the sphincter muscle and to the lower bowel, so that relaxation of this muscle and expulsion of the bowel contents follow. The centre for defæcation in the spinal cord is partially under the control of the brain, so that its action may be either inhibited or augmented.

(c) The *micturition centre.*—This centre acts in a similar manner to that of the defæcation centre, and its centre is also in the lumbar region of the cord. It is stimulated to action by the presence of urine in the bladder or by impulses from the brain, and the relaxation of the sphincter of the urethra is followed by contraction of the bladder and expulsion of its contents.

Fibres also leave the spinal cord by the anterior roots, and join the sympathetic, that regulate the secretion of sweat. Centres subordinate to the principal vaso-motor centre in the medulla are also contained in the cord. Further fibres pass some distance in the cord that pass to the iris and regulate the size of the pupil, and others connected with acceleration of the heart's action.

CHAPTER XV

THE BRAIN

186. The Nervous System.—The nervous system consists of a number of organs, termed *nerve centres, nerves,* and *peripheral end-organs.* The largest and most important nerve centres are the brain and spinal cord, which together constitute the *cerebro-spinal system,* the brain being lodged in the cranium and the spinal cord in the spinal canal. A smaller system of nerve centres consists of a double chain of ganglia on each side of the vertebral column from the cranium to the pelvis, and is known as the *sympathetic system.* It is not, however, a system independent of the brain and spinal cord, as once thought, but an outlying part of the same system and in close connection with it, all its fibres being derived from the cerebro-spinal system. The nerves are white cords or threads traversing the different regions of the body, and connecting the nerve centres with one another and with the periphery or outlying parts of the body. The peripheral end-organs are the minute structures at the extremities of the nerves, and are situated in the skin and other sense-organs, in glands, blood-vessels, and muscles. Under the microscope nervous tissue is seen to consist of nerve cells and nerve fibres in close association, and the structure and function of these two parts have already been in part considered. A previous chapter has also dealt with the spinal cord. We now enter on the study of the complex organ

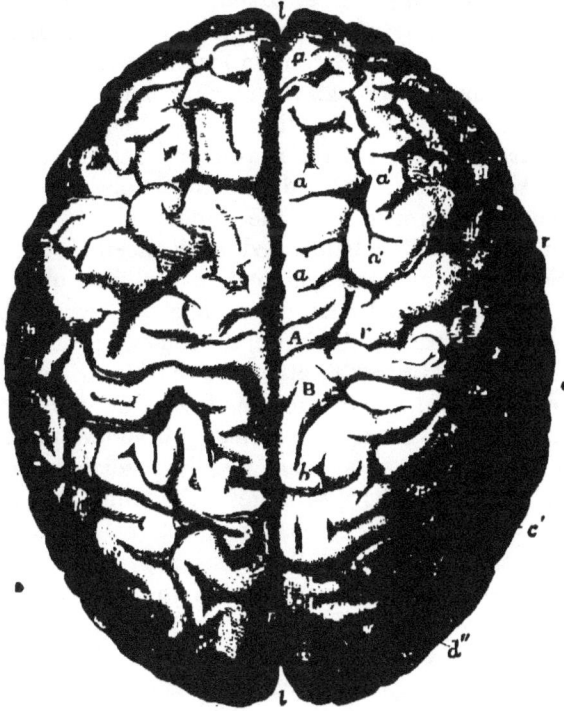

FIG. 178.—The Upper Surface of the Cerebrum, showing its Division into two Hemispheres by the Great Median Fissure, and also the Convolutions.

called the brain, or encephalon. A careful study of the illustrations will greatly aid in understanding the arrangement of the parts of the brain.[1]

187. **General Survey of the Brain** —A human brain removed from the skull and divested of its membranes, when looked at from above, shows nothing but the convoluted surfaces of the two hemispheres that constitute the *cerebrum*,

[1] A good model of the brain will also be of great service.

the hemispheres being separated from front to back by a longi-tudinal fissure. By drawing the hemispheres apart they are found to be connected in the middle half, at about half an inch below the surface, by a transverse band of white fibrous matter, termed the *corpus callosum*. Looked at from the side, the hinder portion of the cerebral hemispheres is seen overlapping the wrinkled *cerebellum*, or lesser brain. The hindmost part of the brain (for the term 'brain' includes all that part of the nervous system within the cranium) is the *medulla oblongata*, which is continuous, through the opening in the occipital bone called the foramen magnum, with the spinal cord. On turning up the base of the brain for inspection, each cerebral hemi-sphere is seen to be subdivided into three lobes, the anterior or frontal lobe separated from the middle lobe by the fissure of Sylvius, the middle lobe divided into parietal and temporo-sphenoidal lobes, and the temporal or occipital lobe. From behind forwards, in and near the middle line, we see, on the under surface or base of the brain, the following parts :—

The *medulla oblongata*, or bulb, pyramidal in shape, overlying the cerebellum, and with its broad end upwards. Certain pairs of the cerebral nerves are seen springing from its surface.

The *pons Varolii*, a quadrate mass immediately above the medulla, showing transverse fibres connecting it externally with the two sides of the cerebellum. Internally it is found to be continuous with the medulla (Figs. 42, 179).

The *crura cerebri*, or peduncles of the cerebrum, striated bundles of nervous matter emerging from the pons and entering the under part of each cerebral hemisphere as they diverge.

The *posterior perforated space*, a small triangular plate of brain tissue traversed by many small arteries, and the *corpora albicantia*, two small white bodies of unknown function, about the size of a pea. Both these structures lie between the di-verging peduncles of the crura.

The *tuber cinereum*, an eminence of grey matter in front of the corpora albicantia, attached to the junction of the optic nerves termed the *optic commissure*.

The *infundibulum*, a hollow conical process passing from the tuber cinereum to a small reddish body of unknown

function, called the *pituitary body*. Being enclosed in the dura
mater the pituitary body is usually detached when a brain is

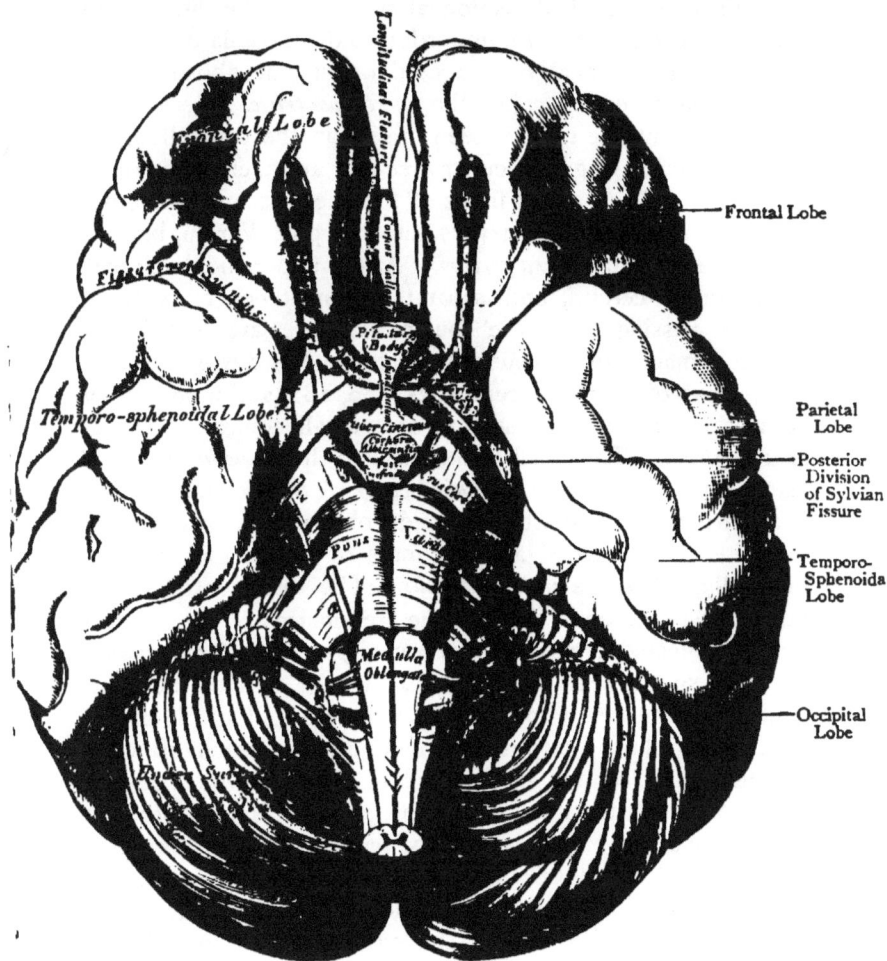

FIG. 179.—Base of the Brain.

removed from the skull. Lying in grooves on the under surface of the frontal lobes of the cerebrum, on each side of the
longitudinal fissure, are the two *olfactory tracts and bulbs*.

To get some idea of the inner relations of the various parts, as well as to see the basal ganglia and cavities of the brain, sections may be made in various directions. Fig. 180 represents a vertical longitudinal section along the middle line. The spinal cord is seen to pass upwards into the medulla oblongata, and the central canal of the cord opens out above into a lozenge-shaped cavity, called the *fourth ventricle*. Overhanging the fourth ventricle there is seen in section the cerebellum, its white matter having a peculiar tree-like arrangement, termed *arbor vitæ*. The upward fibres of the medulla and the cross fibres from the cerebellum pass into the pons Varolii, beyond which are seen the *crura cerebri*. At the upper and back part of the crura cerebri, and separated from them by a small channel passing from the fourth ventricle, are four hemispherical masses of nervous matter, called the *corpora quadrigemina*, two of which are seen in figs. 180, 181, and immediately above them is seen the section of a small conical body, the *pineal gland*. The small channel called the *aquæductus Sylvii* leads from the fourth to the *third ventricle* of the brain. This is a narrow median cavity, each side of which is bounded by the internal surface of an oval mass of matter that projects internally, called the *optic thalamus*. The third ventricle thus lies between the optic thalami, the grey matter of the thalami being connected across the narrow cavity by the *soft commissure* (cut end shown white in fig. 181). In the roof of the third ventricle is the *fornix* (covered by a double fold of pia mater, called the velum interpositum), a longitudinal arch of white fibrous matter, united behind to the *corpus callosum*, and in front to a thin partition called the *septum lucidum*, between the two layers of which is a space termed the *fifth ventricle*.[1] The fornix descends to the base of the brain. An aperture (the foramen of Monro) leads on each side from the front part of the third ventricle by a Y-shaped connection to a *lateral ventricle* in each cerebral hemisphere. By slicing a brain horizontally down to the corpus callosum, and removing this

---

The fifth ventricle is not regarded as a true ventricle, as it has no connection with the others, all of which are in communication and contain a small quantity of fluid, the cerebro-spinal fluid.

sufficiently, the two lateral ventricles are brought to view. Each lateral ventricle is an irregularly curved cavity, extending

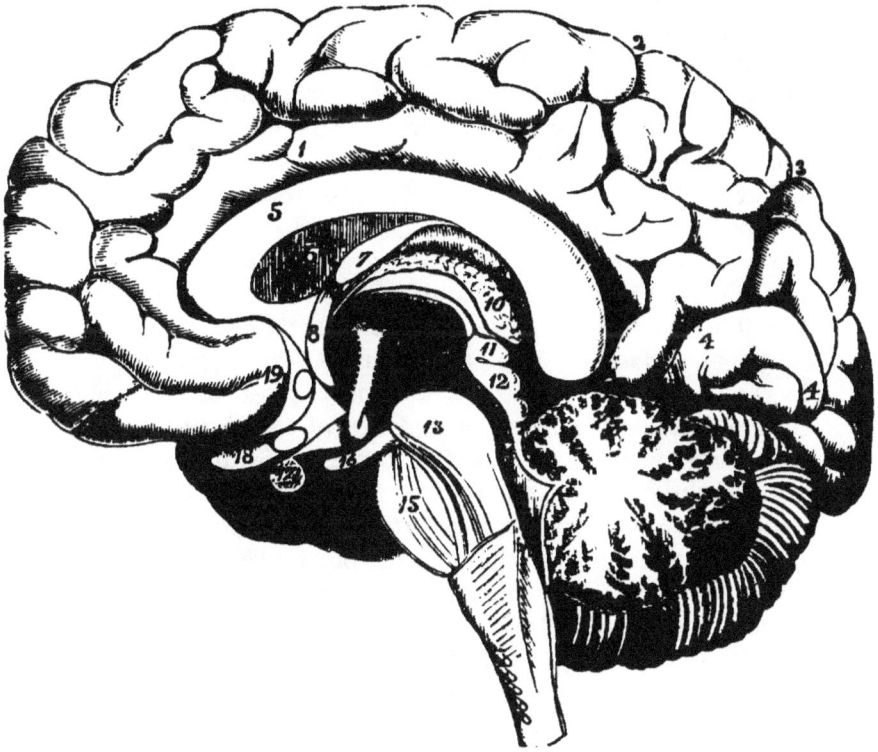

FIG. 180.—Vertical Median Section of the Encephalon, showing the parts in the middle line.

1. Convolution of corpus callosum. Above it is the calloso-marginal fissure, running out at 2 to join the fissure of Rolando.
3. The parieto-occipital fissure.
4, 4 point to the calcarine fissure, which is just above the numbers. Between 2 and 3 are the convolutions of the quadrate lobe. Between 3 and 4 is the cuneate lobe.
5. The corpus callosum.
6. The septum lucidum.
7. The fornix.
8. Anterior crus of the fornix, descending to the base of the brain, and turning on itself to form the corpus albicans. Its course to the optic thalamus is indicated by a dotted line.
9. The optic thalamus. Behind the anterior crus of the fornix, a shaded part indicates the foramen of Monro; in front of the number an oval mark shows the position of the grey matter continuous with the middle commissure.
10. The velum interpositum.
11. The pineal gland.
12. The corpora quadrigemina.
13. The crus cerebri.
14. The valve of Vieussens (above the number).
15. The pons Varolii.
16. The third nerve.
17. The pituitary body.
18. The optic nerve.
19 points to the anterior commissure indicated by an oval mark behind the number.

in the substance of the corresponding cerebral hemisphere for about two-thirds of its entire length, and lined by a prolongation of the ciliated epithelium, which characterises the inner surface of the true brain ventricles.   Each lateral ventricle consists of a central cavity or body, and three small cavities or cornua, the body of each lateral ventricle being separated in front from its fellow by the septum lucidum already referred to.   The roof of each lateral ventricle is formed mainly by the under surface of the corpus callosum.   Into the front portion of the floor of each lateral ventricle a rounded mass of nervous matter, known as the *caudate nucleus* of the *corpus striatum*, projects, a deeper part of the corpus striatum, the *lenticular nucleus*, being embedded in the mass of each cerebral hemisphere.   Part of the upper surface of the optic thalamus also enters into the floor of each lateral ventricle, the inner sides of the optic thalami forming the lateral boundaries of the third ventricle. Each thalamus rests upon, and is connected with, one of the crura cerebri, and has on its outer and hind part two small elevations, termed *corpora geniculata*.   Between the lenticular nucleus of the corpus striatum on the outer side and the optic thalamus and caudate nucleus on the inner side, a tract of white fibres (the *internal capsule*), continuous with the lower or anterior portion (crusta) of the crus, passes upwards to the cortex or outer layer of the cerebrum, some of its fibres diverging on the way in a fan-like manner to form a mass of white matter, known as the *corona radiata.*

The main bulk of the brain consists of the two large ovoid masses called the cerebral hemispheres, separated by the deep median longitudinal fissure at the bottom of which in the middle portion is the connecting corpus callosum.   Each cerebral hemisphere has an outer convex surface in contact with the vault of the cranium, an inner flat surface forming one side of the longitudinal fissure, and an irregular under surface overlapping the basal ganglia in the middle and the cerebellum behind, but resting in front and at the sides on the base of the skull.   All these surfaces are moulded into eminences which form convolutions, or *gyri*, separated from each other by fissures, or *sulci*.   The deeper fissures on the surface mark off each hemisphere into five lobes— frontal, parietal, occipital, temporo-sphenoidal, and central or island of Reil.

The brain, like other portions of the nervous system, contains grey and white matter, and all its parts are more or less closely connected by numerous nerve fibres. The grey matter consists of nerve cells (one pole or process of which is usually found giving off the axis-cylinder of a nerve fibre), a special

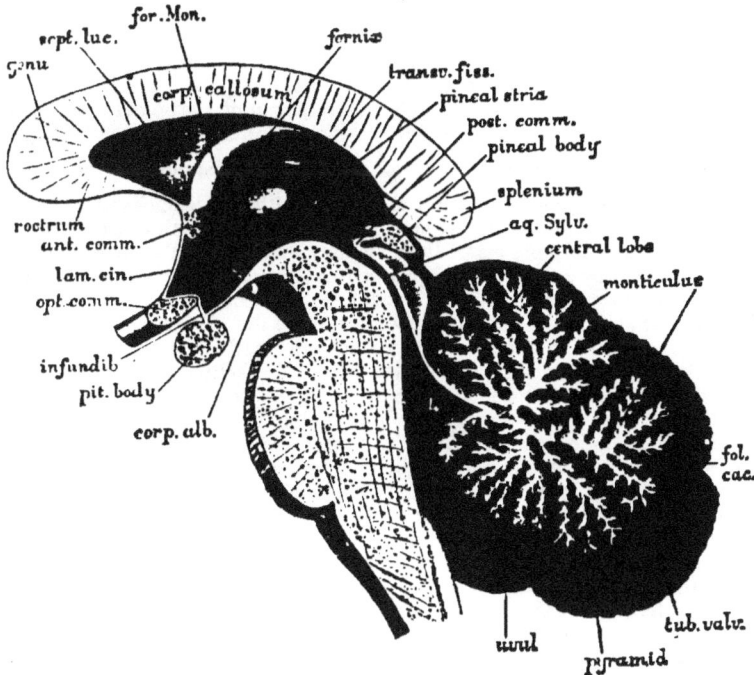

FIG 181.—Portion of a Median Section of the Brain, showing the Corpus Callosum, Third Ventricle, Aqueduct and Fourth Ventricle, Pons, Cerebellum, &c. (G. D. T.)

kind of connective tissue termed neuroglia, and minute blood-vessels. It is found on the *surface* of the cerebrum and cerebellum, and in small masses within the corpora striata, optic thalami, and other parts of the brain. The white matter consists of medullated fibres without the primitive sheath, of various sizes and arranged in bundles separated by neuroglia. These fibres are sometimes arranged into three different systems

U

according to their general course :—(1) Diverging or peduncular fibres (projection fibres) which connect the hemispheres with the lower portions of the brain and the cord, and which are in great measure direct prolongations of the axis-cylinders of the nerve cells of the cortex ; (2) transverse or commissural fibres (including the fibres of the corpus callosum, and the anterior and posterior commissures), which connect the two hemispheres together ; (3) association fibres, which connect different structures in the same hemisphere.

188. The Ventricles of the Brain.—As already mentioned, the brain contains certain cavities or ventricles, and the position and relation of these will now be further described. The central canal of the spinal cord opens out in the upper and posterior part of the medulla into a lozenge-shaped cavity $\langle \ \rangle$, the lower end of which is termed the calamus scriptorius, and the narrow upper end of which is continuous with the small channel termed the aqueduct of Sylvius (fig. 181). The roof of the fourth ventricle in the lower part is simply pia mater lined by epithelium, but in the upper half the roof is formed by a thin layer of grey matter (the valve of Vieussens), which here unites the converging superior peduncles of the cerebellum. The floor in the lower half is formed by the upper portion of the posterior surface of the medulla, and in the upper half by the posterior surface of the pons. The medulla and the superior peduncles of the cerebellum form the lateral boundaries. The Sylvian aqueduct or passage (iter) from the upper end of the fourth ventricle opens in front into the third ventricle, which lies between the optic thalami. The third ventricle is a narrow oblong cavity with the arching fornix covered by the velum interpositum for its roof, the bodies lying in the space at the base of the brain between the diverging crura for its floor and the optic thalami with the anterior pillars of the fornix for its sides. A connecting band of soft grey matter, the middle commissure, passes across the cavity between the optic thalami.

From the front part of the third ventricle an aperture (foramen of Monro) leads on each side into a large curved cavity in each cerebral hemisphere, termed the lateral ventricle. Each lateral ventricle consists of a body or central cavity, and three horns or cornua—an anterior cornu or horn, a posterior horn, and an inferior or descending horn. The roof of the main cavity is formed by a part of the corpus callosum. On the outside is the substance of the cerebral hemisphere, and at the inner side the bodies of the two ventricles are separated by the septum lucidum. ' The anterior horn curves from the foramen of Monro somewhat outwards, with a slight inclination downwards into the frontal lobe ; the body comprises that part of the cavity which extends from the foramen of Monro to its bifurcation into posterior and descending horns opposite the splenium of the corpus callosum, and is separated anteriorly from its fellow of the opposite hemisphere by a thin septum, the septum lucidum ; the posterior horn passes backwards, with a bold curve convex outwards into the occipital lobe ; and the descending horn passes forwards and slightly downwards, also in a bold curve with its convexity outwards, into the temporal lobe, and extends to about an inch from the apex of that lobe.'

The *septum lucidum* consists of two vertical layers of white matter, separating the front part of the lateral ventricles and enclosing the space termed the fifth ventricle. At its posterior part is the arched longitudinal white tract of fibres termed the *fornix*, the anterior pillars of which can be traced forwards and downwards to the corpora albicantia (fig. 181).

189. **The Membranes of the Brain.**—The brain is invested by three membranes or meninges, (1) the *dura mater*, (2) the *pia mater*, (3) the *arachnoid membrane.*

The dura mater is the external dense fibrous membrane closely attached to the inner surface of the skull. It is continuous with the dura mater, that

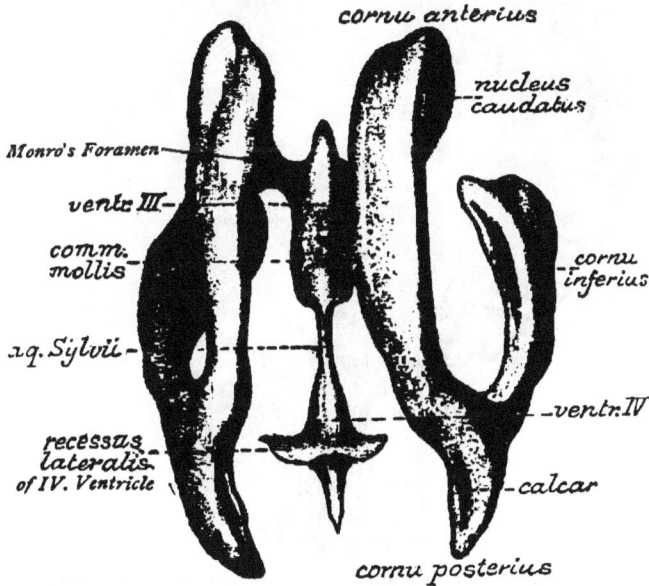

FIG. 182.—View of a Plaster Cast of the Ventricles of the Brain (from above).

forms a loose sheath round the cord, becoming intimately attached to the bony cavity of the skull a little before passing upwards through the foramen magnum. Within the skull it is divisible into two layers, and at various places the two layers separate to form channels or sinuses that contain venous blood. The dura mater also sends inwards three processes or membranous partitions—the *falx cerebri*, lying vertically between the two hemispheres in the longitudinal fissure ; the *tentorium cerebelli*, forming a sloping vaulted partition at the back between the cerebrum and the cerebellum ; and the small *falx cerebelli*, between the hemispheres of the cerebellum.

The pia mater is the delicate fibrous vascular membrane that closely invests the brain, dipping down into the sulci. From its internal surface numerous small blood-vessels pass into the brain substance. At the transverse fissure it is prolonged into the lateral ventricles and over the third ventricle, forming the *velum interpositum* and choroid plexuses. The velum interpositum is the triangular fold of pia mater lying immediately under the fornix, where it can be seen when that structure is cut through and raised. The pia mater is also prolonged over the posterior roof of the fourth ventricle. The arachnoid is a delicate membrane lying outside the pia mater, but it does not invest the brain so closely as the pia mater, as it passes over the fissures without dipping into them. The space between the

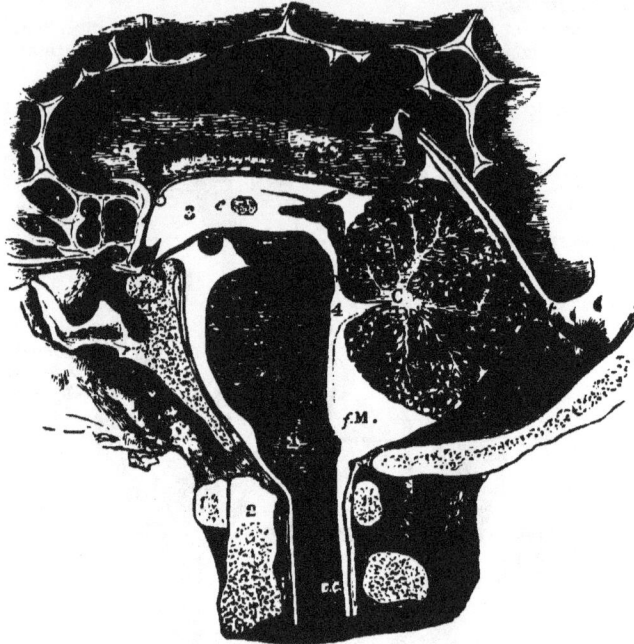

FIG. 183.—Section of the Posterior and Lower Parts of the Brain within the Skull, to exhibit the Sub-arachnoid Space, and its Relation to the Ventricles. (After Key and Retzius.)

The section was made in the frozen state, the cavities having been previously filled with injection.

1, 1′, atlas vertebra ; 2, odontoid process of the axis, 2′ ; 3. third ventricle ; 4, fourth ventricle ; C. C, corpus callosum ; C′, gyrus fornicatus ; C̆, cerebellum ; t, tentorium ; p, pituitary body ; c.c., central canal of the cord ; f.M, in the cerebello-medullary part of the sub-arachnoid space, is close to the foramen of Magendie, by which that space communicates with the fourth ventricle.

dura mater and arachnoid is known as the *sub-dural space,* that between the pia mater and arachnoid, the *sub-arachnoid space.* The sub-arachnoid space is larger in some places than others. At the base of the brain, for example, there is a wide space between the pia mater and arachnoid. In both spinal cord and brain, thin bands of connective tissue connect the pia mater and arachnoid, and in the meshes of the sub-arachnoid space thus formed is lodged the cerebro-spinal fluid, a peculiar clear fluid with but few of the properties of lymph. The sub-arachnoid space communicates with the ventricles of the brain and the spinal canal by a small hole in the pia mater, forming the roof of the fourth ventricle, the foramen of Magendie (fig. 183).

190. **The Medulla Oblongata, or Bulb.**—The medulla oblongata, or bulb, forms the intermediary or bond between the spinal cord and the brain proper. A rearrangement and shifting of the white fibres of the cord takes place, most of them undergoing decussation, *i.e.* crossing over to the opposite side. The narrow central canal of the cord opens out into the wide fourth ventricle on the posterior surface, and the grey matter of the cord becomes separated into four masses on each side by the passage of fibres across the anterior and posterior horns. Additional grey matter is also superadded. Two tracts or strands of fibres can be clearly traced through the bulb to higher parts, the pyramidal tracts to the cerebrum and the cerebellar tract to the cerebellum ; other tracts appear to terminate in groups of cells that act as relays between the fibres of the cord and the white matter of the brain.

The medulla is pyramidal in shape, lying inside the cranium and being continuous with the spinal cord below and the pons Varolii above. Its length is about one inch, breadth nearly one inch, and depth $\frac{3}{4}$ inch. Its outer surface shows fissures and columns, each half consisting of the following parts—an anterior pyramid, olivary body, restiform body, and posterior pyramid. On the anterior or ventral surface is seen an anterior median fissure continuous with that of the cord. The *anterior pyramids* are two bundles of white matter, one on each side of the anterior fissure, and separated from a rounded elevation known as the *olivary body* (see figs. 42 and 184). These anterior pyramids are formed of fibres from the cord, some of the fibres being continuous below with those in the cord forming the *direct pyramidal tract*, but most being derived from the lateral column of the cord of the opposite side, called the *crossed pyramidal tract*, the crossing of the lateral pyramidal fibres seen in the anterior fissure being spoken of as the *decussation of the pyramids*. These two strands of fibres in the anterior pyramid are efferent or motor in function, and are continuous through the crusta, or inferior part of the crura, with fibres in the internal capsule that pass to certain parts of the cerebrum known as the motor areas of the cortex. This is proved by their exhibiting descending degeneration when lesions or injuries of these areas of the cortex are found.

Between the olivary body and the posterior fissure of the medulla are the *restiform body*, and the *posterior pyramid* or *funiculus gracilis* of the medulla. The restiform body contains afferent or sensory fibres from the posterior columns of the cord, and some of these, those forming the direct cerebellar tract, enter the inferior peduncles of the cerebellum. Below the restiform body the posterior median column of the cord appears as the strand of fibres termed the *funiculus gracilis*, the posterior external column appearing as another strand termed the *funiculus cuneatus*. These two strands of fibres terminate within the medulla in grey matter called the *nucleus gracilis* and the *nucleus cuneatus* respectively. Fibres pass across the central grey matter from those nuclei to parts of the brain above, their crossing being known as the *upper* or *sensory decussation*.

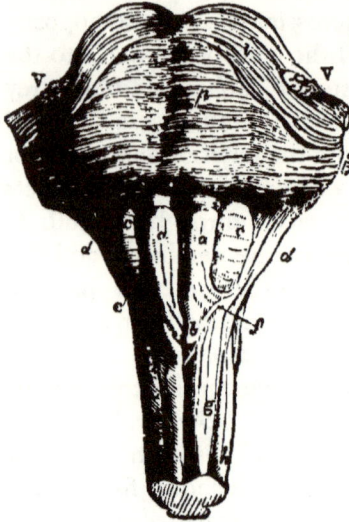

FIG. 184.—View of the Anterior Surface of the Pons Varolii and Medulla Oblongata, with a small part of the Spinal Cord attached.

a, a, pyramids ; *b,* their decussation ; *c, c,* olives ; *d, d,* restiform bodies ; *e,* external arciform fibres, curving round the lower end of the olive ; *f,* fibres described by Solly as passing from the anterior column of the cord to the cerebellum ; *g,* anterior column of the spinal cord ; *h,* lateral column ; *p,* pons Varolii ; *i,* its upper fibres; v, v, roots of the fifth pair of nerves.

The *fourth ventricle* is the space into which the central canal of the cord opens out behind by the divergence of the funiculi or posterior pyramids, which form its lateral boundaries in the lower half. In the upper half its lateral boundaries are formed by the superior peduncles of the cerebellum that pass downwards and outwards. The cavity is, therefore, lozenge- or diamond-shaped, with the widest part in the middle. The pointed lower end is termed the *calamus scriptorius*, from its resemblance to a writing pen. The upper angle reaches to the level of the pons, and communicates by the aqueduct of Sylvius with the third ventricle (fig. 181). Its roof is formed in the upper part by a thin lamina of white matter streaked with grey, termed the valve of Vieussens ; in the lower half the roof consists of a reflection of pia mater from the cerebellum. The floor of the fourth ventricle is formed by the posterior surface of the medulla oblongata and pons. Emerging from a slight fissure in the floor are some white fibres spoken of as *auditory striæ*, though it is doubtful how far they are connected with the auditory nerve. A number of small elevations on the floor correspond to the nuclei of origin of several cranial nerves which arise from collections of nerve cells in the grey matter of the floor of this ventricle. These nerves will be described presently. Additional grey matter is found in the olivary bodies, and in two small detached masses termed the accessory olivary nuclei. The greater part of the central and lateral parts of the medulla consist of grey matter traversed by longitudinal and transverse fibres,

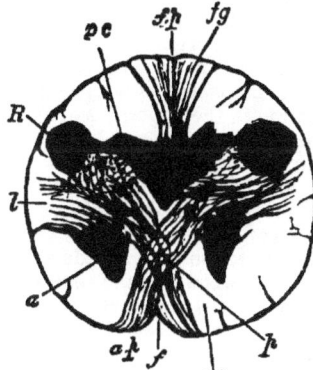

FIG. 185.—Section of the Medulla Oblongata at the Middle of the Decussation of the Pyramids. (Lockhart Clarke.)

f, anterior ; *f.p.*, posterior fissure ; *a.p.*, pyramid ; *a*, remains of part of anterior cornu, separated by the crossing bundles from the rest of the grey matter ; *l*, continuation of lateral column of cord; *R*, continuation of substantia gelatinosa of Rolando ; *p.c.*, continuation of posterior cornu of grey matter ; *f.g.*, funiculus gracilis.

so that this portion is called the *reticular formation.* We now see that, mixed with the nerve fibres conveying impulses to and from the brain, there are in the medulla several ganglionic centres of great importance, as well as the nuclei of origin of important cranial nerves.

191. **Functions of the Medulla.**—The functions of the medulla, or bulb, have already been referred to several times. It acts (*a*) as a *conductor*, (*b*) as a *nerve centre*, or rather a *collection of nerve centres*. (*a*) All the impulses passing between the brain and spinal cord must pass through the

medulla. Efferent impulses travel mainly through the anterior pyramids, where decussation (anterior pyramidal decussation) occurs of those fibres from the cords that have not already crossed in the cord, viz. of the fibres in the crossed pyramidal tract. The fibres of the so-called direct pyramidal tract are believed to decussate at various levels in the cord by the anterior commissure. The afferent or sensory path is not completely made out, but it is probably for the most part along the posterior pyramids, the fibres of which terminate in the nucleus gracilis and nucleus cuneatus of the bulb. From these nuclei fibres pass round the front of the medulla to the opposite side in what is termed the superior pyramidal decussation (sensory decussation). Some fibres may decussate in the pons. At any rate decussation of the fibres connecting the spinal cord and brain is complete in the crura cerebri, so that all impressions to and from the hemispheres of the brain pass across the middle line. Any injury or compression, therefore, of either hemisphere impairs sensation and voluntary motion in the opposite side of the body. A destructive affection of one of the hemispheres usually produces complete motor paralysis and loss of sensibility in the opposite side.

(*b*) The importance of the medulla as a nerve centre is established by the fatal issue that follows its injury or disease. Any sudden displacement of the upper cervical vertebræ, as in hanging, so injures the cord and its connections as to produce instant death. So does injury to the medulla itself, especially in the central part. By experiments on the lower animals it has been shown that the whole of the brain except the medulla may be gradually removed while respiration and life continue some time. The same result follows on removal of the spinal cord up to the origin of the phrenic nerve—a nerve arising from the third and fourth cervical nerves and supplying filaments to the diaphragm. Most of the nerve centres of the medulla are reflex centres, but the reflex actions under its control are much more complicated than those of the spinal cord. Unlike the brain proper it discharges no mental functions, initiating none but reflex movements.

Its most important centres are studied in other parts of the book, and a simple summary is all that can be placed here.

(1) The medulla contains a *respiratory centre*, the centre being bilateral and situated behind the origin of the vagi nerves on each side of the posterior aspect of the *calamus scriptorius*. The efferent nerves are the branches of the vagus distributed to the lungs, and these appear to be stimulated according to the condition of the blood as regards the amount of oxygen and carbon dioxide. The impulse reaching the medullary centre is reflected along the efferent or motor fibres of the phrenics, intercostal and other nerves associated with respiratory movements (see par. 100). Possibly also the venous blood circulating in the medullary centre itself may excite it and lead to its automatic action.

(2) *Cardiac centres* also exist in the medulla, one accelerating the action of the heart through the sympathetic, another inhibiting the action of the heart through the vagus (par. 83).

(3) A *vaso-motor* (vaso-constrictor) centre lies in the medulla, dominating the nerves supplied to the unstriped muscle of the arteries, intestines, &c. The centre is bilateral and lies in the floor of the medulla, a little above the calamus scriptorius. Under ordinary circumstances it keeps the arteries of the body in a state of tonic contraction. Stimulation of the centre leads to contraction of the arteries and a general rise of blood-

pressure ; inhibition of the centre leads to dilatation of the arteries and a great fall of blood-pressure. The vaso-motor centre in the medulla controls subordinate centres in the cord. A special *vaso-dilatator* (vaso-inhibitory) centre, not acting continuously or tonically, is also believed to exist in the medulla (par. 84).

(4) A *centre for mastication* is believed to be situated in the medulla. The afferent nerves are the sensory branches of the fifth or trigeminal, and

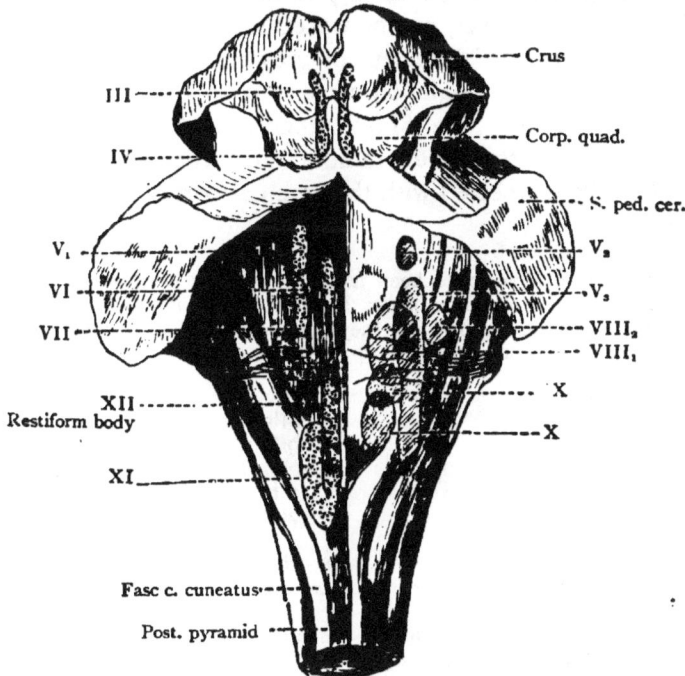

Fɪɢ. 186. Diagram of the Fourth Ventricle, showing Nuclei of Cranial Nerves. (Ziegler.)

III, nucleus of third ; IV, nucleus of fourth ; V_1, nucleus of motor of fifth ; V_2 and V_3, nuclei of sensory of fifth ; VI, nucleus of sixth ; VII, nucleus of facial ; VIII_1, VIII_2, nuclei of auditory ; IX, nucleus of glosso-pharyngeal; X, nucleus of vagus ; XI, nucleus of spinal accessory ; XII, nucleus of hypoglossal.

the tenth or glosso-pharyngeal, and the efferent nerves are the motor branches of the fifth and twelfth cerebral nerves.

(5) A *centre for salivary secretion* is also found in the medulla, and its reflex activity has been described in par. 114.

There are also in the medulla centres for deglutition, vomiting, coughing, and dilatation of the pupil.

192. The Cranial or Cerebral Nerves.—The cranial nerves consist of twelve pairs, and appear to arise from the surface of

the brain in a double series, passing thence through the base of the cranium to their distribution. Their *superficial* origin extends from the under surface of the frontal lobe of the cerebrum to the lower end of the medulla (see figs. 42, 179); and the ultimate distribution of all except the tenth (pneumogastric) and eleventh (spinal accessory) is to some part of the head. Their fibres, however, can be traced into the substance of the brain to some special nucleus of grey matter, which is termed their *deep* or real origin or root.   With one exception (the first or olfactory) the fibres proceeding from the nuclei of origin cross within the cranium, and the nerves are thus functionally connected with the cerebral cortex of the opposite side.

The cranial nerves have been named according to the order in which they pass through the dura mater lining the base of the skull, as well as according to the parts to which they are distributed, or to their functions. They may be thus enumerated :—

1st. Olfactory.	7th. Facial.
2nd. Optic.	8th. Auditory.
3rd. Motor-oculi.	9th. Glosso-pharyngeal.
4th. Trochlear.	10th. Pneumogastric or vagus.
5th. Trigeminus.	11th. Spinal accessory.
6th. Abducens.	12th. Hypoglossal.

The first two pairs differ in their origin and mode of development from all the rest, being in reality actual outgrowths, or processes, of the brain itself.

According to their function the cranial nerves have been divided into sensory, motor, and mixed, thus :—

(*a*) Nerves of special sense.	1st or olfactory, 2nd or optic, 8th or auditory, part of the 5th, part of the 9th.
(*b*) Motor nerves.	3rd or motor-oculi, 4th or trochlear, part of the 5th, 6th or abducens, 7th or facial, 12th or hypoglossal.
(*c*) Mixed nerves.	9th or glosso-pharyngeal, 10th or vagus, 11th or spinal accessory.

The *first cranial or olfactory nerves* are in reality lobes of the brain, and arise by a triple root in the under part of the frontal lobe.   The two olfactory tracts lie in a furrow on either side of the median fissure of the cerebrum.   On reaching the cribriform plate of the ethmoid bone of the skull they expand into bulbs, from the under surface of which ten or twelve true olfactory nerves pass through the cribriform plate to be distributed to the mucous membrane of the nose, there forming the terminal organ of smell. Unlike all other nerves, it is supposed that the sensory impressions do not cross to the opposite side of the cerebrum (fig. 179).

The *second or optic nerves*, distributed to the eyeballs, are connected together at the optic commissure, where a *partial* decussation of fibres takes place. Behind the commissure, under the name of optic *tracts*, the nerves may be traced to their origin in the anterior corpora quadrigemina, the geniculate bodies, and the hinder part of the optic thalami. From their nuclei of origin fibres may be traced to the visual centre in the cortex of the occipital lobe (par. 204).

At the optic commissure fibres from the inner or nasal half of each retina decussate and pass backwards to the opposite half of the brain, the fibres

To OPTIC N.
— *of same side*
— *of opposite side*

FIG. 187.—Course of the Fibres in the Optic Commissure.

from the outer or temporal half of the retina undergoing no crossing. Thus the right optic *tract* contains fibres from the outer half of the right retina and the inner half of the left retina, so that light from objects on the left side of the body is transmitted to the right side of the brain. Similarly, light from objects of the right side passes to the left side of the brain. Hence, section of one optic *nerve* would produce blindness of the corresponding eye, but section of one optic tract would produce a half-blindness of each retina, *hemianopia* as it is called. The distribution of the filaments of the optic nerve in the eyeball is treated of near the end of par. 204 and in fig. 204.

The *third or oculo-motor nerves* have their origin in clusters of cells in the grey matter of the inner part of the crura cerebri on each side of the aqueduct of Sylvius, where the nerves of the two sides decussate. The nerves of the third pair are purely motor, and pass by two branches into the orbits of the eyes, to be distributed to (1) the elevators of the eyelids, (2) the superior, (3) inferior, and (4) internal recti muscles. Fibres also pass to the circular muscle of the iris and to the ciliary muscle, to regulate contraction of the pupil and accommodation.

The *fourth or trochlear nerve* arises from a nucleus of large multipolar ganglion cells immediately below the nucleus of the third in the floor of the aqueduct. It is purely motor in function, and supplies the trochlear or superior oblique muscle of the eye. Its section or paralysis leads to squinting inwards and upwards.

The *fifth or trigeminal nerve* has two roots like a spinal nerve—a large sensory root in connection with the Gasserian ganglion, and a smaller motor root which has no ganglion. Its nuclei are in the part of the pons forming the floor of the fourth ventricle. The sensory part is distributed to the face, the teeth, the mucous membrane of the nose and mouth, and to the conjunctiva of the eye. Motor filaments pass to the muscles of mastication, the tensor tympani, and tensor palati.

The *sixth or abducens nerve* arises from a nucleus in the floor of the pons, about the middle of the floor of the fourth ventricle. It is exclusively motor, and supplies only the external rectus muscle of the eye.

The *seventh or facial nerve* arises in a nucleus in the pons near that of the sixth, and its fibres emerge from the medulla between the restiform and olivary bodies. It is the motor nerve for all the muscles of facial expression, and its paralysis or injury on one side leads to a blank look on that side with drooping of the angle of the mouth. In the chorda tympani branch of this nerve are secretory and vaso-dilatator fibres for the sub-maxillary and sublingual glands (par. 114).

The *eighth or auditory nerve* arises from three nuclei in the floor of the fourth ventricle forming the upper part of the medulla. Passing through the temporal bone by the internal meatus it divides into two branches, one of which passes to the cochlea, and the other to the utricle and semicircular canals of the inner ear. The first is the auditory portion connected with the sense of hearing ; the second branch conveys impulses from the semicircular canals that aid in the reflex maintenance of the equilibrium of the body. Section of the auditory nerve not only causes deafness, but giddiness and other disturbances of equilibrium, that indicate the double function of this so-called eighth nerve.

The *ninth or glosso-pharyngeal nerve* arises from a nucleus in the floor of the lower part of the fourth ventricle. It is mainly a sensory nerve conveying sensations of taste from the hinder portion of the tongue and adjoining mucous membrane of the mouth and pharynx. It has com-munications with the vagus or pneumogastric, with the upper cervical ganglion of the sympathetic, and with the digastric branch of the facial nerve.

'The *tenth or pneumogastric nerve* (*nervus vagus*) has a more ex-tensive distribution than any of the other cranial nerves, passing through the neck and thorax to the upper part of the abdomen. It is composed of both motor and sensory filaments. It supplies the organs of voice and respiration with motor and sensory fibres ; and the pharynx, œsophagus, stomach, and heart, with motor influence. Its *superficial origin* is by eight or ten filaments from the groove between the restiform and the olivary body below the glosso-pharyngeal ; its *deep origin* may be traced deeply through the fasciculi of the medulla, to terminate in a grey nucleus near the lower part of the floor of the fourth ventricle, below and continuous with the nucleus of origin of the glosso-pharyngeal. The filaments become united, and form a flat cord, which passes outwards across the flocculus to the jugular foramen, through which it emerges from the cranium. In passing through this opening, the pneumogastric accompanies the spinal accessory, being contained in the same sheath of dura mater with it, a membranous septum separating it from the glosso-pharyngeal, which lies in front.'

With the ganglion of the root are connected the accessory part of the spinal accessory nerve, a twig from the glosso-pharyngeal, and an ascend-ing filament from the upper cervical ganglion of the sympathetic. The chief branches of the pneumogastric are the laryngeal, cardiac, pulmonary, and gastric.

Most of the functions of the pneumogastric have been already referred to. It carries efferent impulses to the muscles of the pharynx, larynx, œsophagus, stomach, trachea, and lungs ; vaso-motor influences to the same organs, and inhibitory influence to the heart, which is of great importance in the circulation (see par. 83). Among its afferent fibres are those which convey, inwards, impulses from the respiratory and digestive organs, excitor-motor fibres that lead to the reflex phenomena of coughing, vomiting, &c., excito-secretory fibres, and fibres that convey inhibitory influence (depressor) from the heart to the vaso-motor centre.

Section of both vagi in the neck leads to acceleration of heart-beat, difficulty of swallowing owing to paralysis of muscles of pharynx, and

slower but deeper respiration with paralysis of muscles of the larynx.
Foreign bodies accumulate in the insensitive larynx and air-passages,
usually producing fatal inflammation of the lungs.

The *eleventh or spinal accessory nerve* arises in part from a nucleus

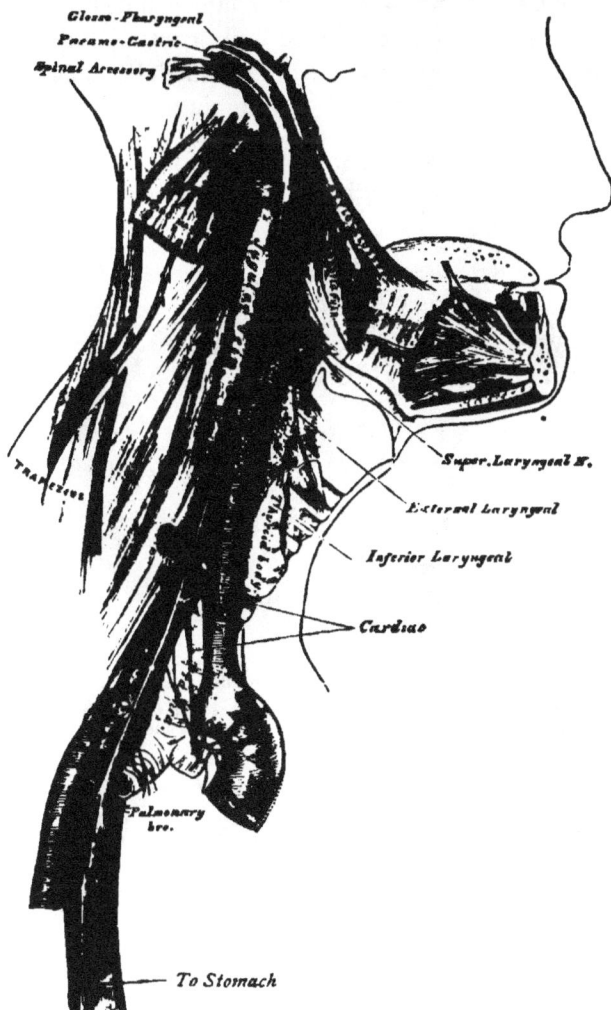

FIG. 188.—Course and Distribution of the Pneumogastric, &c.

below that of the vagus, and in part from the lateral column of the spinal
cord. Motor fibres are distributed from it to the trapezius and sterno-
mastoid muscles, and viscero-motor filaments join the vagus.

The *twelfth or hypoglossal nerve* arises from a nucleus near the middle of the floor of the lower end of the fourth ventricle. It is a purely motor nerve, supplying fibres to the muscles of the tongue and the muscles connected with the hyoid bone.

193. **The Cerebellum.**—The cerebellum, or little brain, consists of two lateral hemispheres united by a central portion called the *vermiform process*, seen beneath the medulla in fig. 179. It lies in the posterior part of the cranium, and is separated

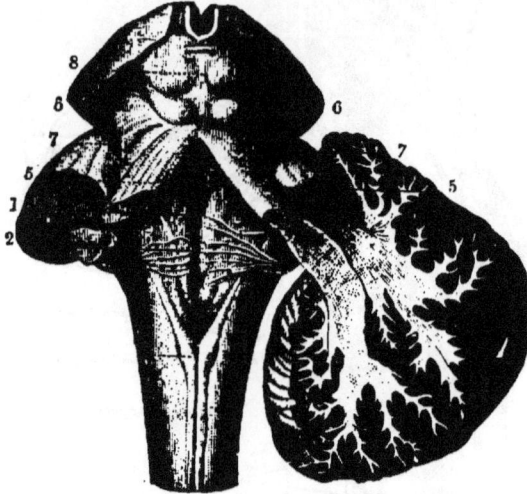

FIG. 189 —Figure showing the Three Pairs of Cerebellar Peduncles. (From Sankey after Hirschfeld and Leveillé.)

On the left side the three cerebellar peduncles have been cut short ; on the right side the hemisphere has been cut obliquely to show its connection with the superior and inferior peduncles.
1, median groove of the fourth ventricle ; 2, the same groove at the place where the auditory striæ emerge from it to cross the floor of the ventricle ; 3, inferior peduncle or restiform body ; 4, funiculus gracilis ; 5, superior peduncle—on the right side the dissection shows the superior and inferior peduncles crossing each other as they pass into the white centre of the cerebellum ; 6, fillet at the side of the crura cerebri ; 7, lateral grooves of the crura cerebri ; 8, corpora quadrigemina.

from the cerebrum above by a partition of the dura mater called the *tentorium*. It is connected with the rest of the brain by three pairs of fibrous stalks termed peduncles, or crura. The inferior or lower peduncles are formed by the prolonged restiform bodies from the medulla. The middle peduncles pass from its two hemispheres transversely to form the transverse fibres of the pons Varolii. The superior peduncles connect it with the cerebrum above. Each superior peduncle, or crus, forms the upper part of the lateral boundary of the fourth ventricle, and is connected with its fellow of the opposite side by the valve of Vieussens, a thin membrane continuous with the white centre of the vermiform process, and forming the roof of the anterior part of the fourth ventricle.

The cerebellum has a laminated or foliated appearance, and is marked on its surface by numerous transverse curved fissures. The cerebellum is composed of white matter and grey matter when seen in section. The grey matter lies outside and forms its cortex, a small nucleus of grey matter being also found near the centre of each hemisphere, termed the *corpus dentatum.* The white matter has a curiously branched arrangement which spreads from a white centre. This arrange-

FIG. 190. Section of Cortex of Cerebellum. (Sankey.)
a, pia mater ; *b*, external layer ; *c*, layer of corpuscles or Purkinje ; *d*, inner or granule layer ; *e*, medullary centre.

ment of white matter is called *arbor vitæ* (tree of life). The minute structure of the grey matter of the cortex or surface

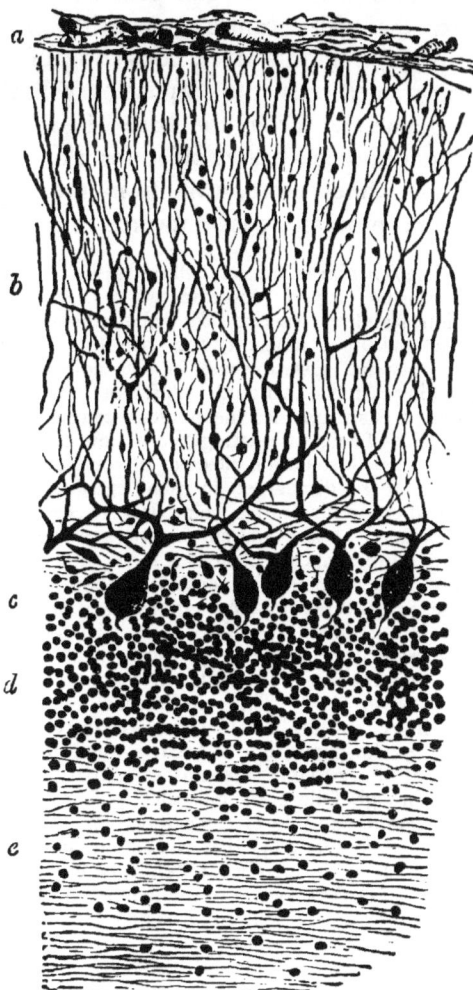

of the cerebellum shows (*a*) an outer layer beneath the pia mater membrane, consisting of delicate fibres with small nerve cells and large neuroglia cells ; (*b*) an inner or granule layer next the white centre, consisting of closely packed granule cells ; (*c*) a middle layer, consisting of a single stratum of large pear-shaped cells (the corpuscles of Purkinje) $\frac{1}{800}$ to $\frac{1}{1000}$ inch in diameter. From the base of each of these cells an axis-cylinder process passes off to form one of the medullated fibres of the white centre, while from the opposite pole of the cells several processes pass outwards into the cuter layer.

The *function* of the cerebellum seems to be that of securing or assisting in securing properly co-ordinated muscular movements, so that in any action, standing, walking, &c., the different muscles employed may each act at the right moment, and with the right force. Other parts of the cerebro-spinal system share in this co-ordinating function, as shown in speaking of the spinal cord. Afferent impulses from the feet, limbs, and other parts stream into the nervous centres during muscular activity, and the office of so co-ordinating the action of the muscles that equilibrium is maintained, locomotion effected, or some definite movement accomplished, appears to be discharged in many cases by reflex centres situated in the cord, these lower centres being largely regulated by higher centres in the cerebellum. These afferent impulses or sensations that lead to co-ordination of movements are of several kinds, and do not always come into distinct consciousness – muscular sensations that form the basis of the muscular sense, visual sensations through the eyes, and those peculiar impulses that arise in the ampullary ends of the semicircular canals (par. 241). All these impulses may reach the cerebellum by one or other of the peduncles, and may in some way contribute to the regular association and action of muscular groups. That the cerebellum is part of the co-ordinating machinery appears clear when we learn—(1) that removal or injury produces, for some time at least, a want of co-ordination of movements ; (2) that injury to one side causes inclination to fall towards the opposite side through failure of muscular power on the injured side ; (3) that excitation of one side of the cerebellum gives rise to muscular contraction of the same side ; (4) that dissection and degeneration show that the connection of the cerebellar hemisphere with the cerebral hemisphere is crossed ; (5) that disease of the cerebellum frequently leads to a staggering gait with loss of muscular power and tone. Hence we may conclude that the cerebellar hemispheres are connected with the same side of the body, but with the opposite cerebral hemisphere, and that the chief function of the cerebellum is to aid in the co-ordination of muscular movements, and in maintaining muscular energy. It does not share in the higher intellectual functions, its destruction not affecting the mental faculties.

194. **The Pons Varolii.**—The pons Varolii lies above the medulla and between the lateral halves of the cerebellum. A section across shows that it consists of transverse and longi-

tudinal fibres intermingled with some grey matter. There are superficial and deep transverse fibres from the middle peduncles of the cerebrum, longitudinal fibres arranged in bundles, that pass down to the anterior pyramids of the medulla and then decussate to form the motor tracts of the cord, other longitudinal fibres from the medulla, and grey matter on the

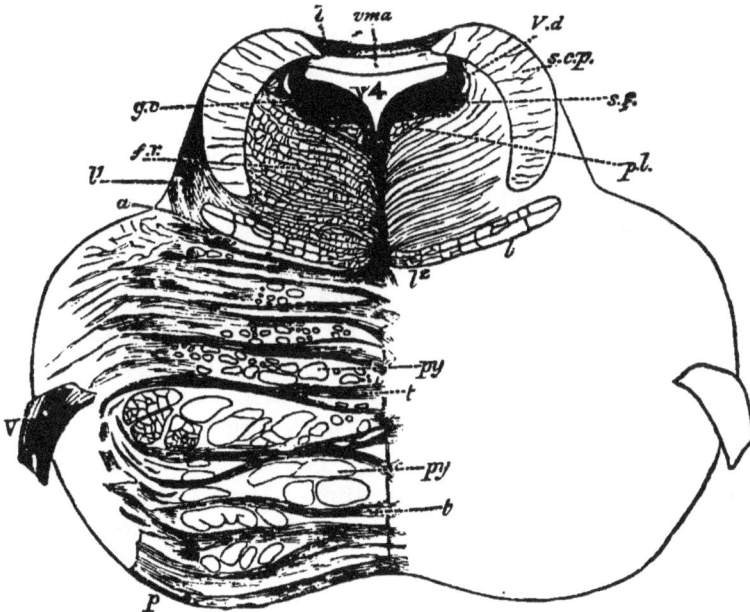

FIG. 191.—Transverse Section through the Upper Part of the Pons. (Schwalbe, after Stilling.) Rather more than twice the natural size.

p, transverse fibres of the pons; *py, py,* bundles of the pyramids; *a,* boundary line between the tegmental part of the pons and its ventral part; *l',* oblique fibres of the lateral fillet, passing towards the inferior corpora quadrigemina; *l,* lateral; *l',* mesial fillet; *f.r.,* formatio reticularis; *p.l.,* posterior longitudinal bundle; *s.c.p.,* superior cerebellar peduncle; *v.m.a.,* superior medullary velum; *l,* grey matter of the lingula; *v* 4, fourth ventricle in the grey matter which bounds it laterally are seen *v.d.,* the descending root of the fifth nerve, with its nucleus; *s.f.,* substantia ferruginea; *g.c.,* group of cells continuous with the nucleus of the aqueduct.

floor of the fourth ventricle forming the nuclei of the seventh or facial nerve, the sixth nerve, and the fifth nerve. The motor fibres from the internal capsule to the facial nuclei are seen to decussate in the pons. Hence injury to one side of the pons in the upper part affects the facial muscles of the *opposite* side, while injury in the lower part, after decussation occurs, paralyses the muscles of the same side. The motor fibres to the limbs decussate in the medulla.

195. The Crura Cerebri.—The crura cerebri diverging from the upper border of the pons pass into the cerebral hemispheres. Each crus on section is seen to consist of two parts separated by dark grey substance termed *substantia nigra*. The *pes*, or *crusta*, is the anterior or lower part, and consists almost entirely of longitudinal fibres, continuous above with some in the internal capsule, and below with some in the pons that go to the anterior pyramids of the medulla. The *tegmentum* of the crus is the dorsal or upper part, and consists of grey matter and fibres continuous with the formatio reticularis of the pons and medulla, separated by transverse arched fibres and some grey matter. Above these fibres pass into the optic thalamus for the most part.

196. The Corpora Quadrigemina.—These are four rounded prominences placed in pairs over the aqueduct of Sylvius, and above the pons and crura. The anterior pair is sometimes called the *nates* and the posterior the *testes*. They are mainly composed of grey matter, with white fibres externally and some internally. From the outer side of each of these eminences a white band termed the arm, or *brachium*, is continued forwards and outwards. The bands from the posterior pair lose themselves beneath two prominences in relation with the posterior part of the optic thalami, termed the *internal geniculate bodies*. Those from the upper pair pass into the *external geniculate bodies* and the optic tract. The superior quadrigeminal bodies are indeed intimately connected with the optic tract and the sense of sight. Their destruction leads to blindness, and they appear to contain centres for the contraction of the iris and for accommodation.

197. The Optic Thalami.—The optic thalami are two oval-shaped masses of grey matter, projecting above into the lateral ventricles of the brain, the under surface resting on the tegmentum of the crus. The posterior and inner end of the thalamus is termed the *pulvinar*, and projects over the arms of the corpora quadrigemina, the geniculate bodies being below and outside. Their inner sides form the lateral bound-

aries of the third ventricle. Their inner surface is united with
a prolongation of the tegmental part of the crura cerebri, and
on their outer side is the white matter of the internal capsule,
formed by fibres from the crusta of the crura that pass into
the cerebral hemispheres without entering the optic thalami.
The optic thalami consist of grey matter with many nerve cells

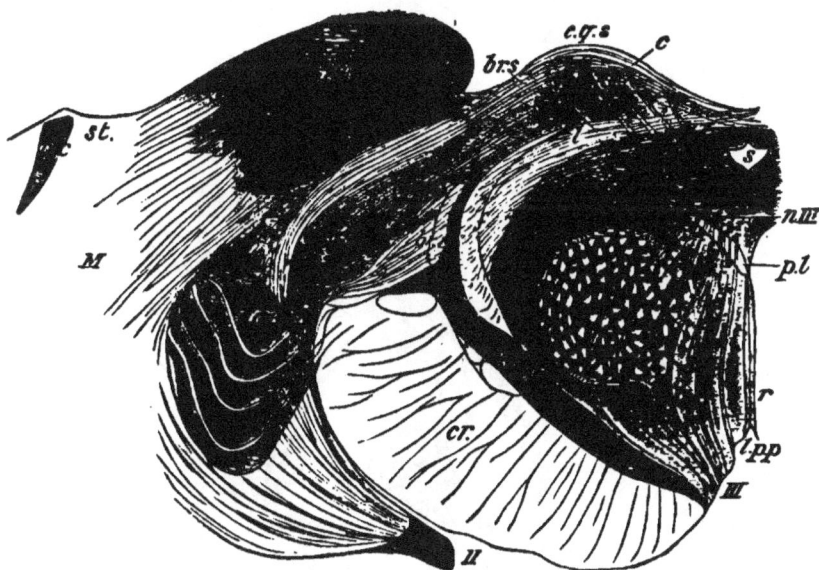

FIG. 192.—Section through the Superior Part of one of the Superior Corpora Quadrigemina, Crus Cerebri and the Adjacent part of the Optic Thalamus. (After Meynert.)

s, aqueduct of Sylvius ; *gr*, grey matter of the aqueduct ; *c.q.s*, quadrigeminal eminence,
consisting of *l*, stratum lemnisci ; *o*, stratum opticum ; *c*, stratum cinereum ; *Th*,
thalamus (pulvinar); *c.g.i*, *c.g.e*, internal and external geniculate bodies ; *br.s*, *br.i*,
superior and inferior brachia ; *f*, fillet ; *p.l*, posterior longitudinal bundle ; *r*, raphé ;
III, third nerve ; *n*. III, its nucleus ; *l.p.p*, posterior perforated space ; *s.n*, substantia
nigra—above this is the tegmentum of the crus with its nucleus, the latter being indicated by the circular area ; *cr*, crusta, or pes. of the crus ; II, optic tract ; *M*, medullary
centre of the hemisphere ; *n.c*, nucleus caudatus ; *st*, stria terminalis.

and white fibres mostly lying on the surface. The optic thalami
are connected with the posterior or sensory paths of the spinal
cord through the tegmenta of the crura, and from their outer
part fibres pass onwards to the cerebral hemispheres.

198. **The Corpora Striata.**—Each corpus striatum consists
of two parts—a pear-shaped part projecting into the lateral
ventricle of the same side in front of the optic thalamus, called

the *caudate nucleus*, and a part embedded in the white substance of the cerebral hemisphere called the *lenticular nucleus*.
Between the two parts is seen in a deep section the white

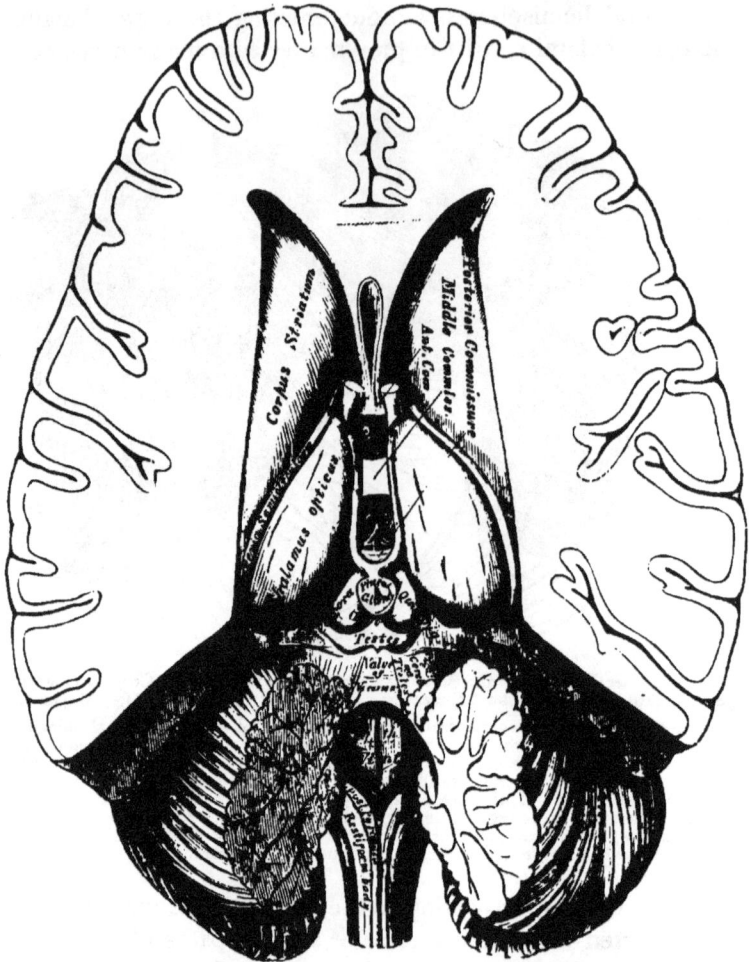

Fig. 193.—Horizontal Section of Brain.

fibres of the *internal capsule*. Outside the lenticular nucleus
is another band of fibres termed the *external capsule*, beyond
which is a thin lamina of grey matter termed the *claustrum*.

The claustrum lies next to a lobe of the cerebrum in the fissure of Sylvius termed the central lobe, or *island of Reil.* In section a corpus striatum shows a striped appearance, owing to diverging white fibres being mixed with grey matter.

199. The Internal Capsules.—The internal capsule is a broad band of white fibres, which for the most part connects the cortex of the brain with the crusta and bulb below. It lies between the lenticular nucleus of the corpus striatum on the outer side and the caudate nucleus and optic thalamus on the inner side, the fan-like expansion of its fibres into the hemisphere being spoken of as the *corona radiata.* In horizontal sections the internal capsule shows a bend termed the knee, or *genu,* a part in front of the genu called the *front limb* and a part behind the knee called the *hind limb,* and the fibres of the capsule curve away in many directions to the various parts of the cerebral surface.

The internal capsules contain :—

(1) The fibres of the pyramidal tract. These can be traced from their origin in the motor areas of the cerebral cortex around the fissure of Rolando through the middle third of the internal capsule into the crusta, or pes, of the crus cerebri, thence into the

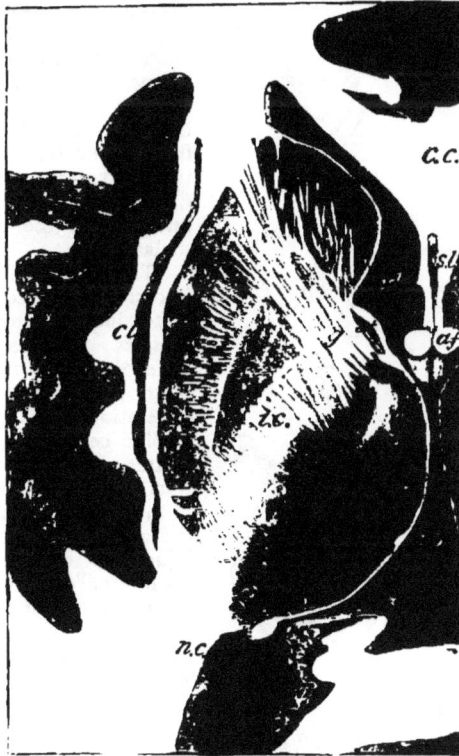

Fig. 194.— Horizontal Section through the Optic Thalamus and Corpus Striatum. (Natural size.)

v.l., lateral ventricle, anterior cornu ; *c.c.,* corpus callosum ; *s.l.,* septum lucidum ; *a.f.,* anterior pillars of the fornix ; *v.* 3, third ventricle ; *th.,* thalamus opticus ; *s.t.,* stria pinealis ; *n.c.,* nucleus caudatus, and *n.l.,* nucleus lenticularis of the corpus striatum ; *i.c.,* internal capsule ; *g.,* its angle or genu ; *n.c.,* tail of the nucleus caudatus appearing in the descending cornu of the lateral ventricle ; *cl,* claustrum ; *I,* island of Reil.

pons and the anterior pyramids of the medulla. At the lower part of the medulla most of the fibres decussate as they pass into the spinal cord, where they form the crossed pyramidal tracts. Some of the fibres of this tract, however, pass to the nuclei of the cranial motor nerves in the pons and medulla. The uncrossed fibres form the direct pyramidal tracts.

(2) Fronto-cortical fibres. These originate in the frontal convulutions anterior to the motor area, and pass down in the anterior third of the capsule through the crusta into the pons, where they appear to terminate in grey matter.

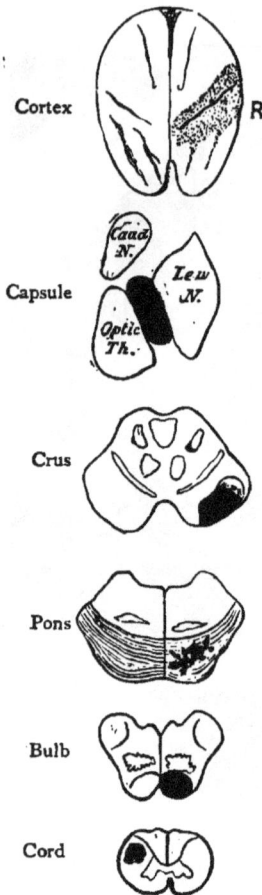

Cortex

R

Capsule

Crus

Pons

Bulb

Cord

FIG. 195.—Diagram to illustrate Degeneration in Motor Pyramidal Tract, through internal capsule, crus, pons, anterior pyramids of medulla, to lateral column of cord on opposite side.

(3) Temporo-occipital fibres. These fibres take origin in the temporal and occipital regions of the cortex, and, passing through the posterior third of the capsule, terminate in the outer part of the pons.

Besides the above tracts of fibres that connect the cortex of the hemispheres with the crus, the internal capsule contains fibres from the nucleus caudatus of the corpus striatum that terminate in the pons, as well as fibres from various parts of the cortex that terminate in the grey matter of each optic thalamus.

The pyramidal tract of the internal capsule is doubtless concerned in conveying voluntary motor impulses from the cerebral cortex to the muscles. The tract is well marked out by the degeneration of nerve fibres that sets in on injury to the motor area, the descending degeneration showing that the fibres have their trophic centres in the cells forming the grey matter of the cortex. As the motor pyramidal tract decussates, the fibres for the face crossing in the pons and those for the limbs just below the anterior pyramids, it follows that lesions or injuries of the motor areas of the cortex or of the internal capsule will be followed by paralysis of the muscles on the opposite side of the face and body. This motor paralysis on one side is termed *hemiplegia*, or 'one-sided stroke.' The muscles affected vary with the extent and situation of the lesion, but it is worthy of note that the paralysis is limited to voluntary movements, and affects the bilaterally associated reflex movements but little. As the posterior part of the hind limb of the internal capsule contains numerous sensory fibres connected with the opposite side of the body, total destruction of the internal capsule gives rise to loss of both motion and sensation on the opposite side of the body. Many cases of ordinary *hemiplegia*, or motor paralysis, of one half of the body involve more or less *hemianæsthesia*, or sensory paralysis of one side.

200. The Cerebrum.—The term cerebrum in the restricted sense describes the two large ovoid masses of grey and white matter called the cerebral hemispheres, that overlap all the rest of the brain.[1] The two hemispheres are completely separated by the great longitudinal fissure, except at about the middle half of their extent, where they are united at a depth of about 1 inch by the transverse fibres of the corpus callosum. The surface of each hemisphere presents a peculiar folded

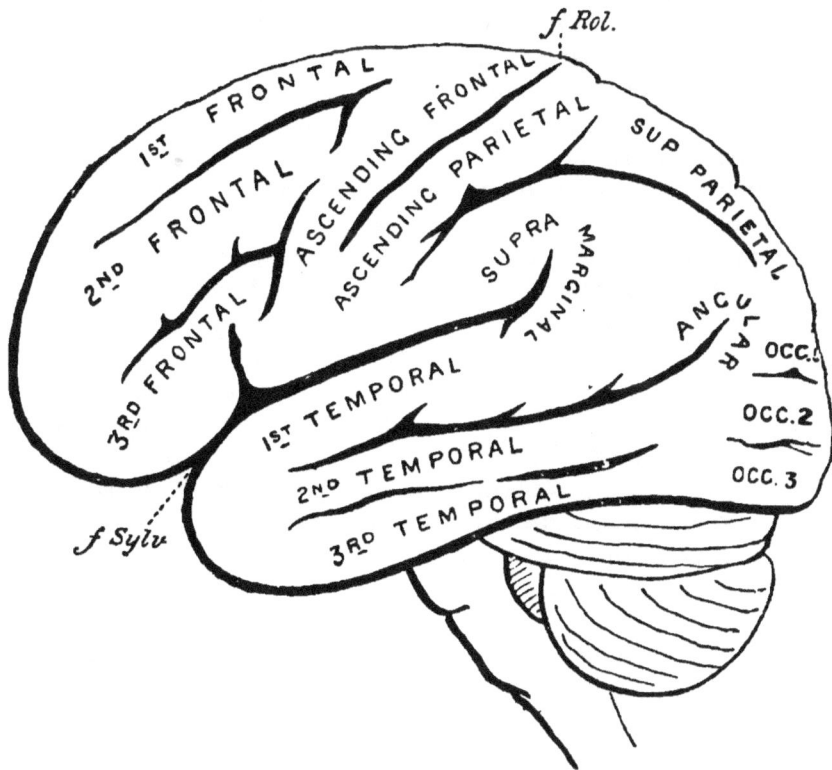

FIG. 196.—Human Brain ; Lateral Aspect of Left Hemisphere. (After Ecker.)

appearance, being thrown into gyri or convolutions by depressions termed sulci or fissures. In this way the superficial area of the hemispheres is greatly increased. The deeper fissures

---

[1] In anatomy the term cerebrum usually includes all the brain in front of the cerebellum and pons, *i.e.* not only the cerebral hemispheres, the corpora striata, and optic thalami, but also the corpora quadrigemina, and the crura cerebri.

divide the outer surface of each hemisphere into five lobes, whilst others separate the convolutions of each lobe from one another.   It is a great help in describing the brain to refer to these lobes and fissures.   The fissure of Rolando begins near the middle of the longitudinal fissure at the vertex, and passes on the outer surface of each hemisphere obliquely downward and forward towards the great fissure of Sylvius, which begins on the under surface of the hemisphere and passes upwards and backwards.   The *frontal lobe* of the brain lies in front of the fissure of Rolando and above the fissure of Sylvius. Smaller fissures divide it into four main convolutions—first or superior, second or middle, third or inferior, and the ascending frontal convolution (fig. 196).   The *parietal lobe* is limited in front by the fissure of Rolando and beneath by a part of the fissure of Sylvius.   It is divided into ascending parietal convolution, superior parietal convolution and inferior parietal convolution, which last is subdivided into supra-marginal convolution round the end of the Sylvian fissure, and angular convolution round the end of a temporal fissure.   The *temporal lobe*, lying below the horizontal part of the fissure of Sylvius, shows three parallel convolutions, termed the first or superior temporal, second or middle temporal, and third or inferior temporal.   The *occipital lobe* is small and lies at the posterior end of the cerebrum, separated from the parietal lobe by the perpendicular parieto-occipital fissure.   It shows three convolutions, a superior (occ. 1), a middle (occ. 2), and an inferior (occ. 3).   The *central lobe*, or island of Reil, can only be seen by pulling apart the edges of the fissure of Sylvius, when the concealed convolutions forming the lobe would be exposed. ·

A view of the inner mesial surface of a hemisphere shows a long fissure beginning below and running backwards some distance above the corpus callosum, to terminate behind the upper end of the fissure of Rolando.   This is known as the *calloso-marginal fissure.*   Between this fissure and the surface of the hemisphere is the marginal convolution, or *gyrus marginalis*, and between the calloso-marginal fissure and the corpus callosum is the callosal convolution, or *gyrus fornicatus*. The posterior end of the gyrus fornicatus comes downwards and then forwards under the name of hippocampal convolution, or *gyrus hippocampi*. At the end of the temporo-sphenoidal lobe is seen the hooked uncinate convolution, or *gyrus uncinatus*. Between the posterior end of the calloso-marginal fissure and the parieto-occipital fissure is the quadrate lobule, while below this, between the parieto-occipital and calcarine fissure, is the wedge-shaped mass termed the cuneate lobule.

201. **Structure of the Cerebrum.**—The cerebral hemispheres consist of white and grey matter, the white pervading

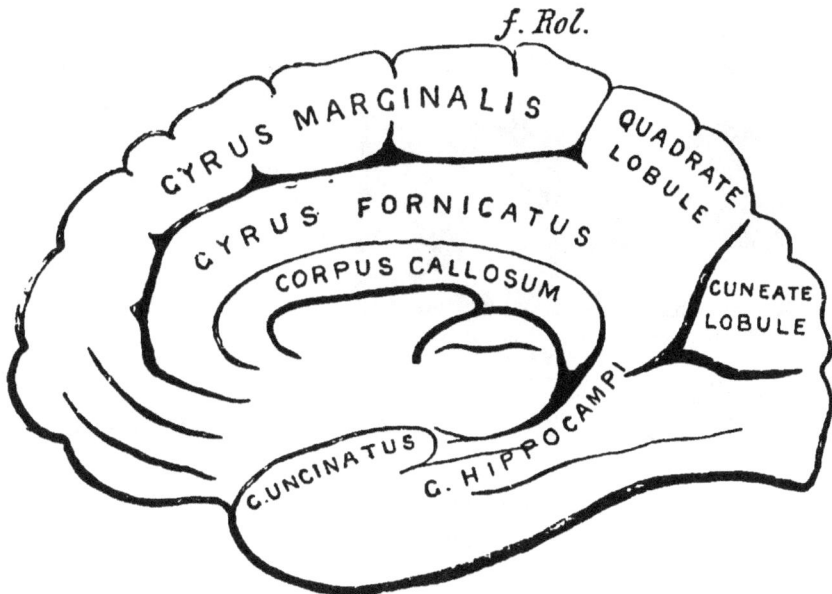

f. Rol.

FIG. 197.—Human Brain; Mesial Aspect of Right Hemisphere. (After Ecker.)

the middle of the hemisphere and extending into the convolutions, the grey forming a layer or cortex from $\frac{1}{6}$ to $\frac{1}{4}$ inch in depth on the outer surface of the convolutions. The white matter consists of medullated nerve fibres arranged in bundles and supported by neuroglia. They vary in size in different parts, but are mostly smaller than those of the cord and medulla. The different systems have already been described. The grey matter on the convoluted surface is arranged in a continuous layer, but divided into strata by lighter lines. A section of a convolution shows to the naked eye the appearance illustrated in fig. 198. There is (1) a thin coating of white matter most conspicuous on the convolutions within the great median fissure ; (2) a layer of grey or reddish-grey matter ; (3) a thin whitish layer ; (4) a yellowish-grey stratum sometimes showing a thin whitish line ; (5) the central white matter of the convolution.

Seen under the microscope, a section of the cortex shows five layers, (1) a superficial layer showing a few small ganglion cells with abundance of neuroglia ; (2) a layer of small pyramidal cells ; (3) a thick layer of large pyramidal cells, each with a process passing towards the surface from the pointed apex, processes passing off laterally, and a process from the centre of the base which becomes continuous with the axis-cylinder of a nerve fibre. These pyramidal nerve cells increase in size towards the deeper part, and bundles of nerve fibres are seen passing downward from this layer into the white matter ; (4) a layer of granules

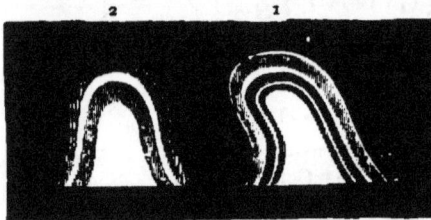

FIG. 198.—Sections of Cerebral Convolutions. (After Baillarger.)

The parts are nearly of the natural size. 1, shows the five layers ordinarily seen in the cerebral cortex when carefully examined with the naked eye ; 2, the appearance of a section of a convolution from the neighbourhood of the calcarine fissure.

and small irregular nerve cells ; (5) a layer of fusiform cells arranged for the most part parallel to the surface (fig. 199). In different regions the various layers bear different relations to each other. Above the fissure of Rolando, in the motor area, the large pyramidal cells are well developed ; in the occipital region the granular layer is well marked and the large cells few ; while the fusiform cells are more abundant in the Sylvian fissure than elsewhere. The axis-cylinder processes of the pyramidal cells pass into the medullary centre to form either association fibres connecting other parts of the cortex of the same hemisphere, or commissural fibres through the corpus callosum to the opposite hemisphere, or projection fibres either to the corpus striatum and optic thalamus, or by way of the internal capsule to the crura, pons, medulla, and spinal cord.

202. Functions of the Cerebrum.—The cerebral hemispheres, especially the grey matter of the cortex, are the seat of consciousness, memory, intelligence, and volition. Various facts may be cited in support of this statement. (1) We find that those races of men that are the most intelligent have the heaviest cerebrum and the most fully developed convolutions. (2) Among vertebrate animals we find that the degree of intelligence and the higher mental faculties increase in the same general proportion as the size of the cerebrum relative to other parts of the brain increase. (3) Imperfect development of the cerebrum in man is accompanied by imbecility and idiocy. (4) A severe injury to the cerebral hemispheres, as in concussion of the brain, suddenly deprives a man of his mental faculties. (5) Removal of the cerebral hemispheres in one of the lower animals deprives it of all intelligent spontaneous action.

Thus, a *frog* which has lost its cerebral hemispheres, though capable of performing many complex movements when the right sensory stimulus is applied, has lost all power of spontaneous movement. It can sit up in its natural attitude, breathing quickly, but if undisturbed will remain motionless for an indefinite time. If placed on a board and the board be lifted, it will crawl up to a position of equilibrium ; if pinched, it will jump away, avoiding any obstacle in its path ; if placed in water it will swim until an object is put before it to rest on, as the contact of the fluid sets up the appropriate reflex actions. But it manifests no hunger, makes no effort to obtain food, and shows no sign of fear. It is a mere machine performing certain purely reflex acts, that may be foretold, on suitable external stimulation. But the complete animal acts very differently, and initiates movements apart from sensory stimuli, going through complex acts spontaneously. Its reactions to outward stimuli vary. ' Led by the feeling of hunger, too, he goes in search of insects, fish, or smaller frogs, and varies his procedure with each species of victim. The physiologist cannot by manipulating him elicit croaking, crawling up a board, swimming, or stopping, at will. His conduct has become incalculable. We can no longer foretell it exactly.' [1]

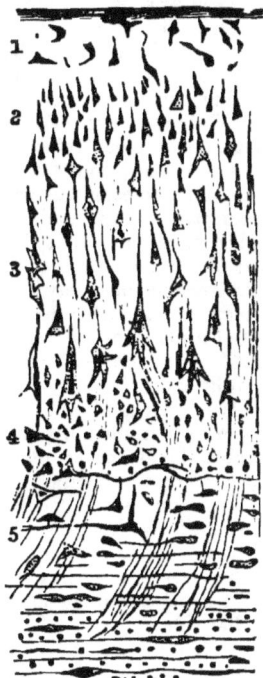

FIG. 199.—Transverse Section through the Cortex Cerebri. Five-laminated or Motor Type. × 50 diam. (Meynert.)

A *pigeon* without cerebral hemisphere behaves much in the same way as a frog. Left to itself, it remains still, though it can be made to fly if properly stimulated. It will starve to death on a heap of corn, though it begins to eat on holding its beak in it. Its emotions and intelligent movements no longer exist.

[1] James, ' Text Book of Psychology.'

In higher animals fewer observations have been made, as removal of the hemisphere often produces a fatal shock. Goltz has described the behaviour of the *dog* after loss of the cerebral hemispheres. After recovering from the surgical operation, the animal walked and moved about in normal fashion, though often wandering restlessly. It slept at night and a loud sound awoke it. A sudden light caused it to close its eyes, and an injury to one foot led it to limp about on three legs, so that co-ordination of both usual and unusual movements was possible. At first it had to be fed, but after some months it would help itself on being started. But it was absolutely wanting in the higher intellectual faculties. No heed was paid to the barking of other dogs or to the caressing of its master. It had no memory, and never seemed to learn that it was being taken out of the case to be fed. Thus the animal was capable of complicated movement, could take nourishment, and had sensation of taste, hearing, and sight. Otherwise it was in a complete state of imbecility.

While the lower centres, therefore, may receive afferent impulses and by an unconscious reflex act discharge efferent impulses to muscles that carry out many complex.acts, the complexity and importance to life of these acts increasing as we pass from the spinal cord to the medulla and lower parts of the brain, yet it is evident from experiment and pathological experience that the hemispheres or cortex of the cerebrum is the part of the nervous system in which sensory or afferent impulses become converted in some way into mental impressions that give rise to conscious perception, and that leave behind vestiges of some kind which as recollected ideas form the basis of intellectual activity. Further, it is evident that from the cortex alone there can issue those impulses termed voluntary or volitional.

It is worthy of note that the mental operations denoted by such terms as perception, memory, imagination, and reasoning, though confined to the cerebral hemispheres, may be carried on with but one hemisphere. Cases have occurred in which one hemisphere has been destroyed by disease, and yet the mental faculties have been apparently unimpaired. Yet in such a case we know, from what has been said about the motor and sensory paths in the brain, there must be complete paralysis of the opposite side of the body, no sensory impulses being received from that side, and the brain being unable by an effort of the will to send impulses to the muscles of that side.

It has already been pointed out that there are certain unconscious reflex actions of the spinal cord and medulla that appear inborn and involved in their very structure. Such reflex actions may be termed *natural* or *original*. Now it is found that by the help of the brain new systems of reflex paths may be set up in the nervous structures and we may become possessed of many *acquired* or *artificial* reflex acts. Actions which in the first instance and for some time require close attention and a continuous effort of the will, become at last so ingrained in the nervous structure that a single sensation or a single impulse from the brain is sufficient to start a whole train of actions. Our voluntary and reflex acts shade into each other gradually, a series being often connected by one or both kinds. Strictly voluntary acts must be guided by idea, perception and volition, but in habitual actions conscious effort is not required beyond giving an impulse that starts a series. A habit is a reflex discharge from some nervous centre, the most complex being, as Prof. James says, 'nothing but concatenated discharges in the nerve centres, due to the presence there of systems of reflex paths, 'so organised as to wake each other up successively—the impression produced by one muscular contraction serving as a stimulus to provoke the next, until a final impression inhibits and closes the whole chain.' And Prof. Huxley says :—' The possibility of all education is based upon the evidence of the power which the nervous system possesses of organising conscious actions into more or less unconscious, or reflex, operations.'

203. **The Fibres of the Brain.**—It has already been noted that there are three kinds of fibres in the medullary centre of the brain : (*a*) *transverse* or *commissural* fibres connecting the two hemispheres together, viz. the fibres of the *corpus callosum* starting in the cortex, and connecting chiefly the frontal and occipital lobes, and the fibres of. the *anterior commissure* connecting chiefly the two temporal lobes ; (*b*) association fibres connecting near or distant parts of the same hemisphere, and (*c*) peduncular, longitudinal, or projection fibres forming the efferent or afferent channels between the cortex of the cerebrum and the lower parts of the brain and spinal cord. These projection fibres are continuous in part with those of the lower or ventral part of the crus cerebri termed the crusta, or pes, and in part with those of the dorsal part termed the tegmentum, ' the latter probably indirectly through the corpus striatum and optic thalamus.' These fibres are in great measure direct prolongations of the axis-cylinder processes of cells in the cerebral cortex, as is proved by the descending degeneration that occurs when injury of a particular area of the cerebral cortex takes place (fig. 195). The trophic centres of the fibres of the cells must therefore be in the grey matter of the cortex. Of the projection fibres passing through the pes or crusta of the crura cerebri, the bundle forming the pyramidal tract is best known. From the parietal region of the cortex around the fissure of Rolando, and known as the motor area, fibres pass downwards through the corona radiata into the internal capsule between the nucleus lenticularis on the outside, and the nucleus caudatus and the optic thalamus internally. From the knee and hind limb of the internal capsule the fibres pass into the crusta of the crus cerebri, thence into the pons, where they are split up by the interlacing fibres of the pons ; from the lower border of the pons they pass into the medulla to form its anterior pyramids.

At the lower part of the medulla most of the fibres cross to the lateral column of the opposite side of the spinal cord to form the 'crossed pyramidal tract.' A few fibres from the medulla do not cross, but are continued on the same side down the cord to form the 'direct pyramidal tract'; but, according to many observers, these cross over gradually in the cord through its anterior white commissure.

At various levels of the cord the fibres make connection with the multipolar ganglionic cells of the anterior cornu of the cord, and through them with the fibres of the anterior roots, which, as we have learnt, are the nerve fibres conveying efferent impulses, especially voluntary impulses, to the muscles. The pyramidal tract thus making connection with the spinal nerves will evidently gradually diminish in size as it passes down the cord. It should also be noted that, in passing through the lower parts of the brain, the pyramidal fibres of the internal capsule give off fibres to the motor cranial nerves that cross the middle line to the respective nuclei of these nerves, as shown in the diagram (fig. 200), where the course of the fibres to the seventh or facial nerve is indicated.

Other fibres pass from the cortex through the internal capsule into the crusta of the crura, but they have not been traced further down than the pons. Still other longitudinal fibres proceed from the cortex of the temporal and occipital regions, and pass round the internal capsule to the tegmentum of the crus. Fibres are also found to connect the tegmentum and the optic thalamus, and from the outer side of the thalamus fibres pass to most parts of the hemisphere. It is not possible to trace the fibres of all the columns of the cord to a termination in the cortex of the cerebrum. While some of the sensory impulses cross over in the cord, others are conducted by fibres which remain on the same side. The fibres of the postero-median column of the cord end in the grey matter of the nucleus gracilis of the bulb; those of the posterior external column end in the nucleus cuneatus. From both these nuclei fibres pass off to the opposite side above by the superior or sensory pyramidal decussation. Thus, all or nearly all the fibres connecting the spinal cord with the brain decussate in some part of their course. So also do the fibres proceeding from the nuclei of origin of the cranial nerves. Hence, injury to or destruction of one hemisphere involves both complete motor paralysis and loss of sensibility on the *opposite* side of the body (par. 199).

204. Localisation of Function in the Brain. — Not only does the cerebral cortex of one side act as the receiver of sensory or afferent impulses, and the dispenser of volitional efferent impulses to the opposite side of the body, but it has been found that particular places in the cerebral cortex are the areas from which particular voluntary movements are set up, or in which particular sensory impressions are perceived and recognised. The evidence of the localisation of function in the brain

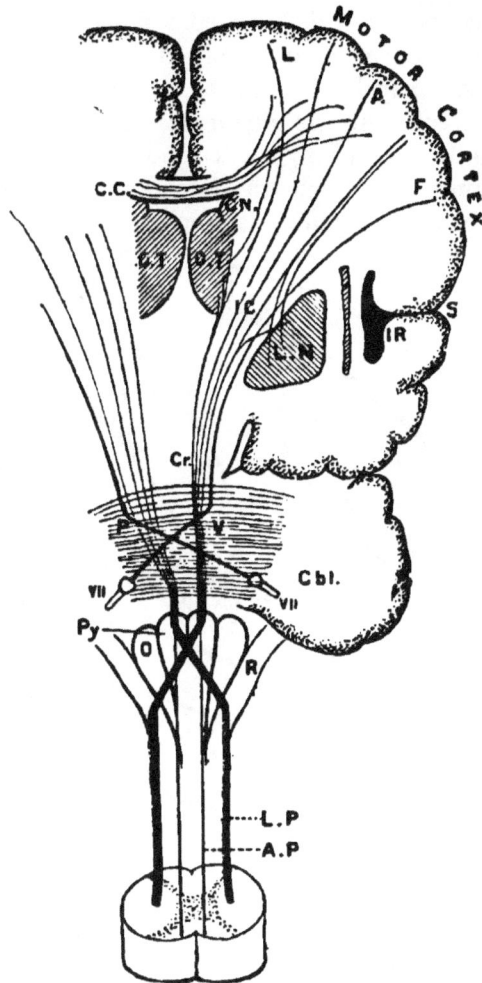

FIG. 200.—Diagrammatic Vertical Section through the Brain, to show the Course of the Pyramidal Fibres, &c.

L A F, centres for leg, arm, and face on motor cortex; I. C, internal capsule; O. T, optic thalamus; C. C, corpus callosum; C. N, caudate nucleus of corpus striatum; L. N, lenticular nucleus; Cr., crus; P. V, pons Varolii; Py, pyramids of medulla in which motor fibres decussate; O, olivary bodies of medulla; R, restiform bodies of medulla; L P, fibres of lateral or crossed pyramidal tract in the lateral column of the cord; A.P, anterior or direct pyramidal tract in the anterior column of the cord; vii, superficial origin of seventh or facial nerve, with fibres passing to opposite side of cortex; I. R, island of Reil at the bottom of s, the Sylvian fissure; cbl., cerebellum.

may be thus briefly stated. For the past ten or twelve years it has been generally admitted that the grey matter of the cortex itself may be excited by certain stimuli, as a weak electric current. At one time it was thought that the stimulus applied to the cortex only excited movement by spreading to the white fibres below, but the latent period is one-third longer than when the stimulus is applied to the underlying fibres, and

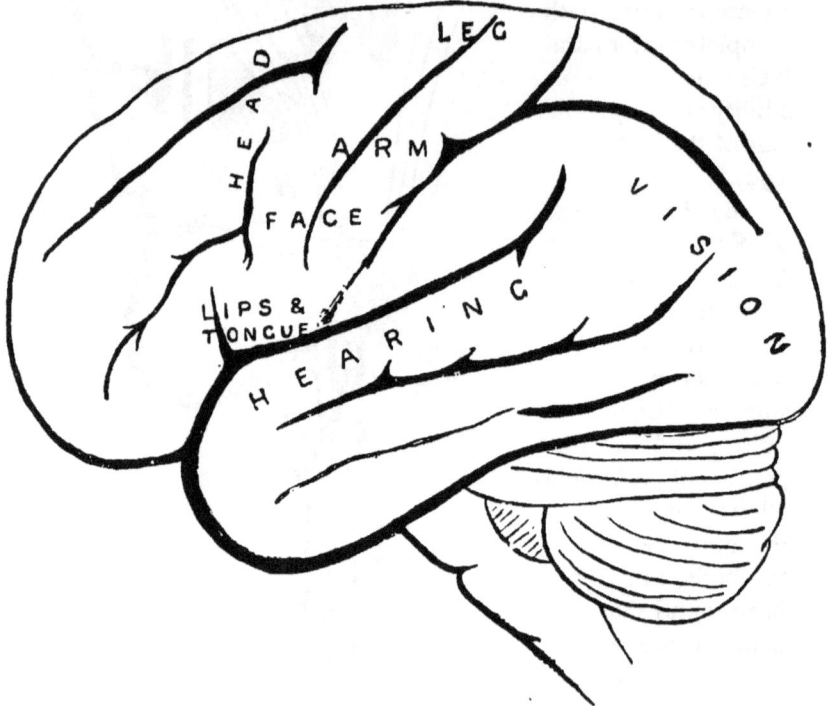

FIG. 201.—Human Brain ; Lateral Aspect of Left Hemisphere. To illustrate Cortical Localisation of Function. (Waller.)

epileptic convulsions are produced more readily by stimulating the cortex than by applying the stimulus to the fibres, while tumours pressing on the surface often produce paralysis of special muscles.

Motor Areas.—By excitation of the convolutions around the fissure of Rolando co-ordinated movements of certain muscles or groups of muscles are produced. Hence this part of the surface is called the 'motor area.' Moreover, the degeneration method proves that from this area motor impulses normally set out (see par. 199), and observation of the variations in the normal phenomena that follow disease of these parts supports the conclusions. These movements occur on the opposite sides of the body, but are often bilateral. Further special movements are associated with particular spots in this region. Thus, on the outer surface stimulation of the upper part around the fissure of Rolando is associated with movements of the *leg*, of the middle part with movements of the *arm*, of the lower parts with movements of the *face* and *mouth*. On the inner surface of each hemisphere movements of the face, arm, trunk, and leg are represented on the marginal convolution (which is merely the median aspect of the frontal and parietal convolutions) from back to front (fig. 202). These areas represent particular movements of the joints rather than of particular muscles, though it should be noted that localisation is not strict,

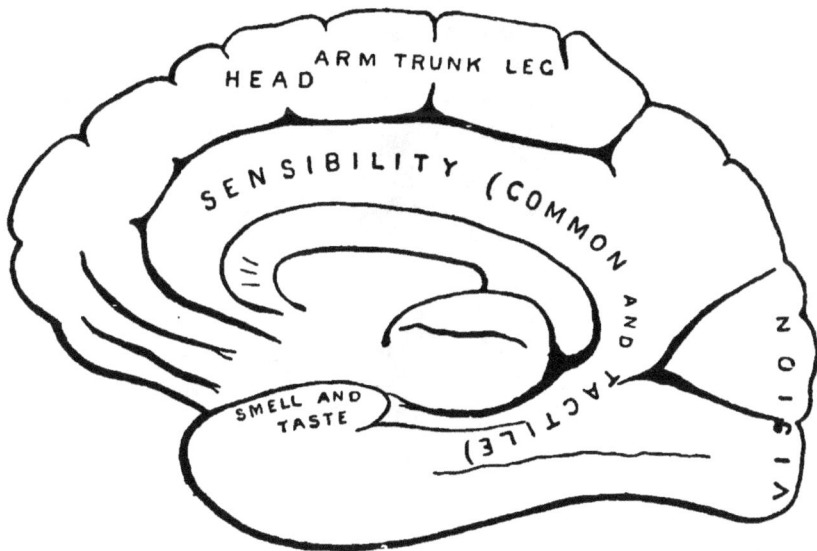

FIG. 202. — Human Brain ; Mesial Aspect of Right Hemisphere. To illustrate Cortical Localisation of Function. (Waller.)

for areas overlap somwhat. Extirpation of any area leads to loss of voluntary movement in the part regulated by that area, though recovery, more or less complete, of this movement may occur after some interval.

In calling these particular parts of the cortex the motor areas for arm movements, leg movements, &c., it is not meant that the voluntary movements of these parts originate in these special centres without any kind of stimulus entering these areas. Sensations of some kind, probably common sensations from the skin, and muscular sensations from the muscles, enter these motor areas of the cortex, and then lead the will to produce movements of a particular quality or force. A particular interest attaches to the *third frontal convolution on the left side*, as a lesion in this part is always associated with the loss of power of speech, or *motor aphasia* as it is called. This part is the same, or close to the motor centre for the muscles of the tongue and

Y

mouth, and is known as the speech centre. It appears that in most persons —right-handed people—the delicate co-ordinated movements of speech are regulated from a particular part of the left hemisphere, the corresponding part of the right hemisphere remaining dormant or uneducated for this office. When this portion of the brain is the seat of injury, there is total loss of voluntary speech. There is no loss of voice, for the patient can laugh and cry, and even sing, but either he is unable to utter any words at all, or he uses wrong ones, speaking incoherently and unintelligibly. He may recognise his mistakes, but cannot avoid them. In this motor aphasia, too, the power of writing is often, though not always, lost, as most people silently articulate the words as they write.

Sensory areas.—Another form of aphasia is termed *sensory aphasia*, the patient being unable to recognise either spoken words (' word-deafness')

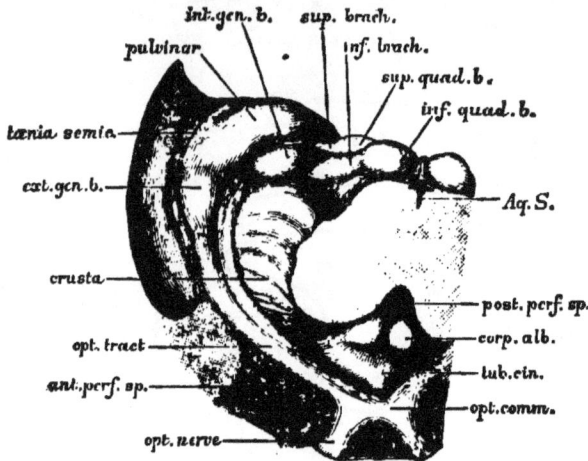

FIG. 203.—Origin and Relations of the Optic Tract. (G. D. Thane.)
The parts are viewed from below, the mid-brain having been divided transversely immediately above the pons, and the pons, cerebellum and medulla oblongata removed. The lower part of the figure is the more anterior.

or written words (' word-blindness '). A person afflicted with ' word-deafness ' may be able to write or speak sensible sentences, but he cannot understand spoken language. This affliction is associated with lesion of the posterior part of the first temporal convolution on the left side. In ' word-blindness,' the patient has lost the power to recognise printed or written words, though he may see quite well, discriminate objects, and speak correctly. This affliction is associated with injury of the cortex of the hinder part of the left parietal lobe near the visual centre.

On destroying the eyes the optic nerves degenerate, being cut off from their trophic nerve-cells in the retina, and a study of the course of degeneration enables us to learn the course and destination of the fibres. Thus it is found that a peculiar partial decussation takes place at the chiasma, the fibres belonging to the temporal half of an optic nerve passing into the optic tract of the same side, while the fibres belonging to the nasal half of an optic nerve pass into the optic tract of the opposite side. Each optic tract therefore contains fibres belonging to one half of each eye, so that what we have called 'corresponding' parts (par. 232) are represented in each optic tract. But the optic tract contains other fibres than those derived from the optic nerves, viz. commissure fibres at its hind part from one internal or median corpus geniculatum to the other, and fibres from the grey matter of the floor of the third ventricle that pass along the optic tracts to the pes, or crusta.

As each optic tract passes along the base of the brain it enters beneath the temporo-sphenoidal lobe, curving round the crusta of the crus and dividing into two roots—an internal root losing itself in the internal geniculate body, and derived from the commissural fibres mentioned above, not connected with the optic nerves, and an external or lateral root passing into the posterior part of the thalamus, termed the pulvinar, the external geniculate body, and into the anterior corpus quadrigeminum by its brachium, or arm. This lateral root, with its triple ending in the three collections of nerve matter just enumerated, contains the true optic fibres from the retina, and the three bodies, external geniculate body, pulvinar, and anterior corpus quadrigeminum, form three 'primary visual centres.'

It is thought that some of the fibres pass from the

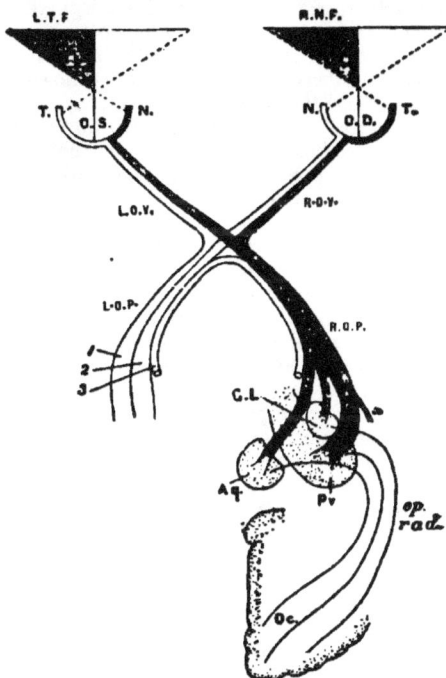

Fig. 204.—Diagram to illustrate the Nervous Apparatus of Vision and the Position of the Visual Centres. The right optic tract, R O P, is shaded, and the left optic tract, L O P, is unshaded. In the left optic tract are seen (1) fibres from the left temporal side of the retina of the left eye, (2) fibres from the right nasal side of the right eye, (3) posterior commissural fibres, having no connection with either retina. The right optic tract is similarly constituted of fibres from each retina and commissural fibres. The fibres from the nasal side of each retina are seen to decussate at the chiasma, x. Each optic nerve, R O V, L O V, contains fibres from each side of the brain. On the right side, the optic tract is represented as sending fibres to G L, the lateral corpus geniculatum, A q, the anterior corpus quadrigeminum, and P v, the pulvinar. OP. RAD indicates the optic radiations or fibres passing to the sensory area of vision in the right occipital lobe, O C. *x* indicates the direct tract to the cortex. L T F, left temporal field of left eye, O S, seen by N S nasal side of left retina, R N F, right nasal field of right eye, O D, seen by T D, temporal side of right retina. Section of one optic *nerve* will evidently lead to complete blindness of one eye; section of one optic *tract* will lead to blindness in the corresponding halves of each retina, or *hemianopia*, as it is called, *e.g.*, section of the right optic tract paralyses the right half of each retina, and blots out the right nasal field and the left temporal field.

optic tract direct to the cerebral hemispheres, and that others are prolonged through the crus cerebri to the oculo-motor nuclei and the cerebellum. But the main course of connection with the cerebral hemispheres is an indirect one through the three bodies forming the primary or lower visual centres, for from these bodies [1] the fibres forming the optic radiations of Gratiolet. may be traced to the cortex of the occipital lobes. Stimulation of the angular convolution produces movement of the head and eyes towards the opposite side, as if to look at an object, and pathological experience shows that the occipital lobes contain *visual areas*. Ferrier and Yeo state that ' the only lesion which causes complete and permanent loss of vision in both eyes is total destruction of the occipital lobes and angular gyri on both sides, and that destruction of the occipital lobe and angular gyrus on one side cause temporary amblyopia (= dulled vision) of the opposite eye and hemiopia of both retinæ on the same side as the lesion.' More definite statements are doubtful. All attempts to localise the sensory areas are difficult, owing to the want of such an indicator as motion, which assists in testing the motor areas, and only very general statements can yet be made. *Hearing* appears to be connected with the upper temporal convolution, as stimulation of this region in some animals leads to movements of the ear on the opposite side, and it is usually atrophied in deafness. The sensory area for *taste* and *smell* is supposed to be in the uncinate convolution, and fibres from the olfactory lobes have been traced to this region.

205. **The Sympathetic System.**—The sympathetic nerve system is the ganglionic chain of nerves lying on each side of the vertebral column together with the branches of this chain. It consists of ganglia united by intervening nerves, and giving off branches which form the main nerve supply of the viscera, glands, and blood-vessels of the body. It is not, as once thought, an independent system, since all nerve fibres are connected with the cerebro-spinal system, and the sympathetic is but a complicated plexus of fibres (with ganglion cells) derived from the spinal cord, the main outflow from the cord being through the anterior roots of the spinal nerves. The ganglia of the sympathetic are divisible into two great groups : (1) *lateral ganglia*, connected by commissural fibres, and lying on the side of the vertebral column and forming the main sympathetic chain, (2) *collateral ganglia* lying on the great branches of the aorta or in close proximity to the viscera. The ganglia are connected with the cerebro-spinal system by white medullated fibres much smaller than ordinary motor fibres, which may be

---

[1] Some authorities hold that the connection with the occipital cortex is through the pulvinar and external geniculate body only, and that fibres from the anterior corpora quadrigemina are in connection with the centres for the movements of the eyes.

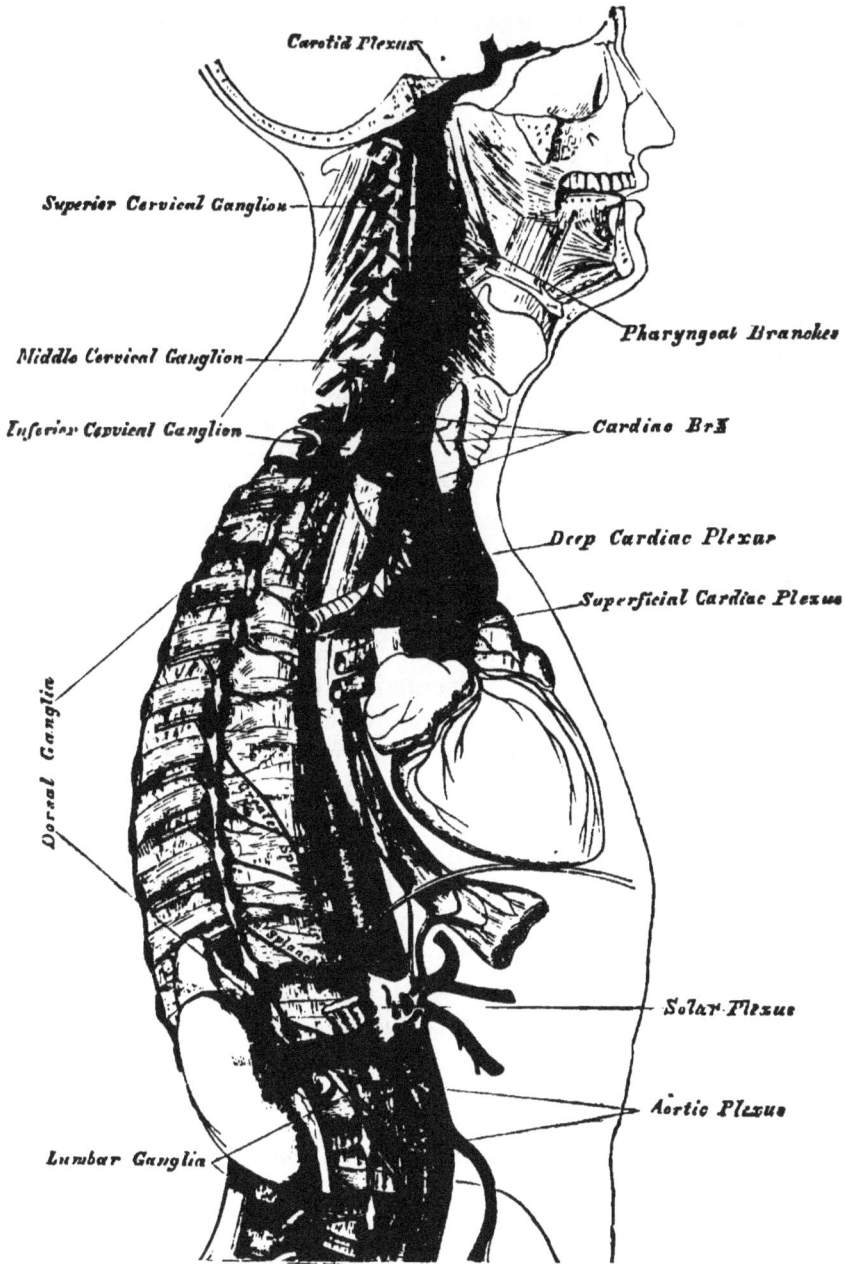

FIG. 205.—The Sympathetic System.

traced from the cord through both roots of the spinal nerves. The 'grey communicating branches' are simply fibres from the ganglia on their way to blood-vessels (par. 38). The ganglia are not centres of reflex action, but simply collections of cells in which nerve fibres lose their medullary sheath, each medullated fibre on reaching a cell being broken up into several naked fibres either on reaching a lateral ganglion or on reaching a collateral ganglion.

By the help of fig. 205 the chief parts of the sympathetic system may be understood. The *cervical* portion of the sympathetic (which is strictly a part of the collateral chain on the carotid artery) consists of three ganglia, superior, middle, and inferior, with connections with the roots of the spinal accessory and the vagus, and branches to blood-vessels, and to the pharynx and heart distributed with the branches of the vagus. Two plexuses of nerves connected with the heart are formed by branches of the sympathetic and pneumogastric, or vagi. The thoracic or dorsal part of the sympathetic consists of a series, usually twelve, of ganglia on each side of the spine opposite the head of each rib. Branches from the dorsal spinal nerves may be seen entering the ganglia externally, while from the ganglia internal branches proceed to blood-vessels and viscera. The internal branches from the lower six dorsal ganglia of the sympathetic unite to form the three splanchnic nerves. Branches from the large and small splanchnic nerves join a large plexus of nerves and ganglia lying in front of the aorta around the origin of the cœliac axis, termed the *solar plexus*. Branches from the solar plexus accompany the blood-vessels to the viscera, and form secondary plexuses on the chief arteries. The lumbar part of the sympathetic consists of four ganglia, and branches pass off to form the hypogastric plexus situated between the common iliac arteries, from which twigs are distributed to neighbouring organs. The sacral part of the ganglionated cord unites at the lower end of the sacrum with its fellow on the opposite side in the *ganglion impar*.

Functions of Sympathetic.—Afferent fibres are said to pass from the lateral ganglia of the sympathetic, and to carry impulses by the posterior roots to the central nervous system. But little definite is known about these afferent fibres. The efferent fibres of the sympathetic system are supplied (1) to the blood-vessels of the body, (2) to the unstriped muscles of the hollow viscera and to the heart, (3) to secreting glands. The vaso-motor part of the system contains two antagonistic sets. The vaso-con-strictor fibres leave the cord as fine medullated fibres by the anterior roots of the spinal nerves from the second dorsal to the second lumbar, and joining the lateral ganglia, where they become demedullated, proceed to their distribution in the arteries as fine non-medullated filaments. The vaso-dilatator fibres appear to leave the cord in the upper cervical and sacral regions, and to pass on to the collateral ganglia and thence to the blood-vessels; but their course is only known in a few cases (see chorda tympani, par. 114). The sympathetic fibres for the viscera leave the spinal cord in all regions, and by their action promote or retard the activity of the involuntary muscles of the alimentary canal and other viscera, and thus aid in regulating nutrition. Accelerating fibres for the heart pass from the upper part of the dorsal sympathetic chain, while cardio-inhibitory fibres arise from the spinal accessory and pass through the ganglion of the vagus trunk to smaller ganglia in the heart, where they then lose their medullary sheath. It is probable that nerves are distributed to most if not all secreting glands to influence their action, but little is known about their course and influence except in the case of the salivary glands (par. 114).

Thus the functions of the sympathetic system include (1) the control of the calibre of the blood-vessels and the regulation of the blood-pressure in the arteries; (2) the control of the contraction of the involuntary muscular fibres in the viscera; (3) the regulation of secretion in the secretory glands.

CHAPTER XVI

TOUCH, TEMPERATURE, MUSCULAR SENSATIONS, TASTE AND SMELL

206. **Touch.**—The sense of touch, which is a sensation of contact or pressure referred to the surface of the body, is located in the skin and mucous membrane near the orifices, the mucous membrane of the tongue and lips being particularly sensitive to touch. The general structure of the skin, which is not only an organ for receiving sense impressions, but also a protective covering, an excretory organ for sweat and seba-ceous matter, and an organ concerned in the regulation of the body heat, has already been described. Cutaneous sensations or sensations derived from the skin, it must be noted, are of various kinds. There are *tactile sensations* or sensations of touch proper, *thermal sensations* or sensations of heat and cold,

and *sensations of pain.* An object placed on the palm of the hand may produce a complex sensation involving a sensation of pressure, of temperature, and, if the pressure be great, of pain. Before describing these we will refer to the various modes of nerve endings of the skin.

207. Nerve Endings of the Skin.—The sensory or afferent nerve-endings of the skin may be thus enumerated :—

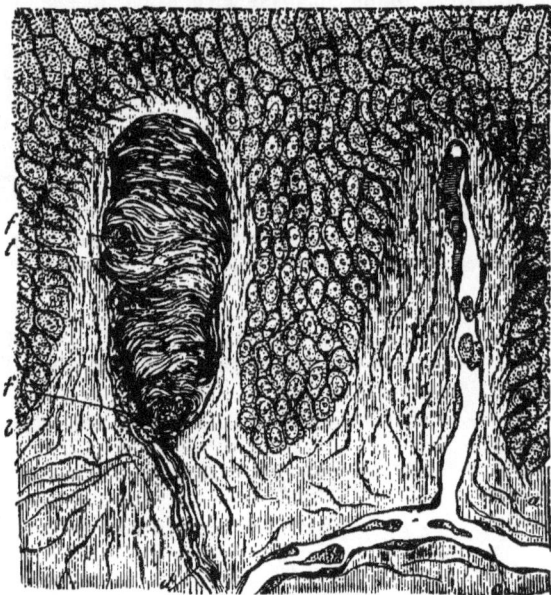

Fig. 206.—Section of Skin showing Two Papillæ and Deeper Layers of Epidermis. (Biesiadecki.)

a, vascular papilla with capillary loop passing from subjacent vessel, *c* ; *b,* nerve papilla with tactile corpuscle, *t.* The latter exhibits transverse fibrous markings. *d,* nerve passing up to it ; *f, f,* sections of spirally winding nerve fibres.

(*a*) *Free nerve endings.*—In all parts of the epidermis and in the cornea fine nerve fibrils are found, derived from the splitting up of the axis-cylinder of a single nerve passing up from the underlying dermis. These nerve fibrils terminate by free ends between the epithelial cells of the upper part of the Malpighian layer, the free ends being sometimes provided with small swellings (fig. 6). In some parts non-medullated nerve fibrils are found to terminate in a small enlargement applied to oval nucleated cells of the epidermis, called 'touch-cells.'

(*b*) *Tactile corpuscles.*—The tactile or touch corpuscles of Wagner are small oval bodies averaging about $\frac{1}{300}$ inch long and $\frac{1}{500}$ inch broad, found in the papillæ of the true skin.   They are very numerous in the palm of the hand and the sole of the foot, especially in the fingers and toes, less numerous on the back of the hand, lips, and tongue, and scanty on the under surface of the arm and eyelids.   In a large portion of the skin they are absent.   Each touch corpuscle is covered outside by layers of connective tissue arranged in transverse layers, within which is the core showing elongated nuclei.   One or more medullated nerve fibres pass to the lower part of the corpuscle and wind round it two or three times, and then losing the sheath, the fibre or fibres enter the interior of the corpuscle, the axis-cylinders of the nerve ending in small enlargements (fig. 51).

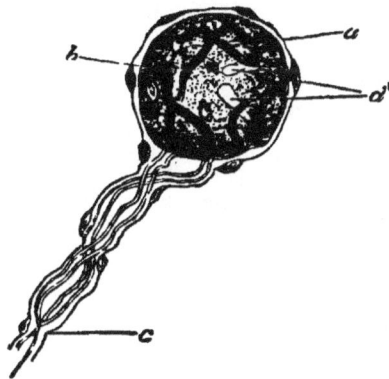

FIG. 207.—End-bulb from the Human Conjunctiva. (Longworth.)

a, nucleated capsule; *b,* core, the outlines of its cells are not seen; *c,* entering fibre-branching, and its two divisions passing to terminate in the core at *d.*

(*c*)   *End-bulbs.* — The end-bulbs of Krause are small oblong or rounded corpuscles from $\frac{1}{360}$ to $\frac{1}{170}$ inch long, into the interior of which the axis-cylinder of a nerve fibre passes, and terminates in a coiled mass or in a bulbous extremity.   There is a capsule of connective tissue with a core of granulated material, and the nerve sheath of Henle with the neurilemma appears to become continuous with the capsule.   End-bulbs are only found in the conjunctiva of the eye, in the mucous membrane of the mouth, and a few other parts (fig. 52).

(*d*) *Pacinian corpuscles.*—These bodies are visible to the naked eye, being from $\frac{1}{12}$ to $\frac{1}{8}$ inch in length, and $\frac{1}{25}$ to $\frac{1}{15}$ inch in breadth.   They do not occur in the skin proper, but in the subcutaneous connective tissue of the palm of the hand and the sole of the foot, including the fingers and toes, along the nerves near the joint, in the nerves of the mesentery, &c. Their structure, resembling much that of an end-bulb, has already been described (par. 49). Being so deep seated it is doubtful whether they can be connected with cutaneous sensations.

As to the function of these various nerve endings but little is known (see par. 211).

208. Touch proper.—Tactile sensations may be distinguished into (*a*) sensations of simple pressure, and (*b*) sensations of locality. (*a*) Mere contact of a body with the skin leads to a slight sensation of touch, and this sensation becomes more acute as the pressure increases up to a certain limit. The least pressure that can be felt, or the smallest difference of pressure that can be appreciated, varies in different parts of the skin. Small weights are allowed to press on the skin of various parts, different weights being used one after another, and the sensations are noted. Measured in this way the greatest acuteness of the pressure sense for a single pressure is found on the forehead, temples, and back of the hand, which detect a pressure of ·002 gram. The skin of the fingers detects a pressure of ·005 to ·015 gram. The greatest sensitiveness to differences of pressure, when small weights are used, is found on the skin of the forehead, lips, and cheeks, which appreciate a difference of pressure $\frac{1}{30}$ of the first pressure ; then come the back of the last phalanx of the fingers, the palm of the hand, and the forearm, which distinguish differences of $\frac{1}{10}$ to $\frac{1}{20}$. Small intermittent variations of pressure, as feeling the pulse, are better noted with the tips of the fingers than with the skin of the forehead. Individual sensations of pressure following each other with sufficient rapidity, as in touching the blunt teeth of a rapidly rotating wheel, fuse into one continuous sensation.

(*b*) When an object touches our skin we not only experience a pressure sensation of greater or less intensity according to the amount of pressure on the part pressed upon, but we are aware of the part that has been touched. This has been called the sense of space or locality. The power of localisation is much finer in some parts of the body than in others, and the parts of the skin most sensitive to the power of discriminating the locality are not the same as those most sensitive to pressure. This power of localisation probably depends on the number of sensory nerves in the part affected ; for the fewer the fibres in a given area, the more likely it is that adjacent points will act on only one and produce but one sensation, and the greater the fibres in a given area, the more likely it is that the different points will be distinguished and the locality determined. The usual mode of testing the power of discriminating points in contact and the sense of locality is to place the two blunted points of a pair of compasses on a part of the skin and determine the smallest distance at which the two points are felt as one impression ; or, the points being kept at the same distance and drawn over the skin, it is ascertained whether the two points give the sensation of approaching or receding. The results obtained in different parts of the skin are set forth below.

Table of variations of the tactile sensations of space or locality in different parts, the measurement indicating the least distances at which two points can be separately distinguished :—

Tip of tongue	$\frac{1}{24}$ in. or 1 mm.
Under surface of end phalanx of third forefinger	$\frac{1}{12}$ in. or 2 mm.
Under ,, ,, second ,, .	$\frac{1}{6}$ in. or 4 mm.
Red part of lip	$\frac{1}{6}$ in. or 4 mm.
Tip of the nose	$\frac{1}{4}$ in. or 6 mm.
Middle of dorsum of tongue	$\frac{1}{3}$ in. or 8 mm.
Palm of hand	$\frac{5}{12}$ in. or 10 mm.
Back of hand	$1\frac{1}{6}$ in. or 28 mm.
Dorsum of foot near toes	$1\frac{1}{2}$ in. or 37 mm.
Upper or lower parts of forearm . .	$1\frac{1}{2}$ in. or 37 mm.
Back of neck near occiput . . .	2 in. or 50 mm.
Upper dorsal and mid-lumbar regions . .	2 in. or 50 mm.
Middle part of forearm	$2\frac{1}{2}$ in. or 62 mm.
Middle of thigh	$2\frac{1}{2}$ in. or 62 mm.

It will be noted that the power of discriminating separate points in space is most acute in those parts of the body that carry out the widest and most rapid movements. Moistening the skin increases this sensitiveness to separate points, but cold and a bloodless condition of the skin blunts this sensibility. Exercise improves it to some extent, though Mr. F. Galton says that the alleged superiority of blind persons in sensitiveness of touch is not great, as ' the guidance of the blind mainly depends on the multitude of collateral indications to which they give heed, and not in their superiority in any one of them.'

The power to localise our sensations with reference to the surface of the body, and to indicate the position of the touching object, enables the brain to construct a tactile field on the surface of the skin to some part of which the body touching us is referred.

209. **Sensations of Temperature.** —Temperature sensations as felt by the skin and certain parts of the mucous membrane, as the mouth, are of two kinds, (*a*) sensations of heat, and (*b*) sensations of cold. These are quite unlike each other as well as unlike those of pressure. Minute areas are found on the skin in which sensations of heat are felt more acutely than

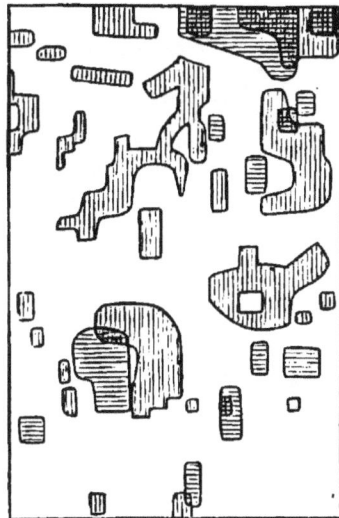

FIG. 208.—Cutaneous ' cold ' spots (vertical shading) and ' hot' spots (horizontal shading) ; anterior surface of the thigh. (Goldscheider.)

in adjoining parts, and such areas are termed ' hot spots ; ' other areas called ' cold spots' are particularly sensitive to cold. These spots seldom coincide, nor do they correspond with the points most sensitive to pressure ; cold spots are more abundant than hot spots, and the spots are generally arranged in lines often somewhat curved, though scattered points between which other sensations may be produced are frequent. A mode of finding these spots is to use a pointed pencil of copper. On dipping it into hot water and touching parts of the skin, points may be found very sensitive to heat, other points not giving this sensation. With the pencil made cold by ice, spots may be found sensitive to cold but not to heat. Such experiments appear to indicate that the nerve fibres from these two kinds of spots are specifically different.

With regard to the two kinds of temperature sensations and the general cutaneous surface, it should be noted that the sensations of heat and cold can only be felt through the nerve terminals in the skin. Direct stimulation of the nerves, as when the epidermis is removed, produces only a sensation of pain. So, also, irritation of the trunk of the ulnar nerve by dipping the elbow into very hot water or a freezing mixture will not only affect the skin of that part with a sensation of heat or cold, but the stimulated trunk of the underlying fibres of ulnar nerve will cause a sensation of pain which, in accordance with a general law, is referred to the peripheral terminations of the fibres in the arm and hand. With regard to variations of temperature it is found that—

    (*a*) Bodies of the same temperature as the part of the skin to which they are applied give rise to no thermal sensations.

    (*b*) The parts of the body having the sense of temperature most acute are, in order, the tip of the tongue, the eyelids, the cheeks, the lips, and the hands.

    (*c*) Small differences of temperature—about $\frac{2}{5}°$ C.—are readily appreciated by the sensitive parts when the temperature lies near that of the body.

    (*d*) Though the power of the skin to recognise changes of temperature is very great, yet our power of estimating absolute temperature by skin sensation is small. Our own feeling of warmth depends on the state of the cutaneous blood-vessels, full vessels leading us to feel hot and empty vessels to feel cold. Hence a body of the same temperature gives a different sensation according as the skin is full or empty of the warm blood.

    (*e*) Illusions in this sense are common, a cold weight feeling heavier than a warm one, a good conductor like metal feeling colder than a piece of wood of the same temperature, &c.

210. **Pain, Common Sensation.**—What is called ' common ' or ' general sensibility ' appears to arise from a number of obscure sensory impulses proceeding from the skin and other parts of the body, and these sensations inform us in a vague manner as to our general condition. If these impulses become intense, we have the sensation of pain, so that painful sensations may be regarded as resulting from the excessive stimulation of the nerves of common sensation, though every violent stimulation of any sensory nerve appears to provoke pain. As every kind of over-stimulation, mechanical, thermal, chemical, or

electrical, may excite pain on the surface of the body, pain may be placed among cutaneous sensations, though it may also occur in almost all other organs. A slight inflammation makes every organ keenly sensitive to pain. Pain may be caused by stimulating a sensory nerve at any part of its course, but the sensation is felt at the nerve termination in accordance with the law of the peripheral reference of sensations. Pains vary in quality and intensity according to the nature and strength of the stimulus and to the excitability of the nerves affected. In violent pain the sensation often appears to spread and to render localisation difficult. Sensations of common sensibility and pain are distinct from other sensations. One great difference between common sensation, which informs us of the general condition of the parts of the body, and of which hunger and thirst are but examples informing us of the condition of the stomach and palate respectively, and the sensations of touch and temperature, appears to be that the latter have special nerve endings.

211. **Functions of Cutaneous Nerves.**—Having learnt that through the skin we experience three different kinds of sensation, those of touch, of temperature, and of common sensibility, rising to pain in certain cases, we may inquire whether there are special nerves for each of these. That there are functionally different kinds of nerve fibres in the sensory nerve trunks of the skin is shown from the following considerations :—

1. In some cases of nervous disease sensibility to touch in certain parts of the skin has been lost, while the power of distinguishing temperature has remained, and *vice versâ*.

2. In other cases common sensibility and pain have been lost over certain areas while the sense of touch has remained ; and, in general, one or other class of cutaneous sensations may be lost while the others remain.

3. The conducting fibres of these different kinds of sensation run, to a certain extent, in different paths in the cord, and have different central endings.

4. The different modes in which the power of appreciating tactile differences, temperature differences, and sensation of pain is distributed on the surface of the skin, indicate that different nerve fibres perform these distinct offices.

5. That the 'hot spots' of the skin do not coincide in position with the 'cold spots' shows that there are points of the skin having nerve fibres sensitive to heat but insensitive to cold, and *vice versâ*. A leg 'sent to sleep' by pressure on its sciatic nerve is found at a certain stage to be more sensitive to cold than to heat, while in some nervous diseases a part is found more sensitive to heat than to cold. These facts point to distinct nerve fibres for these different temperature sensations.

It appears, therefore, that the skin contains nerve fibres of four different kinds, or fibres performing four different functions—pressure, heat, cold, common sensibility or pain. Whether each of these sets of fibres have distinct terminal organs is another question. For the *general* sensations conveyed by the fibres of common sensibility or pain no end-organ for the nerve appears to be needed, for we know that pain can be produced by pinching, heating, &c. the open surface of a wound or the cutaneous nerves themselves. But such stimulation gives no sensation of pressure or heat, only pain ; and this suggests that there must be some special mode of ending for the nerve fibres carrying sensations of pressure, heat, and cold. What has been said above about the different parts of the skin in which these sensations are most acutely developed, and about the loss of one sensation while the others remained, indicates that there are three different sorts of terminal organs for these sensations. Moreover, various experiments and considerations show that the *special* sensations can only be developed by special terminal organs, that stimulation of the terminal organs by their appropriate stimuli (or of the special nerve centres themselves) is needed to arouse the specific sensation, stimulation of the nerve fibres not effecting this.

The various modes of nerve termination in the skin are described in par. 207. As to the part taken by the various end-organs but little is known. The more complex end-organs in which the extremities of the sensory nerves are buried in soft material surrounded by envelopes of connective tissue seem adapted for communicating to nerve fibres slight variations of external pressure and giving to them a rhythmic character. But nothing definite can be said about the part played by end-bulbs and Pacinian bodies. The touch corpuscles being numerous in the papillæ of those parts where the tactile sense is most acute, and few and scattered on the legs and trunk where this sense is dull, would appear to be the end-organs specially helpful and concerned in touch sensations, but their proved absence from certain parts where sensations of pressure can be appreciated shows that they cannot be essential nor contain the only nerves taking part in touch sensations. Probably the fine nerve fibrils terminating among the cells of the Malpighian layer of the epidermis are concerned in the sensation of pressure, and it appears certain that some of these fibrils or the cells connected with them are the instruments by which the altered condition of the skin becomes sensible to temperature sensations, as it is found that heat or cold applied to the nerve fibres underlying the skin only gives rise to a sensation of pain.

212. **The Muscular Sense.**—A special sense called the *muscular sense*, with afferent fibres proceeding from the muscles to nerve centres, is thought to exist, and to enable us to judge of the activity and degree of contraction of the muscles. When we examine our own consciousness, we find that we are aware of the position of any part of our body, even with the eyes closed, and during movements of the limbs we estimate the character and force of the muscular movements. A little consideration teaches us that the afferent sensory impulses from a moving limb, or in lifting a weight, may be of various kinds—

impulses from the stretched or relaxed skin, sensations of pressure when raising a weight or moving a resisting object, as well
as impulses from the contracting muscles and from the joints
brought into action. Muscular sensations are, therefore, closely
conjoined with cutaneous and other sensations, so much so that
some have maintained that there is no muscular sensation independent of the action upon the skin. But this contention fails
when we learn that cutaneous sensations may be lost or greatly
impaired while the power of co-ordinated muscular movement
remains, and that a certain disease of the spinal cord may lead
to 'locomotor ataxia,' or failure of power to use the muscles
with the proper degree of force, though the tactile, temperature
and pain sensations may be unimpaired. Besides, a muscle is
not only supplied with fibres that end in 'motorial end-plates,'
but with afferent fibres, some of which are described as ending
in fine fibrils among the muscular fibres, and these may serve
for the afferent impulses of the muscular sense. At times,
also, we have a sense of fatigue that seems produced by the
condition of the muscles. Besides impulses derived from
the muscle itself, it has been thought that the tendon and
ligaments may supply impulses that enter into the 'muscular
sense.' Tendons are known to have peculiar nerve endings,
termed the 'organ of Golgi.'

The muscular sense contributes largely to our knowledge
respecting the external world. Though the muscular feelings
are rather difficult to localise, they are delicate, and enable us
to discriminate slight differences in the range and force of
movements. Muscular movements are an important factor in
acquiring a knowledge of the space relations between things
generally, though how the muscles help in these space-percep-
tions, whether by their own sensations, or by awakening sensations of motion in the skin, retina, and articular surfaces, is still
undecided. It is in combination with other senses that
this sense plays its most important part, for in almost all
sensations muscular sensations form an element. In taste, the
movement of the tongue is of importance ; in smell, the air
carries the particles into the nose ; in listening, the muscles of
the tympanum contract, or the head is moved in the direction

of the sound. But movement is of greatest importance to the senses of touch and sight. The delicacy of the sense of touch stands in a definite relation to the mobility of the different parts of the body, those parts being most mobile that are most delicate of touch. Combined with tactile sensations, muscular sensations give rise to composite sensations that enable us to estimate small differences of weight and small amounts of resistance, and also render possible the knowledge and relations of the parts of surfaces and solids that enter into the idea of form. Combined with visual sensations (for every position of the eye is accompanied by muscular sensations from the muscle of accommodation and the muscles that move it) they produce changes of the visual field that aid in building up our complex ideas of the objects of the outer world.

213. **The Organ of Taste.**—The sense of taste is located chiefly on the upper surface of the tongue, though it is also found in part of the soft palate and its arches, which receive branches of the glosso-pharyngeal nerve—the principal gustatory nerve. The tongue is a muscular organ covered with mucous membrane. Its base or root is connected behind with the hyoid bone, with the epiglottis, and with the fauces or part of the mouth leading into the pharynx. Its under surface is connected with muscles, but the tip, sides, and dorsum or back are free.

The tongue plays an important part in the actions of chewing, swallowing, and articulate speech, its motor nerve being the hypoglossal. Its sensibility to touch has just been described.

The convex dorsum of the tongue is marked by a slight furrow or raphé along the middle line, that ends in a depression termed the *foramen cæcum*. The mucous membrane of its under surface is thin and smooth, but that of the upper surface is covered with large papillæ, giving it a rough appearance. The epithelium of the mucous membrane is thick with stratified layers of cells, and the dermis, which is highly vascular, is raised in many places to form the well-marked lingual papillæ, some of which are readily seen on examination. These papillæ are of three kinds :—(1) About ten or twelve *circum-*

vallate papillæ, $\frac{1}{20}$ to $\frac{1}{12}$ inch wide, arranged in two rows, and diverging forwards from the middle of the back part of the tongue in the form of a wide V. Each circumvallate papilla consists of a circular projection with a broad free surface, and a smaller attached end, and each is surrounded by a narrow trench or fossa, on the outside of which the mucous membrane is raised to form a wall or vallum. The substance of the papilla consists of corium or dermis, formed of dense connec-

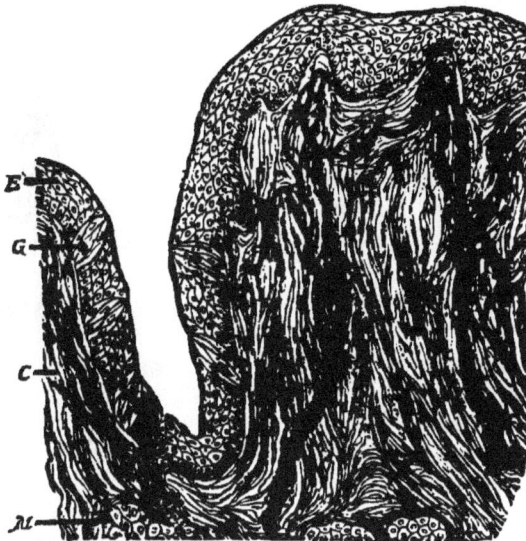

FIG. 209.—Section of Circumvallate Papilla, Human. The figure includes one side of the Papilla and the adjoining part of the Vallum. (Magnified 150 diameters.) (Heitzmann.)

E, epithelium ; *G*, taste-bud ; *C*, corium with injected blood-vessels ; *M*, gland with duct.

tive tissue containing blood-vessels and nerves, and covered by stratified epithelium. The papillæ contain smaller or secondary papillæ. In the epithelium covering the sides of the papillæ, small oval or flask-shaped bodies termed *taste-buds* or *taste-bulbs* are seen in a section under the microscope, fig. 209. (2) *Fungiform papillæ* scattered over the surface of the tongue, but most numerous at the sides and front. These are club-shaped, with a narrow base and of bright red colour, owing to their rich blood supply. (3) *Conical* or *filiform*

z

papillæ scattered over the whole surface of the tongue, and forming the smallest and most numerous papillæ. They are small conical eminences, and are sometimes fringed with epithelial threads or filaments (fig. 211).

The tongue contains numerous small tubular glands opening on the surface, most of which secrete mucus, though some yield a serous secretion.

214. Taste-Buds.—Taste-buds are found in the epithelium on the lateral surfaces of all the circumvallate papillæ, in the epithelium of the

FIG. 210.—Section of Fungiform Papilla, Human. (Heitzmann.)

E, epithelium ; *C*, corium ; *L*, lymphoid tissue ; *M*, muscular fibres of tongue.

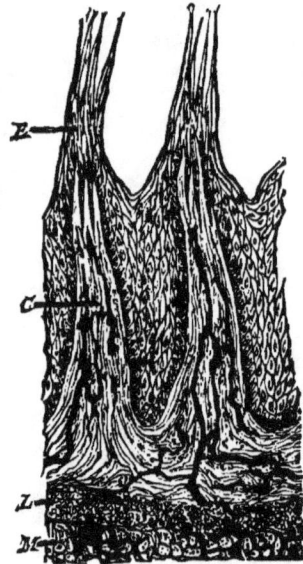

FIG. 211.- Section of Two Filiform Papillæ, Human. (Heitzmann.) (Letters as in previous figure.)

surrounding vallum, in many fungiform papillæ, here and there in the general mucous membrane of the tongue, and on the under surface of the soft palate and epiglottis. They are oval clusters of epithelial cells lying in the epithelium, and set vertically to the surface, and having their broad base resting on the dermis portion of the mucous membrane and their neck opening at a pore on the surface. Each bulb or bud is composed of two kinds of cells, *gustatory cells*, and *supporting* or *sustentacular cells*. The gustatory cells are small spindle-shaped cells with a central nucleus, having an outer process passing from one end to terminate as a fine hair that projects through the gustatory pore, and an inner process, which is believed to be continuous with a nerve fibril. The fibril, in fact, may be considered to take its origin in the gustatory cell, and a similar arrangement will be noticed in the organ of smell. The sustentacular cells are long and flattened, with tapering ends. They are found between the gustatory cells, and also form a sort of covering for the taste-bud.

That the taste-buds by means of the gustatory cells are end-organs of taste appears evident when we learn that they are connected with fibres of the glosso-pharyngeal nerve, that the sense of taste is chiefly found where they are most abundant, and that their cells resemble those of other sensory epithelium. But taste sensations are also distinctly felt near the tip of the tongue where no glosso-pharyngeal fibres have been found, but where fibres from a branch of the fifth cranial nerve are distributed. The lingual branch of the fifth must therefore be considered a gustatory nerve also. Nerve filaments from the chorda tympani also seem connected with the sense of taste, as destruction of this nerve within the tympanum has been followed by loss of taste on the same side of the tongue. But the connections of the nerve of taste are but imperfectly understood.

It would seem that specific tastes have specific nerves, for different parts of the tongue are more sensitive to certain tastes than others: the back to bitter, the tip to sweet and salt, the sides to acid, the middle to hardly any. Further, weak electric currents applied to the tongue awaken sensations of different kinds in different parts, and cocain applied to the tongue in increasing doses is said to abolish sensations of all kinds in the following order: general sensibility and pain, bitter taste, sweet taste, salt taste, acid taste, tactile sensations.

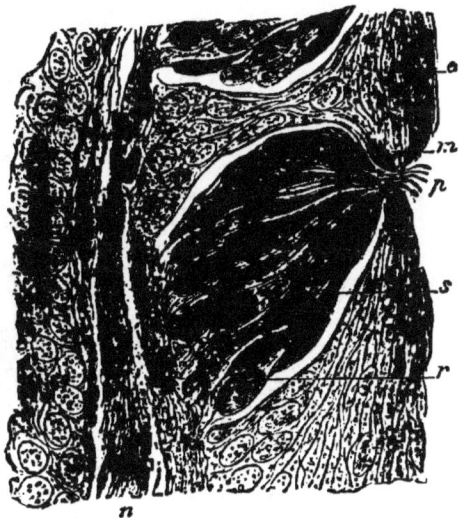

FIG. 212.—Section through the Middle of a Taste-Bud. (Ranvier.)

p, gustatory pore; *s*, gustatory cell; *r*, sustentacular cell; *m*, lymph cell, containing fatty granules; *e*, superficial cells of the stratified epithelium; *n*, nerve fibres.

215. **Sensations of Taste.**—In order that the end-organs of taste may be stimulated, and give rise to a sensation of taste in the brain, sapid substances must be in solution. Such substances are most active at about the body temperature, and when dissolved in fluid or by the saliva, the sensation arises in about $\frac{1}{5}$ of a second on the average. Gustatory sensations may be divided into *bitter, sweet, salt* and *acid*. There are also tastes described as *metallic*. The delicacy of the sense of taste is shown by the power to detect 1 part of sulphuric acid in 1,000 of water. Quinine, common salt, and sugar are less easily detected, but in the order given. Chewing the leaves of an Indian plant (*Gymnema sylvestre*) destroys the

sensibility to bitter and sweet, but leaves the power to discern acid and saline bodies.

The union of taste and smell gives rise to the composite sensation termed the *flavour* of a substance.

216. **The Organ of Smell.**—The olfactory organ is found in a portion of the mucous membrane lining the cavities that are situated between the base of the cranium and the roof of the mouth. The internal part consists of two chief cavities called nasal fossæ, opening in front into the air by the nostrils or anterior nares, and behind into the pharynx by the

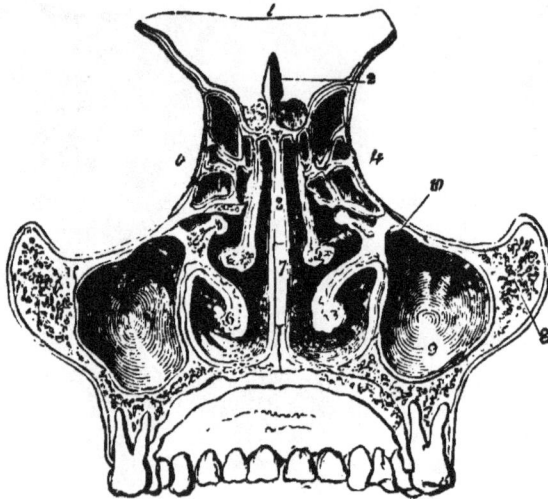

FIG. 213.—Section of the Nasal Cavities, seen from behind.

1, frontal bone ; 3, perpendicular plate of the ethmoid bone ; between 4 and 4, ethmoidal cavities ; 5, middle turbinated bone; 6, inferior turbinated bone ; 7, vomer ; 8, malar or cheek bone.

two posterior nares. The middle wall of each fossa or nostril is formed by a vertical partition, having a smooth surface, but the outer wall is convoluted. The roof of the cavity is formed by the cribriform plate of the ethmoid bone of the cranium. From this bone a central vertical plate passes, and is continued downwards by the vomer and by gristle to complete the partition between the nostrils. The outer or side wall of each chief cavity is formed in part by two scroll-like bones from the ethmoid, and in part by a third similar bone attached to the upper jawbone. These three turbinated bones thus produce three spaces called the superior, middle, and inferior meatus, and these meatuses communicate with small cavities or sinuses in the bone.

The cavities of the nose are lined by mucous membrane continuous with that lining the pharynx and Eustachian tube, and prolonged on each side through the lachrymal canal to the eye. The structure of the nasal mucous membrane (or Schneiderian membrane, as it is called from the anatomist who first explained its secretory function) differs in various parts.

These parts are (*a*) *vestibular region*, or entrances to the air

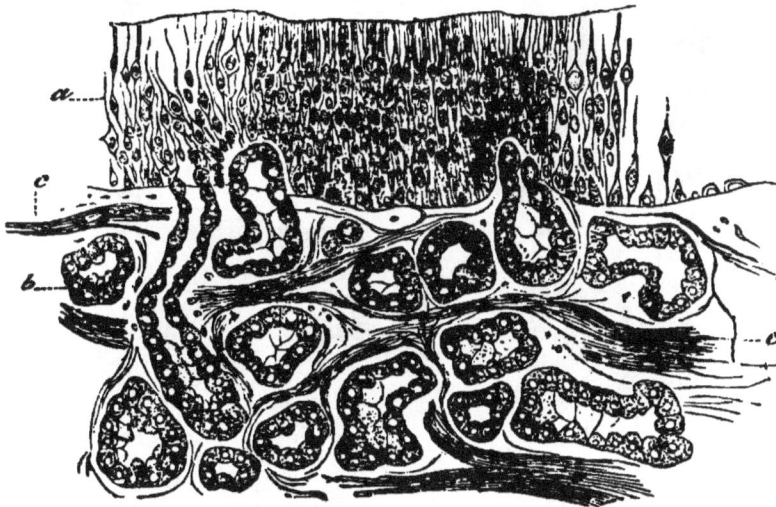

Fig. 214.—Section of Olfactory Mucous Membrane. (Cadiat.)
a, epithelium ; *b*, glands of Bowman ; *c*, nerve bundles.

passages. Its mucous membrane contains numerous sebaceous glands and hair follicles, from which stiff hairs named vibrissæ spring. (*b*) The *respiratory region* includes the lower meatus of the nose, and is lined by thick mucous membrane having numerous mucous glands and a stratified ciliated epithelium. (*c*) The upper or *olfactory region* is the region specially connected with the sense of smell, and is covered with a soft mucous membrane of a yellowish colour. This is found on the anterior two-thirds of the superior meatus, the middle meatus, and upper third of the septum. Fig. 214 shows

a vertical section of the mucous membrane of this region. The dermis portion of the olfactory mucous membrane is very thick, contains numerous blood-vessels and nerve fibres, and a large number of peculiar tubular glands, opening on the surface between the epithelial cells. The epithelium of the olfactory region contains cells of two kinds :—
(1) Long cylindrical epithelium cells with broad nucleated portions coming to the surface, and forked processes stretching to the corium or dermis. These are supporting or *sustentacular cells*. (2) Long spindle-shaped cells with a nucleated central part, from which there passes to the surface a slender filament bearing a free cilium,

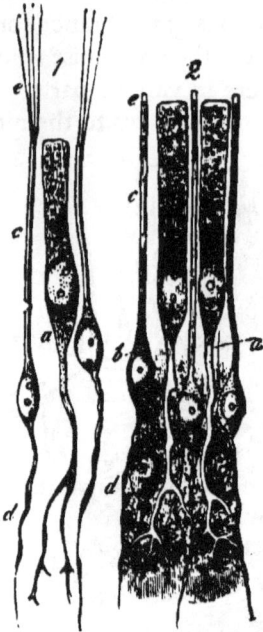

FIG. 215.—Cells and Terminal Nerve Fibres of the Olfactory Region. (M. Schultze.) Highly magnified.

1, from the frog ; 2, from man ; *a*, epithelial cell, extending deeply into a ramified process ; *b*, olfactory cells ; *c*, their peripheral rods ; *e*, their extremities, seen in 1 to be prolonged into fine hairs ; *d*, their central filaments.

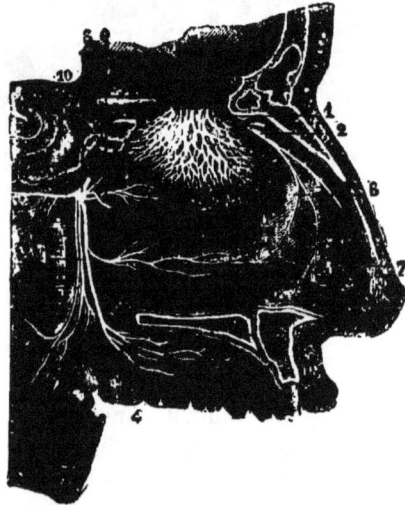

FIG. 216.—Nerves of the outer wall of Nasal Fossæ. 1. Network of the branches of the olfactory nerve.

while another filament passes down to the corium, where it is lost among the nerve fibrils, with one of which it becomes connected. These are the *olfactory* or *rod cells*. Round *basal* cells are also found among the lower parts of the other cells.

The nerve fibrils which are non-medullated at the base of the epithelium of the upper third, or olfactory region of the nose, pass through the cribriform plate of the ethmoid bone to end in the olfactory bulb or first pair of cranial nerves that rest on this plate (par. 298). This is the proper nerve of smell, and its fibres may be seen forming a brush-like expansion on the upper and middle turbinated bone, as well as on the septum before they enter the mucous membrane to become connected with the olfactory cells and the real end-organs of smell. Branches of the fifth cranial nerve also proceed to all parts of the mucous membrane of the nose, but only endow it with common sensibility, for the sense of smell is confined to the parts containing fila-ments of the olfactory nerve, and the sense of smell may be lost while those of common sensation and pain remain.

217. **Sensations of Smell.**—Odoriferous particles carried in the inspired air into the lower nasal cavities pass by diffusion into the upper chambers, and coming into contact with the olfactory epithelium give rise to the sensation of smell. By what physical or chemical process the excitation of the olfactory cells is effected is not understood. The par-ticles must be of extreme minuteness, as will be evident when we remember that air containing an odour may be filtered through cotton wool, and it will still be odorous, and a grain of musk will diffuse its perfume for years without any appreciable loss of weight. It also seems that the particles must be dissolved in the small amount of fluid secreted by the mucous membrane, for if the membrane be too dry or too moist the sensation is not excited. A cold in the head leading to excessive secretion of fluid of altered quality prevents the particles from exerting their action, and smell is abolished. The intensity of an odour varies with the number of odorous particles and the extent of the olfactory epithelium. Sniffing introduces more particles and spreads them with more force over a wider area. The delicacy of the sense is extremely great, as $\frac{1}{30,000,000}$ of a grain of musk can be perceived. Odours are difficult to classify, except into *pleasant*, *indifferent*, and *unpleasant*. Irritating vapours like ammonia not only affect the sense of smell, but stimulate the sensory fibres of the fifth nerve.

CHAPTER XVII

THE EYE AND THE SENSE OF SIGHT

218. **Appendages of the Eye.**—The eyeball is situated in a bony socket termed the *orbit*, which is padded with fatty tissue. Certain structures connected with the eye for its pro-

tection are spoken of as appendages. Of these the eyelids and the lachrymal apparatus are the chief. The eyelids consist of dense fibrous tissue (*tarsus*), covered externally by ordinary skin, and internally by a mucous membrane termed the *conjunctiva*. Beneath the skin are the muscular fibres of the *orbicularis* muscle for closing the lids, and in the upper eyelid there is in addition the *levator palpebræ* for elevating this lid. Beneath the conjunctival membrane near the inner surface of each lid, and embedded in the fibrous tissue, are some parallel rows of a variety of sebaceous glands known as *Meibomian glands*. On turning up the eyelids they may be seen through the conjunctiva like tiny strings of pearls, and from their minute openings at the edge of the lids an oily secretion passes. Along the free edges of the lids curved hairs called eyelashes grow from large hair follicles, to which other sebaceous glands and modified sweat glands are attached. Continuous with the skin at the edge of the eyelids and lining their inner surface is the delicate mucous membrane termed the conjunctiva. From the lids the conjunctiva is reflected over the globe of the eye, becoming adherent to the sclerotic coat, but its epithelial portion only passing over the cornea. The conjunctiva of the lids is thus thicker than the other portions, very vascular, freely supplied with nerve filaments, and has a number of mucous glands where reflection begins. But the chief liquid for keeping the surface of the eye moist is supplied by the secretion of the *lachrymal gland*. This gland is of an oval shape, about the size of a small almond, and is situated at the upper and outer part of the bony orbit, with its under surface resting on the eyeball. The gland is a compound racemose gland, and consists of several lobules, the acini of which are lined by cylindrical granular epithelium, the structure being similar to that of a serous salivary gland (par. 113). Its watery secretion issues by several small ducts that open on the inner surface of the upper lid, spreads over the eyeball, where its overflow is usually prevented by the oily Meibomian secretion on the edge of the lids, and then collects at the inner angle of the eye. The fluid passes off at two small openings, the *puncta lachrymalia*, into small canals that unite to form a sac.

The sac opens below into the nasal duct which runs in a groove of the superior maxillary bone, and terminates in the lower meatus of the nose. The lachrymal secretion is under the control of a centre in the nervous system. Various sensory impulses, as pungent smells or irritating vapours, may produce reflex stimulation of this centre, and lead to such copious secretion that the liquid overflows the lower lid in the form of tears. Certain strong emotions often act in a similar way.

219. **General Structure of the Eyeball.** — The globe-like eyeball consists of segments of two spheres of different sizes. The front portion, forming about one-sixth of the eyeball, is a segment of a small sphere, and the posterior portion, forming about five-sixths of the ball, is a segment of a larger sphere. It is composed of three investing coats or tunics, within which lie fluids and solid bodies

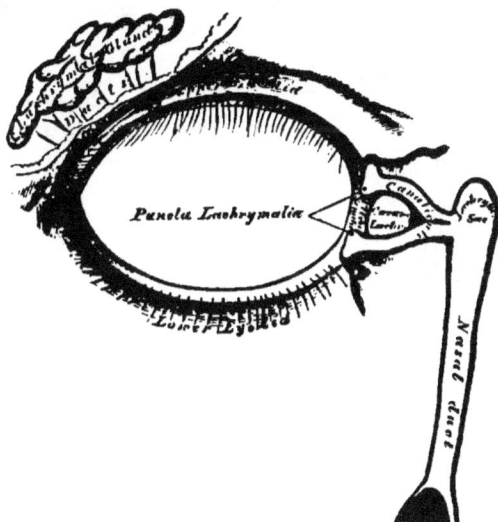

FIG. 217.—The Lachrymal Apparatus. Right Side.

which, from their action on rays of light, are called the refracting media or humours. The three investing layers or tunics are (1) the outer tunic, consisting of sclerotic and cornea ; (2) the middle tunic, consisting of choroid, iris, and ciliary processes ; (3) the inner tunic or retina, spread out on the hinder portion of the choroid. The refracting media or humours are from before backwards : aqueous humour, crystalline lens, and vitreous humour. The iris and crystalline lens serve to divide the interior of the eye into two chief chambers, a small anterior chamber containing the aqueous humour, and a large posterior chamber containing the vitreous humour.

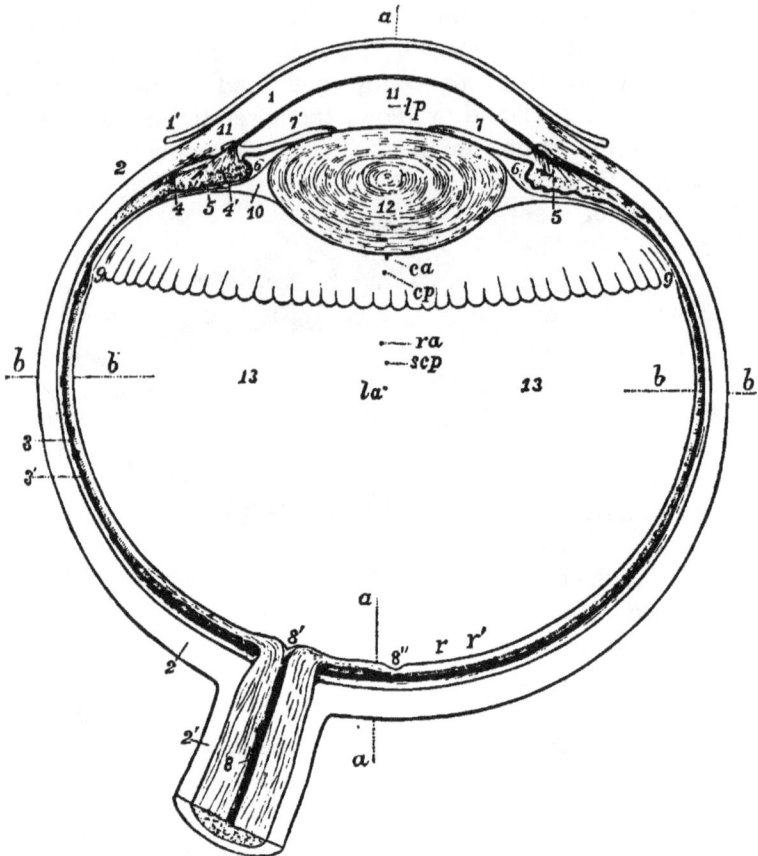

FIG. 218.—View of the Lower Half of the Right Adult Human Eye, divided horizontally through the middle. Magnified four times. (Allen Thomson.)

1, cornea; 1', conjunctiva; 2, sclerotic; 2', dural sheath of the optic nerve passing into the sclerotic; 3, 3', choroid; 4, 4', ciliary muscle; 5, ciliary process; 6, placed in the posterior division of the aqueous chamber, in front of the suspensory ligament of the lens; 7, 7', the iris; 8, central artery of the retina; 8', colliculus of the optic nerve; 8'', fovea centralis; 9, ora serrata; 10, so-called canal of Petit; 11, aqueous chamber; 12, lens; 13, vitreous humour; *a, a, a,* axis of the eye; *b, b, b, b,* equator. It will be observed that from the pupil being placed nearer the inner side of the axis of the eyeball, *a, a,* does not pass exactly through the centre of the pupil; this line falls also a little to the inner side of the fovea centralis. The following letters indicate the centres of the curvatures of the different surfaces, assuming them to be nearly spherical, viz.: *c a,* of the anterior surface of the cornea; *c p,* posterior surface; *l a,* anterior surface of the lens; *l p,* posterior surface; *s c p,* posterior surface of the sclerotic; *r a,* anterior surface of the retina.

220. **The Sclerotic and Cornea.**—The sclerotic coat is the strong, dense, fibrous membrane forming the posterior five-sixths of the external coat of the eye. Its external surface is white (white of the eye), and receives the insertion of the recti and obliqui muscles ; its inner surface is of a brown colour, and connected by fine cellular tissue to the outer surface of the second coat. The sclerotic is composed of white fibrous tissue, with some fine elastic fibres and numerous connective-tissue corpuscles. Behind, it is pierced by the optic nerve a little to the inner or nasal side, the fibrous sheath of the nerve becoming continuous with the sclerotic. Its hinder part around the

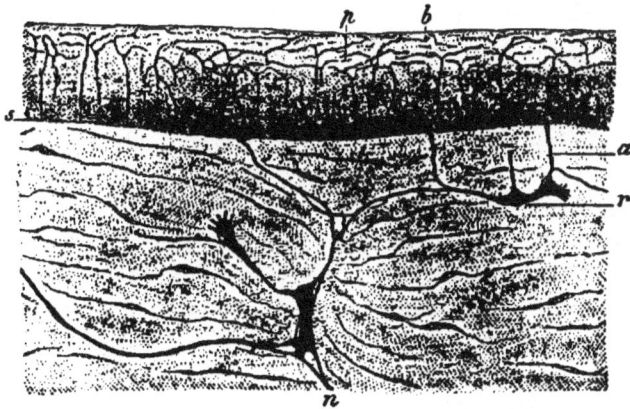

FIG. 219. · Nerve Fibrils in Cornea and Conjunctival Layer.

optic nerve is also pierced by small arteries and nerves—ciliary arteries and nerves—distributed to the sclerotic, choroid, and iris. A small artery for the retina passes in through the middle of the optic nerve, its branches being distributed in the inner layers of the retina only.

The *cornea* is the transparent circular membrane forming the fore part of the outer tunic, and continuous with the sclerotic, which overlaps its margin as the rim of a watch-case overlaps the watch-glass. It projects forward beyond the curvature of the sclerotic, and in old age becomes flattened. Though only about $\frac{1}{25}$ inch in thickness, it is found to consist of five layers :—

(1) A stratified layer of epithelial cells or conjunctival layer, continuous with the epithelium of the conjunctiva lining the eyelids.

(2) An anterior elastic lamina immediately underneath the epithelial cells of the conjunctiva.

(3) The cornea proper, consisting of about sixty thin plates or lamellæ of fibrous connective tissue continuous with that of the sclerotic, and containing spaces in which are branched cells, the corneal corpuscles.

(4) A posterior elastic lamina, forming a thin homogeneous membrane.

(5) A layer of flattened cells lining the anterior chamber of the eye.

The cornea contains no blood-vessels except fine capillaries at the circumference, nutrition being effected by the passage of lymph through the branched spaces in which the corneal corpuscles lie. From the ciliary nerves medullated fibres enter the cornea at the circumference, and losing their sheaths, break up into networks or plexuses of axis-cylinders ; from one of these networks delicate fibrils pass up into the epithelium and end towards the surface between the cells as free fibrils, in a manner similar to the free nerve endings in the epithelium of the skin. It is this nerve supply to the conjunctival epithelium that makes the eye so sensitive to particles of dust, &c. Running round the margin of the cornea in the sclerotic is a small lymphatic channel termed the canal of Schlemm.

221. **The Choroid and Iris.**—The choroid is the tunic between the sclerotic and the retina, and like them found on the posterior five-sixths of the eyeball. It consists chiefly of blood-vessels—the ciliary arteries and veins—connected together by elastic connective tissue, in which lie large stellate corpuscles containing a dark pigment. These pigment cells of the choroid serve to assist in absorbing the light entering the eye. The internal part of the choroid coat contains a dense layer of capillaries derived from the arteries of the outer part, resting on the transparent membrane of Bruch, and these serve to nourish the underlying pigment epithelium of the retina. The choroid coat is continued in front into the iris, but before union it forms a number of radiating folds, or plaits, called the *ciliary processes*. They form a sort of plaited frill behind the iris and round the margin of the crystalline lens. The inner retinal coat terminates in front, where the ciliary processes begin, by a notched edge termed the *ora serrata*, but the retina is represented over the ciliary processes by two layers

of cells termed the *pars ciliaris retinæ*, or ciliary portion of the retina. The *ciliary muscle* is situated between the outer sclerotic at its junction with the cornea and the folds of the ciliary processes. It arises by a small tendon from the inner surface of the sclerotic, and its plain fibres pass partly radially backwards (meridional fibres), to be inserted into the choroid opposite the ciliary processes, and partly towards the iris in a circular course round its insertion. The ciliary muscle forms

FIG. 220.—Choroid Membrane and Iris exposed by the removal of the Sclerotic and Cornea. (After Zinn.) Twice the natural size.

a, part of the sclerotic thrown back ; *b*, ciliary muscle ; *c*, iris ; *e*, one of the ciliary nerves ; *f*, one of the vasa vorticosa or choroidal veins.

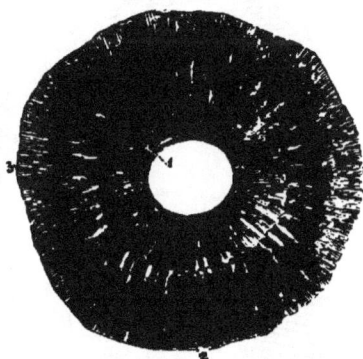

FIG. 221.—Ciliary Processes as seen from behind. Twice the natural size.

1, posterior surface of the iris, with the sphincter muscle of the pupil ; 2, anterior part of the choroid coat ; 3, ciliary processes.

a ring of plain muscular fibres (seen in section in fig. 237) round the eye, between the sclerotic and ciliary processes, and its contraction draws the choroid forwards, and thus relaxes the suspensory ligament of the crystalline lens, so that the lens becomes more convex, and the eye accommodated to vision at different distances (see par. 228).

The iris is the thin, circular, coloured curtain suspended in the aqueous humour behind the cornea and in front of the crystalline lens. It is perforated by a circular aperture, the

pupil, for the admission of light.   At its circumference it is
connected with the choroid, and in front of this it is united
to the cornea by a network of fibres, termed the pectinate

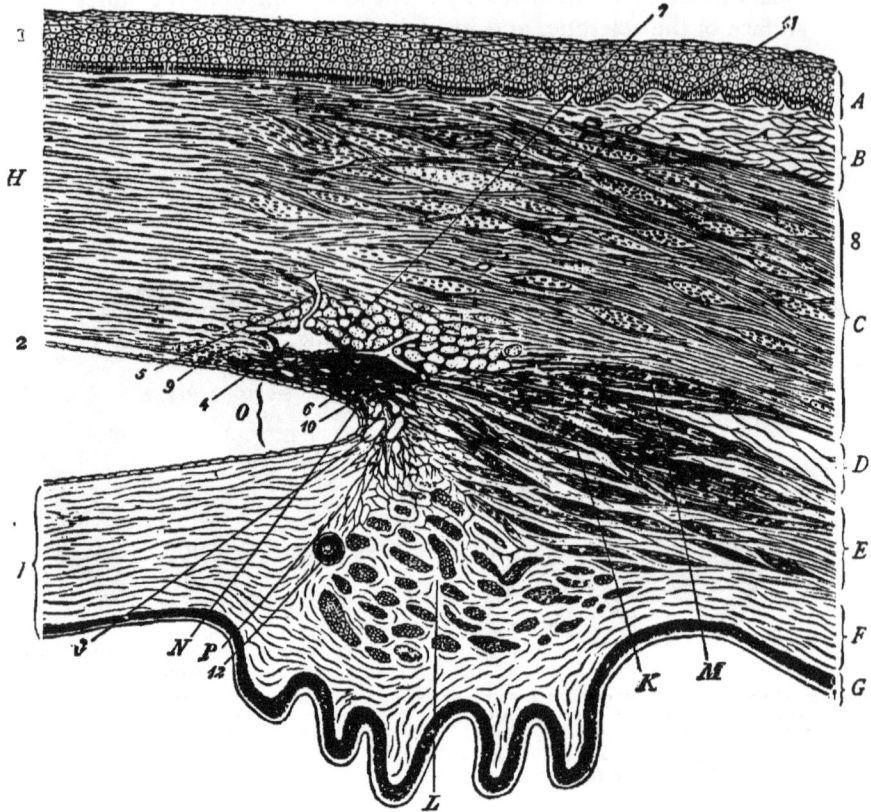

FIG. 222.—Section from the Eye of a Man (aged 30), showing the relations of the
Cornea, Sclerotic, and Iris, together with the Ciliary Muscle, and the Cavernous
Spaces near the Angle of the Anterior Chamber. (Waldeyer.)   Magnified.

A, epithelium ; B, conjunctival dermis ; C, sclerotic ; D, supra-choroid space and laminæ ;
E, opposite the ciliary muscle ; F, choroid, with ciliary process ; G, pars ciliaris
retinæ ; H, cornea ; I, iris ; K, radiating and meridional, and L, circular or annular
bundles of the ciliary muscle ; M, bundles passing to the sclerotic ; N, ligamentum
pectinatum iridis at the angle, O, of the anterior chamber ; P, line of attachment of the
iris ; 1, anterior homogeneous lamina of the cornea ; 2, posterior homogeneous lamina,
covered with epithelial cells which are continued over the front of the iris ; 3, cavernous spaces at the angle of the anterior chamber (spaces of Fontana) ; 4, canal of
Schlemm, with epithelial lining, and with a vessel, 5, leading from it ; 6, other
vessels ; 7, bundles of fibres of the sclerotic having a circular direction, cut across ;
8, larger ones in the substance of the sclerotic ; 9, fine bundles cut across, at limit of
cornea ; 10, point of origin of meridional bundles of ciliary muscle ; 11, blood-vessels
in sclerotic and conjunctiva, cut across ; 12, section of one of the ciliary arteries.

ligament. In structure the iris consists of a fibrous stroma of connective tissue containing muscular fibres, blood-vessels, and nerves. The back of the iris is lined with two layers of dark pigment cells (the uvea), pigment cells being also scattered through its substance, and according as the pigment is more or less abundant the iris has a brown, grey, or blue colour. The pupil is black because this pigment and that of the choroid and retina absorb the light entering the eye, so that little is reflected out. In the stroma or framework of the iris is found a layer of circular unstriped muscular fibres, forming the *sphincter* of the iris, the contraction of which will lessen the pupil ; and some authorities find another layer of plain muscular fibres radiating towards the circumference, the *dilatator* of the iris, the contraction of which widens the pupil.

222. The Retina. —The retina is the delicate transparent membrane that forms the inner tunic of the eye, and upon it the images of external objects are received.

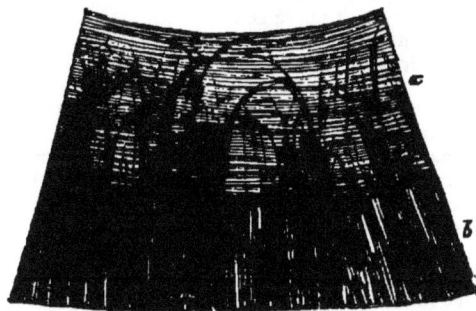

Fig. 223.—Segment of the Iris, seen from the Posterior Surface after Removal of the Uveal Pigment. (Iwanoff.)

a, sphincter muscle ; *b*, dilatator muscle of the pupil.

It extends forwards nearly as far as the ciliary process, where it terminates in an indented border, the *ora serrata*, though beyond this it is represented to the tips of the ciliary processes by a thin layer of different structure containing no nerve fibres, the *pars ciliaris retinæ*. The thickness of the retina diminishes from behind forwards, from $\frac{1}{30}$th to $\frac{1}{200}$th of an inch. In the fresh eye it is translucent and of a pale pink colour, but after death it becomes opaque. On examining the concave inner surface of the retina there may be seen, in a line with the axis of the globe, an elliptical yellow mark about $\frac{1}{20}$ inch in diameter, termed the *macula lutea*, or yellow spot, and in the centre of this a slight white hollow termed the *fovea centralis*. About

$\frac{1}{10}$th of an inch to the inner side of the yellow spot, and there-fore on the nasal side of the axis, is the entrance of the optic nerve, which is marked by a pale round disc, termed the *optic pore*. In the centre of the optic pore is the point from which the central artery of the retina branches to its inner layers.

Though so thin, microscopic examination of vertical sections of the retina shows that there are in it eight layers or strata, together with certain fibrous structures which pass through the membrane and connect the several layers together. These layers, but not the connecting fibres, are shown in fig. 224. The names of the layers from within the eye outwards, that is, starting with the one on which the light first falls, are named as follows :—

1. Layer of nerve fibres.	5. Outer molecular layer.
2. Layer of nerve cells.	6. Outer nuclear layer.
3. Inner molecular layer.	7. Rod and cones.
4 Inner nuclear layer.	8. Pigment cells.

The two series marked limiting membranes are not really membranes, but merely the boundary lines of certain long supporting fibres of the retina, termed the *fibres of Müller*. These fibres begin by expanded bases that form the internal limiting membrane, and pass through the layers of the retinal elements to the bases of the rods and cones. Each fibre as it passes through the inner molecular layer gives off processes, and here too is a nucleated enlargement that shows the original cell nature of the fibre. Other nervous-supporting or neuroglial elements occur as small cells in the various layers.

The optic nerve, composed of more than 500,000 fibres, enters the eyeball a little on the inner side through perforations in the sclerotic and choroid. Its sheaths blend with the sclerotic and choroid, and its fibres losing their medulla pass as naked axis-cylinders to the inner surface of the retina, where they radiate from the optic pore to form the 'layer of nerve fibres' of the retina. They extend as far forward as the ora serrata, some of them bending backwards to end in the inner nuclear layer, but they are absent from the yellow spot. Outside the network of fibres, over the greater part of the retina, is a single layer of ganglionic nerve cells, one process of which is continuous with a nerve fibre. The cells are more numerous, and in several layers at the yellow spot around the fovea.

The inner molecular layer consists of a granular substance in which processes of the ganglionic corpuscles lose themselves. The inner nuclear layer is mainly composed of bipolar nucleated nerve cells, with processes extending into the inner molecular layer in one direction, and through the outer molecular layer as far as the external limiting membrane in the other. The thin outer molecular layer consists chiefly of the inner granules of the rod and cone fibres. Up to this layer the retina may be said to consist of nervous elements ; beyond it is composed of modified epithelium cells.

Layers 6 and 7, termed the *outer nuclear layer* and the *layer of rods and cones*, or Jacob's membrane, are properly one layer, the elements being continuous through the two. It is the sensory or nerve epithelium layer of the retina, and consists of elongated nucleated cells of two kinds. The most numerous cells are rows of rod-like structure set side by side, the rod proper being continued towards the outside in the outer nuclear layer by a fine fibre which swells out near the middle into a nucleated enlargement,

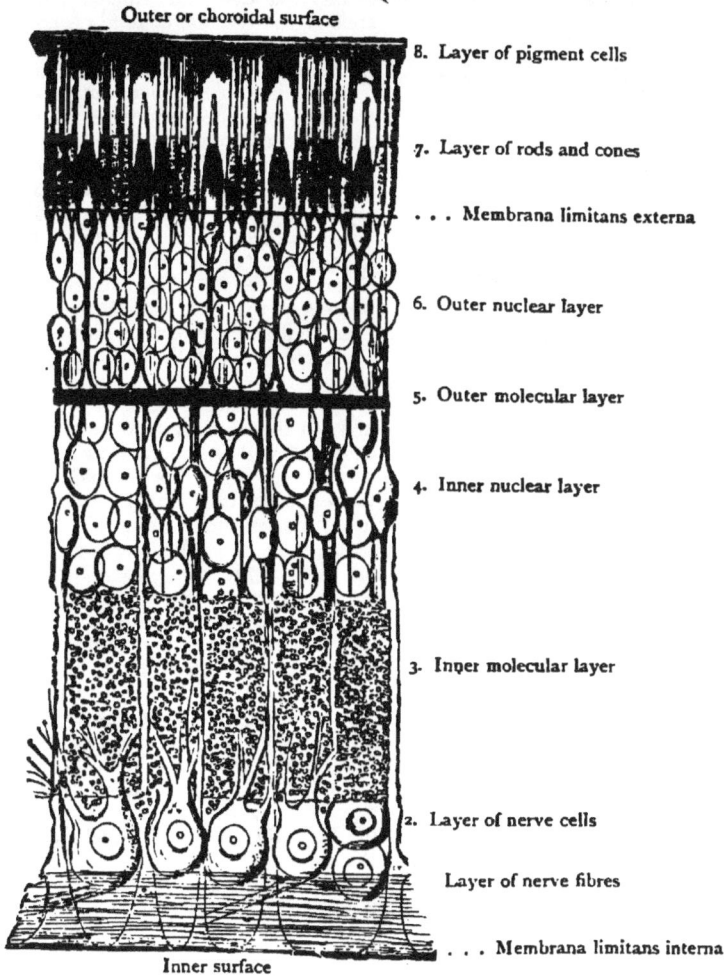

Outer or choroidal surface

8. Layer of pigment cells

7. Layer of rods and cones

. . . Membrana limitans externa

6. Outer nuclear layer

5. Outer molecular layer

4. Inner nuclear layer

3. Inner molecular layer

2. Layer of nerve cells

Layer of nerve fibres

. . . Membrana limitans interna

Inner surface

FIG. 224.—Diagrammatic Section of the Human Retina. (Schultze.

and then passes on to terminate in the outer molecular layer (fig. 224). The rod proper consists of two parts, an *outer* part striated across and cleaving into discs, and an *inner* thicker part striated in part lengthwise. The cone elements consist of a tapering part swelling into a spindle-shaped enlargement, that is prolonged as a thick fibre through the outer nuclear

A A

layer, to terminate by expanded filaments in the thin outer molecular layer. The outer limb of the cones is much shorter than that of the rods, but like that of the rod is transversely striated, and may be made to split into discs. Though the rods are more numerous than the cones, in the centre of the yellow spot there are cones only.

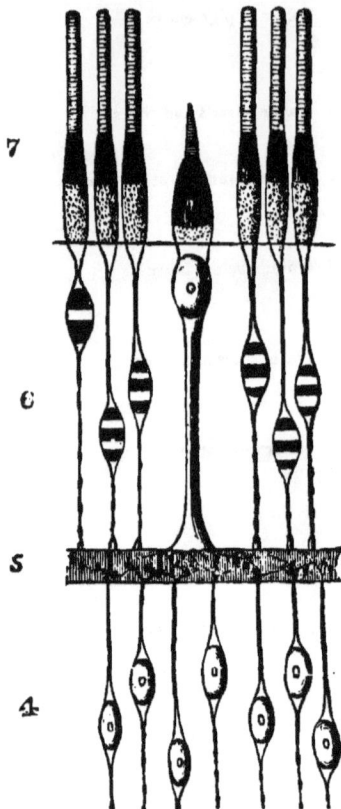

FIG. 225.—Diagrammatic Representation of some of the Nervous and Epithelial Elements of the Retina. (After Schwalbe.)

The designation of the numbers is the same as in fig. 224.

It is believed that the inner ramified ends of the rod and cone fibres come into contact with the fine branches of the fibres from the small bipolar cells forming the inner nuclear layer, and ·that the branched processes of the ganglion cells come into contact with the opposite processes of the small bipolar cells in the inner molecular layer. But no anatomical continuity between these several elements has been found, ' merely an interlacement of ramified fibrils.' The axis-cylinder process of the ganglion cell is, however, in direct connection with one of the nerve fibres of the layer of nerve fibres.

The outermost layer of the retina next to the choroid, and sometimes described with the choroid, as it often comes away when the choroid is detached, consists of a single stratum of hexagonal epithelial cells containing black pigment. They are present in all parts of the retina, except at the entrance of the optic nerve. The outer surface of the cells is smooth and flat, but the inner part is prolonged, on exposure to light, into fine processes that extend between the rods. The pigment granules lie in the inner part of the cell, and, after exposure to light, extend along the prolonged cell processes, where, by their agency, the *rhodopsin* or visual purple of the retina becomes developed in the outer part of the rods. There is no pigment in the cones. Visual purple is bleached by exposure to light, and the function of the pigment cells appears to be to restore the purple colouring matter after being bleached by light. Light falling on the retina causes the processes of the pigment cells to pass inwards between the rods; in the dark the processes are retracted. (The brownish-yellow pigment of the macula lutea lies in the inner layer and not in the cones, so that the central fovea is quite clear.)

Certain *variations in structure in the different parts of the retina* must now be noted. At the optic pore, where the optic nerve enters, fibres

alone are present, and this part is insensitive to light, and called the
'blind spot' (fig. 229). In the central depression, or *fovea centralis* of the
macula lutea (yellow spot), there is nothing below the pigmented epithe-

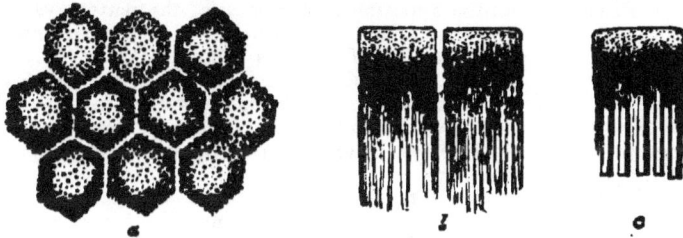

FIG. 226.—Pigmented Epithelium of the Human Retina. (Max Schultze.) Highly
magnified.

a, cells seen from the outer surface with clear lines of intercellular substance between;
b, two cells seen in profile with fine offsets extending inwards; *c*, a cell still in connection with the outer ends of the rods.

lium but long slender cones with their oblique prolongations, the cone
fibres, in the outer nuclear layer, all the other retinal elements having
gradually thinned away (fig. 227). About 7,000 cones are said to exist in

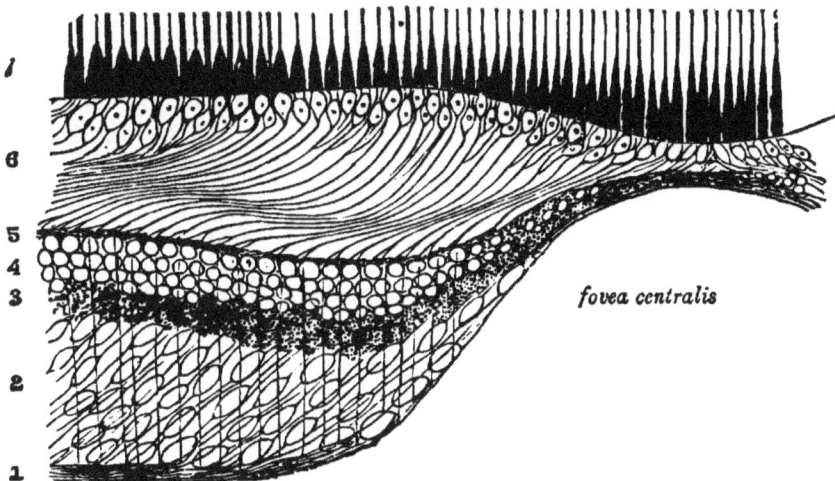

FIG. 227.—Vertical Section through the Macula Lutea and Fovea Centralis; diagrammatic. (After Max Schultze.)

1, nerve layer; 2, ganglionic layer; 3, inner molecular, 4, inner nuclear, and 5, outer
molecular layers; 6, outer nuclear layer, the inner part with only cone fibres forming
the so-called external fibrous layer; 7, cones and rods.

the fovea. The portion of the yellow spot around the macula lutea is
marked by its greater thickness, by the numerous bipolar ganglionic cells
arranged in layers, and by the large number of cones that it contains

compared with rods. Near the yellow spot the retina contains one cone to four rods ; midway to its termination at the ora serrata one cone to twenty-four rods ; at the peripheral part rods only. When the retina is examined from its outer surface, the pigment cells being detached, the ends of the rods and cones present the appearance of a mosaic, the pattern varying with the part of the retina under examination. This is well shown, together with the relative number of the rods and cones, in fig. 228. It has already been noted that the layer of nerve fibres becomes gradually thinner towards the anterior part of the retina. At the ora serrata the nerve fibres, ganglion cells, and rods have disappeared, and over the ciliary process the retina proper ceases, there being only the layer of cells spoken of as the *pars ciliaris retinæ.*

A central artery passes through the optic nerve to supply the retina with blood. On reaching the inner surface it divides into two branches, an upper and lower, and these subdividing pass outwards to be distributed as capillaries in the inner four retinal layers. The outer retinal layers, including the outer molecular layer and layer of rods and cones, have no blood-vessels. The capillaries pass into veins that follow the same distribution as the arteries. The blood-vessels of the retina have no connection with those of the choroid, except near the entrance of the optic nerve (par. 226).

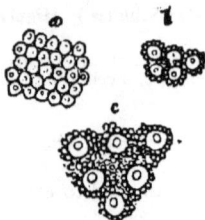

FIG. 228.—Outer surface of the Columnar Layer of the Retina. (Kölliker.) 350 Diameters.

a, part within the macula lutea, where only cones are present ; *b,* part near the macula, where a single row of rods intervenes between the cones ; *c,* from a part of the retina midway between the macula and the ora serrata, showing the preponderance of the rods.

223. **The Evidence that Visual Impression begins in the Rods and Cones.**—Visual impulses begin in the rods and cones on the outer side of the retina, after the rays of light have passed through most of the retinal layers, and the processes started in these sensory epithelial cells of the retina pass back to the layer of fibres on the inner surface of the retina and thence by the optic nerve to the brain. That it is the outer layer of rods and cones that is thus sensitive to light is proved by the following :—

1. At the point of entry of the optic nerve there is abundance of nerve fibres, but light falling on these fibres produces no sensation. It is the 'blind spot.' Its existence can be shown by placing on a piece of white paper a small black cross and a large black spot about 4 inches apart. On shutting one eye and holding the paper 12 to 16 inches away, with the black spot on the outer side, it will be noted that when the cross is looked at the blot disappears, owing to

its image falling on the insensitive blind spot. By using a movable spot, as a quill dipped in ink, and noting exactly the distance of the eye from the white sheet, and the amount of movement of the black spot from visibility to non-visibility, the position of the blind spot with respect to the yellow spot, and its size also, can be calculated.

2. The yellow spot is the region of most distinct vision,

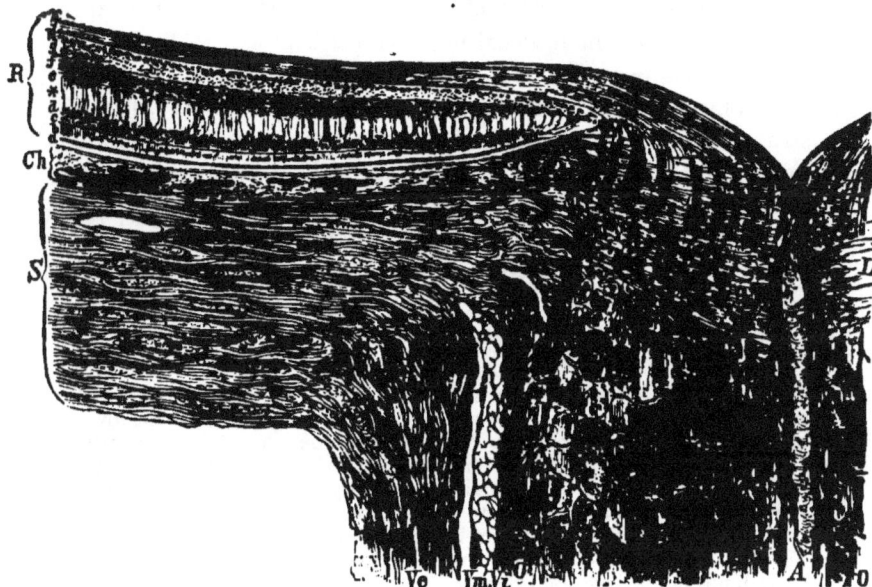

FIG. 229.—Section through the Coats of the Eyeball at the Point of Entrance of the Optic Nerve. (Toldt.)

I'e, dural sheath ; *I'm*, arachnoidal sheath, and *I'i*, pia-matral sheath of the optic nerve, with lymphatic spaces between them ; *O, O,* funiculi of the nerve ; *L,* lamina cribrosa ; *A,* central artery ; *S,* sclerotic ; *Ch,* choroid ; *R,* retina. The small letters refer to the various parts of the retina, *b* being the layer of rods and cones, and *i* that of nerve fibres.

and the spot upon which objects are imaged when they are looked at ; but its central part consists of cones only, no fibres.

3. If a small lighted candle is moved to and fro close to one eye in a darkened room, while the eyes look steadily forward into the darkness, dark branching lines called Purkinje's network are seen on a dull-red ground. These are the

shadows of the retinal blood-vessels thrown upon the sensitive layer of the retina. Various other modes of producing these shadows are known. Now we know that the retinal vessels are distributed in the inner layers (nerve fibres and ganglionic cells) of the retina near the vitreous humour, and the shadows cast behind them must be perceived by something posterior to these vessels. This is a clear proof that the external layers of the retina nearest the choroid, that is, the rods and cones, are the elements in which visual impressions begin.

It thus appears that the real end-organs of vision, the rods and cones, must be in some way connected functionally, if not structurally, with the nerve filaments that pass to the optic nerve, and it is evident that these rods and cones, being backwards from the light towards the sclerotic, must receive the light waves after they have passed through the internal layers of the retina, except at the fovea, where, all the other layers having thinned off, the basal fibres of the cones themselves are directly exposed to the light waves.

It is not possible to say whether the cones or the rods are most essential and important, for although there are only cones in the fovea centralis, these are of a more rod-like character than in other parts. Moreover, cones are absent in some animals ; rods are absent in others. Nor do we know how the undulations of light become converted into nervous impulses that give rise to visual sensations. It has been thought that this may be done by the light bringing about chemical changes in the pigment. Light does produce changes in pigment, and we know that the outer limbs of the rods are tinged with a pigment termed visual purple, or rhodopsin, derived from the pigment cells of the outer layer of the retina. If an animal be killed in the dark and its head fixed in front of a window, the pattern of the window will be bleached upon the retina, the rest of the retina remaining purple. With a 4 per cent. solution of alum this pattern can be fixed, and thus serve as a photograph or 'optogram' taken by the retina. Yet visual purple can hardly be essential to vision, as it is absent from the cones of the fovea, and entirely wanting in some animals that see well.

224. **The Refractive Media of the Eye.**—The **vitreous humour** occupies about four-fifths of the eyeball, its convex hinder surface resting upon the retina, while into its concave anterior surface the crystalline lens fits. The vitreous humour is a jelly-like body, composed of water with a little over 1 per cent. of proteid matter and salts, and the fluid appears to

belong to the same system as the aqueous humour, communication taking place through the suspensory ligaments. It contains no blood-vessels in the adult, and must therefore derive its nutrition from the surrounding vascular structures. A narrow canal passes through it, however, from back to front, terminating at the lens capsule. This *canal of Stilling* takes the place of an artery existing in the fœtus. The surface of the vitreous humour is covered everywhere by a thin, glassy membrane, the *hyaloid membrane*, which separates it from the retina behind and the crystalline lens in front. At the ora serrata the hyaloid membrane divides into two layers, a delicate layer passing over the front surface of the vitreous body beneath the lens, a more fibrous layer extending over the ciliary processes, to be attached to the capsule of the front surface of the lens as a *suspensory ligament*, or zonule of Zinn (fig. 237). Other suspensory fibres are attached to the edge and posterior part of the lens capsule.

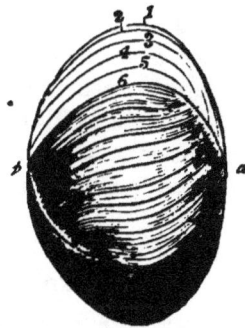

FIG. 230.—Diagram to illustrate the Course of the Fibres in the Fœtal Crystalline Lens. (Allen Thomson.) *a*, anterior; *p*, posterior pole.

The interstices between these fibres form the so-called *canal of Petit*, around the lens.

The **crystalline lens** is a transparent solid body of a biconvex shape, but with the posterior surface more convex than the anterior. It is situated in front of the vitreous body and behind the iris and pupil, and is surrounded by a transparent elastic membrane, strongest in front, termed the lens capsule. To the lens capsule is fused in front the suspensory ligament.

The proper substance of the lens within the capsule consists of a number of concentric laminæ, which, after hardening, may be detached like the coats of an onion, each lamina consisting of elongated ribbon-like fibres with serrated edges. The inner laminæ are closely applied, so as to form a denser core or nucleus. The fibres, which by development are but elongated cells, run from front to back, but are so arranged that no fibres reach from one pole of the lens to the other (fig. 230). The lens contains no blood-vessels, its nutrition being effected by the blood-vessels of the choroid.

The **aqueous humour** is the watery substance found between the cornea and the lens. The space so occupied is divided into two parts by the iris, a large *anterior chamber*, bounded in front by the cornea, and behind by the iris and pupil ; a small *posterior chamber*, formed by the triangular interval at the edge of the lens between the ciliary processes, the iris, and the suspensory ligament of the lens. This posterior chamber is in communication with the anterior chamber between the iris and the lens. Aqueous humour is water with but 1·3 per cent. of solids. It is believed to be a very dilute lymph furnished by the vessels of the ciliary processes, and continually passed from the anterior chamber at the angle of junction of iris and cornea into certain spaces—spaces of Fontana—that communicate with the canal of Schlemm, and thence with the venous system of that region. Both aqueous humour and vitreous humour thus come and go, exercising some nutritive action on the adjacent structures, and being of mechanical use in supporting and keeping tense the coats of the eyeball.

225. **Nerves of the Eye.**—The *optic nerves*, or special nerves of sight, originate from the optic tracts in the mode already described (par. 204), each one passing into the bony orbit of the eye through the optic foramen, and finally terminating in the cup-like expansion of the retina. As each nerve passes through the foramen it receives a sheath from the dura mater, which in the orbit splits into two layers, one forming the periosteum of the orbit, and the other forming the sheath for the nerve until it pierces the sclerotic. Other nerves enter the eyeball, piercing the sclerotic around the optic nerve. These are the *ciliary nerves*, which are mainly branches of the first or ophthalmic division of the fifth cranial nerve. They are accompanied by branches from the sympathetic system. On passing through the sclerotic the ciliary nerves run forward between it and the choroid, and are distributed to the iris and ciliary muscle. Twigs from the first branch of the fifth nerve are also supplied to the cornea, fine fibrils passing between the epithelial cells to the conjunctival layer, which confer on it acute sensibility to foreign par-

ticles. Injury or disease of this branch of the fifth renders the surface of the eyeball insensible to particles of dust and other stimuli. Vaso-motor fibres to the blood-vessels of the choroid, iris, and retina are included in the ciliary nerves.

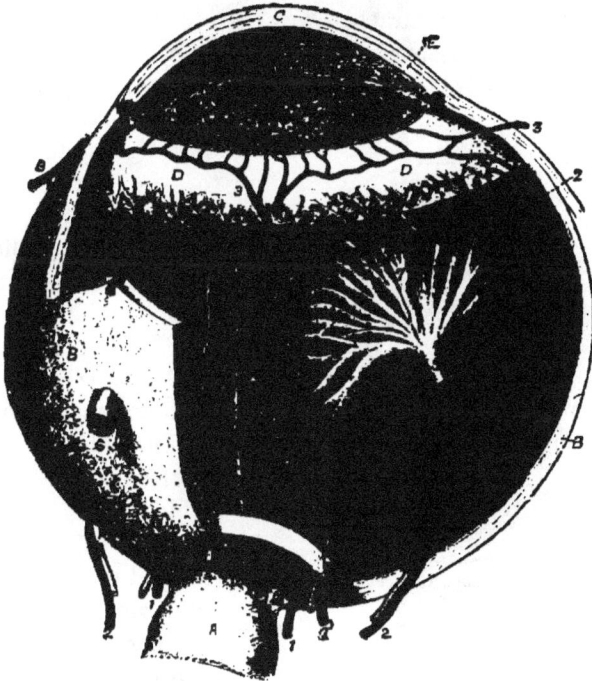

FIG. 231.—Vessels and Nerves of the Choroid and Iris seen from above, the Sclerotic and Cornea being in part removed. (Testut.)

A, optic nerve ; B, sclerotic ; B′, the same in section ; C, section of cornea ; D, ciliary muscle ; E, iris ; F, anterior chamber of the eye ; $1, 1$, short posterior ciliary arteries supplying choroid ; $2, 2$, long posterior ciliary arteries supplying ciliary processes and iris ; $3, 3$, anterior ciliary arteries passing to front of eye with the four recti muscles ; 5, large superior external vorticose vein from the upper and external quarter of the choroid and ciliary processes ; 6, large superior internal vorticose vein as it leaves the choroid ; 7, some of the smaller vorticose veins (the central artery of the retina is not seen in this figure); 4, ciliary nerves passing to blood-vessels, ciliary muscle, and iris.

The chief motor nerves of the eye are the third pair of cranial nerves, or *motores oculi*. Each *motor oculi* supplies all the muscles of the eyeball except the superior oblique and external rectus ; it also sends filaments to the elevator of the eyelid and to the iris and ciliary muscle. Besides controlling the movements of the eyes, therefore, this nerve serves to regulate the amount of light entering the pupil, and brings about accommodation for near objects. A bright light acting through the retina and optic nerve appears to serve as a stimulus to the centre of origin of the third nerve, so that the nerve is thus reflexly excited and the pupil made to contract.

226. **Blood-vessels of the Eye.**—The eye is richly supplied with blood-vessels. The eyelids and glands are supplied from the palpebral and lachrymal arteries, the conjunctival branches passing also to the edge of the cornea. The tunics or coats of the eyeball are supplied by two distinct sets of vessels:—(*a*) The vessels of the sclerotic, choroid, and iris ; (*b*) the vessels of the retina. The first set include the short posterior ciliary arteries, which enter the first part of the sclerotic around the optic nerve, and are distributed to the choroid and ciliary processes ; the long posterior ciliary arteries

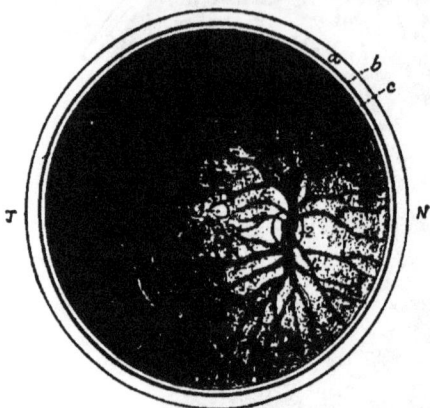

FIG. 232.—Inner Concave Surface of the Retina of the Right Eye.

a, sclerotic ; *b*, choroid ; *c*, retina ; 1, yellow spot (*macula lutea*); 2, optic disc ; 3, 4, 5, 6, branches of the retinal artery, retinal veins coloured blue ; T, temporal side ; N, nasal side.

which enter the choroid behind, and running forward are distributed to the ciliary muscle and to the iris, at the margin of which they form a close vascular circle ; the anterior ciliary arteries, which pierce the front part of the choroid and are distributed to the ciliary processes and iris. From the capillaries numerous veins arise which form a vorticose arrangement on the outer surface of the choroid, and unite for the most part into four large trunks that pass out of the choroid about midway between the cornea and the optic nerve.

The retina is supplied by a central artery which enters at the middle of the optic nerve and gives off branches to supply its inner layers. The capillaries unite to form a central vein. Except near the entrance of the optic nerve the two sets of blood supply to the eyeball are quite distinct.

227. The Formation of an Image on the Retina.—An eye closely resembles a photographic camera filled with water. There is a screen, the retina, at the back of a darkened chamber, from which little light can be reflected to blur the image which is formed by a lens with a spherical surface placed in front. A stop or diaphragm, the iris, cuts off outside rays which would also tend to mar the image. The formation of an image by the eye and by the camera is somewhat roughly illustrated in figure 233.

For a full account of the formation of images by lenses and other bodies, the student must consult a work on Optics. We will, however, give a short explanation of the formation of the retinal image.

As long as rays of light travel through the same substance or medium, they proceed in straight lines. When rays of light pass *obliquely* from a medium of low density, as air, into a medium of higher density, as water or glass, such rays are bent or refracted towards the perpendicular separating the two media. Rays falling *perpendicularly* on the boundary surface of the media are not bent or refracted. Where a spherical convex surface is met by ray of light travelling from a medium of low density into a medium of greater density, all the rays falling normal or perpendicular to the surface pass without refraction through the centre of curvature or nodal point. Parallel rays in such a case are all brought to a meeting-point called the principal focus, and the distance of the principal focus from the boundary surface of the two media varies with the amount of curvature. Other rays diverging from a distant point, and not falling perpendicularly, are brought to a meeting-point or focus by refraction at some distance beyond the principal focus.

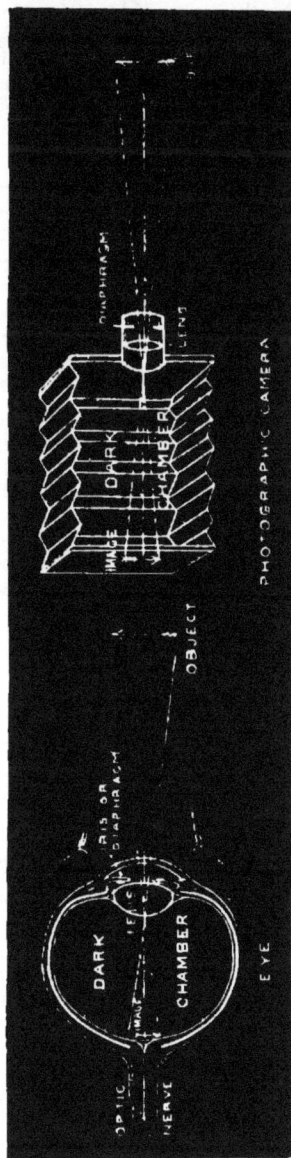

Fig. 233.—The Mechanical Parts of the Eye compared with those of a Photographic Camera.

We may now understand how a simple collecting system forms the image of an object. A simple collecting system consists of two refractive media, separated by a spherical surface. Let P Q represent an object placed in front of the spherical refracting surface seen in section at S S′, C being the centre of curvature, and F the principal focus, or focus for parallel rays. The medium on the right of S S′ is supposed to be the more refractive. The line O C F is called the optic axis. Consider the rays proceeding from the point P. The line P C P′ represents a ray meeting the bounding surface perpendicularly, and passes through the centre of curvature without refraction. The ray P S is parallel to O C F, and will be bent

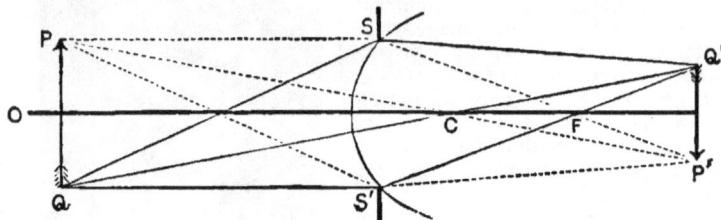

FIG. 234.

so as to pass through F. On prolonging it to meet the unrefracted ray it cuts the latter at P′, where all other rays, as P S′, are also brought to a focus by refraction. At P′, there is then formed an image of the point P by the diverging pencil of rays from P. In a similar way the diverging pencil of rays from Q is brought to a focus, and forms an image of Q at Q′. Points intermediate between P and Q of the object send out rays which are brought to a focus at corresponding points between P′ and Q′. Thus the image P′ Q′ of the object P Q is formed, and might be received on a suitable screen.

In the eye, however, it will be said that there are several different sur-

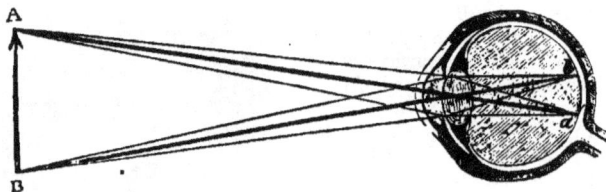

FIG. 235.

faces separating different media, where refraction takes place. This is quite true, for taking the refractive power of the aqueous humour to be the same as that of the cornea, we still have three refracting surfaces—anterior surface of cornea, anterior surface of lens, and posterior surface of lens ; and three media cornea with aqueous humour, the substance of the lens, and the vitreous humour. At each of these surfaces refraction will take place, most at the anterior surface of the cornea, and none at the anterior surface of the biconvex crystalline lens. But by suitable calculation this complicated optical apparatus can be reduced to one mean curved surface of known curvature, and one mean medium of known refractive power. There is thus constructed a simplified or *reduced* eye, which will act and form an image as in the simple collecting system above explained, where S S' may be considered to be the one surface where refraction occurs, and P' Q' the image formed on the retina (see also fig. 235).

It will now be understood how the eye, like a simple collecting system, forms an image of an external object in the field of view, and that this image, owing to the crossing of the rays within the eye, is an inverted image. The size and position of the image of any object depend on the distance of the object, and can readily be found by calculation. But it will easily be seen, without calculation, that as the object is moved the position of the image changes (this may be easily seen if the reader will draw fig. 234. and put the arrow P Q at different distances from the curved surface S S'). In other words, the focal distance, or distance from the refracting surface at which the luminous rays are collected to form an image, varies with the distance of the object from this surface, being less as this is longer and *vice versâ*. Hence there must be some means of adjusting the eye for vision at different distances. A photographer can alter the distance of his sensitive screen from the refracting lens, and thus focus for distant or near objects. This cannot occur in the rigid eyeball, and the minute change of this character that would be required in the eyeball does not seem practicable. Altering the curvature of the refracting surface, however, produces a change in the distance at which luminous rays are brought to a focus so as to form a distinct image, increase of curvature increasing the refracting power, and bringing rays to a focus sooner, and decrease of curvature having the opposite effects.

228. **Accommodation.**—The power by which the eye is enabled to form distinct images on the retina of both distant and near objects is called 'the power of accommodation.' The *need* for accommodation is evident when we consider the mode in which images of objects at various distances are formed, and when we remember that we cannot see distinctly both a near and a distant object at the same time. In the normal eye at rest it is accommodated for distant objects, the rays from which are sensibly parallel. This is proved by the fact that such objects are seen without any effort quite clearly, so that when the mechanism for accommodating the eye is paralysed, and near objects cannot be clearly seen, the images of distant objects still remain distinct. But for distances less than seventy yards, an effort of accommodation is needed. How is this accommodation effected? It is not by altering the position of the retina, as the globe of the eye has been proved not to alter its shape. It is not by altering the curvature of the cornea, as was once thought, for the eye can be accommodated under water, which has practically the same refractive power as the cornea. It has been definitely proved

that the anterior surface of the crystalline lens undergoes a change of curvature when looking at near objects, the increased curvature being necessarily accompanied by increased refractive power. This increased curvature of the anterior surface of the crystalline lens in the act of accommodation is proved by observing 'Sanson's images.' If a lighted candle be held in a dark room, a little to the side of a person's eye, an observer looking at the other side of the eye will see with care three *reflected* images of the flame. (i) A small, erect, and clear image formed by the anterior surface of the cornea. (ii) A larger, but fainter image, also erect, formed by the anterior surface of the lens. (iii) A small, inverted, and faint image formed by the back surface of the lens. If, while the observer

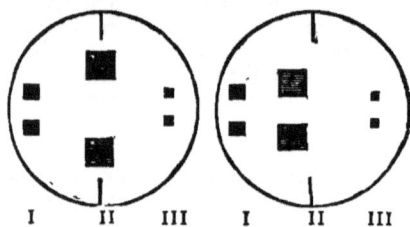

FIG. 236.

With relaxed accom-   With accommodation.
modation.

I, from surface of cornea ; II, from anterior surface of lens ; III, from posterior surface of lens.

is watching these images (it is better to have three pairs of images produced by allowing the light to pass through two holes placed one above the other in a piece of cardboard interposed between the candle and the eye), the person experimented on looks far and then near in one line of vision, it will be noted that with accommodation to a near object II grows smaller and approaches I. This observation can only be explained by the anterior surface of the lens becoming more convex and bulging forwards. Images I and III neither move nor alter, and therefore the cornea and posterior surface of the lens neither move nor alter their curvature.

The change in the anterior surface of the crystalline lens is brought about by the action of the ciliary muscle. When the eye is in the condition of rest or relaxed accommodation, the lens is kept somewhat flattened in front by the tension of the suspensory ligament or zonule of Zinn, which passes forward all round from the ciliary processes of the choroid to its attachment to the margin of the lens. The ciliary muscle springs from a fixed point at the junction of the cornea and sclerotic, and its smooth fibres pass back to be inserted upon the ciliary and front part of the choroid, so that its contraction pulls forward the movable ciliary processes and choroid (but loosely attached to the sclerotic), and thus slackens the suspensory ligament. This lessens the tension of the ligament on the anterior surface of the lens, which then bulges forward and becomes more convex in virtue of its own elasticity.

Various experiments show that this explanation is correct. Thus, if the point of a needle be passed through the sclerotic into the choroid of an eye, it is noticed that when the ciliary muscle is stimulated the eye-end of the needle moves backward, and therefore the point in the choroid moves forward.

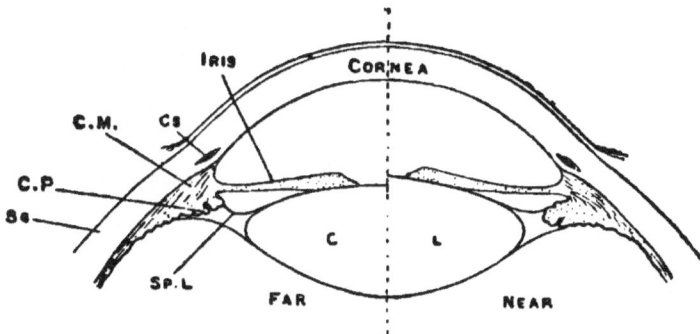

FIG. 237.—Diagram to illustrate Accommodation.
Sc, sclerotic ; C.P., ciliary processes ; C.M., ciliary muscle ; Cs, canal of Schlemm
Sp. L, suspensory ligament ; C L, crystalline lens.

Stimulation of the ciliary nerves, branches from which pass to the ciliary muscle, has been observed to lead to forward movement of the choroid and bulging of the lens.

Accommodation for near objects is associated with convergence of the eyes upon the object viewed, and with movements of the iris that diminish the size of the pupil.

When an object is placed nearer to the eye than 4 inches (10 cm.), it is too near for accommodation, for the rays of light are too divergent to be brought to a focus on the retina. Hence in a normal eye accommodation occurs between an infinite distance, the *punctum remotum*, or remote point, and 4 inches, the *punctum proximum*, or near point. As age advances the near point gets further away, owing to the power of accommodating becoming less in consequence of the lens growing less elastic and the ciliary muscle becoming weaker. Near objects are seen less distinctly, a book being held further and further from the eye. This defect in old people is called *presbyopia.* This kind of long-sighted eye must be distinguished from that due to the eyeball being unusually short.

The luminous rays from all objects beyond 70 yards' distance are virtually parallel. In a normal eye parallel rays are focussed on the retina without any effort. In viewing nearer objects we become conscious of an effort, especially after the distance becomes less than 20 feet, due to the contraction of the ciliary muscle in accommodating. A normal eye is said to be *emmetropic,* or in measure.

In a *myopic* eye distant objects are indistinctly seen, and objects nearer than 5 inches are plainly visible. Such persons are said to be 'short-sighted.' The eyeball is too long—its accommodation being normal—so that distant objects are brought to a focus in front of the retina. In a

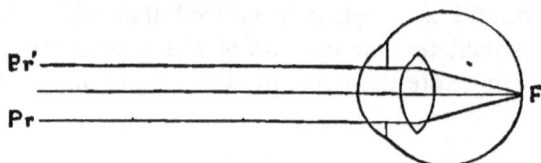

FIG. 238.—Emmetropic Eye. Parallel rays focussed on retina.

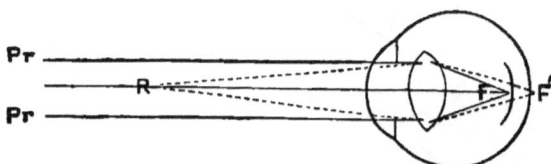

FIG. 239.—Myopic Eye. Parallel rays focussed in front of retina.
ʀ, remote point of distinct V., from which divergent rays are focussed on the retina.

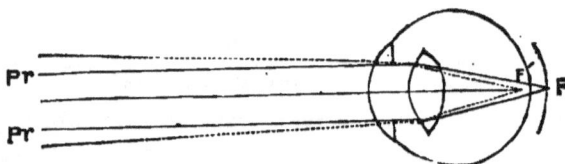

FIG. 240.—Hypermetropic Eye. Parallel rays focussed behind the retina.
Convergent rays focussed on retina.

hypermetropic eye the eyeball is too short, near objects cannot be distinctly seen, and the rays from distant objects are brought to a focus behind the retina unless accommodation is used. Such persons are said to be 'long-sighted.' Spectacles with scattering or diverging lenses (biconcave) are used to remedy *myopia,* or short-sight ; converging or convex lenses are used to remedy *hypermetropia,* or long-sight.

229. Movements of the Iris.—The pupil or central aperture of the iris undergoes variations of size, thus serving to regulate the amount of light entering the eye as well as acting as a diaphragm to cut off marginal rays. No light enters the eye except through the pupil, the dark pigment or uvea absorbing and cutting off others.

The pupil is constricted or lessened, (1) when light falls on the retina, and the brighter the light the greater the constriction, (2) when near objects are viewed to cut off the widely divergent rays that could not be focussed, (3) when the eyeballs are turned inwards, as this is associated with near vision, (4) under the action of certain drugs, (5) during sleep. The pupil is dilated or widened, (1) when the light becomes less, as in passing into darkness, (2) when the eye becomes adjusted for distant objects, (3) when there is an excess of aqueous humour, dyspnœa, or violent muscular effort, (4) when a sensory nerve is strongly stimulated, (5) under the influence of atropin and some other drugs.

Constriction of the pupil is brought about by the contraction of the circular muscular fibres which form the sphincter muscle around the margin of the iris. The contraction of these involuntary fibres is very rapid, unlike that of involuntary muscles generally, and leads of necessity to some extension of the iris inwards with a narrowing of the pupil. Dilatation of the pupil is partly the result of the rebound which occurs when the sphincter fibres relax, and partly the result of contraction of the radiating or dilatator fibres. Some authorities do not recognise dilatator fibres, and attribute dilatation to the removal of tonic constrictor impulses by the inhibitory action of fibres from the sympathetic system. Section of the cervical sympathetic does diminish the aperture of the pupil.

The constriction of the pupil when light falls on the retina is a reflex act. The afferent nerve is the optic nerve, the centre is in the brain in or near the anterior corpora quadrigemina, the efferent nerve carrying the constrictor impulses to the circular fibres of the iris is the third or oculo-motor nerve.

230. **The Muscles of the Eye.**—Besides the *levator palpebræ superioris*, which is the muscle by which the upper lid is raised, there are six muscles of the eyeball, viz. the four straight muscles and the two oblique muscles. The four recti or straight muscles arise behind by a continuous tendinous origin at the bottom of the orbit or bony cavity, and pass forwards, one above, one below, and one on each side of the eyeball, and are inserted by short, membranous tendons into the fore part of the sclerotic coat. The superior oblique or trochlearis muscle springs from the bottom of the orbit, and proceeding towards the front terminates in a tendon that passes through a cartilaginous ring or pulley (*trochlea*) attached to the frontal bones ; the tendon is then reflected backwards and downwards, to be inserted into the upper part of the sclerotic coat, midway between the cornea and the entrance of the optic nerve. The inferior oblique muscle arises at the lower and front portion of the orbit, inclines backwards and upwards, and ends in a tendinous expansion inserted under the external rectus at the outer and posterior part of the eyeball.

B B

The eyeball is turned *outwards* by the external rectus, *inwards* by the *internal* rectus, *upwards* by the combined action of the superior rectus and inferior oblique, *downwards* by the combined action of inferior rectus and superior oblique. Thus, in the lateral horizontal movements the internal and external recti muscles are the sole agents, but the superior and inferior recti, while elevating and depressing the cornea, have both a tendency, from the line of their action being to the inner side of the centre of motion of the eyeball, to produce inward direction as well as movement upwards or downwards. 'The simple action of the superior oblique muscle, when the eye is in the primary position, is to produce a movement of the cornea downwards and outwards; that of the inferior oblique, to direct the cornea upwards and outwards, and in both with a certain amount of rotation. But these movements, caused by the oblique muscles, are precisely those which are required to neutralise the inward direction and rotatory movements pro-

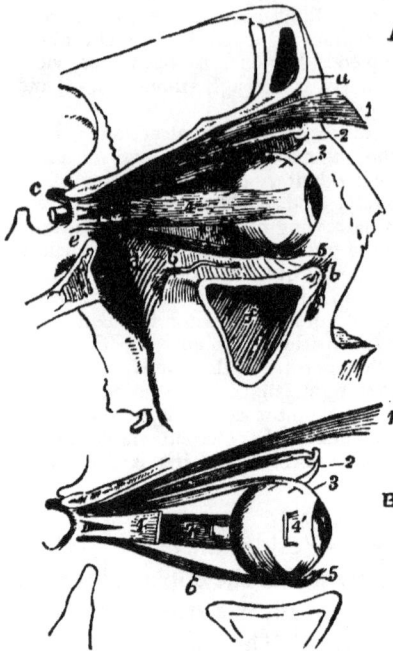

FIG. 241.—A, View of the Muscles of the Right Orbit, from the outside, the outer wall having been removed. (Allen Thomson.) ½.

B, explanatory Sketch of the same Muscles.

orbital arch ; *b,* lower margin of the orbit ; *c,* anterior clinoid process ; *d,* posterior part of the floor of the orbit above the spheno-maxillary fossa ; *e,* side of the body of the sphenoid bone below the optic foramen and sphenoidal fissure ; *f,* maxillary sinus ; 1, levator palpebræ superioris ; 2, pulley and tendon of the superior oblique muscle ; 3, tendon of the superior rectus muscle at its insertion upon the eyeball ; 4, external rectus ; 4', in B, tendon of insertion of the same muscle, a large part of which has been removed ; 5, inferior oblique muscle crossing the eyeball below the inferior rectus ; 6, inferior rectus ; 7, in B, the internal rectus, and near it, the end of the optic nerve cut short close to the place of its entrance into the eyeball.

duced by the superior and inferior recti, and accordingly by the combined action of the superior rectus and the inferior oblique muscles a straight upward movement is effected, while a similar effect in the downward direction results from the combined action of the inferior rectus and superior oblique muscles.'[1] So much for the *straight* movements of the eyeball.

In all oblique movements of direction it is found that two recti act in combination with one oblique muscle. Thus the eye is turned *upwards* and *inwards* by the combined action of the superior and internal recti with the inferior oblique ; *downwards* and *inwards* by the combined action of the inferior and internal recti with the superior oblique ; *upwards* and *outwards* by the combined action of the superior and external recti in combination with the inferior oblique ; *downwards* and *outwards* by the combined action of the inferior and external recti with the superior oblique.

The movements of the eyes are always bilateral. In the upward and downward movements both eyes are turned in the same direction ; in the lateral movements the two eyes may either be directed to the same side, one being abducted and the other adducted, or both may be adducted to bring about the convergence of the visual axes required in near vision. The superior oblique muscle receives its nerve-supply from the fourth or trochlear nerve ; the external rectus is supplied by the sixth or abducens nerve ; all the other muscles are under the influence of the third or common oculo-motor nerve.

231. **Various Definitions and Explanations.**—The *optic axis* is the line which passes through the common centre of curvature (nodal point) and the centre of the cornea. Prolonged backwards it falls on the retina a little to the inner side of the yellow spot.

The *visual axis* or line of vision is the straight line which joins the centre of the yellow spot with the points on which the eye is fixed. These two lines form a small angle with each other.

The *visual angle* is the angle under which an object is seen. It is included between straight lines drawn from the extremities of the object to the nodal point. Its size plainly depends on the size of the object and on its distance from the eye. The visual angle is equal to the retinal angle (angle A *n* B to angle *a n b* in fig. 243 Euclid I. 15), and the base of the retinal angle gives the size of the image formed on the retina. It is evident that objects of different size at different distances may form the same visual angle,

---

[1] Quain's *Anatomy.*

and thus form retinal images of the same size or have the same 'apparent size.' How the real size is estimated will presently be shown. It is also evident that the more distant any object is the smaller will the visual angle and the retinal image become. If we know the size of the object, its distance from the nodal point of the eye, and the distance of the nodal point from the retina, we can calculate the size of the retinal image. Thus if A B, fig. 244 represent a man six feet high at a distance of a mile, the height of the images $a\,b$ on the retina of an observer is given from the properties of the similar triangles $a\,n\,b$ and $A\,n\,B$ thus:—$a\,b : a\,n :: A\,B : A\,n$, that is,

$$a\,b = \frac{a\,n \times AB}{An}.$$

Now $a\,n$ in the reduced eye is about 15 mm. or ·6 inch. Hence we get, reducing all to inches,

$$a\,b = \frac{·6 \times 72}{1760 \times 36} = \frac{1}{1500}$$

inch nearly. In a similar way it may be found that one of these printed letters (about $\frac{1}{20}$th inch in height) will at a distance of one foot form a retinal image of $\frac{1}{100}$th inch, or ·08 mm.

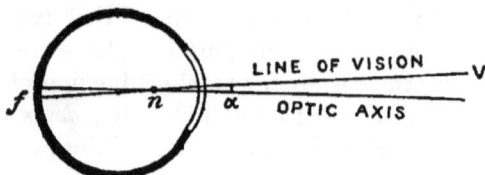

FIG. 242.

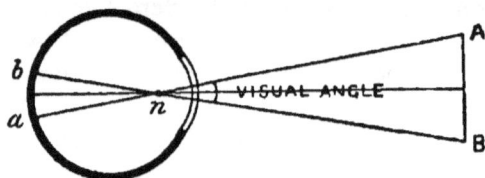

FIG. 243.

Smaller objects than these may be seen at these distances. Acuteness of vision may be tested by ascertaining the smallest visual angle under which two points or two fine lines are distinguished as separate. In most cases the two objects become one when the visual angle is less than 1' or 60". The retinal image corresponding to this is about $\frac{1}{6000}$ inch, or ·004 mm., the distance between two adjoining cones. This implies that the smallest object that could be seen at a distance of 9 inches is $\frac{1}{400}$ inch in diameter, for such an object at such a distance would just form the minimum sensible retinal image. Sharpness of vision diminishes rapidly with distance from

FIG. 244.

the yellow spot. The macula lutea is oval, its long axis being 2 mm., and the fovea centralis has a diameter of ·3 mm.

The *field of vision* is the area taken in by each eye when the head is kept fixed. It is greater for the two eyes than for one, a portion of the field being common to both.

The *horopter* is that imaginary figure in space which comprises all the points images of which fall on 'corresponding' points of the two retinas. Its form varies with different positions of the eyeball. With the head upright and both eyes directed to a very distant point, so that the two visual axes are parallel, it is a plane extending from the observer's feet to the distant point ; with the visual axes converging upon a near object, the horopter is a horizontal circle passing through the object and the nodal points of the two eyeballs.

From any point of the horopter circle, rays are projected on to corresponding points of the retinas shown in the figure (the nasal side of one retina corresponds with the temporal side of the other), and in most cases the image of the object looked at is so small as to lie on corresponding points of the two foveæ. Images of surrounding objects are formed on corresponding

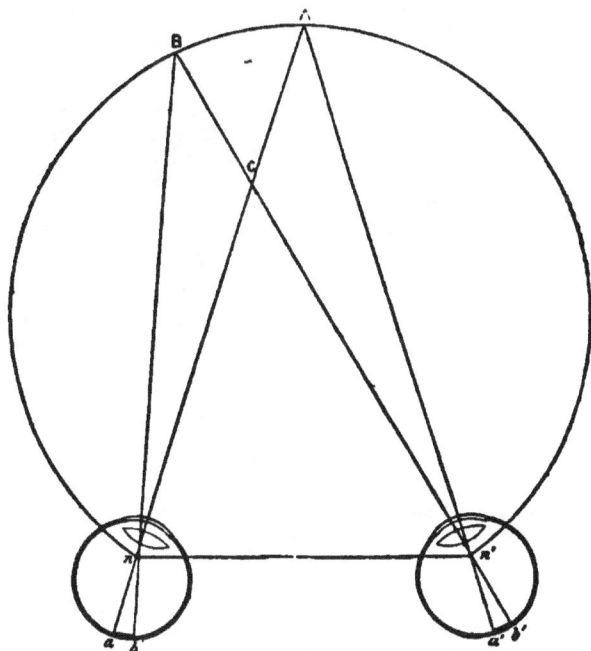

FIG. 245.—Müller's Horopter Circle. (Waller.)

A is the point of regard, and its retinal images are formed on the two yellow spots *a*, *a'*. From B or any other point on the horopter circle images will be formed on corresponding points and will be seen single. By taking a point not on the horopter and drawing lines through *n*, *n'*, the nodal points of the eye, it will be found that the images do not fall on corresponding points, and therefore appear double.

points outside the foveæ, and these being single influence the general view. Other images are formed imperfectly by objects lying in the binocular field but not on any horopter, and not falling on corresponding points. Vision with the fovea centralis is termed 'direct vision ;' vision with the rest of the retina 'indirect vision.' Those images formed within the fovea are alone clearly perceived, others on the surrounding parts being seen indistinctly. Images not falling on corresponding points, or imperfectly focussed images, are neglected or but little heeded by the brain.

On exposing the retina to a very bright object for a time, and then suddenly withdrawing it, the sensation still continues, and is known as the *positive after image.* In a short time the positive image is replaced by another, the *negative after image,* in which the bright parts of the object become dark and the dark parts bright. The positive after image is due to the effect of the stimulus continuing for a short time after it is withdrawn ; the negative after image is caused by the strongly excited portions of the retina or nervous connections becoming exhausted for a time.

It is not possible to say how long a stimulus of light must act on the retina to produce a sensation. It must be extremely short, for the electric flash lasting but a very small fraction of a second enables us to see objects, though rapidly moving objects so seen appear still. The sensation continues after the stimulus is withdrawn for some time, so that if flashes of light follow one another with sufficient rapidity we get a summation of the stimuli, and the successive sensations are fused into one. Thus, whirling a stick with a glowing end gives the sensation of a continuous circle of light. This fusion of visual sensations into one has some resemblance to the mode in which a series of rapidly induced muscular contractions are fused into one prolonged contraction or *tetanus* (par. 36). The duration of a visual sensation after the stimulus is withdrawn is said to be about $\frac{1}{10}$th second, but a strong sensation declines more quickly than a weak one. This is shown by finding that the interval between successive stimuli at which fusion takes place with a faint light is about $\frac{1}{10}$th second, but with a strong light $\frac{1}{40}$th second. A disc with alternate black and white sectors of equal area is made to rotate. With a moderate light it appears of a uniform grey tint when rotated with a speed that causes each pair of sectors to pass a given point in $\frac{1}{10}$th second. If the light be now greatly increased, a flickering is seen owing to imperfect fusion of the sensations, until the speed is increased so as to make each pair of sectors pass much more rapidly.

Certain visual appearances may arise within the eye itself, owing to the fact that the various media through which the light passes are not of uniform transparency, small opaque bodies intercepting the light and throwing shadows on the retina, which are spoken of as *entoptic* pheno-mena. The most common of these are the shadows produced by floating particles in the vitreous humour, the shadows of which produce the impres-sion of small particles floating in space. These are known as *muscæ voli-tantes* (flitting flies). The shadows thrown by the blood-vessels of the inner layers of the retina on the sensitive portion of the retina (the rod and cone layer) behind are entoptic phenomena. (See Purkinje's figures, par. 223.)

We may also note here that although visual sensations are usually brought about by the stimulation of light falling on the retina, other agencies acting upon the retina may give rise to visual sensations. A blow on the eye may so stimulate the retina as to produce flashes of light ; pressure on the eyeball may produce rings of light called *phosphenes* ; and electrical stimulation of the retina awakens sensations of light. Whether irritation of the optic nerve itself produces sensations of light or not is a matter in dispute. However caused, all our visual sensations appear to us to come from external objects, since light from such objects is the normal stimulus. Moreover, the experience gained by this and other senses has taught us to refer the sensation produced to a suitable position in the out-side world. Thus, a luminous impression on the nasal side of the right eye is referred to an object on the outer side of that eye, an image on the lower half of the retina is referred to a source of light above the visual axis, and so on. This explains why we see external objects erect when their retinal images are inverted. We are not aware of the retinal image and do not look at it. The situation of the image on the retina only serves as a hint to the mind to project the various sensations externally in an inverse direction, and the various parts of a retinal image have become associated with muscular sensations set up by moving the eyes and limbs up or down, right or left.

232. Binocular Vision.—A single object forms an image on each retina, and yet we see it as one object under ordinary circumstances. By pressing one eyeball into an unusual position, one object may be made to appear double. Hence 'certain parts of each retina are so related to each other that when an image of an object falls on these at the same time the two sets of sensations excited in the two parts are blended into one.'[1] Parts so related are called 'corresponding' or 'identical' parts,' and the conditions of single vision with two eyes, therefore, are that the images from the various points of an object should fall on corresponding or identical points of the two retinas. These points may be found by tracing the paths of the rays of light from any point to their focus on each retina. By supposing the retinas to be two saucers placed one within the other, so that the two yellow spots coincide, and also the geometrically similar halves (as the right or temporal side of the right retina over the right or nasal side of the left retina), we place all the 'identical' points of one retina over the 'identical' points of the other. The blending of the two sensations which occurs when identical points in each retina are excited by light from an object is probably a mental act due to the experience which we have had of the points acting habitually together, and which has taught us that images falling on corresponding parts arise from a single object. Attempts to explain this remarkable relation of the two eyes by an anatomical arrangement of the optic fibres in their course or termination are not satisfactory. The co-ordinated movements of the eyeballs just described are of such a nature as to bring the images of the objects looked at on identical points of the two retinas, and hence our ordinary vision is single.

The advantages of vision with two eyes are that we get a larger field of vision than with one eye, that we can more exactly estimate the distance and size of objects, and that we are enabled to obtain more easily the perception of depth or solidity. In looking at a fixed point with both eyes one aid in judging of its distance is the greater or less effort required to converge the visual axis upon the point. In looking at a solid object we are assisted in forming our idea of it by the slightly different picture formed on each retina, as well as by the experience gained in estimating the distances of its various points.

Distance is not directly seen by the eye ; it is estimated by means of various visual sensations that serve as signs. With one eye we perceive it imperfectly, as may be shown by suspending a ring in front of a person's face and asking him to put a pencil through it with one eye shut. There is here little but the feeling of accommodation and the size of the retinal image to guide him, and he usually fails. But with two eyes there are additional sensations to help in forming an estimate. There is the muscular feeling caused by the convergence of the eyes, the dissimilarity of the two

---

[1] Foster.

retinal images, the clearness or haziness of the images, all of which serve as signs that we interpret so as to represent to ourselves the amount of movement needed to bring us in contact with the object.

The *size* or *magnitude* of an object is recognised in close connection with that of distance, for our judgment of size depends mainly on the size of the retinal image, and this varies inversely as the distance. Our active touch largely aids in learning real magnitude, and we know that if two objects of the same size have different apparent magnitudes, or are seen under different visual angles, the one with the smaller apparent size is the more distant. If an object of known size intervene, it assists us in forming by comparison an estimate of the size of an unknown object.

Binocular vision is of especial importance in the judgment of solidity. In looking at a solid object the axes of the two eyes are rapidly varying their angle of convergence and their degree of accommodation, in order

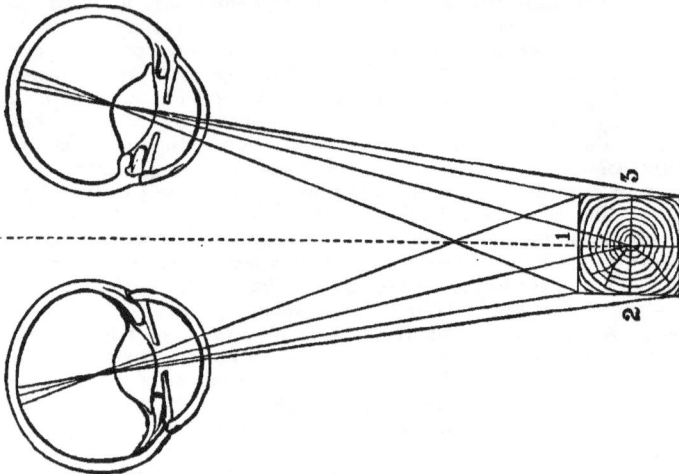

FIG. 246.—Binocular Vision.
The surfaces, 1, 2, and 3, being visible, determine the perception of relief with three diameters.

to see the various points of the object which lie at unequal distances from us. Besides, if we converge the two eyes on the nearest point of the trunk of a tree lying at the distance of a few yards, two dissimilar pictures are formed on the two retinas. ' Thus, in the picture of the right retina there will be imaged some points on the extreme right of the object which are absent in the picture on the left retina. Further, the details of the object, the lines of the bark, which are common to the two pictures, will not be similarly arranged, but will fall on non-corresponding points. Thus, in the case of the right-eye picture the details of the right side of the trunk will image themselves on points lying further from one another than in the case of the left-eye picture. Yet here we do not note the partial dissimilarity of the two impressions, nor do we see double images of the details not imaged on corresponding points. Under these circumstances, which innumerable experiences of active touch have told us answer to the presence of a solid body, we get a new optical effect, viz. *the combination of the two dissimilar impressions in a perception of solidity or relief.* When an object is too far off for the dissimilar retinal effects to come into operation, relief or solidity has to be inferred from other signs.   These include the distribution of light and shade on the surface, or what is known by artists as "modelling."   Thus the prominence of a distant mountain is perceived by the gradations of light and shade.'[1]   That the union of two dissimilar pictures, differing much as retinal pictures do, leads to the perception of a single object in relief is shown by the stereoscope.   In this instrument the rays of light from two such pictures are refracted by wedge-shaped lenses to aid the eye in combining them, so that they are referred to a single object lying midway between, and this is perceived as a solid object.   Even without the stereoscope this combination may be effected. Fig. 247 shows the principal lines of a truncated pyramid held in front, as seen by the right and left eye.   If a card is held between the figures, and

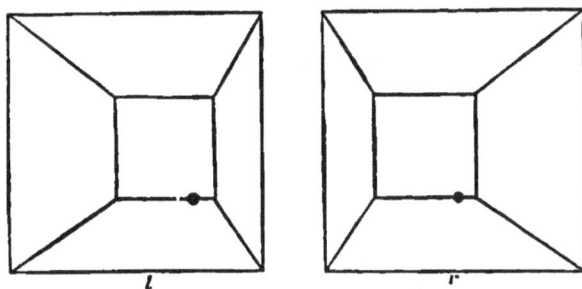

FIG. 247.

they are steadily looked at for a few seconds, they will be seen to blend into one, having an appearance of solidity.

233. **Colour Vision and Colour Blindness.**—By allowing a thin beam of white light to fall on a prism in a darkened room, it is decomposed into a parti-coloured band termed the spectrum.   People whose colour sense is normal recognise in the spectrum the following colours shading off into one another insensibly—red, orange, yellow, green, blue, indigo, and violet. These colours may be recombined into white light in various ways.   If the light thus dispersed by unequal refraction into a colour band be that received from the sun, certain fine black lines are seen passing through the spectrum, and to these lines letters are assigned, B, C, &c.   The cause of these lines does not concern us here, but the letters may be used to indicate

---

[1] Sully, *Outlines of Psychology.*

positions in the spectrum. Thus B and C are in the red, E in the green, &c. Physics teaches us that light is due to undulation or waves in the ether of space, and that just as sound waves of a certain frequency produce in the ear a sound of a certain *pitch*, so light waves of a certain frequency produce in the eye a certain *colour* sensation. White light is the sensation produced by light waves of all visible frequencies or wave lengths striking the retina.

The colour of non-luminous material objects depends upon their power of reflecting certain waves or rays of white light and of absorbing the rest. A red rose, for example, absorbs waves of all frequencies except those giving the sensation of red, and these it reflects. Perfectly transparent objects transmit all kinds of waves in white light ; coloured transparent bodies stop certain of the rays and allow the rest to pass. Colours may be mixed by causing coloured lights from different sources to fall on the retina at the same time, or by making a rotating disc or top with differently coloured segments turn so rapidly that the impressions pass into the eye with such rapidity that the eye or the brain blends them. The sensation of white may be produced, as just remarked, by mixing all the colours of the spectrum. But it can also be produced by certain pairs of colours termed *complementary*. Thus, red and greenish-blue, yellow and indigo-blue, and greenish-yellow and violet are three pairs of complementary colours, each pair producing white. Further experiment shows that all the numerous colours of the spectrum, as well as all shades of grey and white, can be produced by mixing three of the colours termed fundamental. Red, green, and violet, or violet-blue, are usually regarded as the three fundamental colours.

Every colour is distinguished by three qualities—(1) its *hue*, as red, green, &c. ; (2) its *purity*, or measure of its freedom from mixture with white, as expressed by the terms ' pale ' or ' deep ; ' (3) its *brightness*, strength, or luminosity. Colours are identical that possess these three qualities in the same degree, and persons of normal vision appear to have the same kind of colour sense and receive similar colour sensations from coloured objects. But there are persons whose colour sensations are not normal, but different from those of the majority, their power of perceiving and distinguishing colour being defective. Such persons are said to be afflicted with *colour-blindness*. The most common kinds of colour-blindness are red-blindness and green-blindness, as there is a failure to distinguish between red and green. A 'red-blind' person regards certain hues of red as green, and certain hues of green as white. A 'green-blind' person would regard orange as a

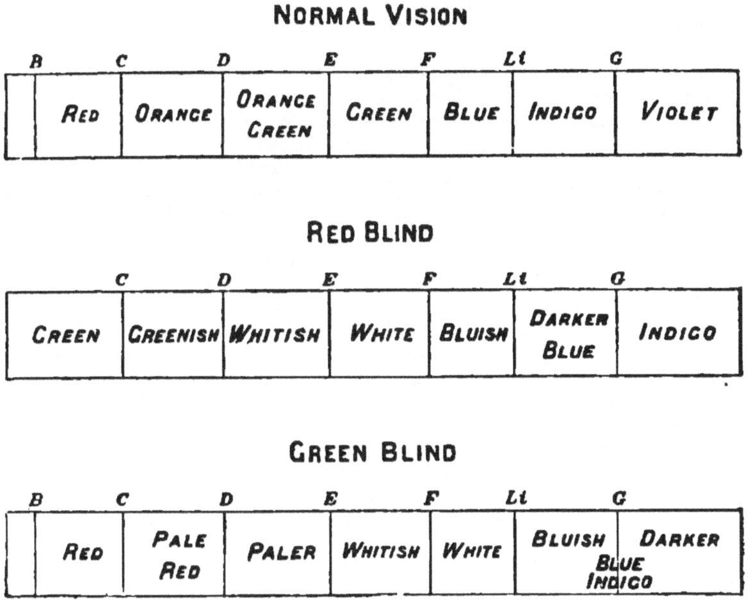

NORMAL VISION

B	C	D	E	F	Lt	G
RED	ORANGE	ORANGE GREEN	GREEN	BLUE	INDIGO	VIOLET

RED BLIND

C	D	E	F	Lt	G	
GREEN	GREENISH	WHITISH	WHITE	BLUISH	DARKER BLUE	INDIGO

GREEN BLIND

B	C	D	E	F	Lt	G	
RED	PALE RED	PALER	WHITISH	WHITE	BLUISH BLUE INDIGO	DARKER	

FIG. 248.

pale red, green as a very pale red with a strip of white, bluish-green would be described as blue diluted with white, and violet as blue. A diagram of the spectrum, though uncoloured, will help to illustrate these two defects. A few cases of 'violet-blindness' have been met with, and in still fewer cases total colour-blindness has been met with, only gradations of light and shade being perceived. Most cases of colour-blindness are hereditary· defects and totally incurable, though colour-blindness may be induced by disease.

Various explanations or theories have been put forth to account for the facts of colour sensation. As all colours can be formed out of three fundamental colours, normal vision is trichromatic, and it has been supposed that there are but three primitive colour sensations. There does not, therefore, appear to be any need for as many varieties of nerve terminals in the retina as there are hues of colour, but the stimulation of the retina by coloured lights may be explained by supposing that the retina contains three different kinds of nerve elements (cones or fibrils) for the perception of the three different primitive colour sensations, one set stimulated by red light waves, another by green light waves, and another by violet light waves. This is the theory of Young, and it has been adopted with some modifications by Helmholtz. The Young-Helmholtz theory of colour vision, therefore, supposes that there are in the retina three kinds of nerve elements, each kind most excitable or most affected by one of the three fundamental colours, but also excitable in some degree by each of the other two. The combination of these primary sensations in varying proportions explains how sensations of various colours may arise, and when the three sensations are equally stimulated white light is perceived. This theory, too, explains defects of colour vision, for, in the case of the red-blind, one set of sensitive elements, viz. those perceptive of red, is supposed to be absent, and only two fundamental sensations felt. Yet red is not totally invisible to such, as a red colour excites feebly his green perceptive nerve elements. It also explains coloured after images, for on staring at any bright coloured object for a time, the fibrils or nerve elements perceptive of that colour become fatigued, and the complementary colour is seen on turning the eyes to a white ground. But it must be remembered that three different kinds of nerve terminals or nerve fibrils have not been discovered, nor does the theory distinguish between darkness, or retinal rest, and the sensation of blackness. The annexed diagram will help to explain the theory still further, the letters B, C, D, &c., simply marking the position of certain dark lines in the solar spectrum.

Helmholtz does not pretend to explain how three kinds of nerve elements, which are sensitive in various degrees to the various simple colours, are excited, though he says we may ' hypothetically refer all these excitations to the presence of three photo-chemically alterable substances in the retina.'

Another theory of colour vision is that of Prof. Hering. He considers that examination of our sensations of light shows that there are six colours quite distinct in nature, which may be arranged in three pairs related to three different parts of the nerve apparatus or ' visual substance.' These pairs are white and black, blue and yellow, red and green, the last two pairs in particular being called ' opposed colour sensations.' He also considers that the sensation of light and shade is a specific faculty, the *totally* colour-blind having but this sensation, and inhabiting a world of grey. To quote the report of the Committee on Colour Vision appointed by the Royal Society :—

' Hering's theory attempts to reconcile, in some such way as follows, the various facts of colour vision with the supposition that we possess these six fundamental sensations. The six sensations readily fall into three pairs, the members of each pair having analogous relations to each other. In each pair the one colour is complementary to the other; white to black, red to green, and yellow to blue.

' Now in the chemical changes undergone by living substances, we may recognise two main phases, an upward constructive phase in which matter previously not living becomes living, and a downward destructive phase in which living matter breaks down into dead or less living matter. Adopting this view we may, on the one hand, suppose that rays of light, differing in their wave length, may affect the chemical changes of the visual substance in different ways, some promoting constructive changes (changes of assimilation), others promoting destructive changes (changes of dissimilation), and on the other hand, that the different changes in the visual substance may give rise to different sensations.

'We may, for instance, suppose that there exists in the retina a visual substance of such a kind that when rays of light of certain wave lengths — the longer ones, for instance, of the red side of the spectrum – fall upon it,

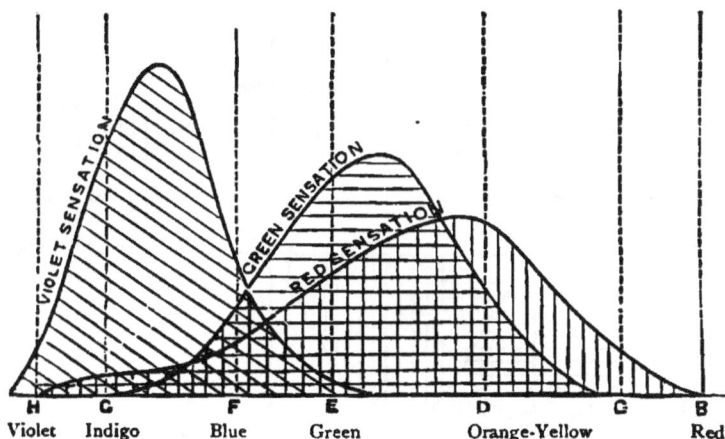

H	G	F	E	D	C	B
Violet	Indigo	Blue	Green	Orange-Yellow		Red

FIG. 249.—Diagram to Illustrate the Young-Helmholtz Theory of Vision.

The curves show the sensitiveness of the three varieties of nerve elements to the different parts of the spectrum, the different degrees of stimulation given to each of the three sensations by every part of the spectrum being indicated by the heights of the curves above the horizontal base.   Thus, at D the heights of the two curves indicate the degrees of stimulation of the two sensations of red and green which produce orange-yellow ; at E we see the degrees of stimulation of the three sensations that produce spectral green.   At the extreme red and extreme violet end the sensations are simple. —(*Report of the Committee of the Royal Society on Colour Vision.*)

dissimilative changes are induced or encouraged, while assimilative changes are similarly promoted by the incidence of rays of other wave lengths, the shorter ones of the blue side.   But it must be remembered that in dealing with sensations it is difficult to determine what part of the apparatus causes them.   We may accordingly extend the above view to the whole visual apparatus, central as well as peripheral, and suppose that when rays of a certain wave length fall upon the retina, they in some way or other, in some part or other of the visual apparatus, induce or promote dissimilative changes, and so give rise to a sensation of a certain kind, while rays of another wave length similarly induce or promote assimilative changes, and so give rise to a sensation of a different kind.

'The hypothesis of Hering applies this view to the six fundamental sensations spoken of above, and supposes that each of the three pairs is the outcome of a particular dissimilative and assimilative change. It supposes the existence of what we may call a red-green visual substance, of such a nature that, so long as dissimilative and assimilative changes are in equilibrum, we experience no sensation, but that when dissimilative changes are increased, we experience a sensation of (fundamental) red, and when assimilative changes are increased, we experience a sensation of (fundamental) green. A similar yellow-blue visual substance is supposed to furnish, through dissimilative changes, a yellow, through assimilative changes, a blue sensation; and a white-black visual substance similarly provides for a dissimilative sensation of white and an assimilative sensation of black. The two members of each pair are therefore not only com-

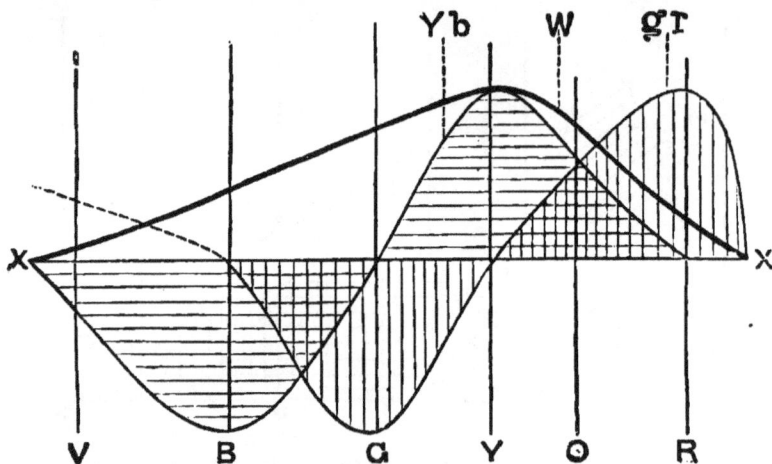

Fig. 250.—Diagram to illustrate Hering's Theory of Colour Vision.

The vertical shading represents the red and green, and the horizontal shading the yellow and blue, antagonistic pairs of sensations. The thick line indicates the curve of the white sensation. All above the line XX indicates destructive changes in the retinal substances, and all below constructive changes. (*Report of the Committee of the Royal Society on Colour Vision.*)

plementary but also antagonistic. Further, these substances are supposed to be of such a kind that while the white-black substance is influenced in the same way, though in different degrees, by rays along the whole range of the spectrum, the two other substances are differently influenced by rays of different wave lengths. Thus, in the part of the spectrum which we call red, the rays promote great dissimilative changes of the red-green substance with comparatively slight effect on the yellow-blue substance; hence our sensation of red.

' In that part of the spectrum which we call yellow, the rays effect great dissimilative changes in the yellow-blue substance, but their action on the red-green substance does not lead to an excess of either dissimilation or assimilation, this substance being neutral to them; hence our sensation of yellow. The green rays, again, promote assimilation of the red-green substance, leaving the assimilation of the yellow-blue substance equal to its dissimilation ; and, similarly, blue rays cause assimilation of the yellow-blue substance, and leave the red-green substance neutral. Finally, at the extreme blue end of the spectrum the rays once more provoke dissimilation of the red-green substance, and by adding red to blue give violet. When orange falls on the retina there is an excess of dissimilation of both the red-green and the yellow-blue substance ; when greenish-blue rays are perceived there is an excess of assimilation of both these substances ; and other intermediate hues correspond to varying degrees of dissimilation or assimilation of the several visual substances.

' When all the rays together fall on the retina, the red-green and yellow-blue substances remain in equilibrium, but the white-black substance undergoes great changes of dissimilation, and we say the light is white.

' According to this theory what are called red- and green-blindness are identical. The yellow-blue and white-black sensations remain, but the red-green sensation is absent in both. The white or grey seen in the spectrum would then be due to the white-black sensation, as it alone is stimulated at that point.'

Other theories have been put forth. It has been suggested, for instance, that ' each of the elements of the visual apparatus is made up of a central structure for the sensation of light and darkness (or light and shade), with collateral appendages for the sensations of colour.'

CHAPTER XVIII

THE EAR AND THE SENSE OF HEARING

234. **Introductory Statement.**—The filaments of the auditory nerve end in peculiar structures buried deeply in the hard portion of the temporal bone of the skull, and special arrangements exist for conducting waves of sound to this deeply seated sensitive part. The external ear assists in collecting sonorous vibrations that pass along a channel termed the external auditory meatus, and impinge against a stretched membrane termed the tympanic membrane, or drum-skin. The vibrations thus set up in the tympanic membrane are transmitted across the tympanic cavity or middle ear by a chain of small bones— the malleus or hammer, the incus or anvil, and the stapes or stirrup—to the inner ear lodged in the temporal bone. The membranous base of the stapes is placed in connection with the inner ear by being fixed into an oval opening in a bony tubular labyrinth consisting of parts termed the vestibule, the

semicircular canals, and the cochlea. Inside the bony labyrinth is a nearly similar labyrinth of membrane filled with liquid, a liquid also lying between the bony and the membranous labyrinth. The vibrations conveyed to this fluid by the movements of the base of the stapes excite the peculiar epithelium of the inner surface of the membranous labyrinth, on and in which are distributed the auditory nerve filaments. Impulses pass from these filaments along the nerve lying in the internal meatus to the brain, and there produce that modification of consciousness which we call the sensation of sound.

The general relation of the parts of the organ of hearing is shown in figs. 251, 252, and 256.

Sound is produced by vibrating bodies such as a tuning-fork, a tense string, or an elastic membrane. Their vibrations are communicated to the air, in which longitudinal waves of alternating condensation and rarefaction are set up, travelling at the rate of about 1,100 feet per second. Irregular vibrations produce noises, regular or periodic vibrations produce musical sounds. Musical sounds differ from one another in intensity or loudness, in height or pitch, and in quality or timbre. Loudness depends on the amplitude or energy of the air vibrations, *i.e.* speaking of the vibration as a wave, on the *height* of the wave. Pitch depends on the number of vibrations that fall on the ear in a second, *i.e.* on the time taken by each vibration. A note of 200 vibrations per second is an octave higher than a note of 100 vibrations. If we divide the velocity of sound per second by the number of vibrations executed in that time, we get the length of each vibration. Hence we may also say that pitch depends on the *length* of the sound wave. Besides intensity and pitch we can distinguish in most musical notes a quality or character which enables us to distinguish the instrument or voice producing the note. To understand this we must bear in mind that musical sounds are either simple or compound. A simple sound is one in which there is only a sound of one pitch, such as is produced by the pendulum-like vibrations of a rod fixed at one end, or a tuning-fork. A compound or complex sound is a sound formed of more than one simple sound, there being a fundamental tone giving the sound its pitch and subordinate higher tones called harmonics. The quality or timbre of a note depends on the number and strength of the upper partial tones that accompany the fundamental tone. When, for example, the string of a piano is struck, besides vibrating as a whole, to produce its fundamental note, it also vibrates in segments that ride, as it were, on the back of the principal vibration, and produce a compound wave of peculiar form, the vibrating segments giving rise to the upper tones. With the same note on a violin or flute, the number and relative strength of the upper tones would be different and produce a wave of different form. Hence we may say the quality or timbre of a musical note depends on the *form* of the air wave produced. A good ear can often pick out some of these overtones in a note, and by the help of resonators compound notes can be readily analysed.

For vibrations to affect the auditory nerve they must have a certain intensity and last a certain time. The rate of vibration frequency within which a sound is audible ranges for ordinary people from 30 to 30,000 vibrations per second, *i.e.* each vibration must last less than $\frac{1}{30}$th second and more than $\frac{1}{30000}$th. Wider limits have been recognised in special cases. A fine ear can distinguish a difference of $\frac{1}{64}$th of a semitone. Taking the range of sounds at ten octaves, each with twelve semitones, we get $(64 \times 12 \times 10 =) 7,680$ as the number of distinct musical sounds for the finest ears.

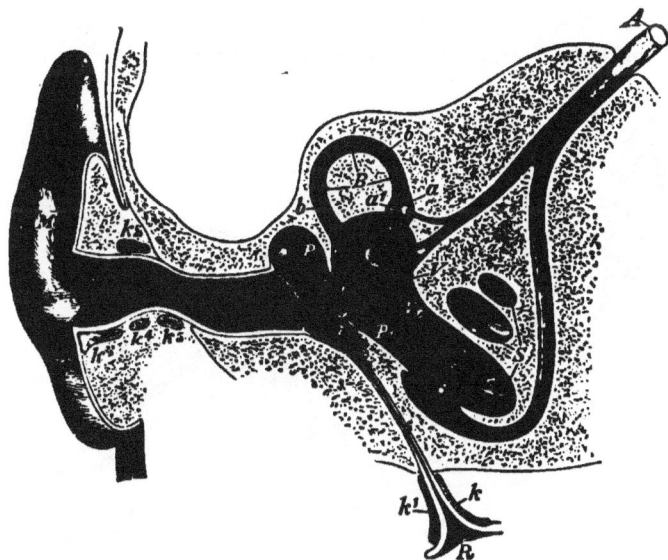

FIG. 251.—Semi-diagrammatic Section through the Right Ear. (Czermak.)

M, concha of external ear; *G*, external auditory meatus; *T*, tympanic membrane; *P*, tympanic cavity in which the auditory ossicles, the malleus, incus, and stapes are placed; *Er*, Eustachian tube leading from pharynx into tympanic cavity; *O*, oval window or fenestra ovalis with stapes fitting into it; *r*, round window or fenestra rotunda; *Pt*, scala tympani of cochlea; *Vt*, scala vestibuli of cochlea; *S*, coils of; cochlea; *b*, one of the membranous semicircular canals lying in its bony canal *B*; *a'*, ampulla of semicircular canal with branch of vestibular nerve passing in; *l*, utricle; *C*, saccule; *A*, auditory nerve passing through internal auditory meatus and dividing into two main branches.

235. **The External Ear.**—The ear is divisible into three parts : the external ear, the middle ear or tympanum, and the internal ear or labyrinth. The first two serve only to collect and transmit vibrations of sound to the nervous apparatus arranged in the internal ear.

The external ear consists of an expanded portion termed the pinna, and the external auditory meatus or canal. The

C C

pinna is composed of a thin plate of irregularly disposed yellow fibro-cartilage covered by skin, the eminences and depressions having special names. It is united to the surrounding parts by ligaments and muscular fibres. The muscles, like the ligaments, consist of two sets, those that connect it with the side of the head, and those which extend from one part to the other. The three muscles connecting it with the head are the *attollens aurem* for raising the ear, the *attrahens aurem* for drawing it forwards and upwards, and the *retrahens aurem* for

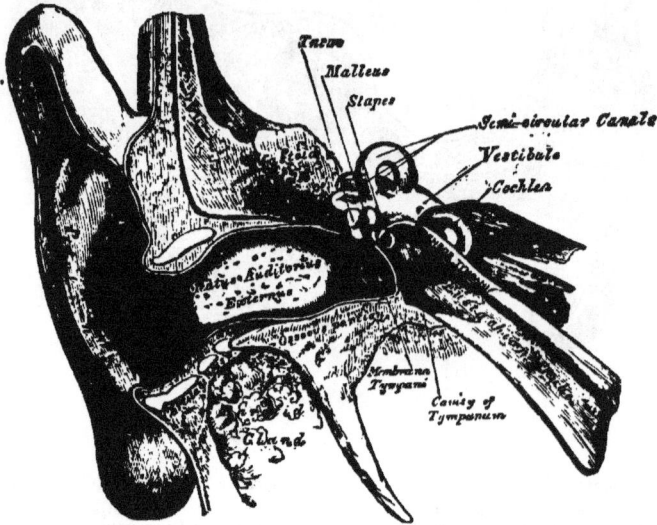

Fig. 252.—A Front View of the Organ of Hearing.   Right side.

drawing it backwards.   In man these muscles have very little action, but in some animals the pinna can, by their action, be moved so as to incline the concavity in the direction of sounds. Even in man this part probably aids in determining the direction of sound, besides assisting in collecting the sonorous waves, as loss of the part interferes with this determination. The external auditory canal is about an inch and a quarter in length, and serves to convey the sound to the membrana tympani which closes its inner end.   Its outer part is cartilaginous, its inner bony.   In the skin of the deeper cartilaginous

part are the *ceruminous* glands, modified sweat glands that
secrete the cerumen or wax of the ear.

236. **The Middle Ear, or Tympanum,** is an irregular cavity,
hollowed out of the temporal bone, and filled with air received
from the pharynx through the Eustachian tube. Its roof is
formed of a thin plate of bone, which separates it from the
cavity of the cranium. A thin layer of the temporal bone also

forms its floor. Its
outer wall is formed
mainly by the mem-
brana tympani and
the ring of bone into
which this membrane
is inserted. Two
small apertures for
the entrance and exit
of the chorda tympani
nerve may also be
seen. (From this
nerve, which is a
branch of the seventh
cranial nerve, fila-
ments pass to the
submaxillary gland
and the tongue.) The
membrane of the tym-
panum is thus placed
obliquely at the end
of the external audi-
tory meatus. It is

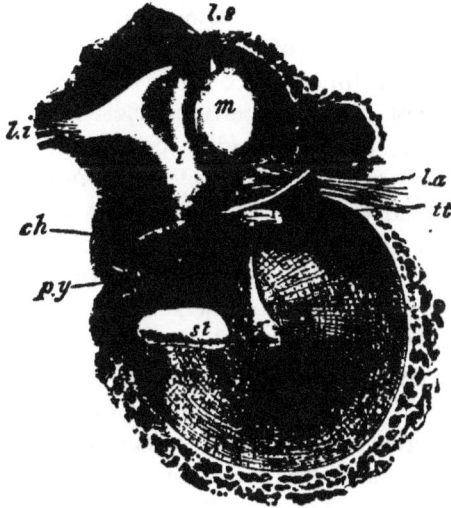

Fig. 253.—View of the Left Membrana Tympani and
Auditory Ossicles from the Inner Side. × 4.
(Quain's Anatomy.)

m, malleus ; *i*, incus ; *st*, stapes ; *py*, pyramid from
which the tendon of the stapedius muscle is seen
emerging ; *tt*, tendon of the tensor tympani, cut
short ; *la*, anterior ligament of the malleus ; *ls*,
superior ligament of the malleus ; *li*, ligament of
the incus ; *ch*, chorda tympani.

nearly circular in shape, about $\frac{1}{3}$ inch in diameter, and is composed of fibres, covered outside by a thin layer of skin, and inside
by mucous membrane. The membrana tympani bulges in
somewhat towards the tympanic cavity, where the handle of
the malleus is attached to it. As the membrane is stretched
across the outer side of the tympanic cavity, it is pressed on
its outer side by the external air entering the external auditory
meatus, and on its inner side by the air entering the tympanic

cavity through the Eustachian tube. Ordinarily the Eustachian tube is closed, and if the closure were permanent, variations of barometric pressure would lead to the bulging in and out of the tympanic membrane, according as the external or internal pressure on it were the greater ; but the Eustachian tube opens during every act of swallowing, and thus the air-pressure on the two sides of the membrane is kept equal. Sound waves throw it into a state of vibration, alternate bendings inwards and outwards, and its peculiar structure and form enable it to take up tones of a great variety of pitch.

The inner wall of the tympanic cavity, or drum, is osseous except at two apertures closed by membrane, the *fenestra ovalis*, or oval window, connected with the vestibule of the inner ear, and the *fenestra rotunda*, or round window, connected with the cochlea of the inner ear. On this inner wall may also be noted a bony prominence between the windows termed the *promontory*, a curved bony canal containing the facial nerve, and termed the *aquæductus Fallopii*, and a conical bone termed the pyramid, out of which a muscle passes to the stapes. The anterior part of the wall also shows a canal containing the tensor tympani muscle, the tendon of which is inserted into the upper part of the handle of the malleus. The upper and bony end of the Eustachian tube opens by an orifice on the anterior wall of the tympanum. The *processus cochleariformis* is a thin plate of bone separating the canal for the tensor tympani muscle from the Eustachian canal. The Eustachian tube, lined throughout with mucous membrane, continuous with that of the pharynx, serves mainly for keeping the pressure of the air in the tympanum equal to that of the external air, the tube being usually closed, but opened in the act of swallowing—an act which is being performed unconsciously at frequent intervals, even when not eating. The whole tympanic cavity, including the chain of bones, is lined with mucous membrane, continuous with that of the Eustachian tube, and like it ciliated except over the tympanic membrane.

237. **The Ossicles of the Middle Ear.**—The three little bones, or ossicles, stretching across the tympanum from the membrana tympani to the fenestra ovalis are the malleus, incus, and stapes. The malleus or hammer

bone has a rounded head articulating with the incus or anvil, and a vertical process called the manubrium, or handle, which is firmly attached to the fibrous layer of the membrana tympani. On the inner side of the upper part of the handle is attached the tendon of the tensor tympani muscle—a muscle whose contraction draws inwards the handle of the malleus, and thus increases the tension of the tympanic membrane. At the top of the handle is a long slender process, the *processus gracilis*, which extends forwards to a fissure in the bony wall, and is connected to

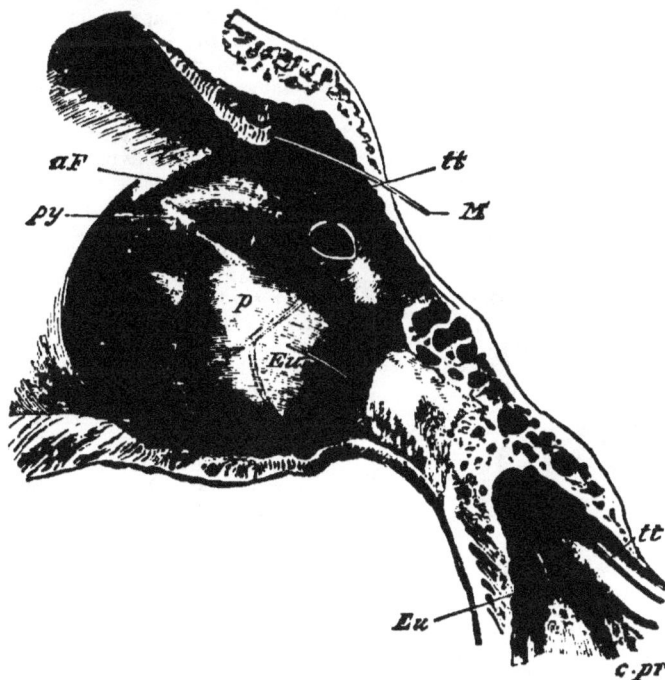

FIG. 254.—Inner Wall of the Right Tympanum. (Quain's Anatomy.)

M, bristle passing into mastoid cells ; *f o*, fenestra ovalis ; *f r*, fenestra rotunda ; *p*, promontory ; *a* F, aqueduct of Fallopius ; *c pr*, processus cochleariformis ; *tt*, bristle passed through the canal for the tensor tympani ; *E* M, bristle in the Eustachian tube.

this by ligamentous fibres. Another ligament descends from the roof of the tympanum to the head of the malleus. The incus, or anvil, is more like a bicuspid tooth with two widely separated fangs. One of these processes, the *processus brevis*, lies horizontally, and is attached to the inner wall of the tympanum by a ligament. The longer process lies vertically parallel to the handle of the malleus, and is articulated with the stapes. The joint between the head of the malleus and the body of the incus is of a peculiar saddle shape, the lower part of the articular surface of each bone having a blunt tooth, so that ' when the hammer is drawn inwards by the handle, it bites the anvil firmly and carries it with it.   Conversely, when the drum-skin with the hammer is driven outwards, the anvil is not obliged to follow it.'   This prevents the stirrup from being pulled out of the oval window by any undue outward push on the tympanic membrane.   The lower end of the long process of the incus bends inwards, and ends in a small flattened bone, the *os orbiculare*, or lenticular bone, which articulates with the head of the stapes.   The stapes lies horizontally at right angles to the long process of the incus, and its bony foot-plate covered internally with cartilage is attached to the margin of the fenestra ovalis by ligamentous fibres.   Into the head of the stapes is inserted the small stapedius muscle, which springs from the inner tympanic wall through a hole near the fenestra ovalis.   It regulates the movement of the stapes, preventing undue pressure in the labyrinth, and may be regarded as antagonistic to the tensor tympani.

The chain of small bones thus jointed together and secured by ligaments may be regarded as a bent lever rotating round an axis passing through the lower end of the neck of the malleus ; the power being applied at the end of the handle of the malleus, and the effect being felt at the foot-plate of the stapes.   Sonorous vibrations pushing inwards the handle of the malleus cause its head, together with the body of the incus, to move outwards, while the outward motion of the body of the incus causes its vertical process to have an inward movement, that presses the base of the stapes into the fenestra ovalis against the liquid of the labyrinth. As the vertical process of the incus is only two-thirds the length of the first arm of the lever, the inward movement of the stapes will be only two-thirds that of the handle of the malleus, though the pressure exerted by the stapes will be, by the

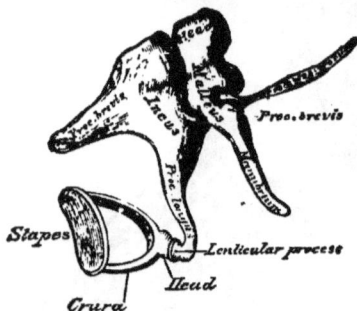

FIG. 255.—The Small Bones of the Ear, seen from the Outside.   Enlarged.

principle of the lever, one-half greater.   Moreover, the area of the tympanic membrane is about twenty times that of the fenestra ovalis.   Thus a relative large movement of small energy is converted by the tympanic membrane and chain of ossicles into a smaller movement of greater energy.

238. **The Internal Ear.** —The internal ear, or labyrinth, consists of chambers and tubes hollowed out in the temporal bone, and enclosed by it on all sides, except for the oval and round windows on the exterior, and apertures for the branches of the auditory nerve and blood-vessels on the interior.   We distinguish (1) the bony or osseous labyrinth, which consists of a series of channels hollowed out of the petrous portion of the temporal bone ; (2) the membranous labyrinth contained in the former.

Between the osseous and membranous labyrinth is a liquid termed the *perilymph*, and within the membranous labyrinth is another liquid termed the *endolymph*.

The *bony or osseous* labyrinth consists of three parts—the vestibule, the semicircular canals, and the cochlea.

The vestibule is the central cavity of the bony labyrinth, and has on its outer or tympanic wall the fenestra ovalis, which is closed in the recent state by the foot-plate of the stirrup bone, and the fenestra rotunda. The interior of its inner wall shows two foveæ or depressions with a ridge between. Behind, it communicates by five orifices with the bony semicircular canals, one orifice being common to two of the canals. In front is the oval opening that leads into the cochlea. The semicircular canals situated above and behind the vestibule lie in three planes, one

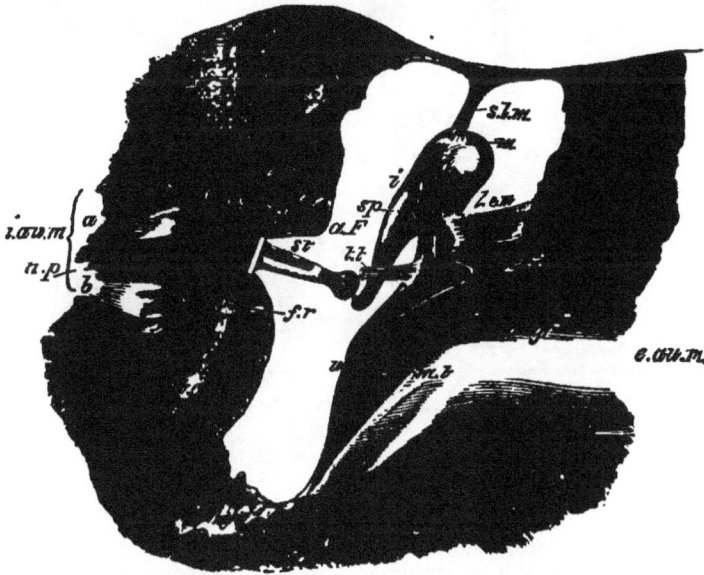

Fig. 256.—Profile View of the Left Membrana Tympani and Auditory Ossicles from before and somewhat from above. Magnified four times. (E. A. S.)

The anterior half of the membrane has been cut away by an oblique slice. *m*, head of the malleus ; *sp*, spur-like projection of the lower border of its articular surface ; *pr. br*, its short process ; *pr. gr*, root of processus gracilis, cut ; *s.l.m*, suspensory ligament of the malleus ; *l.e.m*, its external ligament ; *t.t*, tendon of the tensor tympani, cut ; *i*, incus, its long process ; *st*, stapes in fenestra ovalis ; *e.au.m*, external auditory meatus ; *p.R*, notch of Rivini ; *m.t*, membrana tympani ; *u*, its most depressed point or umbo ; *d*, declivity at the extremity of the external meatus ; *i.au.m*, internal auditory meatus ; *a* and *b*, its upper and lower divisions for the corresponding parts of the auditory nerve ; *n.p*, canal for the nerve to the ampulla of the posterior semicircular canal ; *s.s.c*, ampullary end of the superior canal ; *p*, ampullary opening of the posterior canal ; *c*, common aperture of the superior and posterior canals ; *c.s.c*, ampullary, and *e'.s.c*, non-ampullary end of the external canal ; *s.t.c*, scala tympani cochleæ ; *f.r*, fenestra rotunda, closed by its membrane ; *a.F*, aqueduct of Fallopius.

horizontal and two vertical, and at right angles to each other like three adjacent sides of a cube, the external horizontal canal having two distinct openings into the vestibule. The canals measure about $\frac{1}{20}$ inch in diameter, and have a dilatation at one end termed the *ampulla*. The cochlea, situated in front of the vestibule, somewhat resembles a small snail shell. It consists of a bony canal about $1\frac{1}{2}$ inch in length, winding spirally $2\frac{2}{3}$ times round a central column, termed the modiolus. The interior of the spiral canal of the cochlea in the dry state is seen to be

partially divided into two by a thin bony plate, the *lamina spiralis*, projecting from the modiolus.　From the edge of this bony plate there passes in

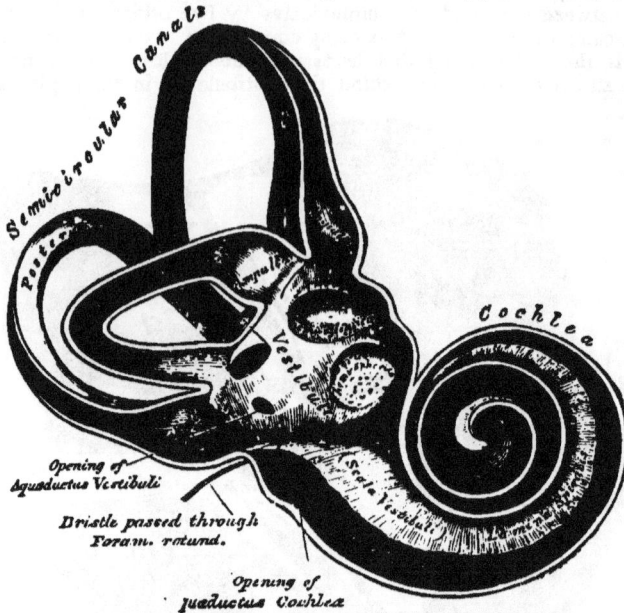

FIG. 257.—The Osseous Labyrinth laid open. Enlarged.

the fresh state a flat membrane, the *basilar membrane*, which completely
divides the coiled tube of the cochlea into two parts termed *scala*. The upper
passage (with the cochlea base down-
wards), starting from the vestibule and
winding round to the apex, is known as the
stairway from the vestibule, or *scala vestibuli*; the lower passage, shut off from the
tympanic cavity at its base by the membrane
of the fenestra rotunda, and communicating
with the upper space at the apex by a tiny
hole, the *helicotrema*, is termed the stairway
from the tympanum, or *scala tympani*. A
small triangular part

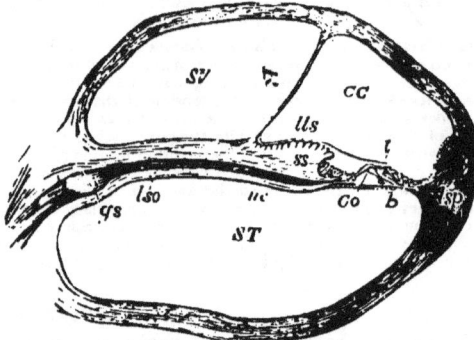

FIG. 258.—Section through one of the Coils of the Cochlea,
diagrammatic (altered from Henle).

ST, scala tympani ; *SV*, scala vestibuli ; *CC*, canalis cochleæ ;
R, membrane of Reissner forming its vestibular wall ; *l s o*,
lamina spiralis ossea ; *l l s*, limbus laminæ spiralis ; *s s*, sulcus
spiralis ; *n c*, cochlear nerve ; *g s*, ganglion spirale ; *t*, membrana tectoria ; *b*, membrana basilaris ; *Co*, rods of Corti ;
l s p, ligamentum spiralis.

of the scala vestibuli is cut off by a fine connective-tissue membrane (membrane of Reissner), passing from the lamina spiralis to the bony wall of the cochlea, and the duct thus cut off is termed the central canal of the cochlea,

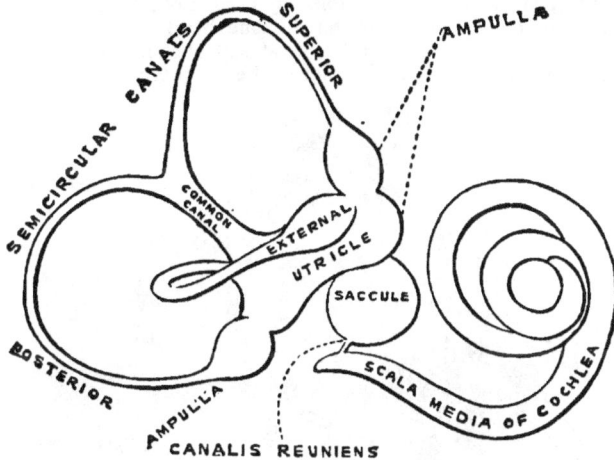

FIG. 259.—The Membranous Labyrinth. Enlarged.

or *scala media*. The bony tubular canal of the cochlea is thus divided into three canals by the membranous structures just mentioned.

The *membranous* labyrinth is the complex closed membranous tube, lined by epithelium and filled with endolymph, that lies within the bony labyrinth just described. It has the same general form as the bony labyrinth, but the portion in the bony vestibule consists of two sacs, the utricle and the saccule. Opening into the utricle are the membranous semicircular canals which are adherent along one side of the bony canals and but one-third their diameter. Each membranous canal is dilated at one of the ends, opening into the utricle to form an ampulla. The utricle and saccule have no direct connection, but communicate indirectly by the *saccus endolymphaticus*, which arises by an isolated limb from each sac, the limbs uniting as in the letter Y, and the terminal sac lying in the skull.

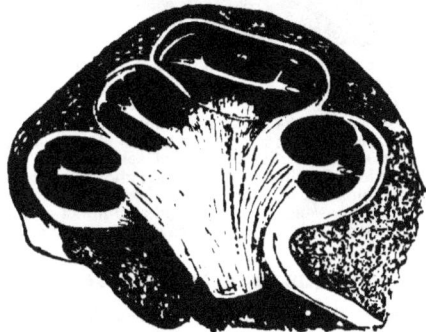

FIG. 260.—Vertical Section of the Cochlea of a Calf. (Kölliker.)

Showing the two scalæ with the intermediate canal of the cochlea, and the lamina spiralis.

From the membranous saccule a narrow tube, the *canalis reuniens*, leads into the small central canal—scala media— of the cochlea, which is the cochlear portion of the membranous labyrinth, and lies between the two scalæ.

The system of spaces formed by the various parts of the membranous labyrinth and filled with endolymph are thus in communication with each other. Around the membranous labyrinth are the portions of the bony labyrinth not cut off by membrane, and occupied by perilymph, the perilymph of the bony vestibule being continuous on the one hand with that in the bony semicircular canals, and on the other hand with that in the scala vestibuli, and through the helicotrema with that in the scala tympani.

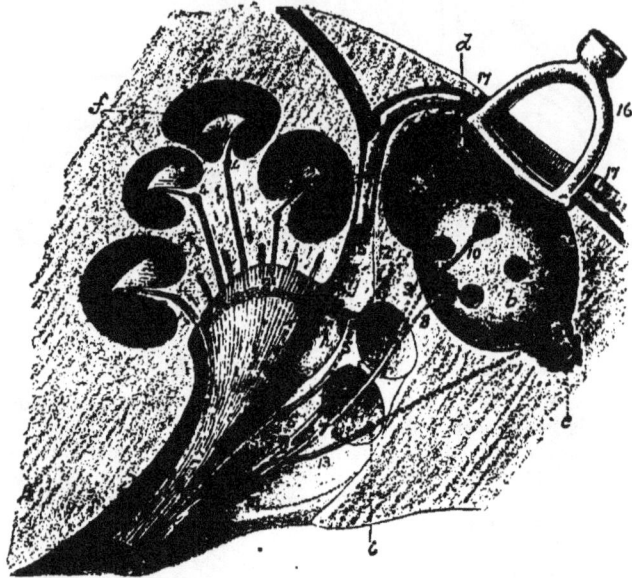

FIG. 261.—Diagram showing the Distribution of the Auditory Nerve. (Testut.)

a, the bony vestibule containing *b*, the utricle, *c*, the saccule, and *d*, the first portion of the canal of the cochlea ; *e*, cut end of a semicircular canal with the ampulla ; *f*, the cochlea ; *g*, the Fallopian aqueduct containing 15, the facial nerve. 1, trunk of auditory nerve ; 2, cochlear branch of auditory nerve ; 3, vestibular branch of auditory nerve ; 4, ganglion at base of spiral lamina ; 5, small branch of 2 passing to first portion of cochlear canal ; 6, ganglion ; 7, superior vestibular branch passing to utricle and ampullæ of superior and horizontal semicircular canals ; 11, inferior vestibular branch passing to saccule and ampulla of posterior semicircular canal ; 16, stapes in fenestra ovalis ; 17, 17, in cavity of tympanum.

239. **The Auditory Nerve.**—Within the internal auditory meatus the auditory nerve separates into two branches, and these, broken up into filaments, pass through perforations of the cribriform plate separating the meatus from the internal ear, to be distributed to the cochlea and vestibule. In both branches are numerous ganglionic nerve cells. Filaments derived from the cochlear branch of the auditory nerve pass up the axis of the cochlea, and give off fibres to the spiral

lamina, at the base of which is the small *ganglion spirale* containing bipolar ganglion cells. From the edge of the spiral lamina the fibres pass on to become connected with the organ of Corti, in a manner to be described shortly. The vestibular branch of the auditory nerve divides into two portions, and its mode of distribution may be learnt from figs. 261 and 262.

240. **The Auditory Nerve Endings.**—The inner surface of the membranous labyrinth consists of an epithelium resting on a basis of connective tissue, and this epithelium is modified in certain places to receive nerve filaments and constitute the nervous end-organ of hearing. The auditory nerve from the brain passes through the temporal bone in the internal auditory meatus and divides into two main divisions, one going to the vestibule and the other to the cochlea. The vestibular nerve supplies terminal twigs to special portions of the utricle, the saccule and the ampullæ of the membranous semicircular canals, and to these parts alone. In the utricle the nerve fibrils come into connection with an area of modified epithelium forming an oval swelling, termed the *macula acustica*. A similar macula acustica is found in the saccule. In each ampulla the modified epithelium where the nerve fibrils terminate is a horseshoe-shaped ridge called a *crista acustica*. The vestibular branch of the auditory nerve thus ends in the macula acustica of the utricle, the macula acustica of the saccule, and the crista acustica of each of the three membranous ampullæ. At these places the modified epithelium is mainly

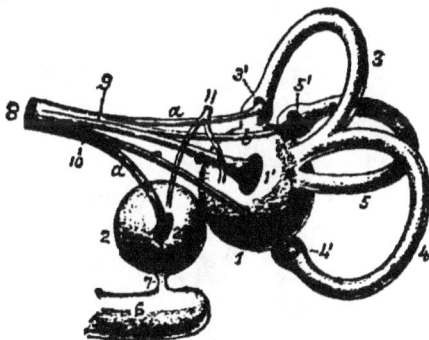

FIG. 262.—The Utricle, Saccule, and Semicircular Canals seen from within, and showing the distribution of the Vestibular Branch of the Auditory Nerve. (Testut.)

1, utricle with 1′, its macula acustica; 2, saccule with 2′, its macula acustica ; 3, 4, 5, the semicircular canals with 3′, 4′, 5′, their cristæ acusticæ: 6, beginning of cochlear canal ; 11, lymphatic canal or saccus endolymphaticus ; 8, vestibular branch of auditory nerve ; 9, superior portion of vestibular nerve with *a*, superior ampullary nerve, *b*, external ampullary nerve *c*, nerve for utricle ; 10, inferior portion of vestibular nerve with *d*, nerve to saccule, and *e*, nerve to posterior ampulla.

formed of columnar cells which are surmounted by stiff tapering processes termed *auditory hairs,* and to these *hair cells* of
the maculæ and cristæ the naked axis-cylinders of the nerve
fibres pass, entering, according to some authorities, the very
substance of the cell itself. Between the columnar hair cells
are a number of thin nucleated rod-like cells (fibre cells), that
are probably of a supporting nature merely. The free ends of
the auditory hairs appear to be imbedded in a cap of mucous
substance floating in the endolymph,
and in the utricle and saccule this
viscid endolymph contains small
crystals consisting mainly of calcium
carbonate, termed *otoconia,* or *otoliths,*
and probably serving as dampers.

The cochlear division of the
auditory nerve passes into a small
bony channel running up the
modiolus, or central column of the
cochlea, and from this branches pass
into the lamina spiralis, to be distributed thence as bare axis-cylinders
to the hair cells that form part of
the organ of Corti in the scala
media, or central canal of the cochlea. The cochlear canal, as already
shown, is a spiral tube triangular in
section, and forming part of the
membranous labyrinth. Its floor is
formed partly by the extremity of
the spiral lamina and partly by the

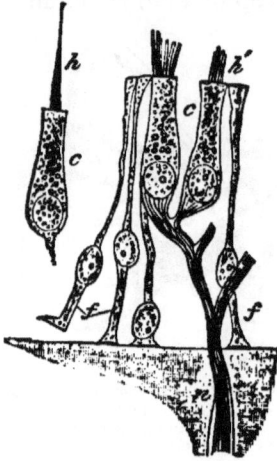

FIG. 263.—Auditory Epithelium
from the Macula Acustica of
the Sacculæ of an Alligator.
(Retzius.) Highly magnified.

c, c, columnar hair cells ; *f, f,* fibre
cells ; *n,* nerve fibre, losing its
medullary sheath and passing
to terminate in the columnar
auditory cells ; *h,* auditory
hair ; *h',* base of auditory hairs,
split up into fibrils.

basilar membrane, which stretches from the end of the bony
lamina to the outer wall of the cochlea, to which it is attached by connective tissue termed the *spiral ligament.* The
basilar membrane, which increases in breadth from the base to
the apex of the cochlea, is composed in its outer part of fibres
extending radially from within outwards. On its upper surface,
inside the central cochlear canal, rests the modified epithelium
termed the organ of Corti.

The organ of Corti thus situated on the basilar membrane consists of—1. The *rods of Corti* about the middle of the organ. These are peculiarly modified epithelial cells arranged along the spiral canal in an inner and outer row, their bases resting apart on the basilar membrane, but their heads leaning against each other like the beams of a house roof. The series

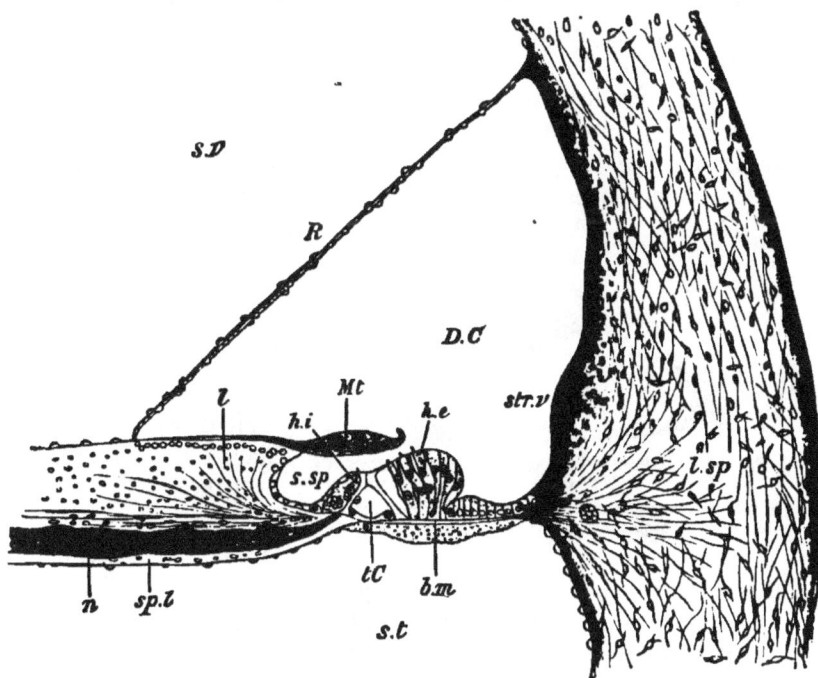

FIG. 264.—Vertical Section of the First Turn of the Human Cochlea. (G. Retzius.)

s.v, scala vestibuli ; *s.t*, scala tympani ; *D.C*, canal of the cochlea : *sp.l*, spiral lamina ; *n*, nerve fibres ; *l.sp*, spiral ligament ; *str.v*, stria vascularis ; *s.sp*, spiral groove ; *R*, section of Reissner's membrane ; *l*, limbus laminæ spiralis ; *M.t*, membrana tectoria ; *t.C*, tunnel of Corti ; *b.m*, basilar membrane ; *h.i*, *h.e*, internal and external hair cells.

of arches thus formed by the two sets of rods enclose a sort of tunnel running along the scala media of the cochlear canal. From the head of each rod a process projects outwards, those of the inner row overlapping those of the outer ; but as the inner rods are about one-half more numerous than the outer rods, the heads of two outer rods fit into three of those in the inner row. Viewed from above by the microscope, the head-plates of Corti's rods resemble very strikingly the keyboard of a piano. 2. The *inner hair cells*, placed on the inner side of the inner rods. These form a single series of columnar cells, and from the upper end there project a number of short auditory hairs arranged in a crescentic form. The pointed base of these inner hair cells rests among long nucleated cells that appear to act as supporting structures. 3. The *outer hair cells*, placed on the outer side of the outer rods. These form in man four rows usually, and are very similar to the inner hair cells, having short hairs arranged in a crescentic form on their upper surface. On the outside of each of these outer hair cells is a supporting cell that does not bear hairs, termed a *cell of Deiters*. 4. A fine *reticulate membrane* extending like a wire net over the outer cells. This is composed of two or three series of fiddle-shaped rings united together by flat bars. Through the holes of the rings of this membrane the heads of the outer hair cells project, while processes from the cells of Deiters are attached to the bars between the rings.

FIG. 265.—A Pair of Rods of Corti, from the Rabbit's Cochlea, in side view. Highly magnified.

b, b, basilar membrane; *i.r,* inner rod ; *e.r,* outer rod The nucleated protoplasmic masses at the feet are also shown.

The organ of Corti is overhung by the *tectorial membrane,* a soft elastic structure that arises from the upper lip of the peculiarly formed piece of connective tissue termed the limbus at the edge of the bony spiral lamina (fig. 264). Some authorities regard it as a damping apparatus for the organ of Corti.

The fibres of the cochlear nerve pass out from the spiral lamina after traversing the ganglionic bipolar cells of a small ganglion (*ganglion spirale*) situated in the bony lamina. As they leave the lower lip of connective tissue tipping the spiral lamina they lose both their primitive and medullary sheaths, and pass as naked axis-cylinders to the hair cells. Most of the nerve fibrils from these axis-cylinders do not appear to pass directly to the hair cells, but to run some distance along the length of the cochlear canal in spiral strands at the bases of the inner hair cells, or in the tunnel of Corti, or along the bases of the

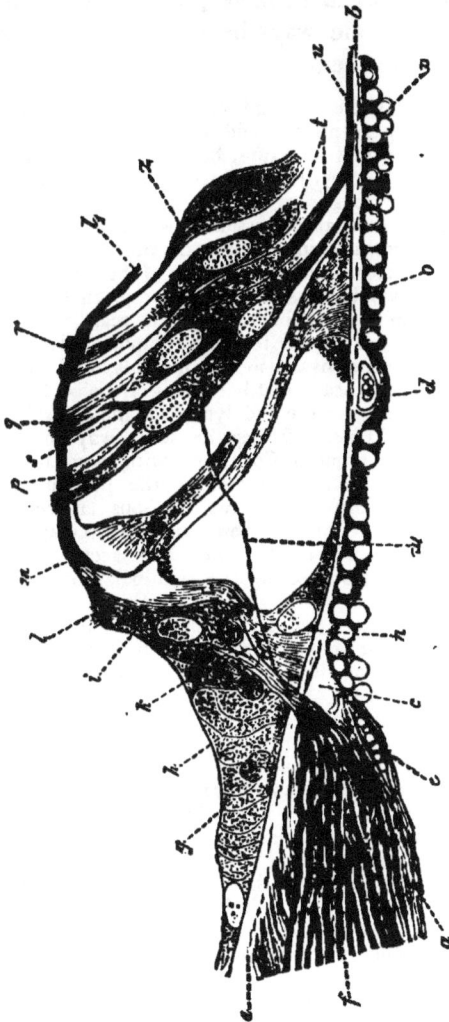

FIG. 266.—Section of the Organ of Corti of the Dog. (Waldeyer. ¾.)

a, a', end of spiral lamina; *b, c,* middle (homogeneous) layer of the basilar membrane; *u,* vestibular (striated) layer; *v,* tympanal (connective-tissue) layer; *d,* blood-vessel; *f,* nerves in spiral lamina; *g,* epithelium of spiral groove; *h,* nerve-fibres passing towards inner hair-cells; *i, k; l,* auditory hairlets on inner hair-cells; *l, l',* lamina reticularis; *m,* heads of the rods of Corti, jointed together; *n,* base of inner rods; *o,* base of outer rod; *p, q, r,* outer hair-cells; *t,* lower ends of hair-cells; *w,* nerve-fibrils passing across the tunnel of Corti; *z,* cells of Deiters.

outer hair cells. Ultimately, however, the fibrils become connected with the hair cells, though it is doubtful whether they actually pass into the cells or merely invest them. It may be regarded as settled that the hair cells, inner and outer, are the true terminal organs of the auditory nerve in which the sensory impulses that pass to the brain are originated, and that the other parts of Corti's organ are merely accessory in function, assisting or guiding in some way the nervous impulses, but not actually giving rise to them.

241. Perception of Vibrations.—Sound-waves collected by the pinna of the external ear pass along the meatus, and set in vibration the tympanic membrane, its peculiar form and structure, being somewhat loosely and unequally stretched and *loaded* by the tympanic bones, enabling it to take up aërial vibrations of various rates. The vibrations of the membrane are transmitted onward by the chain of ossicles swinging as a whole and acting as a lever, their movement being molar and not molecular. With diminished amplitude, but increased force, the base of the stapes communicates the vibrations to the whole mass of the perilymph in the vestibule, each vibration passing from the vestibule along the scala vestibuli, and down the scala tympani to end on the fenestra rotunda, this membrane moving outwards as the fenestra ovalis moves inwards, and *vice versâ*. As the vibrations ascend the scala vestibuli they are also transmitted across the membrane of Reissner to the endolymph of the central cochlear canal and the basilar membrane, affecting in some way the auditory epithelium or hair cells of the organ of Corti, and with them the terminations of the cochlear nerve. The vibrations of the perilymph of the vestibule are also transmitted through the membranous labyrinth to the endolymph of the utricle, saccule, and membranous semicircular canals, to reach the epithelium of the maculæ of the utricle and saccule, and that of the cristæ of the ampullæ, thus affecting the terminal filaments of the vestibular nerve.

It may be noted that vibrations may be transmitted by the bones of the skull to the tympanic membrane, and onwards by the ossicles to the internal ear. The sound from a vibrating tuning-fork held between the teeth may pass into the labyrinth through the skull bones, and so affect the auditory nerve, when the tympanic membrane is injured. An *audiphone* is a plate used in this way.

Confining our attention or the present to the cochlea which contains the organ of Corti, the most highly specialised portion of the auditory apparatus, and therefore the chief agent in hearing, we may inquire further into the mode in which the vibrations reach the hair cells of this organ, and the functions of its other portions.

The note of a musical instrument or the human voice possesses, as we have seen (par. 234), a certain quality or timbre, due to the fact that along with its fundamental tone, determining its pitch, there are upper partial tones or harmonics, of varying intensity and number in different instruments. The complex sound thus produced by vibrations of more than one period travels as a complex or composite air-wave. A composite sound or complex air-wave may be analysed by raising the. dampers of a piano, and allowing the musical note to resound powerfully before it, when a set of strings is brought into sympathetic vibration, namely, those strings which correspond in pitch to the fundamental tone, and to the several upper tones of the note sounded. ' Now, suppose we were able to connect every string of a piano with a nervous fibre in such a manner that this fibre would be excited, and experience a sensation every time the string vibrated.   Then every musical tone which impinged on the instrument would excite, as we know really to be the case, in the ear a series of sensations exactly corresponding to the pendular vibrations into which the original motion of the air had to be resolved.   By this means, then, the existence of each partial tone would be exactly so perceived as it really is perceived by the ear.'   As the ear does possess this power of analysing complex sounds, and can appreciate the pitch and quality of notes, we may inquire what parts of the terminal organ perform the office of the supposed piano-strings, that is, what parts of the terminal organ connected with the ends of the microscopic nerve fibrils can be set in sympathetic vibration by the various complex waves of sound.

Formerly it was thought that the rods of Corti, which vary regularly in the length and span of their arch from the base to the apex of the cochlea, served for the analysis of sound, each pair vibrating in response to a particular simple tone, and stimulating a particular nerve fibril, or group of fibrils.   But their number and variation in size and form do not seem sufficient for the purpose, and in birds, which clearly can appreciate musical notes, there are no rods of Corti in their rudimentary cochlea, though hair cells lie in contact with the basilar membrane.   Hence it is now generally believed that the stretched radial fibres of the basilar membrane, on which the organ of Corti rests, are the vibrating threads which analyse complex sounds, the rods of Corti assisting the transmission of the vibrations of the basilar membrane to the hair cells, and possibly acting as dampers also, as the chain of ossicles serves both purposes.   About 24,000 fibres exist in the basilar membrane, their length varying from ·075 mm. at the base to ·126 mm. at the apex of the cochlea, and their tension probably undergoing changes, owing to the action of what appear to be muscular cells in the outer ligament of the membrane.   Here, then, seems to be an apparatus adapted for the appreciation of pitch and the analysis of complex sounds.   A simple pendular vibration reaching the ear excites by sympathetic vibration the fibres of the basilar membrane tuned to its pitch, the shorter fibres near the base of the cochlea vibrating to high notes and the long fibres near the apex of the cochlea vibrating to low notes ; and the different fibres tuned to differences of pitch affect in some way different auditory cells, which in their turn affect different nerve fibres in order to give rise in the brain to sensations of pitch.   Thus in the case of a complex vibration the basilar membrane resolves it into its elements, one fibre taking up the fundamental tone, and other fibres the various harmonics or partial tones, the synthesis and appreciation of the sound, as having a certain quality, being effected in the nervous cells in the auditory centre of the brain.

Such is one view of the auditory mechanism, though other explanations and theories have their advocates.   Some consider that the sound-wave, on passing into the endolymph of the cochlear canal, impresses the membrana tectoria, and thus sets in vibration the hairs of the hair cells ; but those who regard the basilar membrane as the organ of analysis for pitch and quality believe that the tectorial membrane acts rather as a damper to check vibrations, the auditory hairs serving to convey the damping action to the hair cells when excited in some other way.

If, then, the organ of Corti can appreciate pitch and quality, it must be able to appreciate loudness, as the quality of a note depends to some extent on the relative intensity of the subordinate upper tones. Moreover, as noises involve elements of pitch, and are but partially distinguishable from musical tones, it would appear that the sensory epithelium of the cochlea can also appreciate them.

Having thus found in the cochlea an apparatus sufficient to account for the perception of the intensity, pitch, and quality of sounds, it would appear that the sensory epithelium connected with the vestibular division of the nerve must have other functions than the perception of sound. The semicircular canals, in fact, are believed to be the organs of the 'sense of orientation' which enables us to determine our position in space, and in conjunction with sensations from the skin, muscles, and eyes, furnishes help in guiding the complex coordinated movements by which the body is balanced and equilibrium maintained. Their function, therefore, is probably not auditory but 'equilibrating' Impulses set up in the vestibular nerve endings of the ampullæ of these canals (the cristæ acusticæ) by the varying pressure of the fluid endolymph in them, give rise to sensations that enable us to become aware of the position of the body. As the planes of the three canals lie in three axes of spaces, a movement of the body or a change of the position of the head may lead to changes of pressure or movements of endolymph that affect the ampullæ differently, and so give rise to different impulses in the vestibular branch of the auditory nerve fibres distributed to the ampullæ. Injury to the membranous canals in birds and rabbits does not affect the hearing, but produces, especially in birds, some loss of co ordination of movement, accompanied by movements of the head in the plane of the injured canal and movements of the eyeballs. Menière's disease, of which the chief symptoms are giddiness and staggering, has been found associated with affections of the semicircular canals. It thus appears that the impressions proceeding from the nerve endings in these canals form an important part of the afferent impulses controlling the mechanism of equilibrium (par. 193), and that this part of the internal ear is not connected with the sense of hearing. The sensory epithelium of the maculæ in the utricle and saccule has also been asserted to be connected with the sense of movement, and to have no connection with the sense of hearing   But the latter part of this statement appears doubtful, for certain animals, low in the scale, that have a vestibular labyrinth without a cochlea, or with but a trace of one, undoubtedly hear distinctly.   In certain crustacea the organ of hearing is a mere spherical vesicle, partly lined by auditory hair cells, and some of these have actually been seen vibrating to sounds, the otolith being forced down on the vibrating hairs to act as a damper when the sound was intense. Whether, however, the vestibular nerve filaments of the utricle and saccule distinguish anything but differences of intensity in sound is not known.   The whole subject of the mechanism of hearing is far from being satisfactorily settled.[1]

242. **Auditory Judgments.**--To the sensations of sound the mind adds certain judgments respecting them, forming its conclusions by knowledge previously acquired, and by the aid of other senses.   As a visual sensation is referred to some external object, and not to the retina, so a sound is referred to something outside the internal ear.   Whether the sound is produced inside our own body, or by some object outside ourselves, is mainly decided by noting whether the sonorous vibrations reach us through the external meatus or through the

---

[1] 'The organ which, on structural grounds, we consider to represent that of hearing in animals low in the scale of organisation—as e.g. in the Ctenophora—has nothing to do with sound, but confers on its possessor the power of judging of the direction of its own movements in the water in which it swims, and of guiding these movements accordingly.   In the lowest vertebrates, as e.g. in the dog-fish, although the auditory apparatus is much more complicated in structure, and plainly corresponds with our own, we still find the particular part which is concerned in hearing scarcely traceable.   All that is provided for is that sixth sense, which the higher animals also possess, and which enables them to judge of the direction of their own movements.   But a stage higher in the vertebrate series we find the special mechanisms by which we ourselves appreciate sounds beginning to appear—not supplanting or taking the place of the imperfect organ, but added to it.   As regards hearing, therefore, a new function is acquired without any transformation or fusion of the old into it.   We ourselves possess the sixth sense, by which we keep our balance, and which serves as the guide to our bodily movements.   It resides in the part of the internal ear which is called the labyrinth.   At the same time we enjoy along with it the possession of the cochlea, that more complicated apparatus by which we are able to hear sounds and to discriminate their vibration rates.' Professor Burdon-Sanderson.

bones of the head. If the meatus be filled with water the idea of externality to the body disappears, and the sound that then reaches the tympanic membrane seems to originate in the head. We determine the *direction* of a sound chiefly by the difference of intensity with which it is heard by each ear. If the source of a sound be directly behind or before, our estimate of its position is faulty until the head is turned to one side or the other. The estimate of the *distance* of a sound depends on previous experience of the habitual quality of the sound at varying distances, and any mode of transference, as by a speaking tube, that checks the law of intensity, diminishing according to the square of the distance, leads us to suppose the sound much nearer than it is. Ventriloquists may deceive us by imitating the character of distant sounds. The slight difference, however, with which sounds affect the two ears does not lead to the idea of two sounds, for they are in some way fused into one. Binaural audition is the ordinary mode of hearing, and appears to aid in the perception of space, though in a far inferior degree to binocular vision.

243. **General Remarks on the Senses.**—Each of the special organs of sense may be said to consist of three parts :—(1) A receiving organ situated on the periphery and constituted of modified epithelial cells destined to receive a particular kind of impression or stimulus; (2) a conducting portion consisting of a nervous cord or pathway ; (3) a perceptive organ or central nerve cell where the sensation or particular state of consciousness determined by the stimulus arises.

Although sense stimuli may be mechanical, electrical, thermal, or chemical, each sense organ has its own normal stimulus acting on the special peripheral end-organ of the sense, this end-organ being usually insensible to other stimuli. For sight, the stimulus is the undulations of the ether, which act upon the epithelial cells termed rods and cones, probably through chemical changes induced in pigment; for hearing, the stimulus is of a mechanical nature, and consists of the vibrations of the material air ; for touch, the stimulus is the mechanical pressure on the surface of the skin ; for taste, the stimulus is the molecular vibrations of a sapid substance in solution on the tongue and other parts ; for smell, the stimulus is the odorous particles acting on the olfactory cells in the olfactory mucous membrane ; for the sense of temperature, the stimulus is certain slower ethereal vibrations than those producing the sensations of sight.

The stimulus must have a certain intensity, for there may be sounds so low that we cannot hear them, contact so feeble that it does not affect us, &c. The *least intensity* of stimulus that excites a sensation has been thus estimated for the different senses : --

Sense	Least Appreciable Stimulus
Sight	$\frac{1}{300}$th of the amount of light from the full moon reflected from white paper
Hearing	The sound of a cork ball weighing 1 mm. gram falling from a height of 1 m. metre on a glass plate 90 millim. distant
Touch	A pressure of ·002 gram
Taste	1 part of quinine sulphate in 1,000,000 parts of water
Smell	$\frac{1}{2,000,000}$th milligramme of an alcoholic solution of musk
Temperature	$\frac{1}{4}$° C. when the skin temperature is 98·5° C.

The maximum limit of sensation is difficult if not impossible to state. After reaching a certain pitch of intensity any increase in the stimulus ceases to be perceptible.

As to the smallest appreciable increase of sensation, or the least possible difference in intensity, which gives rise to a sense of difference in succeeding stimuli, much valuable information has been obtained by Weber and others. In the first place it is found that the increase of stimulus necessary to produce an increase of the sensation bears a constant ratio to the total stimulus, *i.e.* that the fraction expressing the least perceptible difference between two succeeding stimuli varies for each sense but is constant for that sense. Thus :—

Sense	Constant of Difference
Sight	$\frac{1}{100}$th of the total amount of light
Hearing	Doubtful (one-third ?)
Touch	One-third of the original pressure
Temperature	Varies with the skin temperature
Muscular sense	$\frac{1}{17}$th of the weight tried

Thus, to perceive an increase in a weight of 17 ounces it would be necessary to add one ounce more, and to perceive an increase in a weight of 170 ounces, *ten* ounces would have to be added, and for 170 lbs. *ten* lbs. would be the least observable difference, the constant of difference being $\frac{1}{17}$th the original weight. In the case of light a sense of different luminosity would be felt by making a difference of $\frac{1}{100}$th in the amount of light.

In the second place we may note that, though the intensity of the sensation increases as the stimulus or exciting cause increases, yet it increases in a less proportion. The law expressing the relation between the sensation and the stimulus may be put thus : 'As the strength of the stimulus increases in geometrical progression, the strength of the sensation increases in arithmetical progression.' Thus, if the stimulus increase in the geometrical progression 1, 2, 4, 8, 16, the sensation only increases in the arithmetical progression 1, 2, 3, 4, 5. This law (Fechner's Law) holds best in regard to light, though even here it is only true within certain limits, *i.e.* when the strength of the stimulus lies within the middle ranges of the scale of intensities.

Time is required for the transmission of a nerve impulse along a nerve to and from the central organ, and we know that the nerve current travels at a rate not exceeding 100 feet per second (par. 46). Time is also occupied in the nerve centres in giving a response, either an immediate motor response as in such a simple unconscious reflex action as winking, in which the reflex time is but 0·01 to 0·015 second, or a response show-ing that the subject has become conscious of, *i.e.* perceives, the sensation. The perception or reaction time for the senses, then, is the interval between the application of a stimulus and the responsive signal that the sensation has been felt. By arranging an electro-magnet so that it marks on a blackened revolving cylinder the moment when a sensory impulse is applied, and the moment when the subject experimented on gives a signal, this interval can be found. It varies with different degrees of attention and health, the varying result being due to the vary-ing time taken by the brain centres, but average values of the reaction time for three of the senses are given in the annexed figure from Dr. Waller.

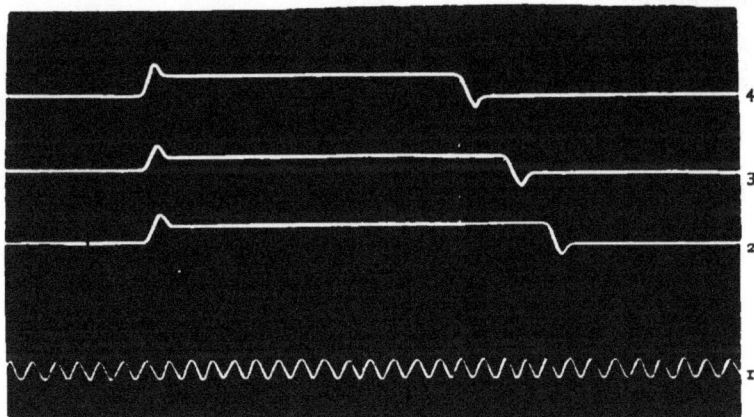

Fig. 267.—1, time indicated in ¹⁄₁₀₀th sec. ; 2, perception time for sight ·18 sec. ; 3, for hearing ·16 sec. ; 4, for touch ·14 sec.

Normally the retina is excited by waves of light which give rise to visual impulses and then to visual sensations. But the retina may also be excited by mechanical or electric stimuli with resulting visual sensations, and it is also said that the excitation of the optic fibres themselves always produces visual sensations, though this last statement is doubted. Hence has arisen a view termed *the specific activity or energy of nerves*, which holds that any impulse in a sensory nerve, however excited, will give rise to a sensation specific to that nerve ; that impulses, *e.g.* of all kinds along an optic nerve give rise to visual sensations, along the auditory nerve to auditory sensations, and so on. But the direct stimulation of the optic fibres is said by some observers not to produce visual sensations, and in the case of the skin it is certain that the direct stimulation of the nerves gives rise to no touch sensation, but rather to painful sensation, so that it would appear that the development of a special sensation must be started in the special terminal organ.

It is admitted, however, that stimulation of the various centres by any means gives rise to the specific sensation of the centre.   The doctrine of the specific energy of nerves thus becomes modified into the doctrine of the 'specific energy of nerve-centres.'

Notwithstanding our knowledge of the structure of the sense organs, of the conditions of sensation, of the relation of stimu⸱ lus to strength of sensation, and of the time taken up in nervous propagation and nervous elaboration, we know absolutely nothing of the mode by which physical energy becomes transformed into mental energy ; nothing of the mechanism occurring in the protoplasm of the cells 'within the book and volume of my brain,' nothing of the mode by which thought is related to the forces of external nature.   Our senses, even taken collectively, respond only to certain forces of nature, and to these only within certain limits, so that our conception of the universe must be but imperfect and incomplete, and all that the most learned can say is what the Soothsayer in 'Antony and Cleopatra' says :—

> In nature's infinite book of secrecy
> A little can I read.

APPENDIX

APPENDIX

APPENDIX

MEASUREMENTS—FRENCH AND ENGLISH

1 lb. avoirdupois .	= 16 oz. av.
	= 7,000 grains
1 oz. avoirdupois .	= 437·5 grains
1 lb. troy .	= 12 oz. troy
	= 5,760 grains
1 oz. troy .	= 480 grains
1 gallon .	= 8 pints
	= 277·25 cubic inches
1 pint .	= 20 fluid oz.
1 gallon of water at 62° F. (16°·7 C.) weighs	70,000 grains
1 cubic foot of water at 62° F. weighs .	62·326 lb. av.
1 metre (m.) .	= 10 decimetres (dm.)
	= 100 centimetres (cm.)
	= 1,000 millimetres (mm.)
	= 39·37 inches
1 foot .	= 30·48 centimetres
1 inch .	= 25·40 millimetres
1 litre (l.) .	= 1,000 cubic centimetres (c.c.)
	= 1·76 pints
1 pint .	= 568 cubic centimetres
1 fluid oz. .	= 28·4 ,,
1 kilogramme (kg.) .	= 1,000 grammes (g.)
	= 2·205 lb. av.
1 gramme .	= 10 decigrammes (dg.)
	= 100 centigrammes (cg.)
	= 1,000 milligrammes (mg.)
	= 15,432 grains
1 lb. avoirdupois .	= 453·6 grammes
1 oz. avoirdupois .	= 28·35 ,,
1 grain .	= 64·8 milligrammes

The unit of microscopic measurements is the micro-millimetre $=\frac{1}{1000}$th part of a millimetre, o·oo1 mm. $=\frac{1}{25000}$ in. nearly. It is denoted by the Greek letter μ.

1 kilogrammetre .	= 7·24 foot-pounds
1 foot-pound .	= 0·1381 kilogrammetre
(1 kilocalorie .	= 424 kilogrammetres)

Centigrade and Fahrenheit Scales.

C.	F.	C.	F.	C.	F.	C.	F.	C.	F.	C.	F.
0°	32°	8°	46·4°	20°	68°	34°	93·2°	38°	100·4°	42°	107·6°
1	33·8	10	50	24	75·2	35	95	39	102·2	43	109·4
2	35·6	14	57·2	28	82·4	36	96·8	40	104	44	111·2
4	39·2	18	64·4	30	86	37	98·6	41	105·8	45	113

Equation of Interchange. $\dfrac{C}{100} = \dfrac{F-32}{180}$

Note on the Microscope and Microscopic Figures.—Microscopes are optical instruments which enable us to see and examine objects which are too minute to be seen by the naked eye. A normal unaided eye can divide an object $\frac{1}{250}$ in. in length at a distance of 10 inches. At 5 inches a smaller object may be divided, but beyond 4 or 5 inches the object is too near for accommodation. With a first-rate compound microscope objects as fine as $\frac{1}{100000}$ in. in length may be seen. Microscopes are either simple or compound. With a simple microscope, as a lens, the object is viewed directly ; with a compound microscope two or more lenses are so arranged that an enlarged image of the object formed by one lens (the objective) is magnified by a second (the eye-piece) and seen as if it were the object. For histological purposes a microscope with two objectives—a low power working at about $\frac{1}{2}$ inch from the object and a high power with a focal distance of $\frac{1}{8}$ inch—should be obtained. Two oculars or eye-pieces of different power are also an advantage. With such an instrument a magnifying power of from 50 to 400 diameters may be obtained. To bring out the minute structure of cells and certain tissues still higher powers are needed. Illustrations of objects seen under such powers will be found in the book, as well as illustrations drawn under low powers and illustrations of objects not magnified at all, and the student should note when examining a figure the conditions under which it is drawn. In this country the size of microscopic objects is often given in fractions of an inch, but the unit of microscopic measurement employed on the Continent is also used. This unit is the *micro-millimetre*, or the one-thousandth part of a millimetre, 0·0000397 inch $= \frac{1}{25000}$ inch nearly. The letter μ (mu) denotes the micro-millimetre. Thus the diameter of a red corpuscle is said to be $\frac{1}{3200}$ in., or between 7μ and 8μ.

It is scarcely necessary to remind the reader that the superficial magnification equals the square of the linear magnification. Thus 500 diameters linear = 250,000 superficial enlargement.

(The student is recommended to consult Prof. Schäfer's ' Essentials of Histology.')

THE CHEMISTRY OF THE BODY

The Chemistry of the Body.—Of the seventy-two elements known to chemists, fourteen enter into the composition of the human body. These are oxygen, carbon, hydrogen, nitrogen, sulphur, phosphorus, chlorine, sodium, potassium, calcium, magnesium, iron, fluorine, and silicon, the first four forming about 85 per cent. of the whole. Other elements, as manganese and lead, have sometimes been found in small quantities. With the exception of oxygen and nitrogen, which are found dissolved in the blood, and hydrogen found as a result of putrefactive processes in the alimentary canal, the elements are always united to form chemical compounds. These compounds, or *proximate principles* as they are called in physiology, are either (*a*) **Mineral or inorganic compounds,** or (*b*) **Organic compounds,** *i.e.* **compounds of carbon.**

The inorganic compounds are water, acids, such as hydrochloric acid in the gastric juice, and salts, such as calcium carbonate and calcium phosphate in bone, sodium chloride in blood and urine, &c. Water is found in greater or less proportion in all the tissues, and forms about two-thirds of the weight of the whole body.

The **organic compounds** are numerous, and include four important classes, **Proteids, Albuminoids, Carbohydrates,** and **Fats,** the last two of which are non-nitrogenous, *i.e.* contain no nitrogen.

Proteids, called also 'albuminous' bodies, form an important class of complex organic compounds containing carbon, hydrogen, oxygen, nitrogen, and sulphur. They occur in a solid viscous condition or in solution in nearly all the solids and liquids of the body. The proteids, with the exception of hæmoglobin, which is really a compound of the proteid globin with hæmatin, are all amorphous or non-crystallisable. They are insoluble in alcohol and ether. Some are soluble in water, others insoluble. Some are soluble in weak solutions of neutral salts, such as sodium chloride and magnesium sulphate, others soluble only in concentrated saline solutions. It is on these varying solubilities that the classification of proteids depends. The following additional account of the proteids is taken with slight alterations from Watts's *Dictionary of Chemistry,* edited by Morley and Muir.

Proteids are never absent from the protoplasm of active living cells, whether animal or vegetable, and they are indissolubly connected with every manifestation of organic activity. A definition of proteids is not possible in the logical sense. Gamgee gives in the following sentences a terse description of these substances, which must take the place of a definition : 'Proteids are highly complex, and for the most part non-crystallisable, compounds of carbon, hydrogen, nitrogen, oxygen, and sulphur, occurring in a solid viscous condition or in solution in nearly all the solids and liquids of the organism. The different members of the group present differences in physical, and to a certain extent in chemical, properties ; they all possess, however. certain common chemical reactions, and are united by a close genetic relationship.'

In vegetables the proteids are constructed out of the simpler chemical compounds which serve as their food. In animals such a synthesis never occurs, but the proteids are derived directly or indirectly from vegetables. By the action of certain digestive juices all proteids are capable of being converted into closely allied substances called peptones, which after absorption undergo reconversion into proteids.

The various proteids differ somewhat in elementary composition within the limits of the following numbers :

	C	H	N	S	O
From	51·5	6·9	15·2	0·3	20·9
To	54·5	7·3	17·0	2·0	23·5

CLASSIFICATION OF PROTEIDS

I. *Native Albumins.* These are proteids which are soluble in water, and not precipitable from their solutions by saturation with sodium chloride or magnesium sulphate. They are coagulated by heat. The important members of the group are egg albumin, serum albumin, and lactalbumin.

II. *Globulins.* These are proteids which are insoluble in water ; they are soluble in dilute solutions of neutral salts, and are precipitated in an uncoagulated condition by saturation with sodium chloride and magnesium sulphate. They are coagulated by heat. The most important members of the group are : serum globulin, or paraglobulin, as it is also called, fibrinogen, myosin, crystallin, and globin.

III. *Albuminates, or Derived Albumins.* This name is applied to the metallic compounds of proteids, and also to acid albumin or syntonin, and alkali-albumin. Restricting the term to the two latter substances, they may be defined as proteids insoluble in water or in solutions of neutral salts, but readily soluble in dilute acids or alkalis. Their solutions are not coagulated by heat. The caseinogen of milk may be put in this class.

IV. *Proteoses.* These are proteids which are not coagulable by heat, and most of them are precipitable by saturation with certain neutral salts. They are precipitated by nitric acid, the precipitate dissolving on the application of heat and reappearing when the solution is cooled. They resemble peptones in being slightly diffusible, and in giving the biuret reaction. They are formed from other proteids as the result of the action of proteolytic ferments on them, being an intermediate stage in the formation of peptones. They are also found in certain animal and vegetable tissues. The best known members of the group are the albumoses, the chief albumoses being proto-albumose soluble in distilled water, hetero-albumose insoluble in water, and deutero-albumose, the most nearly allied to peptones.

V. *Peptones.* These are proteids which are very soluble in water ; they are not precipitated by heat, by saturation with any neutral salt, nor by nitric acid. They are completely precipitated by tannin, by excess of absolute alcohol, and by potassio-mercuric iodide ; incompletely by phosphotungstic acid, phosphomolybdic acid, and picric acid. They give the biuret reaction. Peptones are subdivided into hemipeptones, those which yield leucine and tyrosine as the further result of pancreatic digestion, and antipeptones, those which do not.

VI. *Insoluble Proteids.* This class includes a number of proteids varying in their reactions which cannot be included in any of the foregoing classes, but which all resemble one another in their extreme insolubility in various reagents. This class includes fibrin, coagulated proteid, lardacein, antialbumid, and gluten.

Among the *general* properties of proteids are :

1. *Indiffusibility.* Solutions of proteids are non-diffusible. They belong to T. Graham's class of colloid substances. Peptones, and to a less extent albumoses, are, however, diffusible. This property of indiffusibility enables us to separate proteids from saline admixtures, and also to separate various proteids from one another : *e.g.* if a mixture of albumin and globulin in a saline solution be dialysed, the salts pass out, the albumin remains within the dialyser in solution, while the globulin, which is insoluble in water, is precipitated.

2. *Action on Polarised Light.* Proteids all rotate the plane of polarised light to the left.

3. *Heat Coagulation.* Most of the proteids are coagulated by heating their solutions, especially the globulins and albumins, as in the familiar instance of the solidifying of the white of an egg on boiling. Serum globulin coagulates at a temperature of 75° C., egg albumin at 72° C., myosin and fibrinogen at 56° C.

Among the general *tests* for proteids are the following :

(1) *The Xanthoproteic Test.* On adding strong nitric acid and heating a yellow colour is produced : this becomes orange on the addition of ammonia.

(2) *Millon's Test.* An acid solution of nitrate of mercury gives a white precipitate, which turns brick-red on boiling.

(3) *Biuret Test.* Addition of a trace of copper-sulphate and excess of

potassium hydrate causes a violet colour. In the case of albumoses and peptones the colour produced is a pink one.

The decompositions that proteids undergo in the body lead to their breaking up into simpler bodies, some of which, like urea $CO(NH_2)_2$, form a class of crystalline nitrogenous substances. In the alimentary canal the proteids are converted into proteoses (albumoses) and peptones; this change is probably due to hydration. Under the influence of the pancreatic ferment, a certain class of peptones called hemipeptones are further acted upon, resulting in the formation of leucine, tyrosine, aspartic acid, ammonia, and protein-chromogen (a substance made purple by bromine). Putrefactive processes due to bacteria in the small intestine also occur; these result in the formation of indol, skatol, phenol, and oxy-acids. One of the sources of hippuric acid in the urine of flesh-feeders is the phenyl-propionic acid that results from the putrefaction of proteids in the alimentary tract.

After the proteids have been absorbed from the alimentary canal, they become assimilated by the tissues, and there undergo combustion or metabolism, the chief ultimate products being water, carbonic acid, and urea. It is probable that glycocine, leucine, and creatine and ammonium carbonate are intermediate products in this change. It has also been demonstrated, by experiments on animals, that proteid food gives rise to glycogen in the liver, and to fat in the subcutaneous and other tissues. That proteids can be converted into fats is also shown by the occurrence of *adipocere* in the muscular tissues after death.

Albuminoids.—The term albuminoids is now restricted to certain nitrogenous substances closely allied to proteids and that resemble proteids in many points, but differ from them in others. The chief members of the group are:

Collagen, the substance of which the white fibres of connective tissue are composed. It has been obtained from tendons, ligaments, and areolar tissue. A similar substance derived from bone is spoken of as *ossein*. On boiling with water collagen becomes converted into *gelatin*, a substance that sets into a jelly when the solution in hot water cools. Gelatin gives most of the proteid colour tests, but is not precipitated by acetic acid nor by many of the metallic salts that precipitate proteids. Though it is easily digested, being converted into a peptone-like body which is easily absorbed, it will not entirely replace proteids, acting only as a proteid-sparing food.

The question of the part played by gelatin, which is an easily digestible substance in nutrition, is very important practically, jellies especially being given to invalids. Voit's chief result showed that gelatin will not entirely replace proteids, but that animals rapidly waste which are fed on it alone; but, in conjunction with a certain small amount of proteid, it is capable of maintaining nitrogenous equilibrum as well as if the only nitrogenous food taken was proteid in nature. These results have been since very generally confirmed. Voit distinguishes between circulating and organic albumin; gelatin can never yield the latter, but it may replace the former in so far as it prevents the conversion of organic into circulating albumin. Gelatin also diminishes the waste of fat in the body.

Mucin is the albuminoid that forms the secretion of certain epithelial cells, and the main part of the intercellular substance of connective tissue. It is viscid and tenacious, precipitated by acetic acid and composed of a proteid and animal gum.

Mucin forms the chief constituent of mucus, and gives the sliminess to the secretion of mucous membranes. In mucus it is suspended in an alkaline exudation from the blood and mixed with the *débris* of epithelium cells, and a few white blood corpuscles. The mucin itself is here formed by the protoplasm of certain cells of the epithelium becoming altered, so that it becomes swollen and brightly refracting; the globule of mucin so formed is discharged, leaving a so-called goblet cell. In mucous glands, such as the submaxillary salivary gland, a very similar replacement of protoplasm by mucin (or mucigen, as it is called when inside the secreting cells) takes place. Mucin is also largely contained in the surface secretion of several invertebrate animals, *e.g.* the snail.

Chondrin is the albuminoid formed from cartilage on boiling it with water, and is probably a mixture of mucin and gelatin.

Elastin is the substance of which yellow or elastic fibres of connective tissue are composed.

Nuclein is the chief constituent of cell-nuclei, and resembles mucin in its physical characters, but differs chemically in the percentage of phosphorus that it contains.

Nucleo-albumin is a compound of proteid with nuclein, and is found in the protoplasm of cells.

Keratin is the highly insoluble substance that replaces the protoplasm in the surface cells of the epidermis, in nails, hairs, horns, &c.

Carbohydrates are compounds of carbon, hydrogen, and oxygen, there being two atoms of hydrogen for every atom of oxygen in the molecule. They are derived chiefly from vegetable tissues, and form important foods, but some of them, as *dextrose* (grape sugar), *glycogen*, and *lactose* (milk sugar), are found in animal organisms. Many of them possess the property of causing the rotation of a ray of polarised light, and the direction and amount of this rotation assist in their detection. According to their empirical formula they are arranged into three groups, Amyloses, Glucoses, and Sucroses.

They may be tabulated thus:

Classification of Carbohydrates.

I. Amyloses $(C_6H_{10}O_5)_n$	II. Sucroses or Saccharoses $C_{12}H_{22}O_{11}$	III. Glucoses $C_6H_{12}O_6$
Starch	Cane sugar	Dextrose
Dextrin	Lactose	Lævulose
Glycogen	Maltose	Galactose
Cellulose		

I. **Amyloses** or *Starches*, with the general formula $(C_6H_{10}O_5)_n$, where 'n' varies and is probably large. The chief amyloses are:

1. *Starch*, a body found in microscopic granules in many plants, insoluble in cold water, forms an opalescent solution with hot water that gives a blue colour with iodine, is changed by the ferments of the saliva and pancreas into dextrin and maltose through a process of hydrolysis or taking up the elements of water. With boiling dilute acids starch may be converted into dextrose.

2. *Dextrin*, the intermediate product between starch and sugar, two varieties being distinguished, erythro-dextrin which gives a reddish-brown colour with iodine, and achroo-dextrin which does not.

3. *Glycogen*, the animal starch found in the liver and muscles. The liver appears to receive its carbohydrate material as dextrose, to store it as glycogen, and to reconvert it by the activity of the hepatic cells into dextrose as required.

4. *Cellulose*, a colourless material forming the cell walls and woody fibres of plants, not easily digested by man, and hence starch requires cooking to burst the cellulose envelopes of the grains.

II. **Sucroses** or *Saccharoses* have the general formula $C_{12}H_{22}O_{11}$, and include :

1. *Cane sugar*, a sugar found in the juices and fruits of many plants and forming an important article of food, crystalline and easily soluble in water. Injected into the blood-vessels it is eliminated unaltered, and therefore appears to be non-assimilable by the tissues. Hence, when used as a food, cane sugar undergoes a change before absorption, being 'inverted' or changed into equal parts of dextrose and lævulose by the action of a ferment in the intestinal juice.

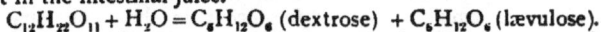

$$C_{12}H_{22}O_{11} + H_2O = C_6H_{12}O_6 \text{ (dextrose)} + C_6H_{12}O_6 \text{ (lævulose)}.$$

2. *Lactose* or milk sugar, a sugar occurring in milk but not assimilated until changed in some way into the glucoses, dextrose and galactose.

3. *Maltose*, the chief sugar formed from starch by the action of the ptyalin in saliva and amylopsin in pancreatic juices, formed also by the action of malt extract (diastase) on starch. Though very soluble in water, it appears to be converted into dextrose before absorption into the portal blood, either by the action of a ferment in the intestinal juice or by the activity of the epithelial cells of the intestines. 'The converting action of extracts of the intestinal mucous membrane is strikingly less than that of the tissue itself ; from this it may perhaps be inferred that the change into dextrose takes place rather *during* than previous to absorption. This fact corresponds closely to the well-known views as to the changes which peptones similarly undergo during their passage through the walls of the intestine into the neighbouring blood-vessels.' [1]

III. **Glucoses** have the general formula $C_6H_{12}O_6$, and include :

1. *Dextrose* glucose or grape sugar is a sugar found in certain fruits, in honey, and in minute quantities in the blood and other tissues. It is crystalline but not so sweet as cane sugar. It is important to note that all the carbohydrates are absorbed from the alimentary canal and reach the blood as dextrose. It is the sugar most readily assimilated. Dextrose possesses the important property of undergoing fermentations, *alcoholic* under the influence of yeast, when alcohol and carbon dioxide are formed ; *lactic* under the influence of another micro-organism. Dextrose is found in large quantities in the urine in diabetes.

2. *Lævulose* is a sugar occurring with dextrose in many fruits, and formed with dextrose from cane sugar by inversion. It is distinguished from dextrose by rotating the plane of polarised light to the left.

3. *Galactose* is one of the sugars formed by hydration of lactose or milk sugar. It ferments with yeast, but not so readily as dextrose.

Inosite or muscle sugar is a crystallisable substance found in muscle and other parts of the body in small quantities. Though possessing the same percentage composition as sugar, it does not really belong to the class of sugars, yielding none of the reactions of the class.

Fats.—Fats consist of carbon, hydrogen, and oxygen, the same three

---

[1] Sheridan Lea.

E E

elements as in carbohydrates, but the oxygen is in less proportion to the hydrogen than in carbohydrates. Fat is found in considerable mass in some tissues and in small particles in most, its average quantity in a healthy well-fed body being a little more than one thirtieth of the body weight. The three fats that occur in man are *palmitin*, *stearin*, and *olein*, the two former, which are solid at the body temperature, being kept liquid by mixture with the olein. Fats are insoluble in water, but soluble in ether and alchohol. Fats are considered as compounds of glycerin with a fatty acid (palmitic, stearic, or oleic), water being eliminated in the act of combination. When heated with a caustic alkali, as sodium hydrate, the fat splits up into glycerin and a compound of the fatty acid with the alkali termed a *soap*, and both glycerin and the alkaline soaps are soluble. This formation of a soap is termed *saponification*. Another change — physical, not chemical — that fats undergo is termed *emulsification*. Shaken up with some liquids fats are reduced to a fine state of subdivision termed an emulsion. Milk is a natural emulsion. Both saponification and emulsification of fats occur in the alimentary canal. The ferment *steapsin* of the pancreatic juice saponifies a portion of the fat of food, and the soap so formed aids the pancreatic juice and bile in forming an emulsion of the remainder. While a large part of the fat laid on by an animal during the process of fattening is derived from the fat of the food, yet it must be noted that fat can be formed from proteids, as is shown by the storage of fat during 'nitrogenous equilibrium,' and also from carbohydrates, as more fat may be stored than could be derived from the proteids.

Ferments. —Certain minute organisms possess the power of inducing definite chemical changes in the fluids or other media in which they live. Thus the cells of the yeast plant can transform sugar into alcohol and carbon dioxide, and to the change thus produced the term *fermentation* is applied, the agent producing the change being called a *ferment*. Other micro-organisms can lead to the formation of vinegar or acetic acid from alcohol, to the souring of milk by the formation of lactic acid, &c. These living organisms are spoken of as *organised ferments*.

In another class of chemical changes the result is brought about not by living organisms, but by chemical substances derived from living cells, these substances possessing the property of inducing chemical transformations in a large mass of certain other substances without themselves undergoing noticeable alteration. Such agents are called *unorganised ferments, soluble ferments*, or *enzymes*. The ptyalin of the saliva which changes starch into sugar is an example. The diastose of malt can also convert an indefinite amount of starch into sugar if the product of its activity is not allowed to accumulate too much. Confining ourselves to the unorganised ferments or enzymes of the human body, they may be classified as follows : —

(*a*) *Amylolytic*, or those which change amyloses as starch and glycogen into sugar. As examples we have the ptyalin of saliva and the still more active amylopsin of pancreatic juice.

(*b*) *Proteolytic*, or those which transform proteids into proteoses (albumoses, &c.) and peptones. As examples we have the pepsin of gastric juice and the trypsin of pancreatic juice.

(*c*) *Steatolytic*, or those which split fats into glycerin and fatty acids, as the steapsin of the pancreas.

(*d*) *Inverting*, or those which convert saccharoses (cane sugar, maltose, and lactose) into the assimilable glucose called dextrose.

(*e*) *Coagulative* or curdling, those which convert soluble proteids into insoluble proteids. As examples we have the rennin or curdling ferment of the stomach, that changes the soluble caseinogen of milk into insoluble casein ; the fibrin ferment of blood, that changes under suitable conditions soluble fibrinogen into insoluble fibrin ; and the myosin ferment of muscle, that leads to coagulation of mysinogen after death.

Most ferment actions are associated with hydrolysis, that is, water is added to the substance acted upon, which then forms a new substance or substances of lower potential energy. In all cases the product of the action ultimately puts a stop to the activity of the ferment unless removed. Ferments, too, are only active within certain limits of temperature, those of the body acting best at about 40° C. The chemical composition of ferments has not been determined, but they appear to be of a proteid nature. In some oases the ferments do not exist in the free and active condition within the secreting cells, but an inactive antecedent called a 'zymogen' occurs. Thus pepsin is not found in the cells of the gastric glands, but a zymogen termed ' pepsinogen ' is found there, and this appears to be converted into pepsin by the action of the gastric juice. Trypsinogen is the zymogen precursor of trypsin in the cells of the pancreas. No zymogen of ptyalin has been found.

(For further details the student may refer to Halliburton's ' Text Book of Chemical Physiology,' or to the same author's 'Essentials of Chemical Physiology.')

The Fœtal Circulation and the Changes that take place at Birth.— It has already been stated that a human being, like other animals, is developed from a cell termed the ovum. When the ovum is fecundated it undergoes a remarkable series of changes, and a description of the origin and formation of the organs in the *embryo*, as the rudimentary animal is called up to the end of the fourth month, is given in works on Embryology. From the end of the fourth month to the full term of nine months the offspring is called a *fœtus*, and it is important to have some acquaintance with the mode in which the fœtus, enclosed within the uterus or womb of the mother, is nourished, as well as a knowledge of the blood circulation in the fœtus. The structure which unites the fœtus to the womb of the mother and establishes a nutritive connection between them is called the *placenta*. The fœtus is united to the placenta by a cord, the *umbilical cord*, attached at the navel. This cord contains the *umbilical vein*, which conveys arterial blood, supplied with nutritive material and oxygen, from the placenta to the fœtus, and two *umbilical arteries* twisted round the cord, which return the blood from the fœtus to the placenta. To understand the course of the fœtal circulation it must be noted (1) that in the fœtal heart there is a communication between the two auricles by means of an opening termed the *foramen ovale* ; (2) that in the arterial system there is (*a*) a short tube, the *ductus arteriosus*, forming a communicating trunk between the pulmonary artery and the descending aorta, (*b*) branches given off from the internal iliac arteries called the *umbilical* or *hypogastric arteries*, which pass out at the navel and return to the placenta the blood that has circulated in the fœtus ; (3) that in the venous system there is a communication between the placenta and the liver by the umbilical vein, and another communication between the umbilical vein and the inferior vena cava by the *ductus venosus*. These peculiarities of structure enable the fœtal blood to gain oxygen and nutrient material from the maternal blood, and to discharge into the same such waste materials as carbon dioxide and urea. Bearing these facts and processes in mind, the student will understand the fœtal circulation by the help of the following figure and account adapted from Gray's ' Anatomy.'

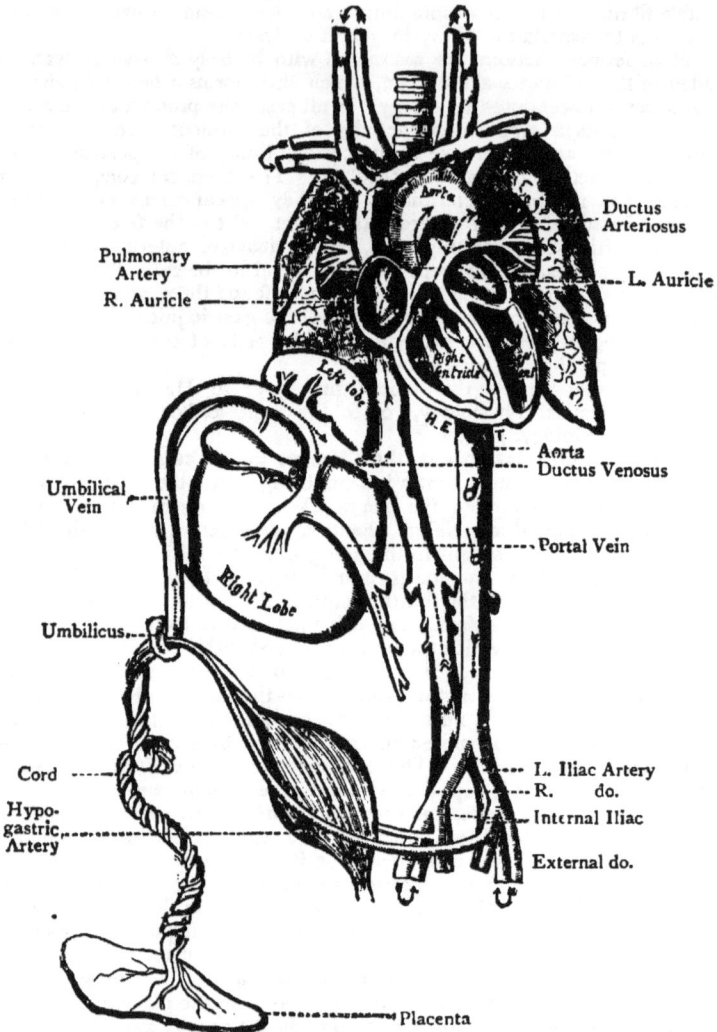

FIG. 268.—Plan of Fœtal Circulation (Gray's ' Anatomy ').

The blood destined for the nutrition of the fœtus is carried from the placenta to the fœtus, along the umbilical cord, by the umbilical vein. The umbilical vein enters the abdomen at the umbilicus or navel, and passing to the liver gives off branches to that organ, a special branch, however, the ductus venosus, proceeding directly to the inferior vena cava. In the inferior vena cava, the blood carried by the ductus venosus and that brought from the liver by the hepatic veins becomes mixed with that returning from the lower extremities and other abdominal viscera. The blood in the inferior vena cava enters the right auricle of the heart, and, without descending into the right ventricle, passes through the *foramen ovale*, guided by the Eustachian valve, into the left auricle, where it becomes mixed with a small quantity brought from the lungs by the pulmonary veins. From the left auricle the blood passes into the left ventricle, the contraction of which sends it into the aorta, to be distributed almost entirely to the head and upper extremities. From the head and upper extremities the blood is returned by the superior vena cava to the right auricle, from which it passes into the right ventricle. The right auricle is thus the point of meeting of a double current, a current from the inferior vena cava which passes through the foramen ovale into the left auricle, and a current from the superior vena cava which descends into the right ventricle. From the right ventricle the blood is sent into the pulmonary artery, but as the lungs of the fœtus are inactive, only a small portion is sent to them along the pulmonary arteries, the greater part passing through a short branch of the pulmonary artery, the *ductus arteriosus*, into the aorta at the point where it begins to descend. Along the descending aorta this blood passes to supply the lower extremities and viscera, a large portion, however, being conveyed by the umbilical arteries back to the placenta. Here this venous blood is brought into relationship with that of the mother, and, giving up waste products, returns again reoxygenated and charged with nutritive material.

At birth the placental circulation ceases, respiration through the lungs is established, an increased quantity of blood passes through the pulmonary arteries to the lungs, and the *ductus arteriosus* contracts and becomes completely closed in a few days. The *ductus venosus* and the *umbilical vein* become obliterated and dwindle to fibrous cords, while the *foramen ovale* is closed by the tenth day, the Eustachian valve being soon reduced to a mere trace. The umbilical arteries, or hypogastric arteries as they are called, within the body, remain in their first portion as the superior vesical arteries to the bladder ; the portion between the bladder and the navel becomes obliterated.

The Distribution of the Spinal Nerves.—There are thirty-one pairs of spinal nerves. The first eight pairs are called *cervical*, the next twelve *dorsal* or *thoracic*, the next five *lumbar*, the next five *sacral*, and the last pair *coccygeal*. The uppermost cervical pair of nerves leaves the spinal canal just above the atlas, and the second and following nerves below the vertebræ in succession, the last leaving below the first vertebra of the coccyx. Each spinal nerve springs from the spinal cord by two roots, an anterior root and a posterior root. These two roots unite in the corresponding intervertebral foramen into a single nerve-cord, and each trunk cord so formed divides immediately into two divisions termed the *anterior primary division* and the *posterior primary division.*

Further, the trunk gives off a small *recurrent* or meningeal branch which passes to the interior of the spinal canal, while the anterior division from the first dorsal to the second lumbar gives off a communicating branch (ramus communicans) to the sympathetic system. A small filament unites the communicating branch and the recurrent branch (fig. 269).

The *posterior primary divisions* of the spinal nerves are in general smaller than the anterior divisions. In most cases each turns backwards

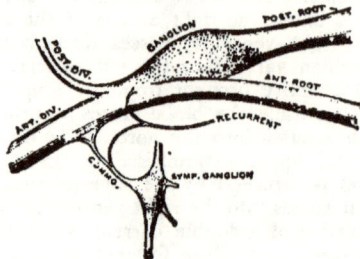

FIG. 269.

at once and divides into two parts, distributed to the muscles and skin of the back behind the spine.

The *anterior primary divisions* of the spinal nerves furnish branches which are distributed to parts of the body in front of the vertebral column, but their mode of division varies in the different parts. Thus the anterior divisions of the first four cervical spinal nerves unite to a complicated network termed the *cervical plexus*. From this branches are supplied to the muscles of the neck, while on each side an important nerve (the **phrenic**) passes to the diaphragm, supplying filaments on its way through the thorax to the pleura and pericardium. The anterior divisions of the lower four cervical nerves and the first dorsal nerve unite to form the brachial plexus (fig. 169), from which nerves pass to the arm. The anterior divisions of the twelve dorsal or thoracic spinal nerves are connected to the gangliated cord of the sympathetic system by short communicating branches (*rami communicantes*), of which there are two kinds, white and grey, both kinds being often united in the same cord. The white *rami communicantes* are mainly composed of fine medullated fibres passing from the spinal nerves to the sympathetic system. The grey *rami communicantes* consist of pale fibres arising wholly from the nerve-cells of the sympathetic ganglia, and passing in the anterior and posterior primary divisions of the spinal nerves to supply vaso-motor nerves to the arteries of the body-wall and limbs, pilo-motor fibres to the muscles of the hairs, and secretory fibres to the secretory glands.

After giving off its communicating branch to the sympathetic system, an anterior division passes forward in the intercostal spaces as an *intercostal* nerve, giving off filaments to supply the skin and muscles of the thorax and abdomen (fig. 270).

The first four lumbar nerves have anterior divisions that unite to form a *lumbar plexus*, from which branches pass to the abdomen and front of the thigh.

A *sacral plexus* is formed by the anterior divisions of the sacral nerves and the first lumbar nerve (fig. 169). From this plexus arises the largest nerve in the body, the *great sciatic*, a neuralgic affection of which is termed sciatica. Passing as continuation of the lower part of the sacral plexus through the great foramen in the pelvis, it descends along the back of the thigh until near the knee joint into the internal and external

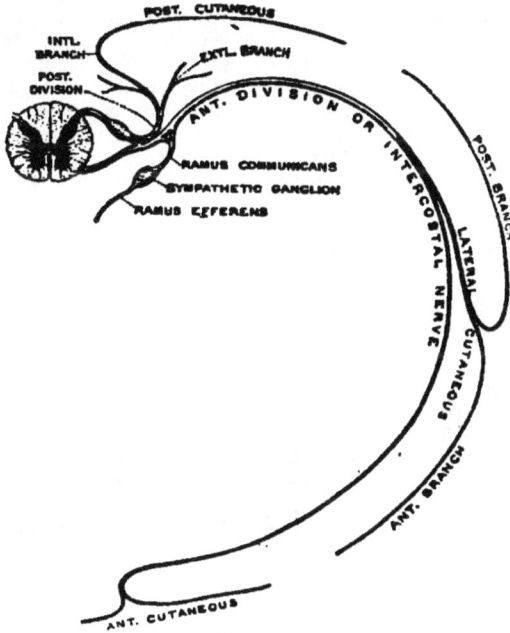

FIG. 270.

popliteal nerves. Its filaments supply nearly the whole of the skin of the leg, the muscles at the back of the thigh, the muscles below the knee, and those of the foot.

NOTE TO PAR. 77

Some important facts regarding the circulation and the blood pressure may be illustrated by the aid of an artificial model or *schema*, such as that shown in the diagram. The heart is represented by an elastic enema syringe, H; the arteries by a piece of elastic tubing leading from it, A; the capillaries by a reservoir C containing several passages; and the veins by another elastic tube leading from the capillaries back to the heart. In the tube representing the arteries there is inserted a manometer or pressure-gauge, M. Another manometer, P, is attached to the veins. The letter L represents a clip or clamp which can be screwed up so as to compress A.

The apparatus being quite full of water, the mercury will stand at the same level in each of the two manometers, since the pressure on the surfaces of the two limbs is the same—viz. the outside atmospheric pressure conveyed directly to m' and p', and through the water to m and p.

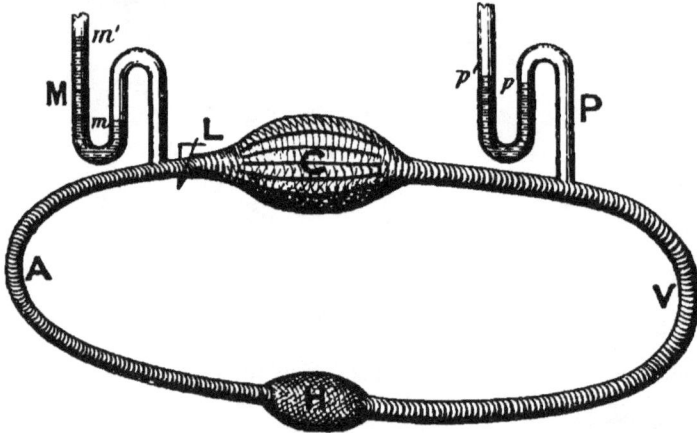

Now work the syringe with the hand to imitate the left ventricular beat that sends the blood into the aorta and its branches. If the connecting pieces are freely open, it will be noticed that after each stroke the increased pressure in A forces the mercury down in the near limb m of the manometer M and up in the far limb m'. The mercury, however, soon falls again, and very shortly afterwards the manometer attached to V shows a similar rise and fall; for the fluid pumped into A easily passes on to V. With each successive pump of the syringe the same rise and fall, owing to the same increase and decrease of pressure, takes place in each manometer.

Now screw up the clamp so as to narrow the tube A and to introduce a considerable resistance to the onward flow into C. This represents 'the peripheral resistance' offered by the contractile action of the small arteries. It will now be noticed that with each squeeze of the syringe the mercury rises higher and higher in the outward limb of M, showing increase of pressure in A, until it attains a certain height, where it remains, merely oscillating a little above and below at each squeeze of H (each heart beat as it were). But the venous manometer shows scarcely any rise of pressure, and but slight oscillation of the mercury in each limb. (The amount of pressure is shown by the difference of the level in the two limbs of the manometer.) For each squeeze of the syringe, or stroke of the pump as it may be called, forces a certain amount of fluid into the arterial tube and into C, but owing to the peripheral resistance this cannot pass on into the capillary lake all at once, so that part of the force of the pump is spent in sending the fluid onwards and part in distending the walls of the elastic arterial tube and increasing the pressure in it. The elastic reaction of the distended tube tends to force the fluid onwards also, and to overcome the peripheral resistance, where the more distended the tube becomes the stronger is this elastic reaction. With a certain rapidity of stroke the elastic reaction becomes so great that the fluid forced out of the arterial tube during the interval between each stroke is exactly equal to that sent in at each stroke, and thus the fluid passes continuously into the venous tube V and flows on steadily at a low pressure, a large part of the arterial pressure having been used up in overcoming peripheral resistance.

Thus an artificial model may be made to illustrate why the blood flows at a high mean pressure in the arteries, why there is an increase of this pressure with a wave of distension at each systole (the pulse), and why there is a continuous flow without a pulse and at a low pressure in the veins. It is also easy to see how increased peripheral resistance would increase the pressure in the arterial tube.

The pressures at certain parts of the circulatory system of man in millimetres of mercury are thus given by Dr. Starling : —

Large arteries (e.g. carotid)	+ 140 mm. mercury	
Medium arteries (e.g. radial). . . .	+ 110 ,,	,,
Capillaries	about + 15 to + 20 ,,	,,
Small veins of arm	+ 9 ,,	,,
Portal vein	+ 10 ,,	,,
Inferior vena cava	+ 3 ,,	,,
Large veins of neck . . .	from 0 to − 8 ,,	,,

SYLLABUS OF SECOND STAGE OR ADVANCED COURSE

(As revised in 1893)

In addition to the elementary stage, a knowledge of the following subjects, as defined below, will be required.

In this stage, in contrast to the Elementary Stage, a practical acquaintance, by the use of a microscope, with the minute structure of the several tissues and organs is essential ; but a knowledge of well-established features only will be required, to the exclusion of all doubtful and controverted details.

a. The Blood and Lymph

The size and structure of the corpuscles of these fluids. The phenomena which they exhibit. Their probable origin, functions, and fate.

The composition of the blood in detail. The nature of the process of blood-clotting.

The quantity of blood in the body, and its distribution.

The microscopic characters and chemical composition of lymph.

b. The Circulatory System

The structure of the heart. The general distribution and arrangement of the principal arterial and venous trunks. The minute structure of the heart, arteries, veins, and capillaries. The disposition of the capillaries in the various tissues.

The detailed analysis of the movements constituting a ' beat ' of the heart ; the changes of form undergone by the several parts of the heart during a cardiac systole, their nature and effects in propelling the contents of the several cavities, and their time relations. The details of the action of the valves of the heart. The sounds of the heart.

The physical properties of the blood-vessels. The mechanics of the circulation. Blood-pressure. The circumstances determining the rate of flow. The phenomena of the pulse.

The nervous arrangements of the heart. The nature of the rhythmic movements of the heart ; how these are maintained, and how they are modified according to circumstances ; the influence of the pneumogastric and of other nerves on the heart.

The vital properties of the blood-vessels. The influence of the nervous system on the blood-vessels ; vaso-motor nerves. The regulation of the blood supply by means of the nervous system. The influences exerted by the walls of the blood-vessels on the flow of blood as seen in inflammation.

The structure of the lymphatic vessels and glands, and the connection of the lymphatic with the blood vascular system. The origin of the lymphatics. The causes which determine the amount of lymph present in the lymphatic spaces, and those which determine the flow of lymph along the lymphatic channels.

c. The Alimentary System

The structure and functions of the salivary glands. The structure and functions of the tongue, the soft palate, uvula and tonsils. The pharynx and the œsophagus, and the structure of their walls. The stomach, its form ; the structure of its walls ; its glands and their functions. The divisions of the intestine. The structure of its walls. Villi. Glands. Peyer's patches. The structure and functions of the pancreas. The peritoneum and the nature of the mesentery.

The phenomena of secretion as observed in the salivary glands, pancreas, and other secreting organs, and the manner in which it is governed by the nervous system. The action of the chorda tympani.

The details of the digestive and absorptive processes.

The structure of the liver, and the course of the blood through it. The arrangements of the ducts of the liver. The composition and functions of the bile. The nature, uses, and fate of glycogen.

The structure of the spleen and of the suprarenal, thymus, and thyroid bodies ; and the probable uses of these organs.

d. THE RESPIRATORY SYSTEM

The structure of the thorax. The pleuræ. The structure of the respiratory organs, and the distribution of the blood through them. The analysis of the respiratory movements in detail. The mechanism by which coughing, sneezing, sighing, and hiccoughing are effected. The gases of venous and arterial blood. The physical and chemical processes involved in the conversion of inspired into expired air, and of venous into arterial blood. The respiratory activity of the tissues and the conversion of arterial into venous blood. The quantity of waste products excreted and of oxygen taken in by the lungs in twenty-four hours. The importance of ventilation.

The manner in which the nervous system maintains and regulates the respiratory movements ; the respiratory centre. How the nervous mechanism of respiration is affected (1) by nervous influences, (2) by chemical changes in the blood. The phenomena of dyspnœa and asphyxia, how they are brought about.

The influence of the respiratory movements on the circulation.

e. THE URINARY SYSTEM

The minute structure of the kidney, ureter, and bladder.

The circulation in the kidney, and the changes which the blood undergoes in passing through it.

The nature of the secretory activity of the kidneys in relationship to the glomeruli and tubules, and the circumstances which affect it. The history of urea.

The quantity and nature of the waste products of all kinds excreted by the kidneys in twenty-four hours.

f. THE SKIN.

The minute structure of the skin, and of the hairs, nails, and glands connected with it.

The quantity and nature of the waste products excreted by the skin in twenty-four hours.

The regulation of sweating by the direct action of the nervous system.

g. NUTRITION

The statistics of nutrition ; the income and output of the body, how they are balanced.

The special effects and uses of proteid, gelatinous, fatty, and carbohydrate food stuffs. The changes in the chemical composition of the body resulting from excess or deficiency (1) of proteid, (2) of non-nitrogenous food ; the effects of starvation. Urea as a measure of the nitrogenous changes of the body. The effects of exercise on the chemical changes and composition of the body. The part played in nutrition by saline bodies.

The energy of the body, its source and expenditure : the amount and character of the daily expenditure.

The production in and the loss of heat from the body ; their balance during health, how maintained and regulated by changes (1) in the amount of loss, (2) in the amount of production. The part in this regulation played by the nervous system. The failure of this regulation of heat ; the temperature of fever.

h. The Muscular System and Animal Mechanics

The minute structure of cartilage, bone, connective tissue, and muscle. Cilia and ciliary action.

The changes, physical, chemical, and structural, which take place in a muscle during contraction. The characters of a simple and of a tetanic contraction. *Rigor mortis.*

The mechanism of standing, walking, running, and jumping.

The structure and working of the larynx.

The mode in which vowel and consonantal sounds and articulate speech are produced.

i. The Senses

The structure of the papillæ of the skin and of tactile corpuscles, end-bulbs, and other nerve-endings in the skin. Pacinian corpuscles.

The localisation of touch. The sense of pressure, pain, and temperature. The muscular sense.

The minute structure and nervous supply of the tongue as an organ of taste. The structure of taste bulbs. Sensations of taste.

The structure of the olfactory organ. The nature and extent of the air-chambers connected with it. The minute structure of the olfactory (Schneiderian) mucous membrane and of the olfactory nerve fibres. Sensations of smell.

The minute structure of the various parts of the eyeball.

The formation of the image on the retina.

The conditions of distinct vision. Accommodation.

Movements of the pupil. Movements of the eyeball. Binocular vision. Judgments of distance, form, and solidity.

Evidence that visual impressions begin in the rods and cones.

Colour-vision and colour-blindness.

The structural arrangements of the organ of hearing. The outer ear. The structure and functions of the middle ear. The minute structure of the inner ear.

Auditory sensations. Use of two ears.

k. The Nervous System

The minute structure of nerves ; medullated and non-medullated nerve fibres. The minute structure of ganglia.

The sympathetic system, its structure and its anatomical and physiological relations to the cerebro-spinal system.

The arrangement and structure of the investments of the brain and spinal cord. The more important structural differences observable in different parts of the spinal cord.

The conduction of motor (volitional) and sensory impulses along the cord.

The general structure of the brain, including the relation and connections of its principal parts, namely, spinal bulb (medulla oblongata), its pyramids ; cerebellum, pons Varolii, crura cerebri, corpora quadrigemina, optic thalami, corpora striata, the fibres of the cerebral hemispheres, and the internal capsule. The relations of the several ventricles. The minute structure of the cerebral cortex.

The relations of the olfactory bulb. The central connections of the optic nerves.

The origins and general functions of the other cranial nerves.

The more important functions of the spinal bulb, or medulla oblongata.

The localisation of function in the cerebral cortex.

The maintenance of equilibrium, and its relationship to the semicircular canals.

l. Reproduction.

The mode in which the fœtus is nourished.

The main features of the fœtal circulation ; and the changes in the circulation which take place at birth.

The lacteal glands and lactation.

Examination for Honours

In this examination, questions will be set at the discretion of the examiner, who will have regard to the state of teaching of physiology in the country, and the means of acquiring information.

The examination will consist of a preliminary written paper, and those who qualify therein may be required to attend a practical examination at South Kensington at a subsequent date.

It does not seem to be sufficiently understood that the standard of examination for honours is very much higher than that for the second or advanced stage. Candidates attempt the honours paper whose answers show that they are but ill prepared even for the advanced stage. Honours will not be awarded unless a candidate shows that he has made physiology the subject of prolonged and careful study.

PROGRESSIVE QUESTIONS (ORIGINAL AND SELECTED).

Questions indicated by an as erisk are taken from papers set by the Science and Art Department.

(*Aid the answers by drawings and diagrams whenever possible.*)

1. Describe the structure of an animal *cell*. How do cells commonly multiply ?

2. What do you know of the composition and properties of *protoplasm* ?

3. What do you understand by the term *tissue* ? How may tissues be classified ?

4. Briefly describe the varieties of epithelium. Where is ciliated epithelium found, and what purpose does it serve ?

5. Enumerate the varieties of connective tissue, and describe the distribution and structure of areolar tissue.

6. Describe the structure of bone as it may be made out by the naked eye. What is the structure as revealed by the microscope? Explain how the various parts of a bone are nourished.

7.* What regions of the body are supplied with ciliated epithelium? Describe the structure of ciliated epithelium, and give an account of ciliary action.

8.* Describe the structure of osseous tissue. In what particulars of construction does the middle of the shaft of a long bone (such as the femur) differ from either of the extremities? What purposes are served by these differences?

9.* Describe the minute structure of adipose tissue, and point out the changes which take place in it during fattening and during starvation. What evidence is there that fat is formed in the body? From what elements of the food is the fat thus formed in the body derived?

10.* Give an account of an ordinary muscle, describing its general appearance, and its structure as far as can be made out without the use of a microscope. State what further details can be made out by the help of a microscope. State briefly what takes place in a muscle (*a*) when a single momentary nervous impulse reaches it along the nerve; (*b*) when a series of such impulses in rapid succession reach it.

11.* In what parts of the adult body is cartilage present? Describe the general characters, and the microscopic structure of cartilage, mentioning the chief varieties of cartilage. To what extent is cartilage supplied with blood-vessels, and how is it nourished?

12.* Describe the structure of an ordinary (voluntary) muscle, and the changes which take place in it during a contraction. Explain the difference between a simple and a tetanic contraction. What is *rigor mortis*?

13. Write out a short description of the structure of the different varieties of secretory glands. What changes does the secretory epithelium undergo in any particular case?

14. What is the general structure of a mucous membrane? Describe that of a bronchial tube, and in particular the nature of a 'goblet-cell.'

15.* Describe the structure, gross and minute, of a large nerve-trunk, such as that of the sciatic nerve. Explain further how such a nerve takes origin from the spinal cord, and give a general description of the manner in which it ends. What do you understand by a 'nervous impulse?'

16. Describe the minute structure of the two varieties of nerve fibres. What changes take place in the structure of a nerve fibre when it is cut off from the nerve cells from which it arises? How may the function of a nerve fibre be ascertained?

17. Explain the nature of reflex action. To what actions may the term *automatic* be applied? What is *reflex time*?

18. What do you know of *tactile corpuscles*, *end-bulbs*, and *Pacinian bodies*? What other modes of peripheral nerve terminations are known?

19. Write out an account of the structure and function of the white and red corpuscles of the blood. What are the probable origin and fate of the red corpuscles? What other microscopic elements are found in blood besides the red and white corpuscles?

20. What do you know of the phenomenon termed *diapedesis*? Under what conditions is it most active?

21.* Describe the phenomena of the coagulation of blood. State what is known concerning the nature of the process, and the reason why blood does not coagulate in the healthy living blood-vessels.

22.* Describe the structure of the red corpuscles of the blood, and give an account of the properties and uses of hæmoglobin.

23.* Blood flows to a muscle by an artery, or by arteries, and returns from the muscle by a vein, or by veins. Explain the chemical differences between the blood in the artery and the blood in the veins ; and state what evidence these differences afford of the changes going on in the muscle. How are these differences affected by the muscle entering into a state of activity, *i.e.* contracting?

24.* What gases are present in arterial and in venous blood, and to what amount in each ? State how the results you give are arrived at. Explain how blood is able to contain such relatively large quantities of oxygen and carbonic acid, and point out the importance of this in the respiratory changes in the lungs and in the tissues. What is meant by dyspnœa and asphyxia ? What respiratory movements are respectively characteristic of these two conditions, and how are they brought about ?

25.* Describe the general arrangement and structure of the lymphatic vessels of a limb. What is the nature of lymph, how does it differ from blood, and what are its uses ? What are the chief causes which drive the lymph along the lymphatic vessels ?

26.* What is the structure of a lymphatic vessel, and of a lymphatic gland ? How do the lymphatics originate ? What are the causes of the flow of lymph ?

27.* Describe the left ventricle of the heart, and give an account of the changes which take place in its walls, its orifices, and its contents during a beat of the heart. State how you know that the events which you describe do really take place.

28.* Describe the structure, attachment, action, and uses of all the valves found on the right side of the heart. Describe in the case of each what disturbance of the circulation would naturally follow from the action becoming imperfect.

29.* Describe the structure of an artery as made out (*a*) without the help of, (*b*) with the help of, a microscope. What are the two chief properties of the arteries in the living body? How do large and small arteries differ in regard to these properties? What part do these properties respectively play in carrying on the circulation of the blood ?

30.* Describe the structure of a vein, as made out (*a*) without the help of, (*b*) with the help of, a microscope. Explain how veins differ in structure from arteries. State, in the order of their importance, the various causes which keep up the onward flow of blood in the veins from the capillaries of the body to the right auricle, and explain how each of these causes acts.

31.* Describe the structure of a capillary blood-vessel, of a very small artery, and of a very small vein, and give an account of the flow of blood through these three kinds of minute vessels, such as may be seen in a transparent tissue like the web of a frog's foot. Explain why the flow, though jerked and pulsatile in the arteries, is steady and even in the capillaries, and why the rate of flow is so much slower in the capillaries than in the arteries or in the veins.

32.* What do you understand by arterial blood pressure? Explain the part which arterial blood pressure plays in the circulation of the blood, pointing out the causes which lead to the establishment of a mean arterial blood pressure, or to a rise or to a fall of pressure. What are the chief results of a rise or fall of pressure? About how much is the mean arterial pressure in a large artery in a large animal, for instance, the carotid of a dog?

33.* What is the 'pulse'? Explain exactly what gives rise to it. How do the characters of the pulse differ in different parts of the body, and in the same part under different circumstances? Explain how these differences in the characters of the pulse are brought about and what they indicate.

34. What is known about the blood pressure within the heart?

35. Explain the mode in which a *kymograph* gives a record of blood pressure in an artery. What may be learnt on examining the blood pressure curve recorded by such an instrument?

36. Trace fully the course of events during a complete cardiac cycle. Describe and account for the sounds of the heart.

37. Give as full a statement as you can about the nervous control of the heart.

38. What is the evidence that the nervous system controls the circulation of any part? Where is the 'vaso-motor centre' situated? Are there vaso-dilatator nerves?

39.* Explain 'blushing' fully, and point out the ways in which the nervous system, by means of 'vaso-motor' nerves, is able to regulate the blood supply of any organ or tissue, according to the wants of the latter.

40.* Give an account of the structure of the lung, pointing out the uses of the several parts in carrying on the work of respiration. State exactly, and explain fully, what happens to a lung when a free opening is made into its pleural cavity through the chest walls, and what are the results of the respiratory movements under those circumstances.

41.* Give an account of the position, attachments, and structure of the diaphragm. State the origin, course, manner of ending, and function of the nerves which are distributed to it. Describe the movements of the diaphragm, and explain exactly how these movements lead to the entrance of air into, and exit from, the chest. What is hiccough? In what acts, other than respiration, does the diaphragm take part?

42.* Describe briefly the structure of a pulmonary alveolus. State the changes which, in life, are taking place in the air of the alveolus, and in the blood which is circulating in the blood-vessels of the alveolus; and explain how these changes are brought about.

43. Give an account of the mechanism by which the cavity of the chest is enlarged and diminished in ordinary quiet breathing – inspiration and expiration. What are the chief types of breathing, and what is *laboured* respiration?

44. Explain how the nervous system maintains and regulates the movements of respiration; describe the position and relations of the respiratory centre, stating what nerves convey afferent impulses to, and what nerves convey efferent impulses from it; and explain how its action is dependent on the amount of interchange which takes place between the air and the blood in the lungs.

45. What are *apnœa*, *dyspnœa*, and *asphyxia*? In what way do the respiratory movements influence the circulation?

46. 'The object of respiration is to give the hæmoglobin of the blood the opportunity of combining with oxygen, and so replacing the oxygen

which it has lost in its circuit of the body, and to allow the carbonic acid, which is partly dissolved in blood and partly in loose combination with its salts, to escape into the air.'—(Dr. A. Hill.) Explain clearly how the various events referred to in this quotation are brought about.

47. What is meant by the 'respiratory quotient'? Why is it less than unity?

48.* Enumerate the several 'food stuffs' or 'alimentary principles,' stating their approximate elementary composition, their most important characters, and the articles of food from which they may severally be obtained. Into what substances is each food stuff transformed before it leaves the body?

49.* What is the chemical nature of the fatty bodies usually serving as food for man? By what agencies, in what manner, and in what condition do fats pass from the interior of the alimentary canal into the blood? What purposes do fats taken as food serve in the economy?

50. What is the average daily loss of material from an adult body in twenty-four hours? Describe a suitable diet to make good such loss. What tissues are most affected during *starvation*?

51. What is the effect of a purely proteid diet? What amount of energy can be obtained from one gram of proteid?

52. The income and output of the body have been put into the form of an equation thus :

Food + oxygen = carbon dioxide + water + nitrogenous waste (urea, &c.) Explain the meaning of this equation.

53. Within the body the energy set free assumes many shapes, but it leaves the body in two forms alone, as set forth in the annexed equation :

. Energy set free by oxidation of food = muscular work + heat given off. Expand and explain the equation.

54. What purposes does water serve in the body? Is more water excreted than taken in or not?

55.* Describe the structure of a salivary gland, and the nature and properties of saliva. Trace out the chain of events by which a flow of saliva follows upon food being placed in the mouth, including an account of the changes taking place in the gland itself.

56.* Describe the microscopic structure of the stomach, pointing out the differences between the mucous membrane at the cardiac end and that at the pyloric end. What appearances are presented by the glands before and during or after secretion, and what conclusions may be drawn from these appearances concerning the act of secretion?

57. Write out a full account of the composition and properties of gastric juice, explaining clearly its action on the different kinds of food.

58. Give an account of the structure, gross and minute, of the pancreas. What changes have been observed in the cells when at rest and after active secretion?

59. State fully the composition and properties of pancreatic juice, and explain its action on the different food stuffs.

60.* Describe the general form, relations, and structure of the small intestine, dwelling chiefly on the minute structure of the mucous membrane. State briefly the changes which an ordinary meal undergoes in passing from the pylorus to the ileo-cæcal valve, and explain how these changes are brought about.

61.* Describe the structure of a villus, and the changes which take place in it during intestinal digestion, dwelling on the manner in

F F

which proteids, carbohydrates, fats, salts, and water are respectively absorbed.

62. What do you know about the position, structure, and functions of the pleuræ and peritoneum?

63.* Give a brief general description of the distribution of lymphatics in the body, and a more careful account of the structure of lymphatic vessels and glands. How do lymphatics begin in the several tissues, and what are the causes which lead to the constant flow of lymph from the tissues to the junction of the great lymphatic trunks with the vascular system?

64.* What is lymph? What is the ultimate source of lymph? What determines its greater or less abundance in a tissue from time to time? In what parts of the tissue is it lodged before it finds its way into the minute lymphatic vessels, and what are the causes which lead to its flowing onward along the lymphatic vessels into the venous system?

65.* Describe the structure of the liver, both as seen by the naked eye and when studied with the help of the microscope. What are the uses of bile in digestion, and on what constituents do these uses depend? What becomes of the bile after it has passed into the intestine? What purpose is served by the gall-bladder?

66.* Give an account of the structure of the liver, dwelling especially on the minute structure of a lobule. What is glycogen? How does it differ from sugar (grape sugar, glucose, dextrose)? What is the evidence that the liver forms glycogen?

67. What is known as to the origin, formation, and destination of *bile*?

68.* Describe the structure of the spleen. Give an account of the changes in bulk which it undergoes under various circumstances. Explain how these are brought out, and what is their probable significance.

69. Write a brief account of the thyroid and thymus glands and the tonsils.

70.* Describe the structure of the kidney. State the nature and the relative amounts of the chief constituents of urine. What are the reasons for thinking that different constituents of urine are secreted by different parts of the kidney?

71. Give a clear account of the renal circulation, and compare it with that in the liver and spleen respectively.

72. What are the composition and properties of urea? What do you know as to its origin and formation in the body? Where is it separated from the blood in the kidney?

73. What do you know of the conditions termed *glycosuria* and *diabetes*?

74. Describe carefully the structure of the skin, pointing out in particular the various nerve-endings in this organ.

75.* State the nature of perspiration, distinguishing between sensible and insensible perspiration. Describe the structure of the organs which secrete perspiration, and explain why profuse perspiration takes place (1) when the body is exposed to warmth; (2) as the result of strong emotion; (3) as the result of dyspnœa.

76.* In what parts of the body, and by what processes, is heat produced, and how is heat lost to the body? Explain the various means by which the bodily temperature of a 'warm-blooded' animal is kept constant under varying conditions.

77. How is it that the temperature of man is nearly the same in arctic and equatorial regions? What evidence is there that the nervous system regulates the production of heat? What is the cause of the increased temperature in fever?

78.* Describe the structure of the larynx, and explain the manner in which the voice is produced and modulated. How do vowels differ from consonants? How does whispering differ from speaking?

79.* Give an account of the appearance, structure, and relations of the chordæ vocales. Describe the changes which take place in the larynx (*a*) when a high note, (*b*) when a low note is sounded, and explain how these changes are brought about.

80.* Describe the manner in which a spinal nerve is connected with the spinal cord, giving an account of the minute structure of the roots, including the ganglion, and stating with what part of the cord each root is immediately connected. State fully the proof of the functions of the two roots, and briefly point out what is known concerning the function of the ganglion.

81.* Describe the appearances presented by a transverse section of a fresh spinal cord as seen by the naked eye, and those of a prepared section seen under the microscope. State exactly the evidence upon which different functions are ascribed to the anterior and the posterior roots of the spinal nerves.

82.* Describe the course of a spinal nerve within the spinal canal, and its manner of attachment to the spinal cord, as far as can be made out with the naked eye, or with a simple lens. Add what further facts can be learnt by the help of the microscope. Describe the structure, and state what is known concerning the functions of the ganglion on the posterior root.

83. By what means has the white matter of the cord been divided into ascending and descending tracts? Name some of these tracts.

84.* State the main parts of which the mammalian brain is composed. Describe what is successively seen when a sheep's brain is sliced away horizontally from above downwards, until the lateral ventricles are well laid open.

85. Give an account of the membranes that invest the brain. What structures could be seen on examining the base or under surface of a human brain?

86. Describe carefully the structure of the medulla or bulb. What nerve centres are found in the grey matter forming the floor of the fourth ventricle?

87. Enumerate the cranial nerves and classify them according to their function. What would be the result of section (*a*) of the left optic tract, (*b*) of the left optic nerve?

88.* What are the corpora quadrigemina, and what are their anatomical relations to other parts of the brain? State what is definitely known concerning the functions of these bodies.

89. Give an account of the nervous fibres or tracts contained in the *internal capsule*. Explain the result of lesion or injury of the internal capsule of one side of the brain.

90. Write out a short account of the cerebral hemispheres, mentioning the names of the chief fissures and convolutions.

91. What evidence have we of the functions of the cerebrum? Describe the microscopic structure of the cerebral cortex.

92.* Give an account of the vagus, or pneumogastric nerve, stating from what part of the brain it takes its origin, and to what parts it is distributed. State in order of their importance the chief functions of this nerve.

93.* Describe the arrangements and structure of the chain of sympathetic ganglia in the thorax and abdomen, explaining how it is connected with the spinal cord. To what structures do nerves pass from these ganglia, and what is known concerning the functions of these nerves?

94.* Give an account of the general structure of the brain, confining yourself to a brief statement of the general nature and form of the chief parts, and of their relations to each other. What actions can a frog, deprived of the cerebral hemispheres, carry out, and what can it not carry out?

95. Give a full account of the sensation of touch. How is acuteness of tactile sensibility estimated? With what nerve terminations is the sense of touch related?

96. Describe the various sensations of temperature, and show why it is thought that special nerves are concerned in these sensations.

97. What nerves and nerve centres are concerned in the maintenance of equilibrium and muscular co-ordination?

98.* Give a general account of the structure of the tongue, with a special description of its mucous membrane and the structures therein found. With what nerves is the tongue supplied, and what are the functions of each?

99.* Give a general account of the nature, origin, position, and distribution of the olfactory nerves. How does the mucous membrane supplied by them differ from the mucous membrane in other parts of the nose? What nerve, besides the olfactory, is distributed to the nasal mucous membrane, and what is its use? Give a physiological explanation of the old custom of pinching the nose when a nauseous draught is swallowed.

100.* Describe the structure of the iris. Under what circumstances is the pupil widened, under what circumstances narrowed? Explain the nervous agency by which the widening and the narrowing are brought about.

101.* What are the various movements of which the eyeballs are capable, and how are these movements effected? Under what circumstances is an object seen single, and under what seen double, with the two eyes?

102.* What are the physical conditions which determine that the visual image of an external object shall be sharp and distinct? What changes in these conditions lead to the visual image becoming blurred and indistinct though the retina remains unchanged? By what changes in the eye are we able to see distinctly, at one time things far off, and at another time things quite near, and what is the proof that these changes do take place, and do give us the power in question?

103.* Give an account of the structure of the retina, including the 'pigment' or 'retinal' epithelium. How does the macula lutea differ from the rest of the retina, and how do we know that the processes of vision begin in the region of the rods and cones?

104.* What is meant by accommodation of vision? How can you prove that you possess this power, and how would you measure the near and far limits of your own accommodation? Explain the nature of the changes in the eye by which accommodation is effected, and how these changes are brought about, giving the evidence on which your statements are based.

105.* We are in the habit of recognising certain distinct sensations, which we speak of as sensations of so many different colours, hues, or tints. What is the evidence that these several sensations are not simple sensations, but are compounded of certain 'simple' or 'primary' sensations, mixed in different proportions? Point out the bearing of this on the explanation of colour-blindness.

106.* Describe the form and position of the tympanic cavity, and explain how air enters into and issues from that cavity. Give an account of the bones of the ear, describing their form and arrangement, and explaining their use and importance in hearing ; state the use of the muscles connected with them.

107. Describe the structure of the cochlea, and give a particular account of 'Corti's organ,' explaining the probable function of its various parts.

108.* Describe the form, position, and structure of the semicircular canals. What is the evidence that they possess functions other than those connected with hearing?

109. Write out a description of the mode of distribution of the auditory nerve, and the different ways in which its two main branches terminate.

110. What tracts of white matter have been observed in the spinal cord? Trace the extension within the central nervous system of the fibres of the two pyramidal tracts.

111. 'The grey matter of the cord may be regarded as a series of segmental centres fused together into a continuous longitudinal mass, each centre giving off a pair of spinal nerves to the two symmetrical halves of a vertebral segment.' Expand and illustrate this statement.

112. What are the phenomena of a typical case of *hemiplegia*, and how may these phenomena be explained?

113. What is meant by 'the localisation of cerebral function'? How has it been proved that the Rolandic area of the cerebral cortex is 'motor' in function? What subdivisions of this 'motor' area have been made?

114. Describe the course of the optic tract. What connections have its endings with the cerebral cortex? What is *hemianopia*, and how may it be produced?

115. What is known of a 'speech centre' in the brain? Describe the condition termed *aphasia*.

116. Give examples of various reflex actions. Mention the afferent and efferent paths taken by the nervous impulses in touching a hot plate, in sneezing, and in winking. What is meant by the inhibition of reflex actions?

117. How have the following been determined? and give the figures as accurately as you can :

(*a*) The velocity of the blood stream.
(*b*) The vital capacity of the lungs.
(*c*) The speed of the nerve current in a motor nerve.
(*d*) The blood pressure in the great veins.

118. Describe the structure of (*a*) a motor end-organ ; (*b*) a Malpighian capsule ; (*c*) the pons Varolii.

119. Discuss the mechanism by which the blood plasma or lymph leaves the blood capillaries. What are the phenomena of *inflammation*?

120. What are the composition and properties of chyle ? What changes does it undergo in passing from the lacteals to the receptaculum chyli ?

121. Classify the proteids and carbohydrates, and explain the changes each class undergoes in the alimentary canal.

122. State what is known regarding the various digestive ferments—their nature, origin, and activity.

123. What is a *nerve-muscle preparation*, and what may be learnt from experiments on such a preparation ?

124. Describe the nervous supply of a submaxillary gland, and explain in detail the action of the *chorda tympani.*

125. Describe the special arrangement of the capillaries of the lung and of striated muscle. What changes does the blood undergo in these two positions, and how are these changes brought about ?

1892

Second Stage or Advanced Examination

INSTRUCTIONS

Read the General Instructions.

You are permitted to attempt only FIVE *questions.*

21. Give an account of the structure and chemical composition of the corpuscles of the blood. State what is known of the life-history of, and the purposes in the economy served by, the red and white corpuscles respectively. (26.)

22. Describe the structure, gross and minute, of the kidney. Explain the nature of the process of the secretion of urine, pointing out the probable functions of the several renal structures. (26.)

23. Describe the structure of a villus, and the changes which take place in it during intestinal digestion, dwelling on the manner in which proteids, carbohydrates, fats, salts, and water are respectively absorbed. (16.)

24. Describe the more important features of the right auricle of the heart, state the events taking place in it during a whole beat of the heart, and point out the uses of the auricle in the work of the circulation. (16.)

25. State the several parts of the nervous system concerned in carrying out an ordinary breath (an ordinary inspiration followed by an ordinary expiration), and explain how the several parts work. Trace out the chain of events through which (*a*) an imperfect supply of pure air, (*b*) powerful emotions, disturb calm breathing. (16.)

26. Describe the structure of a ganglion on the posterior root of a spinal nerve, and state what is definitely known as to the function of such a ganglion. (16.)

27. State the changes which may be observed to take place in an eye when 'accommodation' for a distant object takes place, and explain how these changes are brought about. (16.)

28. Give a brief-sketch of the structure of the larynx and of its con-

dition when easy breathing, in the absence of voice, is taking place. Explain the nature of the changes which take place when vocal sounds of low pitch and of high pitch are uttered. (16.)

1893

Second Stage or Advanced Examination

INSTRUCTIONS

Read the General Instructions.

You are permitted to attempt only FIVE *questions.*

21. Describe the muscular movements which take place in ordinary quiet respiration. Explain exactly how these movements lead to the entrance of air into and exit of air from the lungs. How are the movements changed when by obstruction in the trachea or otherwise the air is prevented from passing into or out of the lungs? How is the change brought about? (26.)

22. What differences exist in the rate of bloodflow and in the pressure at which blood flows and in the character of the bloodflow in the arteries, capillaries, and veins respectively? What is the precise cause of these differences? By what means may the amount of blood sent through any given part of the body be made to vary in accordance with the varying needs of that part? (26.)

23. Where is the chain of structure known as the sympathetic nervous system placed in the body? What is the structural arrangement of this system, and how is it connected with the spinal cord? What are the chief functions of the sympathetic system? (16.)

24. Describe the general structure of the skin and the minute structure as it is seen under a microscope. State what are all the several uses of the skin to the body, and explain with reference to its structure how the skin provides for all these uses. (16.)

25. How does the structure of the mucous membrane of the large intestine differ from that of the small intestine? What changes does ordinary mixed food undergo in the small and in the large intestine respectively? How are these changes produced? (16.)

26. What is the structure of a muscle as it can be made out by mere dissection? What is the further structure which can be made out by the use of a microscope? What changes take place in a muscle when it contracts, and what changes take place when it passes into *rigor mortis*? How would you proceed to ascertain whether a muscle is in *rigor mortis* or no? (16.)

27. In what form is nitrogen taken into the body, and in what form and by what means does it chiefly leave the body? How and where is the change from the one form to the other brought about? (16.)

28. Describe the position, shape, and arrangement of the auditory bones or ossicles of the ear. How do they work in conveying sound from the exterior of the body to the internal ear? Where and by what structures is sound finally made to produce an effect on the nerve of hearing? (16.)

1894

Second Stage or Advanced Examination

INSTRUCTIONS

Read the General Instructions.

You are permitted to attempt only FIVE *questions.*

21. What is blood pressure? How is it produced and maintained? What is its importance in the circulation? How may it be increased or diminished in any region of the body, and what are the chief results of such an increase or diminution? (26.)

22. What parts of the *nervous* system are specially concerned in carrying out the movements of respiration? Explain the working of the parts with special preference to the means, both nervous and other, by which their action may be changed. (26.)

23. The body is always setting free energy in the form of heat and work. From what sources and in what form does it obtain this energy, and by what processes does it get the energy as heat and work out of these sources? How is the waste which leaves the body affected by (*a*) exercise, (*b*) starvation? (16.)

24. Describe the structure of a lymphatic gland. How is a lymphatic gland connected with the lymphatic vessels? What are the chief causes which lead to the flow of lymph along the lymphatic vessels? (16.)

25. Describe the structure of the pancreas. What differences can be observed between the pancreas of a fasting animal and one which has been recently fed? What do these differences teach as to the nature of the processes concerned in secretion? (16.)

26. What is (*a*) the gross and (*b*) the minute structure of a large nerve-trunk such as the sciatic? What can be stated as to what is taking place in a nerve while an impulse is passing along it? How can you tell when an impulse is passing along a nerve? (16.)

27. Describe the general structure and arrangement of the tongue as an organ of taste. What nerves supply the tongue? What conditions are (*a*) essential, (*b*) most favourable, for the tasting of any substance? (16.)

28. What is the structure of the gland by which milk is formed? Explain, by reference to the nature and relative quantities of the substances present in it, why milk is frequently spoken of as 'a perfect food.' (16.)

1895

Second Stage or Advanced Examination

INSTRUCTIONS

Read the General Instructions.

You are permitted to attempt only FIVE *questions.*

21. Describe the minute structure of a lobule of the liver, explaining carefully the means by which bile is discharged from it.

What is glycogen? Under what conditions is (*a*) very little, (*b*) a great deal of glycogen found in the liver? What conclusions may be drawn from these conditions as to the use and fate of glycogen in the body? (52.)

22. Describe the sequence of events which go to make up one complete 'beat' of the heart. State briefly the nervous and other means by which (a) the force and (b) the frequency of the heart-beat may be modified. (52.)

23. What is the relative proportion of the several *gases present in venous and in arterial blood*? What is there peculiar about the conditions in which these gases exist in blood, and what is the importance of these conditions as regards the changes blood undergoes in the lungs and in the tissues?

Why is it always difficult and usually impossible to restore an animal which has been shut up in a room in which a clear charcoal fire is burning? (32.)

24. In what respects does lymph resemble and differ from blood? Where is lymph found in the body, and how does it make its appearance in the places you describe? What are the uses of lymph, and what becomes of it after it leaves the places where it is formed? (32.)

25. Describe the general minute structure of the mucous membrane (only) of the stomach. How does the structure of this membrane near the pylorus differ from that near the cardiac end, and what conclusions may be drawn from these differences? What changes does a piece of fat meat undergo in the stomach? (32.)

26. What is the naked-eye appearance presented by a kidney split open lengthways through the entrance of the ureter? What are the microscopic features of minute structure which correspond to this naked-eye appearance?

State as briefly as possible what are the characteristic constituents of urine, and explain how varying causes, such as food, exercise, &c., lead to variations in the amount of these constituents excreted each day. (32.)

27. Give an account of the structure of the retina and of the retinal (pigment) epithelium. By what means may you observe the blood-vessels of your own retina, and what do you learn from such observations? (32.)

28. Describe the appearance of the brain as seen from a *side* view. What are the most characteristic functions of the several parts of the brain thus seen?

What is meant by the expression 'localisation of function' as applied to the brain? (32.)

1896

Second Stage or Advanced Examination

INSTRUCTIONS

Read the General Instructions.

You are permitted to attempt only FIVE *questions.*

21. Describe the general structure of a lung and the minute structure of a pulmonary alveolus. What changes of pressure take place inside the thorax during ordinary quiet breathing? What is the effect of these

changes of pressure on the heart and large blood-vessels inside the thorax, and hence upon the circulation? (52.)

22. What nervous structures are known to exist in the heart? What nerves connect the heart with (a) the brain, (b) the spinal cord? What is the effect on the beat of the heart of stimulating each of these nerves? Explain the series of events which take place in the body when a person faints. (52.)

23. Give an account of the general chemical composition of the blood, and state briefly the characteristics of the most important substances which enter into its composition. State what is most definitely known as to the part played by the several constituents of blood in causing clotting. (32.)

24. Describe the minute structure of a salivary gland and the general composition and properties of saliva. What differences in structural appearance can be made out between a salivary gland which has been secreting for some time and one which has been at rest for some time? (32.)

25. What is the general structure of an ordinary skeletal muscle? What further details of minute structure can be made out under the microscope? What various changes are known to take place in a contracting muscle? What are the characteristic changes of rigor mortis? (32.)

26. Give an account of the phenomena of reflex action, stating clearly the structures essential to its occurrence and the events which take place in these structures during a reflex action.

Give examples to show the importance of reflex action in (a) the *internal* relationships of the various parts of the body, and (b) the relationship of the body as a whole with its daily *external* surroundings. (32.)

27. What muscles are attached to the eyeball? What is the attachment of these muscles to the skull? What movements is the eyeball capable of executing, and how are they brought about by its muscles? Explain the action of these muscles in connection with binocular vision and squinting. (32.)

28. Describe the minute structure of a typical section of the spinal cord. What evidence is there that certain classes of impulses are conveyed along definite paths or tracts in the spinal cord? State the position of any one or two of the best known of these tracts. (32.)

1897

Second Stage or Advanced Examination

INSTRUCTIONS

Read the General Instructions.

You are permitted to attempt only FIVE *questions.*

21. Describe carefully the minute structure of a medium-sized artery. In what respects is the structure of a large and of a very small artery different from that of a medium-sized artery? What is the exact importance of these differences in connection with the circulation of the blood, and how are they brought into play? (52.)

22. Give an account of the vagus or pneumogastric nerve, so as to explain (i) its origin from the brain; (ii) the outlying parts of the body

to which it is distributed ; (iii) its chief functions in the order of their relative importance. (52.)

23. Describe the arrangement of the blood-vessels of the kidney, pointing out the characteristic peculiarities of this arrangement, and their physiological significance.

What change does the blood undergo in passing through the kidney, how are these changes brought about, and what is their importance to the body ? (32.)

24. Give an account of the general chemical properties of a proteid. What are the changes which proteids undergo in the alimentary canal before they are absorbed ? What structures and processes are concerned in bringing about their final absorption ? (32.)

25. Give an account of the structure, the changes which take place in, and the probable uses to the body as a whole of the spleen. What is the thyroid body, and what is known of its nature and use ? (32.)

26. What is the form and the relationship of the semicircular canals to the other parts of the internal ear ? Describe the minute structure of a semicircular canal. How far are these canals concerned in hearing, and what evidence is there that they possess functions not connected with hearing ? (32.)

27. Describe the structural and muscular arrangements of the larynx so as to explain how it works for the production of high and low notes.

Explain how the sounds characteristic of vowels and consonants are produced. (32.)

28. What is the structural appearance of a cross-section of the spinal bulb (medulla oblongata) at about its middle ? In what respects does this structure differ from that of the spinal cord ? What are the chief functions of the spinal bulb ? (32.)

June 1898

Second Stage or Advanced Examination

INSTRUCTIONS

Read the General Instructions.

You are permitted to answer only FIVE *questions.*

21. Describe the minute structure of the small intestine, with special reference to the structure of a villus. What is the exact use or function of the villi, and how do they work when carrying out this use ? (52.)

22. What do you understand by the expression ' a vaso-motor nerve ' ? Give two or three instances of the action of vaso-motor nerves in the body. Explain in the case of any *one* of these instances exactly (i) what the vaso-motor nerve does, (ii) how it does what you describe it as doing, and (iii) what is the exact use of what it does. (52.)

23. What is lymph ? Where is it found in the body, and what are its uses ? How does it resemble and differ from blood ?

What are the chief causes which drive the lymph along the lymphatic vessels ? (32.)

24. What is the amount of each of the gases which is found in venous

blood and in arterial blood? How has this difference in the respective quantities been brought about?

What is the effect on respiration of (i) breathing a limited portion of air for some time, and (ii) taking active exercise, and how is the effect brought about in each case? (32.)

25. What is the composition and general nature of the fats used as food, and how do they differ from carbohydrates? What changes do fats undergo from the time they are first eaten to the time when they enter the blood, and how are these changes produced? (32.)

26. In what parts of the body do ganglia occur, and how are they connected with the nervous system? Describe the structure of *one* of these ganglia, and state what is most definitely known as to its function or use. (32.)

27. What is the evidence that all our various sensations of colour may be reduced to certain 'primary' sensations mixed in various proportions? What is colour-blindness, and how may it be explained? (32.)

28. With what parts of the brain are the cranial nerves connected, and in what way are they connected?

Taking any *three* (but not more) of these cranial nerves, state the parts of the body to which they are severally supplied, and the general function or use of each of the nerves. (32.)

May 1898

Second Stage or Advanced Examination

INSTRUCTIONS

Read the General Instructions.

You are permitted to answer only SIX *questions.*

21. What are the most obvious and essential facts as to the structure of the corpuscles of the blood? What are the most important facts as to the chemical composition of the corpuscles of the blood? What is known of the origin, uses, and final fates of the corpuscles in the body? (50.)

22. What do you understand by the expression 'a vaso-motor nerve'? Give two or three instances of the existence of vaso-motor nerves in the body. Explain in the case of any *one* of these instances exactly (i) what the vaso-motor nerve does, (ii) how it does what you describe it as doing, and (iii) what the exact use is of what it does. (50.)

23. Describe the minute structure of the large intestine. What are the most important respects in which this structure differs from that of the small intestine?

What changes does food undergo in the large intestine and how are these changes brought about? (25.)

24. What gases are found in arterial and venous blood respectively, and in what amount are they present in each? Explain how it is that blood can contain these gases in the large quantities it does, and how the lungs bring about the differences between venous and arterial blood. (25.)

25. Where and how is heat produced in the body? Where and how is heat lost to the body? By what two means is the balance between production and loss of heat so exactly regulated that the temperature of a

warm-blooded animal is normally nearly constant? How do you account for the high temperature of fever? (25.)

26. Describe the minute structure of a medium-sized medullated nerve.

How are nerves connected to muscles, and what takes place in (i) the nerve, (ii) the muscle, when the nerve supplying a muscle is stimulated or irritated at a point at some distance from the muscle? (25.)

27. What are the several parts of the eye-ball which contribute to the formation of a sharply defined image of external objects on the retina? How and to what relative extents do each of these parts contribute to the formation of the image? Under what conditions may the image be blurred or indistinct? How is this blurring produced in each case, and how may it be naturally or artificially corrected? (25.)

28. What is the general structure and arrangement of the cerebellum? What are its main connections with the other parts of the central nervous system? What is known as to the probable chief functions or uses of the cerebellum? (25.)

GLOSSARY.

Aberration (L. *ab*, from, away, and *errāre*, to wander), a deviation in the rays of light owing to their unequal refraction in passing through a lens (or to unequal reflection from a mirror), so that they fail to unite in a point or focus, and thus produce an indistinct image. *Spherical* aberration is due to the fact that rays from the outer portion of a lens come to a focus sooner than the central rays ; *chromatic* aberration (Gk. *chrōma*, colour) is due to the fact that the coloured rays composing white light are differently refrangible, so that the image is blurred by coloured fringes. The eye is partly corrected for these two defects by its structure and by the action of the iris.

Abnormal (L. *ab*, from, and *norma*, a rule), irregular, contrary to the usual rule.

Accommodation (L. *ad*, to, and *commodāre*, to fit, adapt), the adjustment of the eye for objects at different distances. It usually refers to the adjustment of the eye for near points, and is effected by an increase in the convexity of the anterior surface of the lens owing to the contraction of the ciliary muscle producing a relaxation of the zonular or suspensory ligament, thus allowing the anterior surface of the lens to bulge (par. 228).

Acinus (L. *acīnus*, pl. *acini*, a berry or grape), one of the small sacs forming the terminal expansion of a minute passage, such as the lateral branch of a gland duct ; an alveolus. Acinous glands are those in which the branching twigs of the gland duct end in small saccules —as opposed to the merely tubular glands (fig. 111).

Adenoid (Gk. *adēn*, a gland, and *eidos*, form), glandiform or glandular. Adenoid tissue consists of a network of fine fibrils of connective tissue, the meshes of which contain many leucocytes.

Adipose (L. *adeps*, *adipis*, fat), consisting of fat : as adipose tissue.

Afferent (L. *ad*, to, and *fero*, I carry), carrying to or toward. Afferent nerves are nerves conveying impulses from the outside or periphery to a nervous centre, and afferent impulses are the impulses or currents passing to the centre. Afferent = centripetal. Afferent *vessels* are vessels entering an organ as opposed to those which leave it, termed efferent.

Agminate (L. *agmen*, a group, band). Agminate glands are the groups of small lymphatic glands occurring in patches (Peyer's patches) in the small intestines, as distinguished from the solitary glands or follicles of these parts.

Alveolus (L. diminutive of *alveus*, a hollow, a cavity), plural *Alveoli*, a word used to denote the small cavities found in many parts, as the air-cells of the lung.

Amorphous (Gk. *a*, not, *morphē*, shape), without definite shape ; opposed to crystalline.

Anabolism (Gk. *anabole*, a throwing up, a rising up), assimilation ; the processes by which food substances are converted into protoplasm by the cells of the tissues ; constructive metabolism. *Anabolic* is the adjective.

Anæsthesia (Gk. *an*, without, *aisthēsis*, feeling), diminution or loss of feeling, especially of tactile sensibility.

Analgēsia (Gk. *an*, without, and *algos*, pain), insensibility to pain in a part.

Anastomosis (Gk. *ana*, again, *stoma*, a mouth), inosculation ; the inter-communication of the vessels of any system with one another. To *anastomose* is to unite by anastomosis or intercommunication.

Antiseptic (Gk. *anti*, against, and *sēptos*, putrid), opposing or counter-acting putrefaction ; inimical to the activity of the micro-organisms of disease.

Aphasia (Gk. *a*, not, *phasis*, speech, saying), speechlessness ; impairment of the power to express ideas by speech. *Motor* or *ataxic aphasia* is the form in which the patient remembers the words he wishes to utter but cannot articulate them, although his vocal organs may be sound. It is due to lesion of the third left frontal convolution. *Sensory aphasia* is an impairment of the power to understand spoken words (par. 204).

Aponeurosis (Gk. *apo*, from, *neuron*, a nerve), the white shining membrane or *fascia* of connective tissue covering in the muscles or connecting the muscles and tendons with the parts they move.

Articular (L. *articulus*, a joint), pertaining to the joints.

Asphyxia (Gk. *a*, not, and *sphyxis*, a throb, a pulse), literally stoppage of the pulse, but now used to mean suffocation from interruption of the respiratory movements and extreme deficiency of oxygen in the blood. It is produced by strangulation, drowning, inhalation of coal-gas, &c.

Assimilation (L. *ad*, to, and *similis*, like), the act or process by which living organisms absorb and incorporate food so that it becomes part of the living tissues.

Astigmatism (Gk. *a*, not, and *stigma*, a mark), a defect in the refractive apparatus of the eye, usually due to the cornea being unequally curved in different meridians, so that rays proceeding from one point to the eye are not brought to a focus at one point.

Atrophy (Gk. *a*, not, and *trophē*, nourishment), a wasting away or diminu-tion of a part, owing to defective nutrition.

Automatic (Gr. *autos*, self, and root *ma*, to strive), self-acting, not voluntary, spontaneous ; a term applied more strictly to certain move-ments which are not the immediate result of an exciting impulse from without passing to a nerve-centre, but appear to be due to efferent impulses arising in the nerve-centre itself.

Bacteria (Gk. *bakterion*, pl. *bakteria*, a rod), very minute micro-organisms consisting of colourless protoplasmic cells without nucleus, and of a globular, rod-like, or comma-shaped form. Their origin and the part they play in disease, putrefaction, and fermentation, is discussed in works on Bacteriology.

Bifurcate (L. *bi*, twice, *furca*, a fork), forked, dividing into two branches.

Binocular (L. *bini*, double, *oculus*, an eye), referring to both eyes.

Buccal (L. *bucca*, cheek, mouth), relating to the inside of the mouth. The buccal glands are small mucous glands lying in the substance of the mucous membrane of the mouth formed on the plan of a salivary gland. Some are 'albuminous' and some 'mucous' (par. 112).

Calcified (L. *calx*, *calcis*, lime), hardened by the deposition of lime or its compounds.

Canaliculi (L. *canaliculus*, a little channel), the minute channels radiating from the cell-spaces or lacunæ of bone and joining one lacuna with another.

Cancellated (L. *cancelli*, a lattice or grating), formed of minute cross-bars: as the cancellated or cancellous tissue forming the loose spongy part of bone.

Cardiac (Gk. *kardia*, the heart), pertaining to the heart.

Centrifugal (L. *centrum*, a centre, and *fugio*, I flee), directed away from the centre: as the impulses carried by *efferent* nerves.

Centripetal (L. *centrum*, a centre, and *peto*, I seek), directed towards the centre: as the impulses carried by *afferent* nerves.

Cerebral (L. *cerebrum*, the brain), belonging to, situated in, or affecting the brain or cerebrum.

Cerebro-spinal, of or pertaining to both brain and spinal cord.

Cervical (L. *cervix*, *cervicis*, the neck), belonging or pertaining to the neck.

Cilia (L. *cilium*, pl. *cilia*, eyelid), (1) the hairs growing from the eyelids, the eyelashes; (2) the microscopic hair-like vibratile processes of cells. The adjective *ciliary* means belonging to the eyelids or some other part of the eye, while the adjective *ciliated* refers to cells or other parts provided with the microscopic hair-like processes. Ciliated epithelial cells line the respiratory tract, &c., and the cilia are in constant movement sweeping outward mucous and foreign particles (fig. 13).

Clinical (Gk. *klinē*, a bed), relating to treatment of disease at the bedside of the patient.

Coma (Gk. *kōma*, lethargy), profound stupor, insensibility.

Congestion (L. *con*, together, and *gero*, I bring or carry), the bringing together or accumulation of blood in a part. Moderate congestion takes place in secreting glands during their functional activity, but excessive congestion in a part leads to derangement and disease.

Co-ordination (L. *co*, together, and *ordo*, *ordinis*, order), the capacity of parts to work together in proper strength and order to fulfil some purpose in the body. Thus the nerve-centres co-ordinate incoming impulses with outgoing impulses that produce definite movement.

Corpuscle (L. *corpusculum*, a little body, diminutive of *corpus*, a body), a word applied to certain cells and other anatomical elements of the body: as blood corpuscles, lymph corpuscles, tactile corpuscles, &c.

Costal (L. *costa*, a rib), pertaining to the ribs: as costal cartilage.

Cribriform (L. *cribrum*, a sieve), sieve-like, perforated like a sieve: as the cribriform plate of the ethmoid bone through which the nerve fibrils of the olfactory nerve pass.

Crista (L.), a crest or ridge: as the *crista acustica* on the inner side of each ampulla of the membranous semi-circular canals of the ear.

Crus (L. pl. *crura*, a leg), applied to the diverging structures that form the real or apparent support of a part: as the *crura cerebri*.

Curare, or **Curari,** a drug possessing the power of paralysing the nerve terminals of motor nerves.

Decussate (L. *decusso*, to divide crosswise), to cross in the form of an X.

Defæcation (L. *de*, down, from, and *fæces*, the fæces), the act of expelling the fæces from the bowel.

Deglutition (L. *de*, down, and *glutio*, I swallow), the act or process of swallowing.

Dialysis (Gk. *dia*, through, and *lusis*, a loosening), the diffusion of soluble substances termed crystalloids through animal membranes or parchment. Colloids do not suffer dialysis.

Diapedēsis (Gk. *dia*, through, and *pĕdan*, to leap or ooze through), the oozing of the corpuscles through the walls of the blood-vessels.

Diastolē (Gk. *dia*, through, *stello*, I place), the expansion or passive dilatation of the cavities of the heart during which they become filled with blood.

Dyspnœa (Gk. *dus*, with difficulty, and *pneo*, I breathe), difficulty of breathing.

Efferent (L. *ex*, out of, and *fero*, I carry), carrying out of. Efferent nerves convey impulses out from the centre ; efferent = centrifugal.

Embryo (Gk. *embryon*), the young of an organism before the commencement of free existence ; the fœtus in its earlier stages.

Emulsion (L. *emulgeo*, I milk out), a liquid rendered milky-looking by the suspension in it of fine particles of oil or fat.

Epithelium (Gk. *epi*, upon, and *thele*, a teat or papilla). The term was originally applied to the layer of cells covering the papillæ of the integument of the lips, but it now denotes the superficial layer or layers of cells covering the skin and mucous membranes, lining ducts and serous membranes, &c. Epithelial tissue contains no blood-vessels, the cells being nourished by imbibing the plasma derived from the subjacent blood-vessels. In many cases, however, nerve fibrils are abundant.

Extravasation (L. *extra*, out, beyond, and *vas*, a vessel), the escape of a fluid, as blood, from the vessels containing it.

Extravascular, a term applied to those tissues, as the hair, the nails, and the epidermis, which contain no capillary blood-vessels ; non-vascular.

Filiform (L. *fīlum*, a thread), slender, like a thread.

Follicle (L. *folliculus*, diminutive of *follis*, a bellows or sac), a small tubular or sac-like depression : as the gastric follicles.

Foramen, pl. *foramina*, L., a hole or perforation.

Fungiform (L. *fungus*, a mushroom), having the shape of a mushroom : as certain papillæ of the tongue

Fusiform (L. *fusus*, a spindle), tapering both ways like a spindle, spindle-shaped.

Ganglion (Gk. *ganglion*, pl. *ganglia*, a swelling), a knot-like mass or aggregation of grey nervous matter consisting of nerve cells with nerve fibres running to and from them. They are regarded as the seat of the various nerve centres, and have special names according to their position and function.

Gastric (Gk. *gaster*, the stomach), pertaining to the stomach.

Glomerulus (L. diminutive of *glomus*, a ball of yarn), a cluster of capillary vessels, such as those forming part of the Malpighian bodies of the kidney.

Gustatory (L. *gustus*, taste), pertaining to the sense of taste.

Hæmatin (Gk. *haima, haimatos*, blood), a complex brown amorphous powder containing iron, which, when combined with globin, forms the hæmoglobin of the red corpuscles of blood.

Hæmoglobin, the pigment of the red corpuscles that serves as the oxygen-carrier of the body ; reduced oxyhæmoglobin (see Oxyhæmoglobin).

Hemianæsthesia (Gk. *hemi*, half, *an*, without, *aisthesis*, feeling), loss of sensation in one half of the body, right or left. It may be due to lesion of the internal capsule.

Hemianopsia (Gk. *hemi*, half, *an*, without, *opsis*, sight), or *hemianopia*, a condition in which one-half of the field of vision of each eye is obliterated.

Hemiplegia (Gk. *hemi*, half, *plēge*, a stroke), one-sided stroke, paralysis that affects one lateral half of the body with loss of voluntary movement and sensation on the side affected.

Hepatic (Gk. *hepar, hepatos*, the liver), pertaining to the liver.

Hyaloid (Gk. *hualos*, glass), glass-like, pertaining to the vitreous humour of the eye.

Ingestion (L. *in*, into, and *gero*, I carry), the taking of food into the body.

Inhibition (L. *inhibeo*, I hold in), a restraining or checking upon the action of some organ by the nervous system : as the inhibitory or restraining action of the pneumogastric on the heart.

Intercostal (L. *inter*, between, and *costa*, a rib), between or connecting the ribs.

Katabolism (Gk. *katabole*, a throwing down), the processes by which the protoplasm breaks down into simple products ; destructive metabolism.

Keratin (Gk. *keras, keratos*, a horn), the horny substance forming the chief constituent of epidermal cells, nails and hair.

Kymograph (Gk. *kuma*, a wave, and *grapho*, I write), an instrument for recording the wave-like motions or undulations of blood-pressure.

Lachrymal (L. *lachryma*, a tear), relating to tears, which are secreted and led away by the lachrymal glands.

Lacuna (L. *lacuna*, pl. *lacunæ*, a hollow), a hole or cavity : as the small spaces in bony tissue occupied during life by the bone-cells.

Lamina (L. *lamina*, a thin plate, diminutive *lamella*), a thin plate or layer : as the laminar bony plates.

Lateral (L. *latus, lateris*, the side), situated on one side : as the lateral columns of the spinal cord.

Lesion (L. *læsio*, from *lædo*, I injure), injury, harm or damage, whether visible or not ; any change impairing the function of a part.

Leucocytes (Gk. *leukos*, white, and *kutos*, a hollow vessel or cell), a term applied to white blood corpuscles, lymph cells, and the wandering cells of connective tissue.

Linea alba (L. = white line), the thin white tendinous cord running down the middle of the abdomen and formed by the union of the aponeuroses of the abdominal muscles.

Lobule (L. *lobulus*, a little lobe, diminutive of *lobus*, a lobe), one of the lesser divisions of the cerebrum ; one of the small elementary structures that go to form a lobe of an entire organ, as in the liver, lung, and various glands.

Localisation (L. *locus*, a place), the act of assigning to a particular place ; determination of the place where a function or process is carried on.

Lumen (L. *lumen*, light, window), the opening or passage-way of a blood-vessel or duct.

Malpighian, applied to certain structures first discovered or described by the Italian anatomist M. Malpighi, 1628 to 1694.

Mammary (L. *mamma*, the breast), referring to the breast : as the mammary gland which secretes the milk.

Mastication (L. *mastico*, I chew), the act of chewing, produced by the movements of the lower jaw.

Matrix (L. *matrix*, from *mater*, mother), that which encloses anything ; the groundwork in which the cells of a tissue are embedded ; the portion of corium or true skin lying beneath the root of a nail.

Maxillary (L. *maxilla*, jaw-bone, diminutive of *mala*, cheek-bone), referring to the jaw : as the superior and inferior maxillary bones.

Maximum, a Latin superlative signifying greatest or highest. *Minimum* signifies least or lowest.

Medullary (L. *medulla*, marrow or pith), resembling marrow in structure. The medullary sheath or myelin of a nerve fibre is the clear sheath surrounding the central axis cylinder of protoplasm.

Mesenteric (Gk. *mesos*, middle, and *enteron*, intestine), relating to, or situated on, the mesentery, that fold of peritoneum which attaches the intestine to the abdominal wall.

Metabolism (Gk. *metabole*, a change), the entire series of changes connected with the manufacture of the protoplasm of the tissues and divisible into (1) constructive metabolism or *anabolism*, and (2) destructive metabolism or *katabolism*.

Microbe (Gk. *mikros*, small, and *bios*, life), a minute living organism, especially one of the bacteria ; a micro-organism.

Myograph (Gk. *mus*, a muscle, and *grapho*, I write), an instrument for recording muscular contractions.

Myopia (Gk. *muein*, to shut, and *ops*, the eye), short sight, near sight. The name is due to the habit which myopes, or short sighted people, have of screwing up their eyes.

Myosin, the substance formed when muscle plasma coagulates.

Neural (Gk. *neuron*, a nerve), pertaining to the nerves or nervous tissue.

Neuroglia (Gk. *neuron*, a nerve, and *glia*, glue), the peculiar supporting or connective tissue of the central nervous system.

Nuclear (L. *nucleus*, a kernel), pertaining to a nucleus.

Ocular (L. *oculus*, an eye), pertaining to the eye.

Olfactory (L. *oleo*, to smell, and *facio*, to make), pertaining to the sense of smell.

Optic (Gk. *ops*, eye), relating to the eye or the sense of vision.

Osmosis (Gk. *osmos*, an impulse), the diffusion of a liquid or of substances in solution through membranes or porous partitions.

Osseous (L. *ōs*, *ossis*, a bone), bony, composed of bone.

Otoliths (Gk. *ous*, *ōtos*, the ear, and *lithos*, a stone), the calcareous particles in the membranous labyrinth of the ear.

Oxidation, the process of uniting with oxygen.

Oxyhæmoglobin, hæmoglobin loosely combined with oxygen as it exists in the corpuscles of arterial blood. It is of a bright red colour, and crystallises in rhombic prisms from the blood of many animals when this is suitably prepared. The crystals from human blood are elongated prisms, but they are tetrahedrons from that of the guinea-pig. By mixing defibrinated blood with one-sixteenth its volume of ether, the corpuscles dissolve, the blood assumes a transparent *laky* tint, and after a time the dissolved-out hæmoglobin crystallises. On giving up oxygen in the tissues, oxyhæmoglobin becomes reduced hæmoglobin (or simply hæmoglobin), though, it must be remembered, venous blood always contains some oxy-hæmoglobin. The two kinds of hæmoglobin may be distinguished by their colour and by their different absorption-spectra. Analysis shows that hæmoglobin contains carbon, hydrogen, oxygen, nitrogen, sulphur, and iron, the percentage of iron being 0·4. The oxygen that unites in the lungs with hæmoglobin to form oxy-hæmoglobin, and which is removed as the blood circulates through the tissues, has been called the respiratory oxygen of hæmoglobin, while hæmoglobin is sometimes spoken of as the oxygen-carrier of the body.

Oxyhæmoglobin Crystals Magnified.

1, From human blood ; 2, from the guinea-pig ; 3, squirrel ; 4, hamster.

Papilla (L. *papilla*, nipple), a teat, or something like a papilla, as a papillary process—often very small.

Paralysis (Gk. *para*, and *lusis*, a loosening), abolition of sensation (sensory paralysis), or of motion (motor paralysis), or of both (complete paralysis) in a part, due to interference with its nerve supply. Paralysis of one lateral half of the body is *hemiplegia*, of the lower half *paraplegia*.

Paraplegia (Gk. *para*, beside, and *plēge*, a stroke), paralysis of the lower half of the body and the lower extremities, frequently due to spinal injury or disease.

Parenchyma (Gk. *para*, beside, *en*, in, *cheo*, I pour), the proper tissue or substance of an organ or part as distinguished from the supporting structure or stroma.

Parietal (L. *paries*, *parietis*, a wall), forming or situated upon the side walls of a cavity.

Periphery (Gk. *peri*, around, and *phero*, I carry), the outside, the parts away from the centre. Peripheral nerves are those passing towards the periphery from a nerve centre.

Peristaltic (Gk. *peri*, around, and *stalsis*, a constriction): peristaltic muscular action, seen in the alimentary canals and other tubes provided with both circular and longitudinal muscular fibres, is the rhythmic wave of contraction passing downward, and due to the successive contractions of the circular and longitudinal fibres, whereby the contents of the tube are driven onwards.

Phagocyte (Gk. *phago*, I eat, and *kutos*, a cell), a leucocyte, or cell, that ingests and destroys, or renders harmless, injurious foreign particles, such as bacteria, &c.

Phrenic (Gk. *phrēn*, the diaphragm), pertaining to the diaphragm : as the phrenic nerves from the cervical plexus to the diaphragm.

Plexus (L. *plecto*, I plait, *plexus*, a plaiting), an intricate network, especially of nerves or veins.

Pneumograph (Gk. *pneuma*, breath, air, and *grapho*, I write), an apparatus for recording the movements of the chest in breathing.

Pons (L. *pons*, bridge), a term applied to the nervous structure connecting the medulla and the cruta cerebri.

Process (*pro*, forward, and *cedo*, I go), a word used to signify a projecting part or prominence.

Proteid. A proteid substance is a compound of carbon, oxygen, nitrogen, and hydrogen, with sulphur and phosphorus in very small quantities. Proteids cannot be built up in an animal's body from simpler compounds, and must therefore be supplied in the food. The proteid food-stuffs are called nitrogenous because they are the only kind containing nitrogen. All proteids that enter the body are in the end broken up to form carbon dioxide, water, and urea. The *general* properties and tests for proteids are given on page 414. The property of most proteids found in nature to set or coagulate on heating is seen when white of egg is cooked. The xanthoproteic reaction (heating with strong nitric acid and then adding ammonia) may be applied to show the presence of proteid matter in bread and potato. The pronounced orange colour that results in the case of bread, and the faint orange colour in the case of the potato, indicate the much smaller proportion of proteid in the potato. For the action of gastric juice and pancreatic juice on proteids, see pars 121 and 123.

Protoplasm (Gk. *protos*, first, *plasma*, form), the viscid contractile substance that forms the chief portion of cells ; living matter. Owing to its presence in all organised beings, protoplasm has been called 'the physical basis of life.' Chemically it consists of water, proteids, and small quantities of carbohydrates and mineral salts (see par. 2).

Pulmonary (L. *pulmo*, *pulmonis*, lung), pertaining to the lungs.

Racemose (L. *racemus*, a bunch of grapes), resembling a bunch of grapes on its stalk, having ducts which divide and subdivide and end in small sacs, or acini : exemplified in the salivary glands and pancreas.

Ramify (L. *rāmus*, a branch), to spread out like branches.

Reflex (L. *re*, back, and *flecto*, *flexus*, I bend), bending backward. Reflex action implies the existence of a nervous arc composed of (1) an afferent nerve, conveying a stimulus from its starting-point in the periphery, (2) a nerve centre composed of ganglion cells and capable of converting this impression into an outgoing impulse, and (3) an efferent nerve leading from the centre to the part put in action.

Renal (L. *renes*, the kidneys), pertaining to the kidneys.

Reticulum (L. *rēte*, a net), a network, an interlacement of fine fibres.

Rhythm (Gk. *rhythmos*, measured movement), any regularly recurring motion, as the rhythmic contraction of the heart.

Saponify (L. *sapo*, *saponis*, soap, and *facio*, I make), to convert into soap by combination with an alkali ; to decompose oils or fats into salts of the fatty acid and glycerine.

Sarcolemma (Gk. *sarx*, *sarcos*, and *lemma*, husk or sheath), the delicate sheath surrounding the fibres of striped muscular tissue.

Sciatic (L. *ischiaticus*, from *ischium*, hip bone). The great sciatic, the largest nerve in the body, arises from the sacral plexus (fig. 169), and running through the back of the thigh ends in branches for various muscles of the lower limb. The disease sciatica is marked by severe neuralgic pains along the course of the sciatic nerve.

Sensorium (L. *sensus*, a sense), the part of the nerve centres supposed to receive sensory impression ; the part where sensation resides.

Septum (L. *sepio*, to fence in), a partition, especially a partition between two cavities.

Somatic (Gk. *sōma, somatos*, the body), pertaining to the body.

Sphygmograph (Gk. *sphugmos*, the pulse, and *grapho*, I write), an instrument for obtaining a graphic representation of the blood pressure in the arteries.

Squamous (L. *squama*, a scale), scaly, shaped like a scale.

Starch. A chemical compound of carbon, hydrogen, and oxygen, belonging to the class of compounds called carbohydrates. It is an important food-stuff, and is present in most vegetable products used as food. By itself it is seen to be a white powder consisting of microscopic grains with concentric rings. The granules are enclosed in a coating of cellulose or woody fibre. Starch is insoluble in cold water, but when the water is boiled the granules burst their coating and the starch forms a sort of opalescent solution. The chief test for starch is iodine solution. With iodine solution starch gives an intense blue colour. The presence of starch in bread or boiled potato may be readily seen by pouring a drop or two of iodine solution on these bodies. The various kinds of starch are included under the term Amyloses (see Appendix). For the action of saliva and pancreatic juice on starch, see pars 116 and 123.

Stimulus (L. *stimulus*, a goad or spur, pl. *stimuli*), an agent which is able to cause reaction in muscular or other tissue on which it acts. It may be mechanical (pinching, pricking, &c.), thermal, electrical, or chemical.

Striated (L. *stria*, a furrow or streak), striped, provided with striæ.

Subjective (L. *sub*, under, and *jacio*, I throw), pertaining to the subject or conscious self ; internal ; a term applied to sensations which originate within the organism without the agency of an external object.

Sugar. A chemical compound of carbon, hydrogen, and oxygen, used as a food-stuff. The various kinds of sugar are carbohydrates like the starches, and belong to the two classes Sucroses and Glucoses (see Appendix). Dextrose or grape sugar, one of the Glucoses, is found in small quantities in the blood and certain tissues. It is the form to which all carbohydrates are converted when taken into the circulation.

Sustentacular (L. *sustentare*, to hold up, support), supporting.

Systole (Gk. *syn*, together, and *stello*, I place), the contraction of the heart.

Tactile (L. *tactus*, touch), pertaining to touch or touch sensations.

Tension (L. *tendo*, I stretch), a state of stretching or lightness ; internal pressure or tendency to expand.

Tetanus (Gk. *tetanos*, tension, spasm), a term used to denote the spasm or prolonged contraction of the muscles due to rapidly repeated stimulations.

Tone (L. *tonus*, a sound, from *tendo*, to stretch), a sound of a certain pitch ; a moderate state of tension or contraction, as arterial tone ; a state of healthy vigour.

Trophic (Gk. *trepho*, I nourish, *trophe*, nourishment), pertaining to nutrition. A trophic centre is a group of ganglionic nerve cells presiding over the nutrition of certain nerve fibres.

Turbinated (L. *turbo, turbinis*, a top), convoluted ; rolled like a scroll : as the turbinated bones of the nose.

Ulnar (L. *ulna*, the elbow), situated near or in relation to the ulna. The ulnar nerve is a branch of the brachial plexus (fig. 169) descending on the inner side of the arm to the elbow-joint, and passing thence to the muscles of the palm, the little finger, and one side of the ring finger. Its stimulation by a knock ('hitting the funny-bone') causes the well-known peculiar numb sensation.

Vagus (L. *vago*, I wander, *vagus*, wandering), a term applied to the pneumogastric nerve on account of its length and varying distribution.

Vaso- (L. *vās, vasis*, a vessel), a prefix meaning belonging to a vessel. *Vaso-motor* nerves are nerves regulating the movements of the walls of blood-vessels.

Villus (L. *villus*, a tuft of hair, pl. *villi*), the hair-like processes of the mucous membrane of the small intestine.

Viscera (L. *viscus*, pl. *viscera*, the internal organs of the body), the internal organs of the great cavities of the body.

Vitreous (L. *vitrum*, glass), of a glassy nature or appearance.

Zymogen (Gk. *zūme*, ferment, leaven, and *gennao*, I produce), the peculiar substance of the digestive secreting glands that gives rise to the digestive ferments (see Appendix, p. 419).

INDEX

The numbers refer to Paragraphs, and not to pages

H H

Spottiswoode & Co. Printers, New-street Square, London.

www.ingramcontent.com/pod-product-compliance
Lightning Source LLC
Chambersburg PA
CBHW020904210326
41598CB00018B/1771